Springer-Verlag Berlin Heidelberg GmbH

The Council Members

(as on 1.6.1997)

Prof. Dr. Friedrich O. Beese
Agronomist: Director of the Institute for Soil Science and Forest Nutrition in Göttingen

Prof. Dr. Klaus Fraedrich
Meteorologist: Professor of Meteorology at the University of Hamburg

Prof. Dr. Paul Klemmer
Economist: President of the Rhine-Westphalian Institute for Economic Research in Essen

Prof. Dr. Dr. Juliane Kokott (vice chairperson)
Lawyer: Professor of German and International Comparative Public Law, European and International Law at the University of Düsseldorf

Prof. Dr. Lenelis Kruse-Graumann
Psychologist: Professor of Psychology (specialist in environmental psychology) at the University of Hagen

Prof. Dr. Ortwin Renn
Sociologist: Academy of Technology Impact Assessment in Baden-Württemberg, Professor of Sociology at the University of Stuttgart.

Prof. Dr. Hans-Joachim Schellnhuber (chairperson)
Physicist: Director of the Potsdam Institute for Climate Impact Research (PIK) and Professor of Theoretical Physics at the University of Potsdam

Prof. Dr. Ernst-Detlef Schulze
Botanist: Director of the Max-Planck-Institute for Biogeochemistry, Jena

Prof. Dr. Max Tilzer
Limnologist: Professor of Limnology at the University of Konstanz

Prof. Dr. Paul Velsinger
Economist: Professor of Political Economy at the University of Dortmund, specializing in regional economics

Prof. Dr. Horst Zimmermann
Economist: Professor of Political Economy at the University of Marburg, specializing in public finance

German Advisory Council on Global Change

World in Transition:
Ways Towards Sustainable Management of Freshwater Resources

Annual Report 1997

with 76 illustrations

Springer

German Advisory Council on Global Change (WBGU)

Secretariat at the Alfred Wegener Institute for Polar and Marine Research
Columbusstraße
D-27568 Bremerhaven
Germany

http://www.wbgu.de

ISBN 978-3-642-64171-8 ISBN 978-3-642-59895-1 (eBook)
DOI 10.1007/978-3-642-59895-1

CIP-Data applied for

Die Deutsche Bibliothek – CIP-Einheitsaufnahme

World in transition: ways towards sustainable management of freshwater resources / German Advisory Council on Global Change. [Transl.: Tim Spence]. - Berlin; Heidelberg; New York; Barcelona; Hongkong; London; Milan; Paris; Singapore; Tokyo: Springer 1999 (Annual report/ German Advisory Council on Global Change; 1997) Dt. Ausg. u.d.T.: Welt im Wandel: Wege zu einem nachhaltigen Umgang mit Süßwasser.

Deutschland / Wissenschaftlicher Beirat Globale Umweltveränderungen: Annual report / German Advisory Council on Global Change. - Berlin; Heidelberg; New York; Barcelona; Hongkong; London; Milan; Paris; Singapore; Tokyo: Springer
Früher im Economica-Verl., Bonn; Dt. Ausg. u.d.T.: Deutschland / Wissenschaftlicher Beirat Globale Umweltveränderung: Jahresgutachten 1997. World in transition: ways towards sustainable management of freshwater resources. - 1999

Originally published by Springer-Verlag Berlin Heidelberg New York in 1999
Softcover reprint of the hardcover 1st edition 1999

Translation: Tim Spence, Bremen
Cover design: E. Kirchner, Heidelberg, using the following illustrations
Niagara Falls, USA
Abdullah irrigation canal, Jordan
Water vendor, Marrakech, Morocco
Irrigation with fossil water, Disi, Jordan
Mendenhall glacier, Alaska, USA
Flotsam, Salamis, Greece
(all photos: Meinhard Schulz-Baldes)
Copy deadline: July 1997

SPIN 10654055 32/3136 - 5 4 3 2 1 0 Printed on Workprint 100% recycling paper

Acknowledgment

This Report would not have been possible without the committed and untiring effort of the staff of the Council Members and of the Council's Secretariat in Bremerhaven. The Council wishes to express its fullest gratitude to all those research fellows who participated in the work of the Council when this report was written:

Prof. Dr. Meinhard Schulz-Baldes (Director, Secretariat, Bremerhaven), Dr. Carsten Loose (Deputy Director, Secretariat, Bremerhaven), Dr. Frank Biermann (Secretariat, Bremerhaven), Dr. Arthur Block (Potsdam Institute for Climate Impact Research), Dipl.-Geogr. Gerald Busch (University of Göttingen), Dr. Ursula Fuentes Hutfilter (Secretariat, Bremerhaven), Dipl.-Psych. Gerhard Hartmuth (University of Hagen), Dr. Dieter Hecht (University of Bochum), Andreas Klinke, M.A. (Academy for Technology Impact Assessment, Stuttgart), Dr. Gerhard Lammel (Max Planck Institute for Meteorology, Hamburg), Referendar-jur. Leo-Felix Lee (University of Heidelberg), Dipl.-Ing. Roger Lienenkamp (University of Dortmund), Dr. Heike Mumm (Alfred Wegener Institute, Bremerhaven), Dipl.-Biol. Martina Mund (University of Bayreuth), Dipl.-Volksw. Thilo Pahl (University of Marburg), Dr. Benno Pilardeaux (Secretariat, Bremerhaven), Dipl.-Biol. Helmut Recher (Max Planck Institute for Limnology, Plön).

The Council also wishes to thank the staff members who were responsible for the compilation and editing of this Report:

Vesna Karic-Fazlic (Secretariat, Bremerhaven), Ursula Liebert (Secretariat, Bremerhaven), Martina Schneider-Kremer, M.A. (Secretariat, Bremerhaven).

Moreover, the Council acknowledges the support of the researchers of the projects „Questions" (Potsdam Institute for Climate Impact Research) and „Syndrome Dynamics" (BMBF). Both projects provided invaluable help in the further elaboration and refinement of the syndrome approach:

Dipl.-Geogr. Martin Cassel-Gintz (Potsdam Institute), Dr. Jochen Dehio (RWI Essen), Dipl.-Chem. Jürgen Kropp (Potsdam Institute), Dr. Matthias Lüdeke (Potsdam Institute), Dipl.-Phys. Oliver Moldenhauer (Potsdam Institute), Dr. Gerhard Petschel-Held (Potsdam Institute), Dr. Matthias Plöchl (Potsdam Institute), Dr. Fritz Reusswig (Potsdam Institute), Dr. Hubert Schulte-Bisping (University of Göttingen).

The Council also owes its gratitude to the important contributions and support by other members of the research community. This Report builds on comments and reports in particular from:

Prof. Dr. J. Alcamo, Dr. P. Döll, F. Kaspar and S. Siebert, University of Kassel, Science Center on Ecosystem Research: „Global Mapping of Regional Water Vulnerability".

Dr. G. Bächler, Director, Swiss Peace Foundation, Bern, Switzerland: „Das Atatürk-Staudammprojekt am Euphrat-Tigris unter besonderer Berücksichtigung der sicherheitspolitischen Relevanz zwischen den Anliegerstaaten Türkei, Syrien und Irak".

Dr. N. Becker and O. Leshed, University of Haifa, Israel: „Using Economic Incentives to Mitigate a Potential Water Crisis".

Prof. Dr. E. Brown Weiss, Georgetown University Law Center, Washington, DC, USA: „Prevention and Solutions for International Water Conflicts: the Great Lakes (USA-Canada)".

Prof. Dr. M. Exner, University of Bonn, Hygienic Institute: personal communication.

Dr. W. Grabs, Bundesanstalt für Gewässerkunde, Global Runoff Data Center, Koblenz: personal communication.

Prof. Dr. H. Graßl: World Climate Research Programme, Geneva: personal communication.

Dr. M. Heimann and Dr. E. Röckner, Max Planck Institute for Meteorology, Hamburg: „Modellierung des Wasserdargebots".

Dipl.-Forstw. J. Herkendell, Ministry of Environment, Regional Planning and Agriculture of the Land North-Rhine Westphalia, Düsseldorf: „Literaturübersicht zu allgemeinen gesundheitlichen Aspekten von Wasserproblemen".

Prof. Dr. D. Ipsen, University of Kassel, Working Group on Empirical Planning Research: „Kulturangepaßte Maßnahmen für einen veränderten Umgang mit Wasser".

Prof. Dr. med. J. Knobloch, University of Tübingen, Institute for Tropical Medicine: „Die Ausbreitung wasserverursachter Infektionskrankheiten".

S. Kuhn, ICLEI European Secretariat, Freiburg: „Internationale Übersicht über Aktivitäten zur Lokalen Agenda 21".

Dr. K. Lanz, Hamburg and Prof. Dr. J. S. Davis, ETH Zürich, Switzerland: „Wasserkulturen der Welt im Vergleich – Eine Analyse westlicher Wasserwerte im Lichte fremder Kulturen".

Prof. S. McCaffrey, University of the Pacific, McGeorge School of Law, Sacramento, USA: personal communication.

Prof. Dr.-Ing. Dr.-Ing. e.h. E. J. Plate, University of Karlsruhe, Institute for Hydrology and Water Management: „Wasser und Katastrophen (IDNDR)".

Prof. Dr.-Ing. U. Rott and Dipl.-Ing. R. Minke, University of Stuttgart: „Wassertechnologien: Grundlagen und Tendenzen".

Prof. Dr.-Ing. Dr. rer. pol. K.-U. Rudolph and Dipl.-Ök. Th. Gärtner, Consultants for Water Engineering and Management, Witten: „Die deutsche Wasserver- und -entsorgung im internationalen Vergleich. Schwächen und Stärken eines zukünftigen 'deutschen Modells' sowie umweltpolitische Exportmöglichkeiten".

Prof. Dr. R. Sauerborn, Heidelberg Clinic, Dept. for Tropical Hygiene: personal communication.

Prof. Dr. U. Shamir, Water Research Institute, Technion Israel Institute of Technology, Haifa, Israel: „Sustainable Water Management".

Dr. H.-H. Stabel, Betriebs- und Forschungslabor Zweckverband Bodensee-Wasserversorgung, Überlingen: „Vergleichende Bewertung der internationalen und nationalen Standards für Nutzwasser (Trinkwasser, Irrigation, Industrie, Bergbau u.a.)".

Prof. Dr.-Ing. D. Stein, University of Bochum, Fakultät für Bauingenieurwesen: „Moderne Leitungsnetze als Beitrag zur Lösung der Wasserprobleme von Städten".

Prof. Dr. D. A. Tarlock, Chicago Kent College of Law, Chicago, USA: „The Use of Watermarkets to Reallocate Water to New Demands".

Cand. iur. D. Thieme, University of Düsseldorf, Faculty of Law: „Implementierung der Klimarahmenkonvention: Vorschläge für ein Zusatzprotokoll".

Dr.-Ing. M. Voigt, University of Dortmund, Faculty of Regional Planning: „Was ist die heutige Wasserkultur in Agglomerationsräumen der Bundesrepublik Deutschland und welche Restriktionen und Möglichkeiten zur Entwicklung einer neuen nachhaltigen Wasserkultur lassen sich benennen?"

Dr. R. Wiedenmann, Zürcher Kantonalbank, Switzerland, and Dr. A. Sanchez, Santa Cruz: „Bedingungen, Leistungsfähigkeit und Grenzen von Genossenschaftslösungen für landwirtschaftliche Bewässerungssysteme".

Outline of Contents

Contents

Boxes

Tables

Figures

Acronyms and Abbreviations

ADB	Asian Development Bank
AIDS	Acquired Immune Deficiency Syndrome
BGBl	Bundesgesetzblatt [Official German Collection of Legal Documents and Treaties]
BGW	Bundesverband der Deutschen Gas- und Wasserwirtschaft [Federal Association of German Gas and Water Works]
BMBF	Bundesministerium für Bildung, Wissenschaft, Forschung und Technologie [Federal Ministry of Education, Science, Research and Technology]
BML	Bundesministerium für Ernährung, Landwirtschaft und Forsten [Federal Ministry of Food, Agriculture and Forestry]
BMZ	Bundesministerium für wirtschaftliche Zusammenarbeit und Entwicklung [Federal Ministry for Economic Cooperation and Development]
BOD	Biological Oxygen Demand
BVerfGE	Bundesverfassungsgerichtsentscheidung [Judgment of the German Federal Constitutional Court]
CBD	Convention on Biodiversity
CBO	Community Based Organization
CCD	United Nations Convention to Combat Desertification in those Countries Experiencing Serious Drought and/or Desertification, particularly in Africa
CDC	Centers for Disease Control and Prevention (USA)
CFC	Chlorofluorocarbon
CFS	Committee on World Food Security (FAO)
CGE	Compagnie Générale des Eaux (France)
CGIAR	Consultative Group on International Agricultural Research (USA)
CIMMYT	Centro Internacional de Mejoramiento de Maiz y Trigo (Mexico)
CITES	Convention on International Trade in Endangered Species of Wild Fauna and Flora
CMS	Convention on the Conservation of Migratory Species of Wild Animals
COD	Chemical Oxygen Demand
CSD	Commission on Sustainable Development (UN)
DAAD	Deutscher Akademischer Austauschdienst [German Academic Exchange Service]
DALY	Disability-Adjusted Life Years
DDT	Dichlordiphenyltrichlorethan
DFG	Deutsche Forschungsgemeinschaft [German Research Foundation]
EC	European Communities/Community
ECE	Economic Commission of Europe (UN)
ECHAM4-OPYC	Coupled Atmosphere-Ocean Climate Model (MPI on Meteorology and German Climate Computing Center)
ECMWF	European Centre for Medium-Range Weather Forecast (UK)
ECOSOC	Economic and Social Council (UN)
EIA	Environmental Impact Assessment
EPI	Expanded Programme of Immunization
ESCAP	Economic and Social Commission for Asia and the Pacific

EU	European Union
FAO	Food and Agriculture Organization of the United Nations
FCCC	Framework Convention on Climate Change
FWCW	Fourth World Conference on Women
GAOR	United Nations General Assembly Official Records
GAP	Günedogu Anadolu Projesi
GATT	General Agreement on Tariffs and Trade
GCM	General Circulation Models
GDP	Gross Domestic Product
GEF	Global Environment Facility (UN)
GEMS	Global Environmental Monitoring System (UNEP)
GIS	Geographical Information System
GNP	Gross National Product
GTZ	Gesellschaft für Technische Zusammenarbeit, Eschborn [German Association for Technical Cooperation]
HABITAT	United Nations Conference for Human Settlements (UNCHS)
IAP-WASAD	International Action Programme on Water and Sustainable Agricultural Development (FAO)
ICJ	International Court of Justice
ICLEI	International Council for Local Environmental Initiatives
ICPD	International Conference for Population and Development
ICRISAT	International Crop Research Institute for the Semi-Arid Tropics (India) (CGIAR)
ICSU	International Council of Scientific Unions
IDB	Inter-American Development Bank
IFAD	International Fund for Agricultural Development
IFPRI	International Food Policy Research Institute
IGBP	International Geosphere Biosphere Programme (ICSU)
IHDP	International Human Dimension of Global Environmental Change Programme (ICSU)
IIASA	International Institute for Applied Systems Analysis (Austria)
IIMI	International Water and Irrigation Management Institute (Sri Lanka) (CGIAR)
IITA	International Institute of Tropical Agriculture (Nigeria) (CGIAR)
ILA	International Law Association
ILC	International Law Commission (UN)
ILM	International Legal Materials
ILO	International Labour Organisation
IPCC	Intergovernmental Panel on Climate Change (WMO, UNEP)
IPF	Intergovernmental Panel on Forests (CSD)
IRRI	International Rice Research Institute (Philippines) (CGIAR)
IS92a	Carbon dioxide emission scenario (IIASA)
KA	Kritikalitätsabschätzung [Criticality appraisal]
KI	Kritikalitätsindex [Criticality index]
LA21	LOCAL AGENDA 21
LAWA	Länderarbeitsgemeinschaft Wasser [Länder Working Group for Water]
LCA	Life Cycle Analysis
LED	Lyonnaise des Eaux-Dumez (France)
MAB-UNESCO	Man and the Biosphere Programme (UNESCO)
MPI	Max Planck Institute
NATO	North Atlantic Treaty Organization
NGO	Non-Governmental Organization
OAS	Organization of American States
OAU	Organization of African Unity
OCP	Onchiocerciasis Control Programme (WHO)
OECD	Organisation for Economic Co-operation and Development

OSCE	Organization for Security and Cooperation in Europe
PAHO	Pan American Health Organisation
PAI	Population Action International
PCB	Polychlorinated Biphenyl
PEEM	Panel of Experts on Environmental Management for Vector Control (WHO, FAO, UNEP)
PIK	Potsdam-Institut für Klimafolgenforschung [Potsdam Institute for Climate Impact Research]
PKK	Partîya Karkeren Kurdistan (Kurdische Arbeiterpartei)
PLO	Palestine Liberation Organization
POP	Persistant Organic Pollutant
PRA	Participatory Rural Appraisal
QELRO	Quantified Emission Limitation and Reduction Objective
Ramsar	Convention on Wetlands of International Importance especially as Waterfowl Habitat (signed at Ramsar, Iran)
RRA	Rapid Rural Appraisal
SAUR	Societé d'Amenagement Urbain et Rural (France)
SEI	Stockholm Environment Institute
SHIFT	Studies of Human Impact on Forests and Floodplains in the Tropics (BMFT)
SRU	Rat von Sachverständigen für Umweltfragen [Council of Environmental Experts]
T & D	Towns & Development
TRIPS	Trade-Related Aspects of Intellectual Property Rights
UBA	Umweltbundesamt, Berlin [Federal Environment Agency]
UN	United Nations
UN DTCD	United Nations Department of Technical Cooperation for Development
UN IN-STRAW	United Nations International Research and Training Institute for the Advancement of Women
UNCED	United Nations Conference on Environment and Development, „Rio Conference 1992"
UNCHS	United Nations Centre for Human Settlements (HABITAT)
UNCLOS	United Nations Convention on the Law of the Sea
UNCTAD	United Nations Conference on Trade and Development
UNDP	United Nations Development Programme
UNEP	United Nations Environment Programme
UNESCO	United Nations Educational, Scientific and Cultural Organization
UNFPA	United Nations Population Fund
UNGA	United Nations General Assembly
UNICEF	United Nations Children's Fund
UNIDO	United Nations Industrial Development Organisation
UNPD	United Nations Population Division
UPOV	International Convention for the Protection of New Varieties of Plants
VIP	Ventilated Improved Pit
WaterGAP	Water - Global Assessment and Prognosis (Model)
WBGU	Wissenschaftlicher Beirat der Bundesregierung Globale Umweltveränderungen [German Advisory Council on Global Change]
WEU	West European Union
WHG	Wasserhaushaltsgesetz [Water Management Act, Germany]
WHO	World Health Organization (UN)
WMO	World Meteorological Organisation (UN)
WRC	Water Resources Committee (ILA)
WRI	World Resources Institute
WTO	World Trade Organisation
WWI	Worldwatch Institute
WZB	Wissenschaftszentrum Berlin für Sozialforschung [Berlin Science Center]

Executive Summary A

Summary of individual chapters 1

Never have I held back the waters of the Nile,
never have I barred the water its way,
never have I dirtied the Nile.

PHARAONIC INSCRIPTION
IN THE VALLEY OF KINGS (RAMESES III)

Water resource management – the harvesting, distribution, utilization, purification and control of water – has shaped the history of human civilizations to a major and permanent degree. Management of water resources is also one of the greatest challenges facing the present generation. Today, around 2 billion people have no access to clean drinking water and sanitation, and only 5% of the world's wastewater is treated or purified. As a result, one person in two in the developing countries suffers from a water-related disease, and 5 million people die each year after drinking contaminated or infested water. Freshwater is the most important factor limiting food production, with agriculture already accounting for 70% of global water use. Worldwide, as many as 40,000 dams are in operation to secure and increase the supply of water, with a new dam being added daily. The total volume stored in reservoirs is five times that found in all the world's rivers. International conflicts are expected to arise from the growing scarcity of this crucial resource in many parts of the world. Referring to Ethiopia's plans for dam projects on the Blue Nile, President Sadat, the former Egyptian president, once threatened that "Anyone who plays with the waters of the Nile is declaring war on us!"

The dimensions and implications of today's freshwater problems, the source of a potential major crisis of global society and the environment, have prompted the Council to focus this year's Annual Report on this burning issue. The Council analyzes and evaluates the total complex on the basis of facts and interrelationships, describing in detail the available instruments for freshwater management and outlining ways to prevent a global crisis from unfolding. The solutions put forward by the Council are based on two elements. The first key element is generated from the Council's *"guard rail" model*, which is an attempt to resolve the dilemma between social, environmental and economic goals by setting clear priorities. A robust paradigm for the "sound management of freshwater resources" is generated in the process.

The guiding principle developed by the Council can be summarized as follows: TO ACHIEVE THE GREATEST POSSIBLE EFFICIENCY WHILE OBSERVING THE IMPERATIVES OF EQUITY AND SUSTAINABILITY. This principle takes account of the fact that water, like no other environmental asset within the global commons, is both a *scarce* and a *crucially important resource*. Water is not only a commodity, but also a foodstuff. Its essential properties define the sociocultural and ecological framework and the non-sustainable limits (the "guard rails") within which water must be used efficiently in order to optimize the general welfare of humans everywhere. The very scarcity of water resources requires that, within the guard rails, there are as few obstacles as possible to an efficient search for beneficial freshwater use. However, efficiency can only be achieved if appropriate institutional, technical and educational conditions are met.

From this paradigm, the Council develops ways to solve the water crisis and addresses these to specific policy and research fields. The second key element therefore involves a *global strategy* for putting this paradigm into practice. The strategy is sub-divided into three components: creating an international consensus, instituting a World Water Charter and drawing up an international Plan of Action against the freshwater crisis.

Essential foundations for analyzing the global water crisis

BIOLOGICAL AND PHYSICAL FOUNDATIONS

A description of the natural state serves as the basis for further analyses. The first step is to describe the various freshwater habitats and the threats to limnetic biodiversity. This is followed by a description of the abiotic factors of key importance for the hydrological cycle. Such a description must take account of the interactions which occur between the atmosphere and vegetation. In what ways can key elements of the water balance and the hydrological cycle be altered by climate change? To answer this question, the Council presents an analysis in which characteristics of the hydrological cycle under present climatic conditions are compared to those in a simulated climate with CO_2 doubling (equivalent to twice present-day levels). Here, the Council draws on calculations made with the ECHAM/OPYC coupled atmosphere-ocean model developed by the German Climate Computing Centre (DKRZ) and the Max Planck Institute for Meteorology (MPI). Simulations with the model show that more precipitation falls on land masses in a warmer climate, especially at high latitudes and in parts of the tropics and subtropics, while other regions have less rain. The latter include large areas of Brazil, southwest Africa, as well as Western and northern Australia. The cumulative effect of human-induced climate changes will cause a forcing of the hydrological cycle, although with substantial regional variances. This means that there will be losers and winners.

WATER NEEDS AND WATER DEMAND

The Council has predicted future trends in global water withdrawals by agriculture, industry and private households in a scenario, the basic elements of which were developed at the Potsdam Institute for Climate Impact Research and the Environmental Research Center at the University of Kassel. These computations are based on the future development of core trends relating to water, such as variations in water supply as a result of climate change, consumption levels in relation to demographic and economic trends, and the efficiency of water use. Water prices, cultural influences and institutional conditions for water withdrawals were not taken into consideration. The predicted figures show that total water withdrawals by agriculture will increase by 18% in the thirty-year period between 1995 and 2025. Despite this increase, the proportion of agricultural withdrawals in relation to the total global figure falls to 56%, 19% less than in 1995. This decline is attributed to water withdrawals by industry, which will treble by the year 2025, i.e. at a rate much faster than population growth. Household water consumption will increase substantially, especially in Africa and Asia, but falls are expected in Europe and Latin America.

WATER QUALITY

In order to define environmental and social guard rails for water quality, it is necessary to carry out monitoring operations as comprehensively as possible. However, current data on water quality are distributed very unevenly in geographical terms. Human impacts on water quality are impairing the natural and cultural functions of water, primarily through direct interference by agriculture and through pollutant loads emanating from point and non-point sources in settlements, the small business sector, agriculture and industry. Too little is known about the behavior of substances that enter water through human activities, about their decomposition and conversion, and about the impacts they have on ecosystems and humans. The most important factors influencing global water quality include acidification, eutrophication, salinization, and pollution caused by organic and inorganic trace compounds (pesticides and heavy metals, for example). Quality standards such as those governing agricultural and industrial uses have yet to be defined for many other types of use. Those standards already in operation tend to vary considerably from one country to the next, one example being drinking water, for which the highest quality criteria must, of course, apply. Setting limits can provide only relative safeguards against damage to health. If water stress levels are to be kept below the critical threshold, quality targets must be defined on the basis of expert knowledge and appropriate efforts made to meet such targets.

FLOODS

The greater part of the Annual Report addresses problems that arise from shortages or the poor quality of water resources. However, too much water can also lead to major problems and even disasters. Floods are the natural disasters which cause the greatest economic damage worldwide, often with great loss of human lives. The Council examines above all the mechanisms by which floods originate, how global change influences the incidence and severity of floods, and how the risk of floods can best be mitigated. The next Annual Report produced by the Council will focus in detail on risks and risk management.

Impacts of global change syndromes on the freshwater crisis

In its various Reports to date, the Council developed a concept for the holistic analysis of global environmental changes (WBGU, 1994–1996). This approach enables the most important global environmental problems to be described in the form of 16 "clinical profiles" or syndromes afflicting the Earth System. The Council now applies this systems approach to the crisis of freshwater resources. Of these 16 syndromes, the Council has selected three that are particularly relevant to water and which therefore require detailed study: the Green Revolution, Aral Sea and Favela Syndromes.

Analysis centers, firstly, on the role played by water within the "Global Network of Interrelationships", a method developed by the Council for organizing the complex interactions within global change into a form suitable for further analysis. By applying this method to global water problems, it is possible to examine how typical trends in the hydrosphere (such as freshwater scarcity, groundwater depletion or changes in the local water balance) are linked to other trends of global change. The interactions are described and graphically portrayed as a water-centered Global Network of Interrelationships.

Significance of the regional freshwater crisis

The regional importance of the freshwater crisis is emphasized further by the criticality index developed by the Council. This approach involves assessing the water crisis using a composite indicator that combines natural water stocks and the drain on water resources caused by humans, while also taking society's problem-solving capacity into consideration. On the basis of detailed scenarios for water supply and water withdrawals, which were developed and computed at the level of subnational catchments by the Center for Environmental Systems Research at the University of Kassel and linked to national problem-solving capacities by the Potsdam Institute for Climate Impact Research, we derive world maps that show the present "hot spots" of the freshwater crisis and other relevant aspects. With the help of additional scenarios for population trends, a climate scenario developed by the MPI in Hamburg, and assumptions about future water withdrawals, the regions which will face severe problems in the future are identified and presented.

The Green Revolution Syndrome

The Green Revolution Syndrome circumscribes the extensive, centrally planned and rapid modernization of agriculture with imported, non-adapted agricultural technology, whereby negative side-effects on geographical conditions of production and the social structure can occur, and indeed are put up with. The successes of the Green Revolution are primarily achieved in irrigated agriculture; within the space of a few years, however, typical water-related problems can arise. The "evolution" of the Green Revolution Syndrome is characterized by a particular combination of geopolitical, biological, population and economic trends (the interplay of national interests, the "seed revolution" in agriculture, population growth and impoverishment respectively). The Green Revolution was forced upon the people through "from above" within the framework of large-scale plans, and on a global scale through the transfer of technology and know-how "from the rich to the poor".

The syndrome analysis approach illustrates that the food security problem cannot be reduced to food shortage alone. Poverty is often accompanied by chronic malnutrition and famine. Close links must therefore be forged between rural development and increased production. The Council recommends initiating a "New Green Revolution", i.e. enhancing food production while at the same time ensuring the growth of the small business sector, the craft trades and market practices. Secure land tenure rights are essential if farmers are to have the capacity to plan the utilization of their resources on a long-term basis. Enhancing security of legal certainty for small farmers is thus a contribution to resource protection and a better means of realizing the right to food and water laid down in the International Covenant on Economic, Social and Cultural Rights. Water rights should therefore be specified in greater detail, and special institutions should be established with responsibility for implementing and enforcing such rights. Environmentally sound management methods that protect resources, such as agroforestry (combined agriculture and forestry practices) and multiple cropping are very difficult to implement on a large-scale without start-up assistance. States must therefore become involved in the field of rural development and assist in the adaptation of agricultural practices. The "debt for food security swaps" recommended by the World Food Summit are supported by the Council as an important instrument in this respect.

The analysis of the Green Revolution Syndrome with special reference to water problems reveals that current irrigation systems are urgently in need of improvement, as almost two thirds of all land irrigated worldwide is in need of rehabilitation. Subsidies should be dismantled, but without endangering the subsistence of small farmers. One way to achieve this result would be to introduce a special form of "water money" for specific target groups, whereby those most vulnerable to crisis would have to be identified. Water resource development projects and water management systems must form an integral part of regional development programs, with preference given to local, small-scale solutions.

The Aral Sea Syndrome

The Aral Sea Syndrome refers to the problems associated with centrally planned, large-scale projects involving water resource development. Such projects are ambivalent – on the one hand, they provide the additional resources that are required (water for food security, renewable energy), or they protect existing structures and people (flood control); on the other hand, they can have severe impacts on the environment and society. The effects of these large-scale installations are rarely confined to the local or regional area, but can assume far-reaching and even international proportions.

The various manifestations of the Aral Sea Syndrome are illustrated in two case studies. Attention is directed first and foremost to the greatest environmental catastrophe ever caused to regional water resources by mankind – the desiccation of the Aral Sea that lends the syndrome its name. The second study concerns the Three Gorges Dam that China is currently constructing on the Yangtze River, and describes the benefits derived in the form of electrical power and flood control, as well as the serious problems engendered in the form of compulsory resettlement of more than a million people and major environmental impacts.

How can the "susceptibility" or "vulnerability" to the Aral Sea Syndrome on the part of the various regions be measured? To do this, a complex global indicator is being developed for assessing the anthropogenic changes in surface runoff caused by large-scale projects. A second indicator mirrors the vulnerability of the various regions to the occurrence of the syndrome as a function of various geographical and societal factors. Combining these two data sets produces a global indicator of the intensity of the syndrome.

Applying the syndrome approach gives rise to the general imperative to preserve the integrity and function of catchment areas and to prevent the degradation of ecosystems and soils. The Council attaches enormous weight to the reduction or avoidance of the disposition to large-scale water resource development projects with severe environmental or social consequences. If large-scale installations are nevertheless essential, they must be subjected to a cautious assessment in which all environmental and social costs are first internalized. The Council specifies guard rails that may not be crossed and puts forward recommendations regarding the assessment procedure.

The Favela Syndrome

The Favela Syndrome refers to the progressive impoverishment and environmental degradation brought about by uncontrolled growth of human settlements. Due to the sheer speed of such informal urbanization and the failures evident in many policy fields, states become incapable of controlling further settlement (e.g. by means of development plans and building schemes) or of constructing water supply and wastewater treatment facilities. Uncontrolled urban agglomerations have a very high level of water demand and in most cases an inadequate system for sewage disposal. Most people living there have no access to clean drinking water or adequate sanitation. This explains the diseases typical of this syndrome (e.g. cholera), which can spread to other regions of the world as a result of global mobility.

How can the Favela Syndrome be mitigated? Firstly, it is essential to combat the basic underlying causes, such as rural exodus, which give rise to the Favela Syndrome in the first place and which ultimately produce the water-related problems. To eradicate the latter, the Council recommends establishing the prerequisites for integrated treatment of water-related problems in the urban agglomerations, for example by capacity-building in the local government sphere and through closer cooperation between public administration and the informal sector. Water prices are too low in most cases and lead to wastage (frequent when water supply companies are state-owned); conversely, however, water prices can often be much too high (where private-sector water traders operate) and impose a particularly heavy burden on the poor. The system for pricing water in urban agglomerations should therefore be changed in such a way that prices minimize wastage without, however, depriving the poor of access to water. Here, too, it may be necessary to consider paying "water money" to the needy. The Council also recommends a series of technical measures for mitigating water crises. A very practical method could be the institution of inter-city partnerships focusing on solutions to the water crisis in the favelas and in the surrounding areas from which people migrate to the favelas.

Key issues in the freshwater crisis

Certain problems are common to all syndromes and are dealt with by the Council as cross-cutting "key issues" of the freshwater crisis.

Conflicts

One such issue concerns the potential for political conflict ensuing from water resource problems. Are international "water wars" conceivable? Under what conditions are water wars especially likely? What options are available for the peaceful settlement of international conflicts over water resources? These questions are examined for four conflicts with very different trajectories. Disputes over the Great Lakes in North America have generally been resolved through cooperation, and in the case of the conflict between Hungary and the Slovakia, both parties accepted the jurisdiction of the International Court of Justice. There are no signs of an agreed solution to the conflict between Turkey, Syria and Iraq over the waters of the Tigris-Euphrates basin. Some observers see the possibility of a renewed escalation of political conflict between Israel, Jordan, Syria and the Palestinian administration of the West Bank over the allocation of water resources.

Health

Medical aspects form an important dimension of the freshwater crisis. In the first half of this century, many vector-borne diseases appeared to be in decline. However, these diseases are increasingly commonplace again in many developing countries. Such infections have acquired greater significance in industrialized countries as well, especially through highly resistant strains of pathogens. There are manifold reasons for this trend: human settlements with high population density even in the vicinity of forests and swamps, growth in world trade with greater mobility of people and goods, excessive use of pesticides and antibiotics, the adaptation of pathogens to environmental conditions, social and political collapse, rapid population growth and regional climate disturbances. Waterborne infections are one of the main causes of disease and death worldwide. At present, diseases transmitted through water or water-related vectors afflict about half the world population. Regulating the supply of water and the treatment of wastewater according to the quality criteria drawn up by the WHO is therefore the most effective precaution against disease. Investments in this area are likely to provide one of the highest possible "health gains". The Council therefore recommends, inter alia, that drinking water and wastewater treatment projects be given greater levels of support within the framework of development cooperation, and that food security programs be linked to infrastructural improvements for drinking water supply. The construction of reservoirs and open irrigation installations should no longer be supported as long as their health impacts have not been examined and countermeasures implemented. Vaccination against waterborne diseases should be improved and distributed more widely; this also requires greater investment in the development of vaccines.

Food

In irrigated regions, the issues of food and water supply are intimately linked. In the large river basins such as the Nile, the Euphrates or the Tigris, the use of water in agricultural irrigation systems enabled the rise of the oldest civilizations over 5000 years ago. Although there have been quantitative and qualitative improvements in the supply of food to humans over the last thirty years, the situation in regions with water scarcity and large fluctuations in rainfall continues to be highly problematic. In many developing countries, economic stagnation, climatic and pedological disadvantages, distributional problems and population growth are causing a dramatic deterioration in the food situation. Whereas undernutrition is no longer a serious problem in the growth economies of Southeast Asia, states in sub-Saharan Africa as well as South Asia give cause for concern. One person in three in sub-Saharan Africa is chronically undernourished.

At the same time, the area of cropland available for growing staple foods is in decline. Today, there are 16 million hectares less land in use for cereal production compared to the 1981 figure. Even though the area of irrigated land is increasing by 1% per annum, this corresponds to a per capita decrease of 12% in real terms by the year 2010. The trends for cropland are even worse on the whole, as the available land per capita will fall by a total of 50 million hectares (21%) by the year 2010, despite increases in the area of land used by agriculture. The response recommended by the Council is to increase irrigation efficiency, reduce the amount of water wasted through pumping, diversion or delivery to plants, and make greater use of salt-tolerant plants. Rainfed cropping should also be improved. Improvements should be made in the cultivation of locally adapted crops and varieties. Another option would be to optimize aquacultures and develop strategies for multiple use of water resources.

Degradation of freshwater habitats and neighboring biotopes

The Council examines in close detail the degradation of freshwater habitats, i.e. the damage to waterbodies by physical, chemical or biotic factors exceeding the former's stress-bearing capacity. Degradation reduces the quality of the natural areas affected and impairs their usefulness for humankind. Every reduction in water quality alters the composition of the biota; in most cases, the number of species declines. In cases of severe damage, species diversity is reduced to a small number of common species with high resilience. Increasing salinity produces similar impacts. Sulfur and nitrogen compounds released through the burning of fossil fuels are transported by air currents across large distances and later deposited as "acid rain", the most important factor causing acidification of waterbodies. Eutrophication – enhanced nutrient loads to waterbodies – is accelerating the production of primary organic material and biological decomposition processes in particular. In numerous industrialized countries, the ecosystems of the major rivers have been largely destroyed through the construction of dams for hydroelectric power generation. The Council recommends that no untreated wastewater be fed into non-flowing waterbodies, that the shores of lakes be placed under special protection and that action be taken to prevent the erosion of slopes near the shores of lakes. The importation of unlisted exotic species should also be stopped. Wetlands perform a special ecological function and should no longer be drained; renaturalization measures are needed here instead.

WATER TECHNOLOGY

Technological solutions for supplying households, agriculture and industry with water, for efficient water use and for purifying wastewater play a key role in the sustainable management of this precious resource. Due to the mounting contamination of surface water and groundwater, increasingly costly treatment processes must be deployed in order to supply people with drinking water. Pollution from industrial sources should therefore be reduced as far as possible by integrating environmental protection into production. Leaching of chemicals and other problematic substances from agriculture should be avoided. Because of the high costs involved in treating sewage, the wastewater of one third of humankind is discharged without treatment, even in OECD countries. In developing countries, where the majority of people have neither access to clean drinking water nor sewage systems, there is an urgent need to develop and implement culturally and locally adapted technologies for water supply and sewage treatment. In many cases, existing technological potential for efficient water use is not exploited to the extent possible, despite the fact that substantial reductions in water consumption by irrigation schemes, industry and private households could often be achieved with suitable technical equipment.

Ways out of the global water crisis

On the basis of the new paradigm and the guiding principles for "sound management of water resources", the Council describes the sociocultural and individual foundations for water resource management. The ways in which people use water depend not only on environmental and economic conditions, but also on manifold cultural factors, in other words the specific "water culture" of a society. Water resource management was often the starting point for human civilizations: ancient civilizations such as those in Egypt, the Indus Valley or along the Huang He in China arose from regional water cultures. The water culture in a specific type of society is multidimensional, featuring scientific, technical, economic, legal, administrative, religious, symbolic and aesthetic dimensions, for example. Another important aspect is how people perceive water: for instance, water is barely acknowledged as a resource in many industrialized countries, where it flows "straight from the tap" at a relatively low price.

A central pathway out of the water crisis is therefore to enhance environmental education and public discourse. The Council has developed a series of concrete solutions in this respect. Possible responses include media campaigns to save water, information on specific options, or pilot projects in selected urban districts. Communication between those involved, thus triggering learning processes that can subsequently lead to changes in behavior patterns, is essential here. In the industrialized countries, discursive forms of planning and conflict resolution are gaining in popularity – examples are round table discussions, citizen involvement, alternative procedures for settling conflicts, or LOCAL AGENDA 21 initiatives. In general, water-related problems should be made easier for people to perceive. The extent to which individual behavior affects water resources must be made clear to all, as must the successes that can be achieved by modifying behavior. An effective step would be, for example, to publish the water consumption figures of a particular community on a noticeboard so that a local "water-saving culture" can be promoted.

The instruments with which policymakers could shape society's management of water resources in a viable way are described in detail. This requires a highly differentiated analysis, since there is hardly a single resource that is used in as many different ways as water. To arrive at an optimal distribution of water resources, attention is focused primarily on institutional solutions. However, precisely because of the different uses to which humans put water, no institutional solution can provide a convincing response by itself. The Council therefore recommends that consideration be given in any case to a combination of different instruments, whereby the criteria of efficiency, equity and sustainability must be complied with as optimally as possible in each specific case. Market-based solutions generally hold out the promise of more efficient water use. However, the state must support them by establishing an appropriate framework and by implementing measures in order to conform to the criterion of equity and the secure coverage of basic needs for water as a basis for life (e.g. through anti-trust law, water money for the needy and similar).

The main focus of the section dealing with the role of law in the sound management of water resources is on multilateral aspects of water use. A distinction is made between two areas in which international cooperation is essential: firstly, riparian states bordering on inland waterbodies, i.e. which utilize and share a river basin or an inland lake, must engage in cooperation. In such cases, international freshwater law requires that riparian states share the utilization of water resources in an "equitable" and "reasonable" manner. This aspect is regulated in detail by the Convention on the Non-Navigational Uses of International Watercourses recently adopted by the Sixth Committee of the United Nations General Assembly. Secondly, the Council's view is that the global water crisis also demands international partnership over and beyond the riparian states of an inland waterbody. The entire international community is called upon to support all those states that are affected or directly threatened by a water crisis. The Secretary-General of the United Nations has called for a global consensus on international freshwater policy, which the Council expressly supports. From the viewpoint of the Council, this consensus could form the basis for a variety of institutional solutions: states could agree on a new Plan of Action, or go a step further and adopt a World Water Charter establishing behavioral standards for states, multilateral organizations and non-governmental bodies that would be non-binding in international law but nevertheless representing a political commitment on the part of signatories. A third step would be to use the Desertification Convention as a model for negotiating an international convention on the protection of freshwater resources, in the form of a legally binding Framework Convention on Freshwater Resources, for example. The Council believes, however, that the time is not yet ripe for this latter step. Nevertheless, Germany should make a concerted effort to initiate negotiations for a World Water Charter, for which the Council has drafted a basic outline in its Report.

Potential instruments within an international and national strategy for sustainable management of water resources could take a variety of forms. Optimal "instrument mixes" for responding to various problems of freshwater policy are presented, for example in the fields of water supply, wastewater treatment, protection of health, irrigation, human nutrition, disaster prevention and control, and for the settlement of disputes at national and international level.

Recommendations to the German Federal Government

A number of specific recommendations to policymakers and on further research needs can be derived from the "ways out of the water crisis" developed by the Council. The basic guiding principle for efficient, equitable and sustainable management of freshwater resources as applied by the Council must be operationalized in specific contexts and given shape and form through practical action. Germany can help resolve global water problems primarily by asserting its influence in various fields of international policymaking. These include international development cooperation, foreign trade, the transfer of knowledge and technology, and support of existing and forthcoming international regimes in the environmental and development field. Furthermore, by implementing a national water policy complying with the guiding principles outlined by the Council, Germany can strive for an enhanced role as a "model" of sound water resource management for other regions to follow.

The Guard rails

Sound management of water resources requires a definition of the sociocultural and ecological "guard rails". It is crucially important in this context to give equal consideration to environmental and development standards and to elucidate in sufficient depth the repercussions of water-related projects. Specifically, the Council recommends that:

1. minimum standards be defined for the supply of drinking water and water-related sanitation facilities to individuals,
2. the resultingant country-specific and culture-specific demand for freshwater be ascertained in respect of quantity and quality, giving special consideration to health aspects,
3. general safety standards be defined in respect of natural water-related disasters,
4. the geographical and sociopolitical pattern of vulnerability and the resultant need for precautionary action be determined in light of 3. above,
5. international principles of equity governing access to national and transboundary freshwater resources be agreed upon,
6. global groundwater reserves in fossil aquifers and the renewal and self-purification rates of recent groundwater reservoirs be determined,
7. the global stock of ecosystems dominated or influenced by freshwater and in need of protection be identified and classified,
8. the respective stress-bearing limits of the semi-natural systems identified under (7) above be determined in respect of water stocks, water quality and water variability, and that
9. the methods for integrated analysis and assessment of water-relevant projects in the private or public sector be developed further.

A fundamental consensus between competing users, societal groups or states on the specific guard rail criteria for sound management of freshwater resources does not automatically mean that these limits will be respected, however. This would require agreement on institutional regulations that can be enhanced by technical, educational and economic programs.

International regimes and international law

In connection with the further development of international law and international regimes formation, the Council recommends that the Federal Government support negotiations for a World Water Charter and for a comprehensive Global Plan of Action for "Sustainable Water Management". Additionally, water-relevant standards should form a more integral part of international trade and credit agreements (WTO, World Bank programs, Hermes credit guarantees, etc.), and sound management of water resources should always be taken into consideration more as a cross-cutting task in sectoral regimes for sustainable development (examples being the Climate Convention, negotiations on the protection of forests, the Biodiversity Convention and the Desertification Convention). International cooperation should also be stepped up with regard to water-relevant aspects of the International Covenant on Economic, Social and Cultural Rights and the relevant responsibilities of the United Nations High Commissioner for Human Rights.

Finally, it is important to improve the coordination of international organizations and programs in the field of "sustainable development"; here, the Federal Government should exert political pressure to ensure integration of the latter within a single "Organization for Sustainable Development". In particular, it would be possible to integrate the UNEP, the CSD and the UNDP in one body, while closer links could be forged between it and the World Bank, the International Monetary Fund, the World Trade Organisation and UNCTAD.

Regarding the amendment of the United Nations Charter, which Germany supports (German membership of the Security Council), the Federal Government should also lend support to the inclusion of clause articles on sustainable development, for example by including environmental protection in Article 55, and the goal of sustainable development in the Preamble and in Article 1, or 2 or 55 of the UN Charter.

Negotiations on the Convention on the Non-Navigational Uses of International Watercourses have been successful. The Council recommends that the Convention be ratified as soon as possible.

Foreign policy, foreign trade policy and development cooperation

As far as foreign trade policy and development cooperation are concerned, the Council recommends that securing a basic supply of water for nutrition and sanitation purposes be given greater consideration, in addition to environmental aspects, in multilateral agreements on development cooperation, whereby agreement must be reached with the partner countries in question. Preference should be given to recycling water as opposed to primary withdrawals, whereby withdrawals from fossil aquifers should be seen as a last resort. Local cultural traditions of protecting waterbodies and the environment, as well as indigenous knowledge, must be respected and supported as a matter of principle. It is essential to ensure public participation on the part of those affected, as this is the only way to guarantee the social acceptability and effectiveness of development policy measures and to determine the real needs of users. These aspects should be taken into consideration above all in the debate over a "New Green Revolution"; it is precisely here that efforts should be made to bring about a greater diversity of seeds and breeds in agriculture, and to promote rainfed cropping in particular, more intensively. A second main focus of water-specific development cooperation should be the improvement of water supply to poor sections of the urban population. In general, integrated water resource management in cities should be carried out to a greater extent by examining quantity and quality in combination only, by linking supply issues to wastewater treatment issues and by basing planning on whole catchment areas rather than district or national boundaries.

The Council recommends, in particular, that better support be given to states affected or threatened by water crises, especially for the modernization of existing agricultural irrigation systems, for the repair and expansion of water supply networks, and for establishing or improving systems for pumping drinking water, treating wastewater and recycling water. These activities should be carried out within the framework of bilateral development cooperation as well as through close collaboration with international organizations such as the FAO, the WHO, the UNDP or the World Bank.

In addition, environmental and development projects promoting and advancing the cause of peace should be vigorously supported and promoted in areas suffering from water crisis (the Middle East, for example). Another important activity concerns the transfer of technology and expertise to maintain sociocultural and ecological water standards, especially in areas affected by water crisis, and to protect the world's natural heritage, whereby special weight must be attached to water-saving and environmentally, culturally and locally compatible methods. Macroeconomic externalities (such as long-term impairment of water body quality as a result of industrial activities) should be taken into consideration by means of appropriate operationalization of the liability principle. The ecological guard rails can be complied with in an effective way by granting tradable emission certificates, for example. Improvements should be made to the conditional framework for efficient management of scarce freshwater resources; to this end, rights of tenure and disposal should be secured as far as possible, available water resources should be subjected to economic valuation, and limits imposed on subsidies that reduce competition. Where effective competition and anti-trust laws are in place, international water markets should be furthered in various regions of the world. A basic supply of freshwater in water-scarce countries must be secured by appropriate forms of direct assistance ("water money" rather than large-scale water resources projects).

Environmental education must similarly be advanced, also in relation to LOCAL AGENDA 21 initiatives. Special efforts should be made here to increase awareness of the interactions between individual behavior and damage to the environment, to provide the population with feedback on successful modification of behavior (such as information on consumption, and implications for water charges) and to enable learning from models.

FINANCING ASPECTS

As far as financing activities are concerned, the Council believes that greater efforts need to be made in order to increase Germany's financial contribution to the in support of water management policies in countries with insufficient resources, particularly in light of the UN Secretary-General's estimate of US$ 50 billion per year to meet global drinking water needs over the 1990–2000 period. All opportunities for reducing the debt servicing burden on developing countries threatened by water crisis should be considered and exploited to this end, whereby links to water policy programs should be examined (debt for water security swaps). The Council also recommends exploring the possibility of assistance to financially overburdened countries from a global Water Fund replenished via robust international financing mechanisms (for example, a "World Water Penny" levied on water consumption).

INTERNATIONAL RESEARCH COLLABORATION

With regard to international research collaboration, the Council recommends: enhancing the international transfer of knowledge about physiological, epidemiological and environmental factors of relevance to water resources, and on all aspects of sound management of freshwater resources, whereby special emphasis should be placed on communicating scientific and technological interrelationships (in the fields of hydrology, hydraulic engineering, water treatment or hygiene, inter alia), tried-and-tested regulations of the institutional organizations, and methods for efficient management of scarce environmental resources; developing integrated and participatory mechanisms for maintaining water-specific standards in private- and public-sector projects (water audits, water impact assessments, etc.), and disseminating information on same.

Five years after Rio – an initial assessment

In addition to the key focus on water resources, the Council assesses the follow-up process to the 1992 UN Conference for Environment and Development in a further section of the Report. In the Rio Declaration of 1992, almost all states agreed on "the goal of establishing a new and equitable global partnership through the creation of new levels of cooperation". Growing institutionalization of international environmental and development policy can indeed be identified, following the coming into force of the Montreal Protocol on Substances that Deplete the Ozone Layer in 1989, the Biodiversity Convention in 1993, the Framework Convention on Climate Change and the Convention on the Law of the Sea in 1994, and the Desertification Convention in 1996. The first follow-up documents have meanwhile been signed, for example the Draft Agreement for the Implementation of the UN Convention on the Law of the Sea (focusing on straddling and highly migratory fish stocks) or the 1997 Kyoto Protocol to the Framework Convention on Climate Change. The next steps could be a Protocol to the Framework Convention on Climate Change, a Biosafety Protocol to the Biodiversity Convention and a Forests Convention.

The intervening period has seen a whole series of UN summits closely related to the goals of AGENDA 21, such as the United Nations Conference on Human Settlements (HABITAT II) in Istanbul, the World Summit on Social Development in Copenhagen and the World Food Summit in Rome. Even though these major UN conferences ended in non-legally binding "Declarations" and "Plans of Action", they perform a key function at the symbolic level of politics, where the agenda for international policymaking is determined and where general expectations are formulated for policymaking at national level. This is the case, for example, with the so-called 20/20 compact, according to which developed and developing country partners agreed to allocate 20% of the ODA provided by the donor countries and the development banks and 20% of the national budget, respectively, to basic social programs.

Despite the various "Earth negotiations" that have been conducted recently, one must not lose sight of the fact that AGENDA 21 can only really be implemented through participation and initiative on the part of every individual. The Rio Conference set in train a process of implementing AGENDA 21 at local level, which is an important addition to the processes operating at international level. The "LOCAL AGENDA 21" initiatives form an indispensable part of the overall effort to preserve global environmental assets. Both trends are equally important constituents of effective political action to protect the global environment.

The Council's overall conclusion is that the Rio Conference in itself was a major step forward. For the first time in history, the overwhelming majority of nations adopted the guiding principle of sustainable development. The countries represented at UNCED acknowledged their responsibility for the global environmental and development system and accepted the need for action at the global level. However, the negative trends that led to the Rio Conference in the first place continue unabated and in some cases have worsened still further. The strategies adopted in Rio must therefore be pursued with added vigor and determination. Moreover, every effort must now be made to ensure that the pressures exerted by national problems and the greater financial constraints now operating do not lead to declining involvement in global issues. The fact that difficulties at national level are often linked through globalization to global environmental problems and development concerns, and experience feedback from the latter, means that national and global tasks can only be tackled through joint action. Germany has a special obligation and responsibility to bear in this context. As one of the major causal agents of global environmental problems and as one of the most powerful economic nations in the world, Germany should display a special level of commitment in the field of global environmental and development policy.

Key recommendations for action

The Council's analysis shows that the emergent global crisis of freshwater resources could become even worse in the future. The political domain should therefore take action without delay; national and international action programs must be designed and implemented as rapidly as possible in order to minimize risks and reverse current trends. The sheer complexity of the freshwater crisis calls for detailed and case-specific recommendations on activities and research, which the Council presents in the separate chapters of this Report and in condensed form in Section E. In line with the Council's criteria for "sound water resources management", these recommendations can be summarized in the form of four central and three syndrome-specific demands:

1. Increasing efficiency and effectiveness

Water is a scarce resource that is becoming more and more scarce for humankind and nature as a result of population growth and rising individual demands. It is therefore all the more important that any assessment of water resources be oriented to water scarcity around the world.

- The Federal Government should therefore exert political pressure to ensure that reliable and efficient water supply and wastewater treatment systems are established in all countries; these systems should involve water prices that reflect the scarcity of water resources, on the one hand, while, on the other, ensuring that the right of access to sufficient drinking water to meet basic needs is safeguarded and that environmental standards are met. The Council's view is that this demand can best be met by establishing water markets based on competition as well as proprietory rights to water supply and wastewater treatment systems. Another option at local or regional level would be the creation of cooperatives.
- The regulation of water supply and water demand should be governed by the principle of subsidiarity. Decentrally organized systems and regulations for water supply are usually more efficient than inflexible, centrally planned solutions; they are also easier for those concerned to understand and utilize, and tend to be better adapted to the specific characteristics of the region in question.

2. Compliance with the social "guard rails"

Efficient management of scarce water resources provides benefits for all humankind. However, the distribution of water must also conform to the principles governing the individual's right to a livelihood as well as – especially in international conflicts – the principle of distributive justice. There should also be adequate protection against severe droughts and floods. The Council puts forward the following key recommendations in this respect:

- The Federal Government should play an active role in enforcing the right to water worldwide. Efforts must be primarily geared at ensuring not only that all countries have the technical means to provide free access to water, but also that a (regionally defined) minimum supply of water for individuals is provided to low-income strata of the population throughout the world. This should be achieved by allocating "water money" (akin to housing benefit in Germany), or by appropriate water charges, i.e. inexpensive rates for the quantity of water defined as the minimum consumption level for individuals.
- Combating the freshwater crisis by means of national and international action programs requires the provision of financial support to those regions facing severe water scarcity, even when major increases in water efficiency are achieved. The UN Secretary-General estimates that annual investments of US$ 50 billion until the year 2000 are necessary to meet the global requirement for drinking water, an amount far beyond the financial resources of many developing countries affected by severe water crises. The Council therefore recommends that the Federal Government do everything it can to increase Germany's contribution towards supporting water policy efforts in

countries with insufficient financial resources of their own. Consideration should be given here to providing assistance from a global Water Fund, replenished through robust financing mechanisms (such as a "World Water Penny" levied on water consumption).

- Educational programs centering on the interrelationships between water, health and the environment are as essential as full public participation at local level if the subsidiarity principle is to be complied with and the supply of water is to be adapted to regional lifestyles and cultures. What are also needed are equitable decision-making processes for determining the amount of water use and the minimum level of environmental protection to be provided for waterbodies and surrounding areas of land. This is another area where the traditions, lifestyles and role patterns (gender roles, for example) of those affected have to be respected and taken into account. The Council therefore recommends that the Federal Government lend support to culture-specific educational work and to appropriate means of public participation (like the "water parliaments" in France).
- Another basic problem concerns the inequitable utilization of water resources by upstream and downstream riparians of river basins, or by joint users of waterbodies. This problem is the source of many international conflicts, and these are likely to escalate in the future. The Council therefore recommends that the Federal Government promote pilot projects for the equitable use of transboundary rivers, provide international mediators for settling such conflicts and insist on compliance with equity principles as a criterion in development cooperation.

3. Compliance with the ecological "guard rails"

Utilization of water resources by humans is placed under natural constraints wherever essential ecological functions are disturbed or where valuable biotopes are threatened. The basic principles that should operate include protection of species diversity in freshwater ecosystems, ensuring that water quality does not deteriorate beyond an environmentally acceptable level, and conservation of all major wetland areas. The impacts of water withdrawals and waterbody utilization on surrounding areas of land (land consumption in particular) must be taken into consideration here, as must the indirect human-induced effects operating through the media of soils and air on water-based habitats.

- The Federal Government should therefore carry out activities for preserving and re-establishing the structural and functional integrity of water-based ecosystems (including neighboring habitats) and exert political pressure to ensure that such measures are promoted in other countries as well. Such activities are of paramount importance for preserving and re-establishing the vital habitat functions that freshwater resources perform. The Federal Government can contribute towards this goal through the transfer of knowledge and technologies and by supporting specific rehabilitation projects.
- The core principle of sustainable use of water resources by human societies defines a crucial environmental "guard rail", or boundary zone, for preserving the natural resource base on which present and future generations depend. This implies that the annual withdrawals of (ground)water in a catchment basin must not exceed the renewal rate. The Council recommends that states restrict the delivery of water and/or water rights whenever a critical level is exceeded. To safeguard the quality of water, substance and organism loads may not over-tax the capacity of the medium to purify itself. The Council therefore recommends the definition of quality targets in accordance with the precautionary principle.
- The Federal Government should continue to provide financial and research support for protection of the biotopes included in the UNESCO World Heritage List. It should also promote the inclusion of additional, globally important and significant freshwater habitats.

4. Building the capacity of international institutions

In addition to national activities aimed at a model leading role and geared to bilateral economic, development and finance policy, it is absolutely imperative to codify the goals of sustainable water use in the form of international conventions and treaties.

- The Council therefore recommends that the Federal Government launch a "World Water Charter" to which all governments, local authorities, international organizations and non-governmental bodies may become signatories. The Charter would operate as a global code of conduct behavior committing all concerned to take the requisite political action to combat the freshwater crisis.
- Cooperation among the international community is made overly difficult by the hypertrophy endemic to the international system of institutions and organizations. This is the background to the Council's recommendation to improve the coordi-

nation of international organizations and programs in the field of "sustainable development" and to integrate the separate bodies in a high-level "Organization for Sustainable Development" endowed with extensive monitoring powers. This organization could unite existing institutions and programs such as the United Nations Environmental Programme (UNEP), the Commission for Sustainable Development (CSD) and the United Nations Development Programme (UNDP) within a single body. It should engage in close collaboration with institutions such as the World Bank, the International Monetary Fund, the World Trade Organisation (WTO), the World Health Organisation (WHO), the United Nations Food and Agriculture Organisation (FAO) and the United Nations Conference on Trade and Development (UNCTAD).

5. Mitigation of water-relevant syndromes

Besides the key recommendations directly derived from the guiding principle of sound management of water resources, the Council has identified three syndromes in which increasingly negative trends in the water-centered network of interrelations are concentrated. They therefore play a crucial role in exacerbating the global water crisis and hence require rapid and effective strategy responses. Key recommendations giving special consideration to the systemic nature of the freshwater crisis can be derived in this area as well.

- The analysis of the Green Revolution Syndrome shows that the food security problem cannot be attributed solely to food shortages in specific regions. More specifically, poverty and the severe lack of resources and capacities are primary determinants of chronic malnutrition and famine. The Council's recommendation to the Federal Government is that it take steps within its development projects to ensure that farmers have clearly defined water rights and fair competitive conditions with water traders so that they can plan ahead with confidence and thus achieve a modicum of local sovereignty. Adequate training and education programs must be implemented to improve awareness of the interrelationships between agricultural activities and the environment and to build the capacities of local communities to solve their water problems.
- The analysis of the Favela Syndrome shows that health and hygiene problems in the slums of major cities have reached extremely alarming proportions. The Council recommends that the damage to health caused by contaminated water be seen as the priority issue in the field of development policy and that action be taken to combat the root causes. Additional steps should involve the development of affordable wastewater treatment systems and the provision of support for essential health care (e.g. simple forms of disinfection and hygiene education).
- To cure the Aral Sea Syndrome, the Council recommends that environmental and development strategies be designed in such a way that large-scale water development projects be granted financial and otheror non-material assistance only on condition that the social and environmental costs are taken into consideration as far as possible. Construction of large-scale facilities should be dispensed with entirely if the environmental and social guard rails are overstepped.

Introduction

B

Ἄριστον μὲν τὸ ὕδωρ

Water is the best of all things

(Pindar, 552–446 B.C. (?), Olympic Odes 1,1)

In the Valley of Kings, scenes from the ancient Egyptian books of the dead adorn the walls in the tomb of Pharaoh Rameses III. The deceased ruler raises his hand in oath to the godhead Osiris and swears the following: never has he held back the waters of the Nile in the season of floods, never has he dirtied the waters of the Nile, never has he been cruel to any animal working in the far-reaching system of water use.

Today, the River Nile has been dammed by the gigantic Aswan Dam ("Sa'ad el Ali"), the river is a canal for transporting waste and pollutants of all kinds, while the flora and fauna in the river valley suffer the consequences of a society in transition driven by a preoccupation with growth.

Are we to conclude that the 3000-year-old message from the Pharaoh's tomb – admonishing purposeful, equitable and environmentally sound management of the most valuable resource a high civilization can possess – has become irrelevant for our time? The precise opposite is the case: in terms of its importance for sustaining life, freshwater ranks a close second behind the very air we breathe. It is at once the medium for the most elementary physiological processes and for evolution itself, the cohesive force for cultural organization and the source of individual well-being. However, unlike air, the freshwater medium is distributed very unevenly in time and space due to its physico-chemical properties and the geography of the world. As a result, many regions on our planet have little or no share in the overall supply of this virtually inexhaustible and continuously regenerated life-giving substance. Even within well-endowed regions, the volume and quality of the freshwater to which humans have access can vary considerably. This combination of indispensability and scarcity make freshwater the most precious raw material that our environment provides.

Water resource management – the harvesting, distribution, utilization, purification, control and defense of water – has shaped the history of human civilizations to a major and permanent degree and signifies a challenge of immense importance for present generations. Today, around 2 billion people have no access to clean drinking water and sanitation, and only 5% of the world's wastewater is treated or purified. As a result, half the population in the developing countries suffers from a water-related disease, and 5 million people die each year after drinking contaminated water. Between 1992 and 1995, almost 800 million people were victims of floods or landslides, and the number who have died in droughts over the last thirty years is beyond estimation. Freshwater is the most important limiting factor for food production, as seen by the fact that agriculture accounts for 70% of global water use today. Worldwide, as many as 40,000 dams are in operation to safeguard and increase the supply of water in space and time, with a new dam being added daily. The total volume stored in reservoirs is currently about 10 trillion liters, five times the total in all the world's rivers. Dams of all kinds influence more than three quarters of all natural runoff in North America, Europe and Northern Asia. Competition for this precious resource is hard and often ruthless. The legal system in India, for example, is dealing at present with a series of water conflicts between a number of states, e.g. Gujarat and Madhya Pradesh; many observers view these conflicts as a serious destabilizing factor for India as a nation-state. Were the Ethiopian government to implement its plans to dam the Blue Nile, thus diminishing Egypt's water supply, escalation into military conflict would become increasingly likely. The threat pronounced by President Sadat, the former Egyptian president, that "Anyone who plays with the waters of the Nile is declaring war on us!" remains as topical as ever.

For a full understanding of today's global water problems, it is essential to analyze freshwater as an integral factor within the specific causal webs that generate the dominant crises of environment and development – the *syndromes of global change* (WBGU, 1997). It is only within the context of uncontrolled urbanization (the Favela Syndrome), the environmental and social impacts of large-scale projects aimed at "taming nature" (the Aral Sea Syndrome), or the political and economic offensive to increase food production by importing inappropriate techniques (the Green Revolution Syndrome) that one can identify the reasons for poor management of water resources – as the direct or indirect results of (improper) human behavior.

The future of the international community's freshwater resources is even more foreboding, due to the unrelenting and mutual enhancement of the major driving forces behind water-specific syndromes: the world population continues to grow at a rapid pace, for example, and will stabilize at 8–10 billion after 2050, according to "best case" scenarios. However, it is not only the sheer numbers of people in the next century that pose severe problems, but also their invariable concentration in megacities and large-scale urban agglomerations, coupled with the growth in individual demands associated with the transformation of lifestyles across the globe. In order to realize the sheer scale of resource demand, one need only compare present-day water consumption in India (25 liters per inhabitant per day) to water consumption levels in the tourist centers around the Mediterranean (1,000 liters per visitor per day). If this tourist standard were extrapolated for the expected world population, then humankind would empty the total content of all the rivers on the planet within the space of six months!

This illusory scenario contrasts starkly with the realistic forecast that the number of chronically undernourished people (currently more than 900 million) will increase still further unless global food production is raised by about 60% by the year 2010. This can only be achieved by increasing the area of irrigated land – already 17% of the total cultivated area and accounting for almost 40% of global food production.

However, this might only reinforce a vicious circle, with greater pressure to produce leading to intensified use of natural resources and hence to the degradation of soils, ecosystems and landscapes, and thus to further and further deterioration of the basic conditions for agricultural production. One of the greatest threats is that the transformation of river catchments by human settlements could gather additional momentum, with all the negative repercussions this would involve; even today, the sediment load transported by the Earth's rivers has risen five-fold as a result of land-use changes (approx. 45 billion tons).

These future prospects are further overshadowed by the expected climate changes due to human activities, which will lead in all likelihood to modified precipitation patterns on the continents and thus to severe pressures on humans and nature to adapt accordingly. The international community is at a crossroads: unless the right environmental and development policy measures are put into force, the developing countries in particular will experience dramatic water problems that could escalate to a worldwide crisis through long-distance mechanisms such as migration, infection, exported conflicts or normal trade links. That said, there are ways to prevent such a trajectory, in that the freshwater problem is exceedingly responsive to strategic policy action. There is hardly another sector within the entire environmental-developmental complex that promises a comparable humanitarian dividend per US$ or DM spent. Furthermore, there are economic, institutional, technological and educational potentials worldwide for better management of freshwater resources. These potentials must be mobilized quickly, however, because many countries in the world are already verging on a developmental crisis as a result of water-related factors. For states in the Middle East and in North Africa, in particular, time is running short.

The dimensions and implications of today's freshwater problems, the source of a potential major crisis of global society and the environment, have prompted the Council to focus this year's Annual Report on this burning issue. In Section D, the overall complex is firstly analyzed and evaluated in terms of basic facts and interrelationships. This is followed by a detailed description of the available instruments for freshwater management, before ways to prevent a global crisis from unfolding are outlined. The solutions put forward by the Council are based on two elements. The first main element is the Council's "*crash barrier*" model, which is an attempt to resolve the dilemma between social, environmental and economic goals by setting clear priorities. A robust paradigm for the "sound management of freshwater resources" is generated in the process. The second main element consists of a *global strategy for implementing this principle*; the strategy is sub-divided into three components: creating an international consensus, instituting a World Water Charter and drawing up an international Plan of Action to combat the freshwater crisis. The latter Plan should be guided by the basic awareness that water is a scarce resource and must be priced accordingly, with only minimal exceptions.

National and international institutions play a special role in the prevention and mitigation of crises. The various treaty regimes, rules and authorities responsible for ensuring "sound management of water resources" should be endowed with greater flexibility, which has been the case so far, and should promote the principle of public participation. However, there is also a general need to improve international cooperation in the field of freshwater resource management. Despite the long tradition of intergovernmental agreement regarding the use of transboundary water resources, the level of cooperation has been inadequate in many regions. One good sign is that the Framework Convention on the Non-Navigational Uses of International Watercourses, which has been in preparation for twenty years, is likely to be adopted by the United Nations General Assembly in the near future.

The Council particularly welcomes the fact that the integrated management of water resources was made a key sectoral issue at the 6th Session of the UN Commission on Sustainable Development (CSD). The "Water 21" initiative of the European Union, which was proposed at the 5th Session of the CSD in April 1997, should be given strong support by the Federal Government. The Council expressly supports the priority attached to this issue, not only by the Secretary-General of the United Nations in his call for a "Global Consensus", but also by the EU in its Community policies and in its common international policy.

The freshwater issue must be understood more clearly than hitherto as a key component of global environment and development policy in the so-called Rio process.

The Council dedicates the standard section of the 1997 Annual Report to a review of this latter process over the last five years. The aim is an initial assessment of the extent to which the objectives and measures defined in AGENDA 21 have been implemented through efforts at national level as well as through collective international operations. Within this context, the issue that must be raised is whether there is evidence of an integrated concept, adequate for tackling a cross-cutting problem such as freshwater supply, to link the institutional facets of the Rio process.

Such an integrated concept would have to center around three core elements, namely

1. a common principle for controlling the environment and development process,
2. a strong, independent international organization as the driving force behind this process,
3. a robust global financing mechanism for supporting and strengthening this organization.

In the recommendations section, the Council will put forward its specific suggestions.

The central concerns in this year's Annual Report are the *global dimensions* of the freshwater problem; the aim is not to provide an in-depth water management or limnological analysis for Germany. There are two main reasons for this: firstly, Germany is a "country of surplus" as far as water is concerned, and, secondly, the Council of Experts on Environmental Issues (SRU) shall be presenting a study on key national aspects in 1998. Aspects of freshwater resource management pertaining specifically to Germany, such as concepts for waterworks, are referred to only when model solutions for the international community can be derived from them. The most important recommendations to the Federal Government regarding policy action and research will not relate to the national management of water resources, but to *measures within global environment and development policy* that would have to be coordinated, in the ideal case, by the main federal bodies with responsibility in this field (BMU, BMZ, BMBF, etc.).

Even if certain aspects of the freshwater problem are filtered out, it remains an issue of confusing breadth and scope. The reader will need a good deal of motivation and patience to digest this Report. Only rarely are there simple answers to complex issues, and these are invariably wrong.

Five years after the UN Conference on Environment and Development in Rio de Janeiro

C

Introduction 1

Establishment of the CSD – Special Session of the UN General Assembly

The dramatic escalation of global problems in the 1970s and 1980s showed that they are often too great and complex to be solved at the purely national or regional level. This realization led the UN General Assembly in 1989 to call a Conference on Environment and Development (UNCED) in Rio de Janeiro in 1992. Attended by more than 100 heads of state and government leaders, as well as around 1,400 non-governmental organizations, the Earth Summit turned out to be the biggest conference in the history of humankind, a demonstration of the broad consensus among the international community that is reflected in AGENDA 21.

The basic message of the Rio conference was that environment and development can no longer be viewed as discrete policy areas, but are indissolubly linked through the new principle of "sustainable development". Besides the two conventions on climate and biodiversity, as well as the non-legally binding declaration of principles for a global consensus on the sound management, preservation and sustainable development of all types of forest (the Declaration of Forest Principles), the most important document to come out of the summit was AGENDA 21, the United Nations Programme of Action for the 21st Century. A Commission on Sustainable Development (CSD), reporting directly to the Economic and Social Council of the United Nations (ECOSOC), was also established to monitor, assess and continue the Rio process.

Although foundation of the CSD by the Rio Conference fell far short of more far-reaching proposals (enhanced status for UNEP, a new UN special organization, an Environmental Security Council), a major discussion forum for implementation of the Rio process was created. A total of 53 nations are represented in the CSD; central to its work are the annual meetings of the Commission, which focus on specific chapters of AGENDA 21 and draw up recommendations for ECOSOC. These are then passed on to the UN General Assembly.

The fifth Session of the CSD in April 1997 was used to prepare the Special Session of the UN General Assembly in June, which was aimed at assessing the implementation of AGENDA 21, including possible recommendations for improvement. It can only be hoped that the international community will utilize this opportunity to bring its awareness of problems up to date and to accelerate the Rio follow-up process.

Five years after Rio, an initial appraisal is called for. In the section that follows, we look at the various international activities relating to specific core problems of global change (Section C 2; on core problems see WBGU, 1995), and the important role played by local implementation of AGENDA 21 (Section C 3), in order to produce a summary analysis of the Rio follow-up process as it has unfolded to date (Section C 4).

2 International policymaking in response to Global Change

The Montreal Protocol – UN Framework Convention on Climate Change – Land-based Marine Pollution – Convention on Straddling Fish Stocks – Biodiversity Convention – Declaration of Forest Principles – Plant genetic resources – Desertification Convention – World Population Summit in Cairo – World Conference on Women – World Summit on Human Settlements – Human Rights Summit – GATT/WTO – World Food Summit – World Summit for Social Development

2.1 Atmosphere

2.1.1 The Montreal Protocol

Strictly speaking, the Montreal Protocol on Substances that Deplete the Ozone Layer is not part of the Rio process, in that the first version was agreed upon as early as 1987, following adoption of the underlying Vienna Convention in 1985. However, the Montreal Protocol has acted in important ways as a model for other regimes, and is closely connected to the Rio follow-up process through the various amendments and adjustments of 1990, 1992 and 1995. For the industrialized countries, the definitive ban on most ozone-depleting substances has been implemented, with only a few essential CFC uses, such as asthma sprays, being excepted.

In 1987, the developing countries were granted an extension period of ten years; most of these countries will have fulfilled their commitments by the respective deadline. This was made possible above all through the industrialized countries' promise, in the 1990 amendment to the Protocol, to meet "all agreed incremental costs" in this area and to set up a special Multilateral Fund to finance implementation of the Protocol. A welcome provision in the amended version of 1990 concerns the balanced representation of the Parties in decision-making, according to which further-reaching measures require the simple majority of the developing countries and the simple majority of the industrialized nations.

However, further action still needs to be taken. Given existing chlorine concentrations in the stratosphere, the reduction schedule for the groups of substances that have only been included since 1992, especially methyl bromide and partly chlorinated fluorocarbons, should be speeded up as a matter or urgency in order to bring forward the target for restoration of the ozone layer. In the light of experience in North-South negotiations at the last two Conferences of the Parties, however, this will probably require additional contributions to the Multilateral Ozone Fund.

2.1.2 UN Framework Convention on Climate Change

The UN Framework Convention on Climate Change came into force in early 1994; since then, however, the Parties have failed to set specific time scales and reduction targets for the industrialized countries' commitment under the Convention to reduce their greenhouse gas emissions. At the first Conference of the Parties, held in Berlin in 1995, the signatory states agreed to continue negotiations (the so-called Berlin Mandate).

That said, the Berlin conference succeeded in establishing some key foundations for the current negotiation process. The outcome of the latter is to be an international legally binding instrument that commits the signatories to greenhouse gas reductions within specific time frames. New commitments for the developing countries going beyond the general, non-specific climate protection commitments in Article 4.1 of the Convention have not been an item of negotiation to date, because the per capita emissions of the industrialized countries exceed those of the developing countries many times over. However, to gain experience with joint implementation of activities to meet reduction targets, a pilot phase for this new instrument was started in 1995.

A binding legal instrument might possibly be adopted in December 1997 at the third Conference of the Parties in Kyoto, either in the form of an additional Protocol to the Convention, or as an amendment of the convention. According to currently available drafts by some individual states, it is likely that two different kinds of reduction commitment will be agreed for industrialized states for the period after 2000: negotiations center on proposals that the industrialized countries commit themselves in their post-2000 climate policies to so-called QELROs (Quantified Emission Limitation and Reduction Objectives). A number of industrialized countries are calling here for differentiation of commitments between individual industrialized countries, which could be based on indicators such as GDP, energy efficiency or per capita emissions of greenhouse gases. The USA, in particular, is demanding that trading in emissions permits between states be allowed and that such trading be made an integral component of the new instrument. Another issue under debate is to commit industrialized countries to joint policies and measures to combat climate change, such an agreement on certain technical standards. The EU has put forward a comprehensive proposal in this regard. However, there is no telling to what extent these political measures will be legally binding; one conceivable approach could be that only some categories of measures are binding, or that the states are granted the option to select measures from a "menu".

What is clear in any case is that developing countries will not take on any commitments that involve specific reduction targets. Uncertainty continues to surround the future status of eastern European countries, which are allowed a "certain degree of flexibility" in Article 4.6 of the Climate Convention in order to enhance their ability to address climate change. Other important areas of negotiation relate to the financial support for developing countries in the field of climate protection, the financial compensation of developing countries that suffer economic disadvantages as a result of future climate policy, and future dispute-settlement procedures in connection with implementation of the convention.

It is difficult to tell from the current state of negotiations whether a legally binding instrument will be adopted at the Kyoto conference. The Council considers it imperative that the parallel debate on harmonization of policies and measures does not detract from the central purpose of climate protection policymaking, namely the agreement of quantitative and legally binding reduction targets for greenhouse gases.

2.2 Hydrosphere

2.2.1 Protection of the Seas from Land-based Pollution

Progress has been achieved since 1992 regarding protection of the oceans (Chapter 17 of AGENDA 21). In 1995, as part of the Rio follow-up process, a new instrument was created to limit land-based marine pollution, which accounts for up to 80% of total marine pollution – the Washington Global Programme of Action for the Protection of the Marine Environment from Land-based Activities. This non-binding action program is based on older UNEP guidelines that were recognized as inadequate in the course of UNCED. The global action program is now being translated by the states into national action programs, but the developing countries are being given too little financial assistance from the Global Environment Facility (GEF) (WBGU, 1996).

The Washington Global Programme of Action and the Washington Declaration call on the international community to negotiate a legally binding convention on twelve groups of persistent organic pollutants (POPs). The intention is to subject particularly hazardous POPs, which are mostly prohibited in Germany, to a globally binding reduction regime or to ban them entirely. A model for such a regime, to which reference was often made during preliminary negotiations, is the Montreal Ozone Protocol.

The Council welcomes both the Global Programme of Action as well as the planned POP Convention as a first significant step towards the International Convention on Protection of the Seas recommended in the 1995 Report (WBGU, 1996). Consideration should be given to widening the scope of the POP Convention to include not only those groups of hazardous substances currently under discussion (the "dirty dozen"), but other groups of substances as well; the Council sees major research gaps in this area. As with the Montreal Protocol, it is essential that the POP Convention enable rapid adjustments to be made as scientific understanding advances.

2.2.2 Overfishing

Another key outcome of the Rio follow-up process concerning protection of the seas is the Agreement for the Implementation of the Provisions of the United Nations Convention on the Law of the Sea of

10th December 1982 relating to the Conservation and Management of Straddling Fish Stocks and Highly Migratory Fish Stocks. The aim of this new agreement on fisheries is to limit states' authority over fisheries zones and to lay down some general obligations regarding cooperation and consultation. The Council continues to view overfishing of the seas as a serious challenge for biodiversity and marine protection policy, and lays particular stress on the social dimension of the problem. Experience with market-based measures to combat overfishing – e.g. fishing ground licenses – indicates that such instruments can have positive impacts. Another instrument could be international regulations for reducing driftnet and similar fishing methods, which cause excessive damage to marine biodiversity due to the other organisms which are caught in the nets.

2.2.3 The International Tribunal on the Law of the Sea in Hamburg

An important role in international dispute settlement could be performed by the new International Tribunal on the Law of the Sea, which commenced work in Hamburg in 1996 and which decides on disputes over interpretation of the UN Convention on the Law of the Sea – provided, of course, that states bring their conflicts before the tribunal. This makes it the second major UN institution, along with the Climate Convention Secretariat, to be sited in Germany.

On the whole, the world's seas have not been assigned the position they deserve in the international "hierarchy" of problems, especially in light of probable trends in the future (WGBU, 1996a). Nor is protection of the seas granted adequate consideration as yet in multilateral financing programs such as the GEF.

2.3 Biosphere

2.3.1 Convention on Biological Diversity

The Convention on Biological Diversity (Biodiversity Convention) was agreed in Rio de Janeiro, came into force in 1993, and has since been ratified by more than 160 states.

It is a cross-cutting convention aimed not only at preserving biodiversity, but also at the sustainable use and equitable sharing of the benefits deriving from such use (WBGU, 1996). Given this range of tasks, it is inevitable that conflicts will arise between the protection and use of biodiversity resources, as in agriculture and forestry, for example.

The three Conferences of the Parties (1994 in Nassau; 1995 in Jakarta; 1996 in Buenos Aires) have set special emphases in this respect – decisions have already been adopted on the preservation of biodiversity in agriculture (especially on combating genetic erosion), on coastal and marine biodiversity and on forests.

The convention has also established a new legal basis for regulating access to genetic resources, which are no longer deemed to be commons property to which there is free access; instead, states have sovereign rights over and explicit authority to govern access to their genetic resources. The interests of industrial enterprises seeking sources of genetic resources, the rights of local and indigenous communities with their traditional knowledge about this diversity, and the sovereignty of states must be harmonized with each other (Henne and Loose, 1997). Some states have passed legislation implementing the provisions of the convention as national law, in order to ensure this balance of interests and above all equitable sharing of the benefits derived from genetic resources.

Negotiations are currently being conducted on a supplementary protocol relating to biosafety. The Convention has a financing mechanism, which the GEF is administering on an interim basis. However, the funds required exceed the financial resources of the GEF to such an extent that the Conference of the Parties is now compelled to find new sources of finance.

Integration into existing multilateral structures has been improved by building bridges to the other Conventions and organizations of relevance to biodiversity (e.g. CITES, the Ramsar Convention, the FAO; see Fig. C 2-1). In addition to specific work on implementation, the negotiation process itself is operating as a forum for activities aimed at the preservation and sustainable use of biological diversity. This function is further supported by a clearing house mechanism for promoting the exchange of information and technology.

The convention process has led to open and constructive discussion, stimulating fertile debates and activities within nations as well. Germany took an active part in setting up the clearing house mechanism and in promoting cooperation with the private sector. The Council welcomes the efforts being made to encourage environmentally compatible forms of tourism, since this promotes the economic valuation of biological diversity. On the whole, however, Germany is lagging behind in implementing the Convention. Neither the re-enactment of the Federal Nature Conservation Act, which has been pending for a long

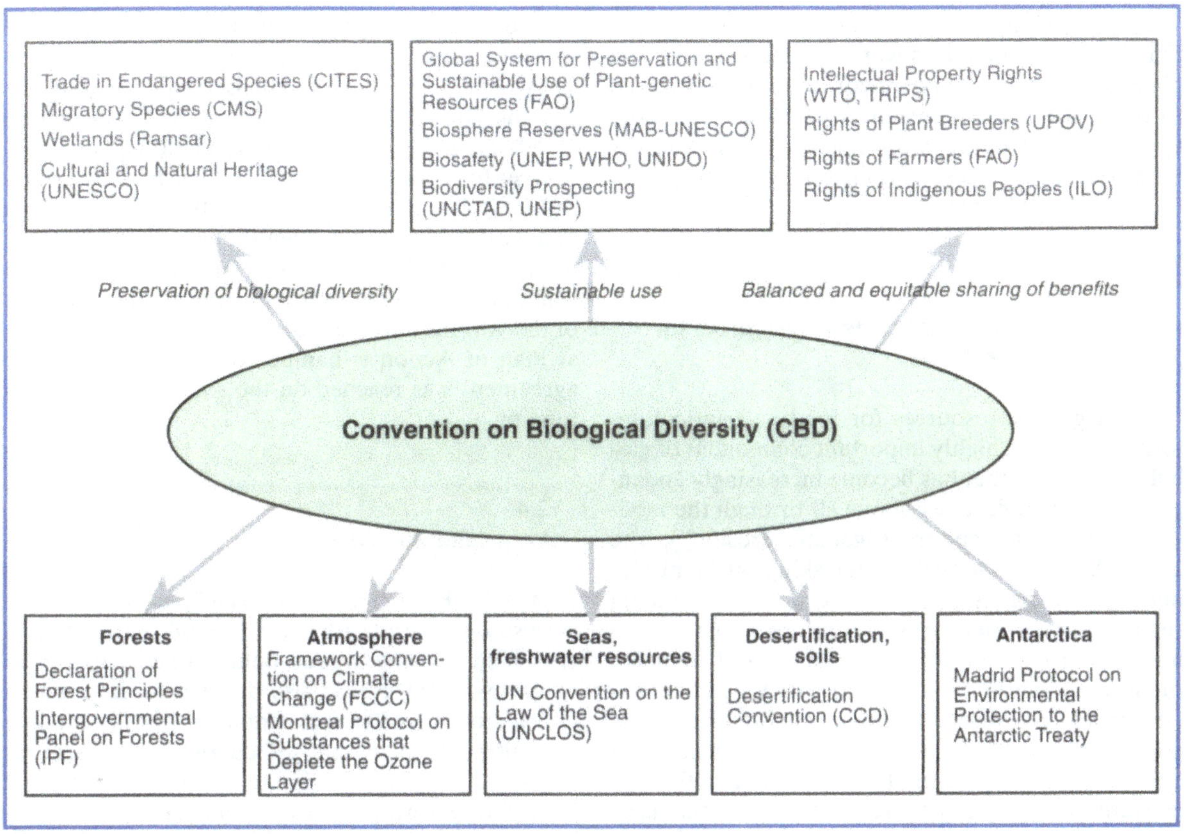

Figure C 2-1
Linkages between the Biodiversity Convention and other global environmental regimes.
Source: Gettkant et al., 1997

time now, nor the development of a national strategy for implementing the Biodiversity Convention have been completed. Politicians and researchers in Germany are realizing only slowly that the preservation and sustainable use of biodiversity is not a problem peculiar to nature conservation policy, but is a cross-cutting task in environmental and development policy.

2.3.2
Intergovernmental Panel on Forests

In the field of forest protection, efforts to engineer a Forests Convention in the run-up to UNCED met with failure. All that was agreed was a "non-legally binding authoritative statement of principles for a global consensus on the management, conservation and sustainable development of all types of forest" (the Declaration of Forest Principles). In an attempt to revive discourse, the CSD established the Intergovernmental Panel on Forests (IPF) in 1995. The IPF was asked to deliver a report to the Fifth Session of the CSD in April 1997, containing recommendations on the management, preservation and sustainable use of forests.

The IPF's efforts to institute a Forests Convention continue to be dogged by the opposing interests of stakeholders. For precisely this reason, many non-governmental organizations consider this to be an inopportune time to be negotiating a Convention. They argue that the final outcome might be based on the lowest of common denominators, and that the negotiation process itself is liable to delay any form of effective action.

Very little real progress has been made so far. Neither conceptual issues (interpretation of "sustainable forestry") nor procedural issues (continuation of negotiations beyond the current mandate of the IPF) have been clarified. The work of the IPF has more to do with the preparation and management of discussions between governments, non-governmental organizations and the public. Forests have been put back on the political agenda. In many countries, debate has received new stimulation above all through the improved technical knowledge generated by the IPF process and through the discussions on certification and indicators.

In view of the IPF's problems in renegotiating a Forests Convention, it is worth asking whether – in place of a new, independent convention – a protocol to the Biodiversity Convention would be the more appropriate instrument for regulating, in a legally binding form, the preservation and sustainable use of forests (WBGU, 1996).

2.3.3 Negotiations on Plant Genetic Resources for Food and Agriculture

Plant genetic resources for food and agriculture form a small but highly important component of global biodiversity that has become increasingly endangered in recent decades, above all through the modernization of agriculture ("genetic erosion"). The 1983 FAO International Undertaking on Plant Genetic Resources represents a first, albeit non-binding regulation to protect these resources as part of the human heritage, to collect, research and to make them accessible as commons property for all interested parties (e.g. research into plant breeding). The coming into force of the Biodiversity Convention in 1993 has produced a need for adjustment, in that the convention places genetic resources under the jurisdiction of sovereign national states (see Section C 2.5), although not without specifying certain framework conditions governing access to such resources. The necessary reform of the International Undertaking and some complex issues associated with it (e.g. farmers' rights, access to ex-situ collections) are currently being dealt with in a new negotiation process administered by the FAO. The result may be a legally binding protocol on plant genetic resources within the framework of the Biodiversity Convention; however, the outcome is still uncertain.

The Fourth International Technical Conference on Plant Genetic Resources organized by the FAO in Leipzig (June 17–23, 1996) marked an important step forwards towards a global system for managing these resources. In the run-up to the conference, the FAO presented the first report on the state of global plant genetic resources for food and agriculture (FAO, 1996a).

Access to genetic resources and equitable benefit-sharing are controversial issues, since the Biodiversity Convention does not regulate access to existing ex-situ collections (e.g. gene banks and botanical gardens). The issue of farmers' rights, meaning the recognition of the services performed by local farming communities that have long been breeding plant varieties of their own, has a direct bearing in this context.

The most important outputs of the conference were the consensus on a common policy message to the World Food Summit – the Leipzig Declaration – and the Global Plan of Action on the Conservation and Sustainable Utilization of Plant Genetic Resources for Food and Agriculture. The Leipzig Declaration states that crop biodiversity (including gene banks) is under threat, even though this is the basis for food security and hence for the "survival and well-being of humanity". However, implementation of the various technical measures listed in the Global Plan of Action is hampered by the fact that no agreement was reached on the source of additional funding.

2.4 Lithosphere/Pedosphere

At UNCED, the lithosphere and pedosphere were discussed above all with respect to soil degradation in arid, semi-arid and dry sub-humid zones (Chapter 12 of AGENDA 21). It was also decided to draw up an international convention containing legally binding commitments to combat soil degradation in arid regions and to reduce the threat of droughts. Two years later, the Desertification Convention (UN Convention to Combat Desertification in Countries Experiencing Serious Drought and/or Desertification, Particularly in Africa – CCD) was adopted; the Convention came into force in 1996. The first Conference of the Parties will be held in September 1997. The objective of the Convention is to combat soil degradation in arid zones and to mitigate the effects of drought. The treaty has regional implementation annexes for Africa, Asia, Latin America and the Caribbean, as well as the northern Mediterranean region. Priority is attached to Africa, which is particularly exposed to droughts (see WBGU, 1995 and 1996 on the processes which led up to the Convention).

The most important component of the Convention concerns the recognition of participative strategies, the enhanced role of non-governmental organizations and so-called bottom-up approaches as the basis for implementing the action programs, although implementation is still in its early days as far as practical development policy is concerned. Another new aspect is the call to use local, culturally specific knowledge (Article 16). Research is mainly needed to identify the conditions on which successful participation depends.

Another major advance signified by the Convention is that it treats soil degradation in arid areas and droughts as primarily a socioeconomic problem. The negotiation process created a stronger awareness of the need for support and coordination mechanisms.

Special reference was made to the need for coordination with other conventions on environmental and developmental issues.

In Article 10, states commit themselves to introducing long-term action programs. The most important measures concern the promotion of adequate livelihoods, the improvement of national economic environments with a view to strengthening programs aimed at the eradication of poverty and at ensuring food security, sustainable management of natural resources, improved institutional and legal frameworks, improved monitoring, national capacity-building and environmental education. Another important step was the establishment of the Committee on Science and Technology, the advisory body for the CCD that will report on scientific and technological progress in combating soil degradation in arid areas.

The Convention has already produced some initial results. In Morocco, the Gambia, Botswana and South Africa, national action programs are in the making or undergoing further development. France is supporting national action programs in Burkina Faso, Senegal, Chad, Cape Verde and Mauritania. In 1997, Germany earmarked DM 2 million for sectoral activities to combat desertification, sponsored the Pan-African Conference of Ministers in March 1997, and took over the lead agency function for Mali as provided for in the Convention (see WBGU, 1996 on the German contribution in previous years).

The Convention attracts criticism for its lack of funding commitments; the Parties could only agree to improve administration of existing funds under a Global Mechanism (Article 21). In its 1995 Annual Report, the Council has already drawn attention to Germany's declining contribution to agricultural research for the developing countries and noted with concern that the financial cutbacks in international agricultural research since 1991 have not been reversed. The location of the Secretariat, which shall commence operations in 1999, will be decided at the first Conference of the Parties in Rome in 1997. Spain and Canada have applied, as has Germany.

The Desertification Convention covers only part of the wider, global problem of soil degradation. In order to give soils in general a similar degree of protection under an international regime, the Council recommended in its 1994 Annual Report that the Federal Government examine whether a global Soil Convention would be a feasible option.

2.5 Population

2.5.1 UN Conference for Population and Development (Cairo)

The International Conference for Population and Development (ICPD) was held in Cairo in September 1994 (WBGU, 1996). The summit debated the perspectives for future population policy in the context of sustainable development, measures to enhance women's reproductive rights and reproductive health. The departure from the narrower concept of family planning at the Cairo conference signified a change of paradigm in population policies for developing countries.

The central focus shifted to health care and health promotion, with special reference to the availability, gender-related accessibility and culture-specific acceptance of health care services and facilities. The central role of family and kinship structures was confirmed, and special attention to the needs of young people established as a major principle. In the view of many observers, women's rights were articulated in a stronger form in the Global Plan of Action of the Cairo conference than at the World Conference on Women (Beijing).

A key element of the document is that it gives special consideration to human rights as well as the cultural and religious frameworks in which the Global Plan of Action shall be implemented. Of equal importance is the section on the special rights and needs of children, adolescents and youths, e.g. with regard to information on birth control and AIDS prevention.

Intervention by the Vatican and some Islamic states caused debate to revolve around the abortion issue in many respects (Section 8.25), to the detriment of the two main conference issues (namely the relationship between population and development, and between population and environment). That aside, the Global Plan of Action endorses the "right to development" referred to in the final documents of other UN summits. The conference also acknowledged that solving the population problem is directly linked to sustainable development. In June 1996, ECOSOC called on the international community to speed up implementation of the Cairo plan.

2.6 Social organization

2.6.1 United Nations World Conference on Women (Beijing)

The Fourth World Conference on Women (FWCW) ended on September 15, 1995 in Beijing. The Conference adopted as its final act the Programme of Action based on Chapter 24 of AGENDA 21 (Global Action for Women Towards Sustainable and Equitable Development), in which the special role of women in the field of environment and development is emphasized. The Platform of Action was summarized in the 38 sections of the Beijing Declaration.

The central statement of the Platform is that although the situation of women has advanced in some important respects since the Third World Conference on Women held at Nairobi in 1985 and the many UN conferences of the intervening period, major obstacles remain which stand in the way of women's equal participation in development. The Platform of Action therefore stresses the importance of recognizing and implementing the human rights of women at all levels and in all areas of life, and addresses issues that have been on the international political agenda of recent years, namely poverty, migration and the environment. In addition to the realization of human rights, the empowerment of women is another central element of the recommendations, which are addressed not only by governments, but also by banks, international financial institutions, non-governmental organizations, research institutions, women's organizations, the private sector and trade unions.

The final act of Beijing affirms that poverty and environmental degradation are closely interrelated, and identifies unsustainable patterns of consumption and production as the major cause of the continued deterioration of the global environment. In view of the lack of recognition and support for women's contribution to environmental protection, women's perspectives and concerns should be integrated into all environmental decision-making. As in the Convention on Biological Diversity, governments are called upon to encourage the protection and use of the specific knowledge, innovations and practices of women in indigenous and local communities. The feminization of poverty and the lack of women's participation in decision-making are criticized, as is the fact that women lack or are denied access to education.

Many issues raised by the women's movement were first put on the international political agenda at the World Conference on Women. The conference evidenced intense polarization along religious and cultural lines among conference participants and government delegations, and a North-South cleft in the setting of priorities. The Platform of Action was adopted by consensus, but various states hurried in the aftermath to express their reservations. The emphasis on national and sociocultural autonomy and traditional values played a major role for developing countries especially, whereas individual and equal human rights for men and women were highlighted by participants from the industrialized world.

National governments, in consultation with non-governmental organizations, should have developed implementation strategies by the end of 1996. The Conference on Women did not produce the breakthroughs that many observers had hoped for, particularly in respect of an all-embracing concept of human rights, but led to further sensitization regarding the rights and participation of women in world politics and at local and national level. Issues that had not been debated at this level before (e.g. marital rape, pornography) were "internationalized" and put on the agenda for the first time in Beijing. The conference failed to adopt effective financial arrangements for the measures called for, despite the fact that the financial resources deployed so far for the advancement of women are clearly inadequate. The demand that 0.7% of the GNP of the industrialized countries be provided as official development assistance was endorsed.

A national follow-up conference to the World Conference on Women was held in Germany in March 1996 under the auspices of the Federal Ministry for Women and Youth. In line with the obligations arising from the Platform of Action, national strategies were developed in conjunction with the relevant ministerial departments, the *Bundesländer* and non-governmental organizations. The main foci related to the equal access of women to decision-making positions at all levels of society, improving the situation of women in the economy and the employment market, human rights and the eradication of violence against women.

Germany ranks 18th among the industrialized countries on the UNDP Human Development Index scales for "Women and capabilities" and "Women and political and economic participation". This highlights the need for action at national level as well, if full sexual equality is to be accomplished. The Council has repeatedly referred to the key function of women in society and politics, and has identified the policy action and research that is required (WBGU, 1995 and 1997).

2.6.2
United Nations Conference on Human Settlements (Habitat II) (Istanbul)

Sustainable settlements development and the eradication of poverty (Chapters 3 and 7 of AGENDA 21) were the focus of the second United Nations Conference on Human Settlements (HABITAT II) in Istanbul in 1996, which commenced with a global report on human settlements. The key issues at HABITAT II were the right to adequate housing laid down in the International Covenant on Economic, Social and Cultural Rights of 1966 and sustainable settlements development. The right to adequate shelter was endorsed as a central element in the final documents of HABITAT II (the Global Plan of Action and the Istanbul Declaration).

The conference debated whether and in what form the constitutional position and the operational capacity of local authorities could be strengthened through the structural reform of the United Nations system. Local self-government, non-governmental organizations and decentralization were recognized in an unprecedented way as key prerequisites for urban development. Other major features of the HABITAT conference included the redefinition of cities as partners in the United Nations, the emphasis on social aspects of sustainability, gender issues, the proposal to transfer development assistance directly from local authorities, and enhanced cooperation between urban areas. HABITAT II gave a clear signal that partnership between all levels, especially for implementation of LOCAL AGENDA 21 (see Section C 3 below), is more important for the future of cities than central government activities.

Not only international organizations, but also the private sector and local governments are called on to mobilize the requisite financial resources for the follow-up process. There was substantial agreement that more development assistance is needed from the public sector; the Group of 77 also proposed that development aid be transferred from towns and cities as well. Another topic at the conference concerned the introduction of *debt for shelter swaps* as a means of financing settlements development.

During the run-up to HABITAT II, governments were expected to include quantitative data on 46 key indicators of urban development in their national reports. Germany was among the many states which submitted a national report on settlements development and policy prior to the conference. The states participating at HABITAT II committed themselves to monitor global conditions and trends in human settlements development on a regular basis. These data shall be fed into a new database at the United Nations Centre for Human Settlements (UNCHS) and shall form the standard basis for subsequent reporting to the United Nations. The UN General Assembly shall review progress in implementing HABITAT II in 2001.

The BMZ (Federal Ministry for Development Cooperation) has initiated a number of projects as part of the HABITAT II follow-up. Their main foci relate to decentralization in urban development projects and alternative implementation models for settlements management at local authority level. Continued support is also given to the Urban Management Programme (see WBGU, 1997).

2.6.3
World Conference on Human Rights (Vienna)

Sustainable development is inseparable from the protection of human rights, meaning civil and political rights as well as social, economic and cultural rights. AGENDA 21 refers at several points to the importance of human rights in implementing the Programme of Action.

A World Conference on Human Rights was held in Vienna in 1993, at which the international community committed itself, in the Vienna Declaration and Programme of Action, to strengthening respect for human rights. The conference reaffirmed that all human rights are universal. A large proportion of the final documents relate to civil and political rights. Another important element, precisely with regard to the Rio follow-up process, was the emphasis on economic and social rights. The Vienna Declaration, albeit non-binding, reaffirms the right to development as a "universal and inalienable right" that should be fulfilled in such a way as to meet equitably the developmental and environmental needs of present and future generations. The international community is called on, for example, to "make all efforts to help alleviate the external debt burden of developing countries, in order to supplement the efforts of the Governments of such countries to attain the full realization of the economic, social and cultural rights of their people". "Immediate alleviation and eventual elimination" of extreme poverty are also essential for the enjoyment of human rights. Another social right is the right to water, which is seen as part of the human right to food, on the one hand, but also debated as part of the right to life (see Box D 3.3-3).

One important step that was supported by Germany especially is the establishment of a United Nations High Commissioner for Human Rights, who shall strengthen and promote the political, civil, economic, social and cultural rights of humans, including the right to development. The Vienna conference also encouraged the examination of optional protocols for the International Covenant on Economic, Social and Cultural Rights, similar to those already developed for the International Covenant on Political and Civil Rights.

2.7 Economy

2.7.1 General Agreement on Tariffs and Trade / World Trade Organization

The interaction between the GATT/WTO regime and the problems of global change have received increased attention in the scientific and public debate in recent decades (Kulessa, 1995; Helm, 1995). Firstly, environment and development organizations argue that the current liberalization of world trade can have not only welfare-enhancing, but also negative impacts on the environment and on development in some regions and fields. Secondly, there is the problem of how to shape national environmental policy in such a way that competition is not distorted at international level.

There is no overlooking the need for better coordination of international environment policy and trade policy. On the other hand, it is important to have safeguards against covert protectionism in the form of unilaterally defined environment standards, as ruled by the GATT panel in the well-known Tuna-Dolphin Case (GATT, 1991 and 1994; Dunhoff, 1992). The accusation of protectionism is leveled above all by developing countries against the environmental policies of the industrialized countries. The first Conference of the Parties to the WTO – in Singapore in 1996 – failed to produce a final ruling on this issue.

In specific cases, provided a number of conditions are met, the Council supports the imposition of trade restrictions – e.g. those agreed upon in Article 4 of the Montreal Protocol on Substances that Deplete the Ozone Layer (WBGU, 1996): the trade barriers must be an instrument of last resort, and may only be introduced after prior negotiations with non-member parties; they must also relate in a direct way, as far as possible, to the environmental problem at issue, and the revenues generated should be used for the benefit of the environment. Care must be given to ensure that trade measures against non-member states are based as far as possible on a broad consensus within the international community, in other words that they impact on only a small minority of states. The issue should therefore remain on the political agenda within the WTO, so that unanimous arrangement is reached and unilateral trade restrictions minimized.

2.7.2 United Nations World Food Summit (Rome)

Debate at the 1996 World Food Summit in Rome was concentrated on food security, rural development and the eradication of poverty (Chapters 3 and 14 of AGENDA 21). The final acts of the conference, the non-binding Rome Declaration and the Global Plan of Action, acknowledge poverty as the main cause of famine and underline the necessity of a sustainable increase in food production. The right to food is recognized as a basic human right, as was proclaimed in the non-binding Universal Declaration of Human Rights of 1948, in Article 11 of the International Covenant on Economic, Social and Cultural Rights (1966) and at the World Food Conference of 1974. The Conference did not provide a specific description of the right to food; instead, the UN High Commissioner for Human Rights is called on to define this in operational terms.

The summit identified a stable and effective political, economic and social framework, including good governance and a transparent and enforced legal environment, as crucial for sustainable food security. Essential foundations include the peaceful resolution of conflicts, respect for human rights and basic freedoms, democracy, as well as transparent and responsible government. Trade is also viewed as vital to food security, but equal emphasis is placed on the necessity of mitigating the potential adverse impacts of trade liberalization. The document also contains a clear rejection of unilateral trade measures. The role of civil society in achieving food security, and especially the contribution of women, is mentioned in numerous places. There were no negotiations on how to finance the Rome decisions. One instrument referred to was *debt for food security swaps*.

Many countries presented national reports on their food situation to the Conference. These reports are an important basis for the follow-up process, particularly since standard indicators for monitoring global food security are to be developed under the auspices of the FAO Committee on World Food Security (CFS).

International implementation of the Plan of Action adopted by the World Food Summit is being integrated into the existing UN mechanisms for implementing the action plans of other major summits. The FAO will have a central role to play in this context, in that the CFS was nominated as the body responsible for implementing and monitoring the follow-up process to the World Food Summit at governmental level. The Plan of Action also stipulates that its implementation must be coordinated with other international organizations, including the Bretton Wood institutions and the regional development banks. By implementing the declarations of Rome, the number of people suffering from starvation is to be halved by the year 2015. An interim assessment shall be made in 2006, and a review shall be carried out in 2010 to determine whether the target can be reached.

2.7.3 United Nations World Summit for Social Development (Copenhagen)

Chapter 3 of AGENDA 21 is concerned with the eradication of poverty, the main issue at the World Summit for Social Development held in Copenhagen in 1995. Although the summit failed to adopt a new Social Charter, it managed to establish a new quality in the way that social security issues are approached. Whereas "people-centered" development was a contentious issue at the fourth UNCED preparatory conference in 1991, this approach has now become a key element in the final documents of the World Summit for Social Development.

The Global Plan of Action consists of 10 commitments. One of the most important outputs of the Copenhagen summit was the so-called 20-20 approach, according to which developed and developing countries allocate 20% of official development assistance and 20% of the national budget, respectively, to basic social programs. The voluntary nature of this commitment remains a weakness, however. The industrialized countries reaffirmed the non-binding target of spending 0.7% of GNP on overall official development assistance. The subject of debt relief was dealt with at length, with the conference calling for implementation of debt relief measures as soon as possible.

The Global Plan of Action stresses the central role of women in the development process. For the first time at international level, the promotion of full employment was defined as a priority goal of economic and social policies. The states committed themselves to ensuring that structural adjustment programs include social development goals. They are also called on to monitor the impact of structural adjustment programs on social development, especially with regard to their impacts on women.

The most important task of governments is now to translate the "people-centered" policies debated in Copenhagen into practical activities and to set up national programs of action.

3 Local government implementation of Agenda 21

Global change as a behavioral problem – "Bottom-up" approaches – LOCAL AGENDA 21 initiatives – Participation culture – Public participation at local community level – Potentials and barriers of LOCAL AGENDA 21

3.1 Importance of local-level political processes for sustainable development

Global change, as the Council has repeatedly emphasized in its Annual Reports, is ultimately a behavioral problem, and the crisis of the environment is a crisis of society. Behavior relevant to global change occurs at various levels of society, from the individual through the family, the place of work and the local community to national and international organizations. People always act in local contexts that are spatially and temporally specific, and are influenced by these in turn. This applies to the causes of global change problems, as well as eradication of or adaptation to them. It is therefore essential to integrate measures for sustainable development into local sociocultural contexts and to shape them according to the specific problems and target groups concerned.

Local and regional initiatives for achieving sustainable development are therefore of utmost importance, particularly since the top-down policymaking approach (multilateral negotiations on conventions and protocols at global level) is proving to be a strenuous and protracted process with uncertain outcomes. Activities involving local authorities and/or local communities are particularly appropriate, because at this level there is a kind of "hinge" between politicians and community interests. On the one hand, people are most likely at the community level to have opportunities to bring their interests into the policymaking process, while, on the other, it is here that they can best be shown the positive and negative impacts of their activities. Furthermore, local authorities can play a key role in sensitizing people to sustainability issues. Society's acceptance of far-reaching national and international decisions on sustainable development is critically dependent, in turn, on such sensitization.

3.2 The Local Agenda 21 concept

AGENDA 21 stresses the importance of social groups for sustainable development. Chapter 28 requires local authorities to undertake a consultative process with their populations in order to reach consensus on a LOCAL AGENDA 21 for the community. Each local authority should enter into a dialogue with its citizens, local organizations and private enterprises in order to specify and implement a LOCAL AGENDA 21.

AGENDA 21 is thus adopting a perspective that is new in international policymaking: multilateral policies for sustainable development (top-down processes such as negotiations on global conventions) must be accompanied by bottom-up approaches that address specific local problems and which actively involve the local community in the formulation and implementation of local government policies ("think globally, act locally"). A LOCAL AGENDA 21 (LA21) is therefore to be understood as a "participatory, multisectoral process to achieve the goals of AGENDA 21 at the local level through the preparation and implementation of a long-term, strategic action plan that addresses priority local sustainable development concerns" (ICLEI, 1997).

Many municipalities and communities setting out to implement AGENDA 21 at local level, in particular the goals relating to the environmental, economic and social aspects of sustainable development, already have programs in place on which they can base such activities. However, there are some differences between this process and traditional local government policymaking on environmental, economic, developmental and social welfare issues that make the adoption of an LA21 program a daunting political task. The most important difference concerns the integration of what are often competing policy areas to develop a cross-cutting action program guided by the sustainability principle. As well as this *integration* of different policy fields, it is essential to involve all local stakeholders and groups with their different interests, concerns and know-how into the LA21 process, in both target-setting and the implementation of activities. The concept of *communication and participation* at the heart of AGENDA 21 goes much further in this context than traditional models for public participation in local government planning. Citizens should not only be informed and consulted, but should take an active part themselves in political decision-making. This requires the creation of suitable forms of consultation and consensus building. The principle of sustainable development also implies new, long-term time frames for local authority planning, which runs counter to the prevailing focus on electoral periods or on short-term economic thinking.

3.2.1 Local government participation in the LA21 process

Developing a LOCAL AGENDA 21 is a long-term process involving activities that are often slow to get off the ground and which have to cope with a diversity of problems. This explains the failure to meet the schedule laid down in AGENDA 21, according to which most local authorities in each country should have achieved a consensus on a LOCAL AGENDA 21 by 1996.

There is no agreement on criteria for LA21s, due not least to their specifically local character. It is proving difficult, despite the efforts of various institutions (ICLEI, 1996; Kuby, 1996, for the Council of European Municipalities and Regions; Rösler, 1996, for the German Association of Municipalities), to arrive at a clear assessment of progress in implementing Chapter 28 of AGENDA 21. According to a survey carried out by the World Secretariat of the International Council for Local Environmental Initiatives (ICLEI), approximately 1,800 local governments in 64 countries were involved in LOCAL AGENDA 21 activities in 1996 (ICLEI, 1997; as a comparison, around 170 countries signed AGENDA 21). European LA21 initiatives are mostly concentrated in Great Britain, The Netherlands and Scandinavia. In Sweden, for example, virtually all local governments are in the process of developing their own LA21s, thanks to active support from the environment ministry.

Many of the local authorities working on LA21s are members of international networks. After the Rio Conference, ICLEI centered its activities on the worldwide promotion of LA21. ICLEI was also a key player in setting up the European Sustainable Cities and Towns Campaign in 1994, an association of around 290 cities, towns and counties in Europe that have committed themselves, by signing the Aalborg Charter, to engage in LOCAL AGENDA 21 processes. Towns & Development (T & D), the international North-South network of local authorities and NGOs that has led to around 2,000 partnerships based on the *Berlin Charter* between local authorities throughout the world, is similarly involved in LA21 activities.

3.2.2 LA21 activities – an international comparison

Local government activities aimed at developing and implementing LA21s can utilize the specific sociopolitical contexts in the respective country – e.g. the presence of a local environment authority or a "participatory culture". The political, legal, economic and social frameworks for LA21 processes in the various countries display considerable variation. It therefore makes little sense to perform an international comparative assessment. However, an international comparison of LA21 activities brings to light the enormous range of structures, methods and strategies deployed by local governments.

Almost all local authorities active in this area consider a policy resolution on LA21 to be an essential first step towards establishing political commitment. The various forms of commitment range (with ascending binding and motivational force) from the signing of the Aalborg Charter, to adoption of a LOCAL AGENDA 21 and provision of adequate funds, to the designing of a local development concept including key targets and a program of activities.

Most towns and cities assign responsibility for implementing LA21 to the local environment authority. While this means a contact address within local government, often well-motivated at that, this type of approach fails to appreciate the real aim of the LA21 process, namely the integration of different issues and sectors for the preparation and implementation of an overall strategic concept. It is interesting in this context that wherever responsibility for coordination is decentralized, the process is organized in a very targeted and efficient manner.

Involving as many local stakeholders as possible from the very outset is almost impossible within existing structures. Sooner or later, all active municipalities have established special structures for involving citizens and stakeholder groups. These include open community forums, many of which meet on a regular basis, stakeholder groups to discuss principles and targets, as well as smaller working groups and expert forums to develop more specific proposals.

Many local governments confine themselves at the beginning of LA21 processes to traditional consensus-building methods, such as public meetings and public awareness campaigns. In contrast, almost all cities where the process is already further advanced are working with more up-to-date and sophisticated methods like planning cells and visioning workshops, or at least with some form of professional coordination.

As far as the integration of issues – the key innovation of AGENDA 21 – is concerned, the industrialized countries in particular continue to display an almost total bias towards environmental quality. Wherever efforts are made to solve existing structural problems with the help of the LA21 process, economic and social criteria have been incorporated as a matter of course. Nevertheless, many cities now have focus groups or specialist forums on non-environmental issues, which suggests that an increasingly integrated approach is being taken towards cross-cutting aspects.

Two main strategies can be observed regarding the translation of AGENDA 21 aims to the local government level. In the first case, the primary focus is on setting development targets and developing the relevant indicators, so that specific activities can be successively derived or assessed. The second strategy relies on projects and isolated activities, without participatory processes for target-setting. The latter approach may lead more quickly to visible successes, and in that sense have a motivational impact, but it easily creates an illusion that sustainable development can be achieved in the relatively short term provided one operates the right collection of measures, without having to make further-reaching and ongoing changes to behavioral and decision-making patterns.

3.2.3 LA21 initiatives in Germany

According to a survey conducted by the European Secretariat of ICLEI in late 1996, only about 60–70 local governments in Germany – out of a total of more than 16,000 – are involved in developing an LA21. This would put Germany in the middle ranks on an international comparison. Other sources identify as many as 200 local communities (Kuby, 1996; Rösler, 1996), while some refer to an "outbreak" of LA21 initiatives in the second half of 1996 (de Haan et al., 1996).

In addition to local environment authorities, who mostly provide the initiative, the main actors in the German LA21 process are above all the environmental and North-South NGOs, the adult education centers, and organizations affiliated to the churches. Involving local enterprises is proving difficult in many cases. Mobilization of citizens who are not members of the groups just mentioned is similarly a problem in many local authorities.

Many German municipalities are highly active in the environmental field especially. Existing structures and policies, some of which are already traditional, provide an initial basis here. A hearing on LA21 before the Enquete Commission on Protection of Humanity and the Environment revealed that the activities adopted or already implemented by German municipalities with LA21 initiatives are primarily related to urban and transport development, energy, nature conservation and public awareness campaigns (Enquete Commission, 1996). In contrast, however, they tend to neglect the social, development policy and economic dimensions of sustainable development. There are also shortcomings with regard to other key elements of LA21, such as long-term targets and public participation. Many local authorities (not only in Germany) wrongly believe that active policies to protect the environment are synonymous with the goals of a LOCAL AGENDA 21. When LA21 activities are assigned to the environment authorities, as is regularly the case, this frequently operates as an obstacle to any cross-cutting approach.

Although some specific LA21 activities adopted by local governments in other countries (e.g. Great Britain, Southern Europe) are very similar to those implemented by local governments in Germany 10–15 years ago (e.g. environmental reporting, separation of waste, recycling, environmental awareness-raising), these local governments are explicitly linking what in some cases are very young (environmental) policies at local level to the new concept of LA21, thus using the opportunity to work under the Local Agenda philosophy from the very outset (long-term approach, inclusion of global aspects, partnerships, public participation).

No local authority in German has yet managed to produce a collection of measures commanding a broad consensus in society. Most LA21 initiatives are still concerned with achieving agreement on general principles for LA21, or with analysis of the status quo in their areas in relation to these general principles (de Haan et al., 1996).

LA21s involve specific activities and procedures that are less typical nowadays in the operations of German local governments. The idea of integrated urban development, for example, was already on the agenda as far back as the 1970s. The negative experiences of that time, due especially to organizational shortcomings, are frequently a barrier to extensive and early involvement of stakeholders by local governments.

As well as reviving the idea of citizen involvement in local community development, it is essential to promote modern forms of citizen participation, balancing of interests and conflict resolution, to propagate them and to train those charged with coordination and moderation of participation schemes. Rudimentary forms of citizen participation, as established so far in the German laws on planning and approval, but which have been increasingly reduced again in the recent past (the so-called procedural acceleration laws) are inappropriate models to emulate. They are biased too much towards acquiring acceptance for planned measures, and function more as a legal safeguard than as active participation of citizens in the policymaking process.

The Scandinavian countries and The Netherlands, where environmental protection at local government level and a high degree of citizen involvement have traditional roots and are brought into LA21 processes, provide excellent models for local authorities in Germany to emulate. Furthermore, as a comparison with Great Britain, The Netherlands and Sweden shows, a national coordination body is very helpful for disseminating information, exchanging experiences between local governments with LA21s, and especially for preventing duplication of labor.

3.2.4
Towards sustainability with LA21 – Potentials and barriers

The LA21 concept is especially important on account of its sizable potential for achieving sustainable development at local government level.

- For all the novelty of the concept, LA21 initiatives can build on existing policies (relating primarily to the environment in most cases), on existing communication links between local authorities and NGOs, on the environmental awareness and commitment of the population, and on any related experience that has already been acquired.
- Global aims may, indeed must be linked to the concrete situation on the ground (environmental situation, labor market, locational issues, etc.) in order to arrive at a real consensus.
- At local level, most of the relevant actors are known. Responsibilities can be assigned to persons or groups, thus enabling the complexity of problems to be kept within certain limits at least.
- Setting up new forms of popular participation can enhance the oft-neglected social and humane components of sustainable, environmentally and socially acceptable development, and can spark off a learning process among all involved.
- "Environment and development" is becoming a cross-sectoral challenge at local level, too, and one which takes on clear contours through LA21. This requires new criteria for planning and decision-making on the part of local governments.
- By addressing people in their specific lifestyle contexts at local level, there is greater potential for sensitizing them to the somewhat abstract principles and aims of sustainable development, and for motivating them to adopt sustainable patterns of behavior. To that extent, LA21 processes are an important channel for educating the population on environmental and development issues.

The potential strengths of LA21 are countered, however, by a whole set of barriers that can greatly impede the success of LA21 processes:

- Five years after Rio, there are still severe gaps in information among the population and public administration with respect to the basic concepts of "AGENDA 21" and "sustainable development".
- Even at local level, the issues involved in LA21 are already of such complexity that many potential actors are hampered or prevented from taking part in the process. Another aspect is that few if any actors have ever learned "networked thinking" to the extent needed for the LA21 process.
- In many cases, the attempt to operationalize environment and development as cross-sectoral issues within the local context turns out to involve severe complications and cannot work without institutional reforms.
- All local initiatives aiming at sustainable development rely on international or national top-down processes that frequently impose a framework of critical relevance.
- The leeway open to local governments is often restricted in so many respects (legal, financial and human resource constraints) that ambitious goals are often followed by humble deeds. On the other hand, lack of financial resources can compel authorities to find creative solutions.
- Decision-makers in politics and public administration often see themselves as being an upper tier compared to the general population, an attitude which makes agreement, never mind cooperation based on confidence-building and consensus, a difficult enterprise. If key political personages in a community fail to give their support to the LA21 process, then the latter has little or no chance of success.
- There is a lack of experience on all sides with new forms of participation and communication, relating to group dynamics or conflicts, for example, and how these can be handled.
- Taking part in the LA21 process consumes so much time and energy, especially for volunteer workers, that people are discouraged from getting

involved, or those who do are prevented from carrying out their other tasks.

- As a rule, local government planning processes have little impact on the general public and take up considerable amounts of time. There is always a risk of plans being forgotten and – intentionally or not – winding up in a dead end.
- The prioritization of short-term as opposed to long-term planning in local government policy-making runs counter to the time scales needed for LA21.

Implementation of AGENDA 21 at local level is still a very young process, the results of which will become apparent as time goes on. Only time can tell whether the methods in operation can actually lead beyond defining a local AGENDA 21 to real sustainable development at local community level. Developing a local action plan is no guarantee in itself that sustainable development will be achieved.

Summary and prospects

The main points in the Council's assessment of the Rio process are as follows:

- In itself, the Rio conference was a major step forwards, in that it marks the point at which the principle or goal of sustainable development was first adopted by the overwhelming majority of the international community. The countries represented at UNCED acknowledged their responsibility for the global environmental and development system and accepted the need for action at the global level.
- Given the paucity of time and the complexity of global change, it is not surprising that there are no clear signs yet of a deliberate reversal of trends as a consequence of the Rio process.
- The negative trends that led to the Earth Summit are unabated, indeed intensifying in some cases. The strategies adopted in Rio must therefore be pursued with added vigor and determination. The serious problems that nations are experiencing within their own borders, and the growing shortage of financial resources, must not detract from their global commitments.
- The fact that severe difficulties at national level are often linked through globalization to global environmental problems and development concerns, and experience feedback from the latter, means that national and global tasks can only be tackled through joint action.
- Germany has a special obligation and responsibility to bear in this context. As a "global player", one of the major causal agents of global environmental problems and as one of the most powerful economic nations in the world, Germany should display a special level of commitment in the field of global environmental and development policy (WBGU, 1997).
- If the process initiated in Rio is to bear fruit and the problems associated with global change are to be brought under control, it is essential to promote and push forward the horizontal process of self-organization among nation-states – i.e. *global governance*.
- Where conventions, protocols and action programs are essential instruments for achieving this goal, they should be embedded in a rational way within the existing system of institutions so that no duplication of effort occurs. In view of increasing international institutionalization, the coordination and refinement of existing institutions is becoming an issue of ever-greater importance.
- International organizations and departments already existing within the United Nations system should be committed more strongly than before to the principles adopted at UNCED. The Council therefore advocates that the UN Environment Programme (UNEP) be enhanced in its function as catalyst and initiator of international environment and development policy based on the resolutions of the Rio Summit.
- Given the obligations that nations enter as part of the Rio process and which will grow in future, the sometimes drastic levels of developing country debt represent a restriction on implementation that the international community should take very seriously. According to AGENDA 21, implementation in the developing countries will require an additional US$ 600 billion per annum, of which the international community is supposed to provide US$ 125 billion (per annum in the 1993–2000 period) (see AGENDA 21, 33.18). Taking Germany's 8.93% quota of the United Nations budget in 1993 (1997: 9.5%), this means an annual contribution by Germany of US$ 11.16 billion. For 1993 – the first year of the AGENDA 21 planning frame – this would correspond to 0.59% of Germany's GNP (US$ 1,908 billion, 1993). Meeting such commitments would come very close to the agreed 0.7% of GNP target for development assistance endorsed by the various summits of the Rio follow-up process. Since economic cooperation with the developing countries comprises more than the "pure Rio follow-up costs", the real obligation amounts to signficantly more than the 0.7% target. Germany's current expenditure on development cooperation is equivalent to 0.31% of GNP (1995 figures). In contrast, comparable industrialized countries such as The Netherlands, Denmark

and Norway are achieving or exceeding the 0.7% target.

- With regard to support for developing countries, a diversity of experience has already been gained with financial mechanisms and instruments such as the Global Environment Facility jointly administered by the World Bank, UNDP and UNEP (Jordan, 1994), the Multilateral Fund for the Implementation of the Montreal Protocol (Biermann, 1997), or the somewhat older Fund for the Protection of the World Cultural and Natural Heritage of Outstanding Universal Value (the World Heritage Fund) (Birnie and Boyle, 1992). "Debt swaps" are a new instrument involving debt relief for developing countries in return for environmental measures ("debt for nature swaps") or food security ("debt for food security swaps") (Jakobeit, 1996). Alongside direct funding of environmental projects through separate mechanisms, environmental protection must be integrated more as a cross-sectoral task into the work of the other "traditional" financing mechanisms (e.g. multilateral mechanisms such as the World Bank, or bilateral programs like the Hermes export credit guarantees in Germany). More work needs to be done to examine which instruments are particularly appropriate in specific cases.
- The Council stresses the necessity of sanctions against states which fail to comply with their commitments. Direct sanctions on states that do not comply with the conventions at all or inadequately is envisaged in most convention texts, but in practice are often unenforceable. The Council refers in this connection to the positive experience with new kinds of non-confrontational dispute settlement mechanisms (Victor, 1996). "Weaker" commitments of states in such regimes, mandatory reports for example, have an important role to play here as well (Levy, 1993). In certain problematic areas where global compliance with quantified environmental commitments is impracticable or unenforceable within the specified time frames, reporting obligations may serve as a surrogate.
- Non-governmental organizations have acquired a key role in responding to global change. As global "trustees of the environment", NGOs exert a crucial influence during diplomatic negotiations; they are also key actors in the national process of implementing AGENDA 21. The Rio conference marks the starting point for the involvement of non-governmental organizations at summit conferences, as reflected in their informational and consultative rights at conferences. As the Council emphasized in 1996, consideration should be given to extending these participatory rights of environmental and development organizations within the UN system (see also Sands, 1989).
- Responding adequately to global change is only possible if lifestyles and consumption patterns are changed. Education and raising awareness of the global crisis are crucial factors in coping with environment and development problems. The Council's recommendation here is to strengthen education in the field of environment and development, both in and outside schools (WBGU, 1996). Germany should maintain or expand upon its existing policy of promoting NGOs in Germany and abroad that work in the environment and development field. In a similar vein, UNEP's educational and public awareness campaigns at local level should be enhanced and extended.

Since the Earth Summit in Rio, there have been many initiatives to develop and implement a LOCAL AGENDA 21 at community or regional level. Their bottom-up approach, addressing the concrete lifestyle and behavior of people in their immediate surroundings, makes them an ideal complement to international responses to global change. The Council recommends that federal and *Länder* support be given to LOCAL AGENDA 21 initiatives and that efforts be made to promote their interlinkage at national and international level.

The Council's overall conclusion is that further efforts are essential, however much progress has been made in the Rio follow-up process. Responding to global change is a central task for any future-oriented policy at domestic and international level, and one which must not be pushed into the background by the present-day focus on national problems.

Focus: Water

D

The freshwater crisis: Basic elements 1

1.1 Water functions

Resources of drinking water and process water: a minute proportion of global water stocks – Water as food, habitat and regulator – Utilization and pollution of water resources by people

The entire stock of water on the "blue planet", Earth, is estimated at approx. 1.4 billion km³ (Fig. D 1.1-1). The world's oceans, which cover 71% of the Earth's surface, account for 96.5% of that total. The rest is found as polar ice masses and glaciers (1.77%), as groundwater (1.7%) and as the waters of lakes, swamps, rivers, permafrost and the atmosphere (0.03% all told). Freshwater stocks account for only 2.5% of the total volume of water in the hydrosphere (35.1 million km³), of which 69% is stored in glaciers and permanent snow cover and around 30% as groundwater. A mere 0.3% of all surface water is freshwater. Although the water volume on the planet has been constant for a long time, distribution among the three different water phases (ice, water and water vapor) has been subject to major fluctuations during Earth's history.

The water resources of crucial importance for humankind and which are continuously renewed through rapid turnover are the small amounts of water in lakes, streams and "active" groundwater. Annual runoff to the oceans is about 41,000 km³, of which 28,000 km³ per annum flows directly as surface runoff, and 13,000 km³ through groundwater to rivers.

Around 8% of this renewable resource is used at present, 69% being agricultural use, 23% industrial use and 8% domestic use. Regional variations are very large – for example, 96% of all industrial water use occurs in North America and Europe.

Before we deal in detail with freshwater stocks, freshwater turnover and freshwater use, an overview of the manifold *functions of water* on Earth, the "water planet", should first be given. A description of water's functions shows not only the significance of water for life and the balance of nature, but also the role of water as a renewable resource and as a cultural asset for people. A list of these functions is found in Box D 1.1-1.

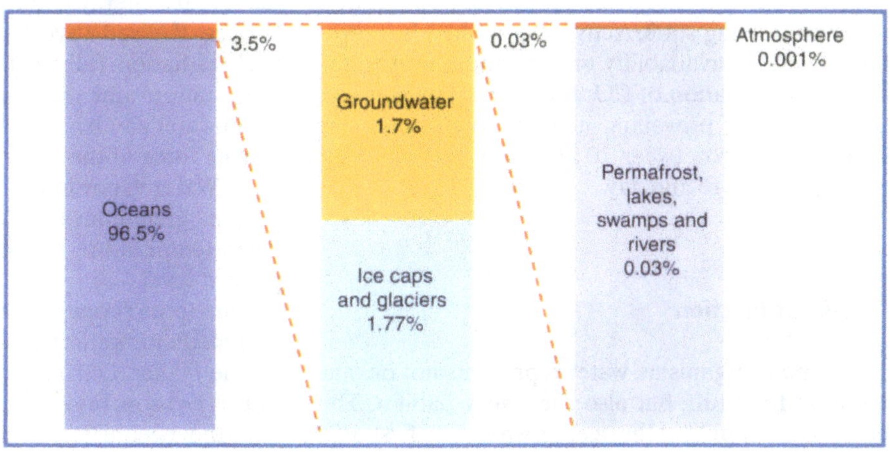

Figure D 1.1-1
Global distribution of water resources.
Source: Gleick, 1993

BOX D 1.1-1

Functions of water

NATURAL FUNCTIONS

- Life-sustaining function
- Habitat function
- Regulatory functions
  - Energy balance
  - Hydrological cycle
  - Matter balance
    - Morphological function (floods, erosion, sedimentation)
    - Solvent and transportation medium
    - Self-purification of nature

CULTURAL FUNCTIONS

- Consumption, withdrawals
  - Foodstuff function
    - Drinking water
    - Preparation of meals and beverages
  - Cleansing function
  - Production functions
    - Raw material (chemistry, food production)
    - Process water (transport medium, detergent, solvent, extinguishing agent, etc.)
    - Crop and animal needs (irrigation, aquaculture, watering places, etc.)
- Utilization
  - Therapeutic function
  - Energy carrier function
  - Transportation function
  - Recreation function
  - Design function
  - Aesthetic functions
  - Religious functions
  - Supply function (fisheries)
- Pressures
  - Disposal function
  - Self-purification function

1.1.1 Natural functions

1.1.1.1 Life-sustaining function

Not only do all organisms consist primarily of water (plants and animals 50–95%, humans 60%), but their physiological processes require an aqueous milieu as well. Terrestrial organisms therefore need a regular input of water due to the unavoidable water losses to their surroundings. Lack or scarcity of water can mean the loss of vital functions within a short space of time, which means death if the organism is unable to survive protracted periods in a suspended, water-conserving state. Active life on land is inseparable from the availability or consumption of water, and the assimilation of CO_2 for biomass production. Depending on prevailing climate conditions, crop stands consume between 10 and 100 cubic meters of water per hectare and day.

1.1.1.2 Habitat function

For many organisms, water represents not only an essential foodstuff, but also their very habitat. This holds true not only in the oceans, where life first originated, but also in surface waters such as rivers, lakes and swamps, groundwater, and the interstitial water of the pervious zone (soils). The planet could never have acquired its present state without the many different life forms that evolved in the various aqueous habitats. In addition, aquatic habitats are in close interaction with terrestrial ecosystems and the atmosphere. Habitats exert mutual influence on each other, especially through substance loads and exports.

1.1.1.3 Regulatory functions

Water's different phases and its specific phase transition properties make it a critical regulator of the Earth's energy and mass balance. Ice surfaces influence the radiative balance on account of their back radiation (albedo). The energy consumed by evaporation and thawing and the energy released when water condenses and freezes are a fundamental driving force of the Earth's weather and climate systems. Water vapor in the atmosphere is the most important greenhouse gas, without which the annual mean temperature would be significantly below the present level.

Due to its properties as a solvent and transportation medium, water also influences the mass balance on the planet. During heavy rains and floods, water causes erosion, moving soil material to sink areas or into the ocean. Rainwater infiltrating the ground chemically alters rock and soil (weathering). In doing so, dissolved ions and other components are flushed into groundwater, waterbodies or the sea. Biological communities in waterbodies, soils and in groundwater may absorb and decompose nutrients and energy carriers (organic substances) that are dissolved in water. This is of major importance for the biogeochemical cycles of carbon, nitrogen and other nutrients.

1.1.2 Cultural functions

Clean and fresh water is quintessential for life on Earth and human activities. In many regions of the world, however, people have virtually no access to clean drinking water or water for industrial use. This results not only from very uneven spatial distribution of water resources, but also and especially from the irresponsible way in which they are managed.

Box D 1.1-1 lists the manifold functions that water performs for humans. In order to use water, it must first be withdrawn from surface waterbodies or groundwater. It is then either consumed, i.e. converted from the liquid to the gaseous or chemically bound state, or it is used in some way and thus changed in its quality. The wastewater that ultimately results is channelled in most cases to surface waters or to groundwater after passing through the soil. Water resources perform a function here as a medium for disposal, transport and distribution.

In order to survive, people require between 3 and 8 liters of water per day, depending on prevailing climate, in order to cover their needs and to prepare their food. People consume additional quantities of water to varying extents. Consumption occurs when liquid water is evaporated, for example in crop irrigation or when used as a coolant. Water is also consumed, however, when it is used as a raw material in the chemical and food industries. Water use refers to all those activities in which water undergoes a marked change in quality and loses some of its original characteristics. Such changes occur in cleaning, through use as process water, in crop and animal production and in aquaculture.

Waters thus contaminated are returned to the environment, where they pollute soils, surface waterbodies and groundwater. The disposal and purification functions of waterbodies are exploited in order to dispose of waste. However, critical loads are often exceeded, with the result that the habitat functions of lakes and streams are severely impaired. The filter and cleansing function of water in the unsaturated (vadose) zone is used in a similar way when unclean water is purified during its passage through soil, after which it is clean groundwater and ready for re-use.

Water is exploited by humans in various ways that involve no change in the actual volume of water. For example, water is utilized in power generation, while waterways serve on a large scale for the transportation of goods and passengers. The therapeutic and recreational functions of water are commercially exploited. In addition, water is used as a design element in the human environment, e.g. in fountains and landscaping. Even when these types of use do not result in volume reductions, they may involve pollution that in turn impairs the habitat functions of waterbodies.

Because of water's mobility and properties as a solvent, contaminants may be transported along with water into neighboring ecosystems, where they can cause severe damage. Examples include acid rain, salinization of soils, as well as nitrate and pesticide loads to groundwater. Waterbodies are also used on a large scale for waste disposal, with negative impacts on the environment.

All forms of water resource management, be it withdrawal and consumption, use or contamination, are influenced to a greater or lesser degree by the values operating in the respective sociocultural context. This specific context, which may be termed water culture, can be classified in terms of several features. These, in turn, are interdependent and can be observed in different societies, separated in space and time, in specific "mixes" (see Section D 5.2). The most important dimensions of water culture include:

- the scientific and technological dimension,
- the economic dimension,
- the legal-administrative dimension,
- the religious dimension and
- the aesthetic and symbolic dimension.

These dimensions of water culture overlie the way in which a society manages its water resources, as manifested on different levels of social action. This generates a kind of matrix, with the cultural functions of water and the sociocultural dimensions as the two axes, and the cells of the matrix having a different content depending on the society in question.

This description shows clearly that water resource management can be sustainable and environmentally sound only if due consideration is given to the manifold and interrelated functions of water, and existing knowledge is used to develop wise and appropriate strategies for water use.

1.2 Water as habitat and its importance for neighboring environments

Uneven global distribution of freshwater lakes – 10–400 liters of water in 1 cubic meter of soil – Functions of wetlands – 66% of all extinct species lived in freshwater ecosystems

Freshwater ecosystems (lakes and streams) cover 2.5 million km^2 in total, or less than 2% of the Earth's surface (Wetzel, 1983). Together with soils, they contain only 0.014% of the Earth's total water stock. Freshwater ecosystems are highly diverse in their physical structure and species composition (Hutchinson, 1957, 1967 and 1975; Hynes, 1970; Wetzel, 1983). Through the exchange of water and the material dissolved or suspended in it, freshwater habitats are linked to the surrounding terrestrial ecosystems, and via atmospheric pathways are exposed to impacts from distant regions.

1.2.1 Standing waters

The properties of inland lakes are closely linked to the way in which they were formed, the size and characteristics of their catchments, geological factors and climate (Hutchinson, 1957). Most lakes are young in geological terms (10,000–20,000 years; Wetzel, 1983). The exceptions are lakes of tectonic or volcanic origin (Lake Baikal, lakes of the East African rift, lakes of volcanic origin). Geologically old lakes often exhibit high species diversity and high endemism, i.e. they are host to species that are found nowhere else in the world (Snimschikova and Akinshina, 1994) (see also Box D 1.2-1).

The water stored in freshwater lakes is unevenly distributed over the Earth's surface. Approximately 18% of all liquid freshwater is stored in Lake Baikal alone (23,000 km^3). The Great Lakes in North America hold the same volume of water in total.

The unique physical properties of water – its high density (780 times that of air), heat capacity and optical density – are overriding determinants of the biotic conditions in aquatic habitats. Standing waters have vertical gradients of temperature and light, as well as varying concentrations of gases. The amounts of dissolved and particulate substances as well as organisms vary with water depth, and in many cases are subject to daily or seasonal changes.

The cycles of many chemical elements and nutrients in lakes are closely linked to the biological communities present. Producers (especially green plants) form living organic material that is used by fauna (consumers), and finally decomposed to its inorganic constituents by bacteria and other microorganisms (decomposers). Organic material that is less easily decomposed is deposited as sediment on the lake floor. The manifold biological functions operating within an aquatic ecosystem are maintained by a large number of species. Bacteria play a particularly important role in nutrient cycling and biological self-purification.

Biological communities in the benthic zone (the floor of lakes and oceans) have certain features in common with terrestrial communities. However, the high specific density of the water as compared to air allows a diverse community in the pelagial zone to which there is no equivalent in non-aquatic habitats (Wetzel, 1983). Open-water biological communities are composed primarily of microscopically small organisms called plankton, which are able to remain passively suspended in water. Plankton consists of predominantly unicellular cell algae (phytoplankton), bacteria (bacterioplankton), as well as small animals (zooplankton). Phytoplankton forms its organic substance through photosynthesis. With the help of pigments (especially chlorophyll), light is absorbed and used as an energy source to form biomass from carbon dioxide and nutrients (primary production). Because the intensity of light declines exponentially as depth increases, primary production in inland lakes is confined to water layers near the surface (less than 20 meters in most lakes). Zooplankton, in turn, feeds on phytoplankton, suspended bacteria and protozoa, which it takes up and assimilates (secondary production). In inland lakes, the main zooplankton are cladocerans ("water fleas"), copepods and rotifers. Cladocerans, which are at most only a few millimeters in size, are the preferred food source for plankton-feeding fish. Most bacteria live on organic substances dissolved in water. Bacteria and zooplankton secrete nutrients, which in turn can be taken up by phytoplankton. The regeneration of inorganic nutrients (especially phosphorus and nitrogen) allows for efficient use of these resources, which are often in short supply. The maximum possible biomass depends on the amount of available nutrients, which determines the trophic state of a lake and influences the species composition of the biological communities (Lampert and Sommer, 1993). The state of inland waterbodies is closely linked to the characteristics of the drainage basin, and is subject to anthropogenic influences on the supply of nutrients (Section D 4.4).

At greater depths where there is insufficient light for primary production, organisms depend on the organic material that is present. Oxygen is then consumed when such material is decomposed. Where there is a high concentration of biomass, as is often the case in nutrient-rich (eutrophic) lakes, the dissolved oxygen may be depleted fully at deeper layers that receive no light, thus depriving many organisms of life support.

The characteristics of the biological communities of the benthic zone differ markedly between shallow areas in the vicinity of lakeshores, and the lake floor. Sufficient light for plants to grow (emergent, floating-leaved and submersed macrophytes, sessile algae) is available only near lakeshores, where species-rich biological communities can develop. On the lake floor, organisms depend for nutrition on organic material that sinks to the bottom. This habitat is colonized by various species such as horsehair worms, oligochaets (*Tubifex*), insect larvae, mussels and bacteria. In cases where there is no dissolved oxygen in bottom waters, colonization of the lake floor is usually limited to bacteria. Freshwater fish feed on plankton, or draw nutrition from the benthos. Many of them, especially salmon-like species (e.g. trout), are highly sensitive to low oxygen levels and elevated water temperatures.

1.2.2 Running waters

Only about 0.004% of the total liquid freshwater on Earth is found in streams and rivers (Brehm, 1982; Hynes, 1970). Communities in running water are dominated by fish, sessile organisms or species (such as freshwater snails and blackfly larvae) that are suited to life in water currents (e.g. through adaptations such as flattened body shape and attachment mechanisms). Plankton are found only in slow-moving lowland rivers (Schwoerbel, 1987). Sessile algae and higher aquatic plants (e.g. amphibious buttercup) predominate in the riverbed flora. Among the animals, water insects play a dominant role. In running waters, the most important source of nutrition is the organic material deposited from neighboring terrestrial areas. In addition to predatory invertebrates and fish, the dominant organisms are those which take up decomposing organic matter (detritus eaters).

Along the course of running waters, the conditions for life change gradually (Niemeyer-Lüllwitz and Zucchi, 1985). Water temperature generally increases from the spring to the estuary, with daily fluctuations decreasing in amplitude. The flow of water enhances exchange processes with the atmosphere. The upper reaches of rivers with intense mixing are therefore saturated with oxygen as a rule. Here, cold-adapted animals with high oxygen requirements (larvae of mayflies and stoneflies) predominate. In lower reaches, the river depth increases and gas exchange with the atmosphere declines. Turbidity increases, the substratum is more and more finely grained, and the concentration of dissolved and particulate organic substance increases. Oxygen is consumed when organic material is decomposed by microbes. If oxygen is not replenished in organically or thermally stressed waterbodies, fish and other water organisms may perish (see Section D 4.4).

Along the course of European rivers, a distinction is made between four regions with declining oxygen demand. These are designated according to the typical fish species present there (trout zone, grayling zone, barbel zone and bream zone) (Fig. D 1.2-1). This zoning is paralleled by invertebrate communities, and is comparable to the zoning patterns of other continents (Illies, 1961). Water quality usually declines along the course of running waters due to civilizatory factors such as waste discharges, despite dilution and self-purification (Schmitz, 1961).

1.2.3 Soil and groundwater

Depending on its type, 1 cubic meter of soil may contain between 10 and 400 liters of water. In the unsaturated zone above the groundwater, water and air fill the pore spaces. Soil water and its biological communities are immensely important for the stability of terrestrial ecosystems and the maintenance of nutrient cycles. Water is the solvent and transportation medium for many substances, and provides plants with the water and nutrients they need to grow. Soils manifest steep vertical physico-chemical gradients and differences in colonization. Photosynthesis is not possible due to the absence of light. With the exception of bacteria that can use special chemical reactions to produce their own energy, living organisms in soil and interstitial water are therefore heterotrophic, i.e. they are dependent on organic substance as their source of energy and carbon. These include fungi and invertebrates such as horsehair worms, earthworms, freshwater isopods, wingless insects and larvae of winged insects. Loads of organic carbon to topsoil may amount in productive ecosystems to 5–10 tons per hectare and year, falling to a mere 10–100 kg under the rooting zone. By far the greatest proportion of this organic carbon is biologically utilized and thus decomposed. Microorganisms, in particular, play an important role in the self-purification of soils and groundwater. Bacteria and fungi decompose organic compounds. Fungi have coenzymes which enable them to break down even complex molecular structures. These processes are influenced by the quantity of soil water that is present. If there is water scarcity in topsoil, the biotically most active layer, then the

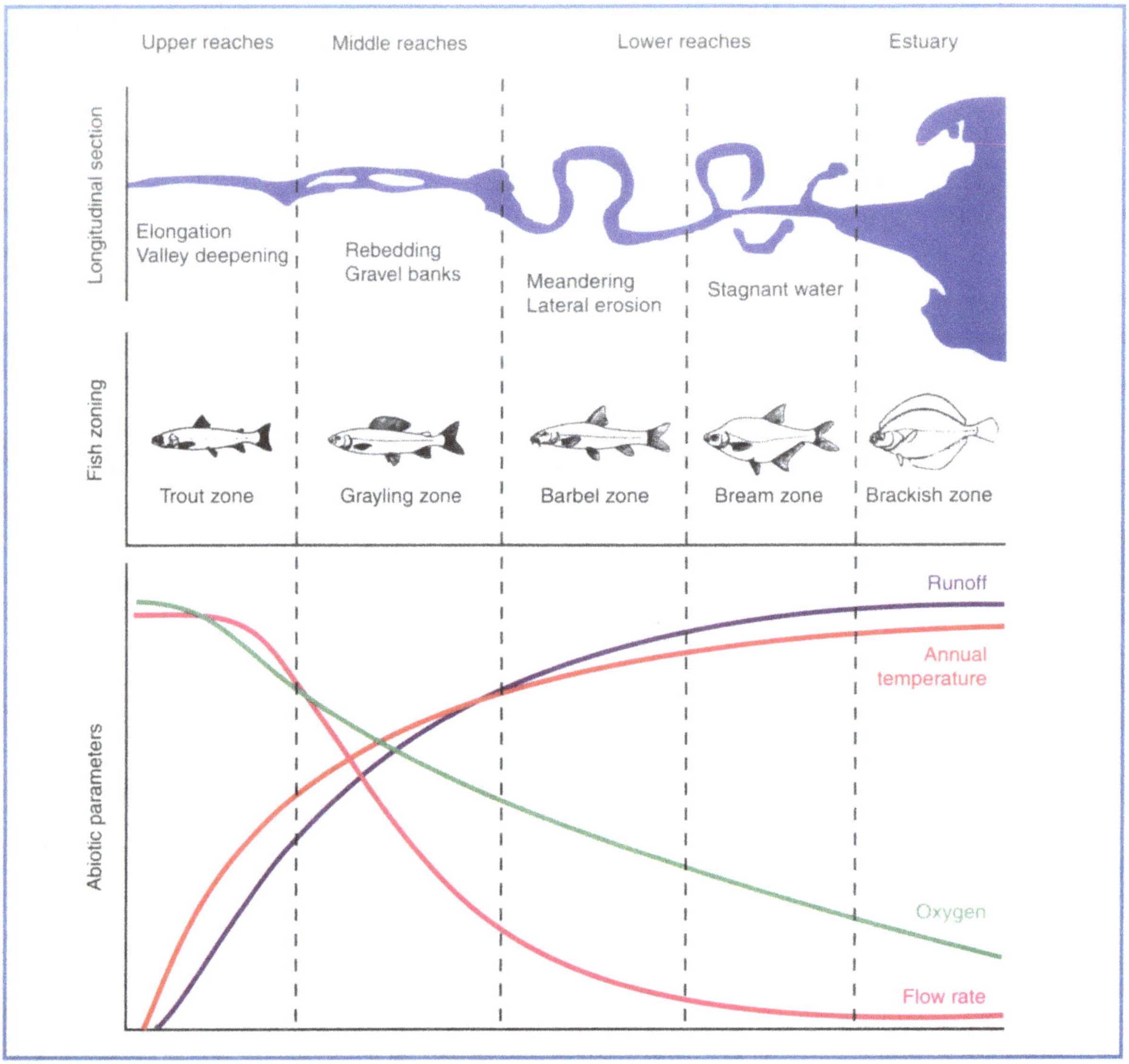

Figure D 1.2-1
Diagram of a river course, showing zoning of fish fauna, physical gradients and oxygen profile.
Source: modified from Niemeyer-Lüllwitz and Zucchi, 1985

turnover capacity of organisms is severely diminished or even stopped altogether.

The zone of permanent groundwater saturation has a constant environment. Temperatures in the groundwater are stable and at depths of only a few meters have reached the mean annual air temperature. This zone is characterized by the total absence of light, a confined habitat and low oxygen level, but is colonized nevertheless by surprisingly diverse communities of organisms that have adapted to these conditions. They include protozoa, flatworms, rotifers, threadworms, seaworms, waterbears, mites and various types of crustacean. The predominant life forms are microscopic, worm-like and eye-less fauna, which have very low metabolic rates and comparatively low growth rates.

Through feeding, groundwater fauna regulate bacterial density, on the one hand, and contribute through the processing of larger particles to higher bacterial activity, on the other. Organic substances can therefore be decomposed faster. In this way, intact communities provide for self-purification of groundwater. If larger amounts of organic material are flushed into groundwater, the environment that groundwater fauna require to survive is impaired. Their slow life cycle makes subterranean biocoenoses vulnerable to environmental disturbances. Even though these interrelationships are still rather poorly known, it probably takes some considerable time for the original state to be reestablished (Schminke, 1997).

1.2.4 Wetlands

Wetlands cover an estimated 5.6–8.6 million km^2 of the Earth's surface – about 4–6% of the total (Mitsch et al., 1994) – and can be found in all climate zones. The sheer number of designations and definitions is evidence of the many different forms in which wetlands occur. The most important freshwater wetlands in the world include wetland areas in river deltas, floodplains, siltation areas of standing and running waters, swamps, bogs, moors, humid forests, freshwater springs and oases. Very large wetlands can be found worldwide in the estuaries of major rivers, in lake districts and in boreal moors, mainly in North America (Canada and Alaska), South America, the territory of the former USSR and in Asia (Table D 1.2-1).

As transitional habitats between terrestrial and aquatic ecosystems, wetlands belong to the world's most productive ecosystems. They are eminently important for the water balance and are unique habitats for specific flora and fauna, e.g. for amphibians, waterfowl and shorebirds. Many wetlands are populated by endemic species. Some species threatened with extinction, such as the Bengal tiger, the jaguar and several species of crocodile, are found in wetlands only.

Continental wetlands receive water from precipitation, groundwater and surface waters, e.g. from flooding. Biological productivity may vary considerably, depending on the nutrient content in the catchment. "Mature" wetlands are formed gradually through the slow growth of special microclimatic conditions and soils. Because of the sheer variety of wetland structures and colonizing species from terrestrial and aquatic habitats, wetlands display a great richness of species and ecotypes that enables complex nutrient cycles and food webs to be sustained.

Wetlands perform numerous functions of global and regional significance (Dugan, 1993). One study on the value of the world's ecosystem services and natural capital arrived at a figure of more than US$ 14,000 per hectare and year provided by wetlands. This value far exceeds that of forests and agricultural land, and is surpassed only by coastal ecosystems (Costanza et al., 1997). Wetlands perform numerous ecological functions:

- They release runoff to groundwater, thus recharging groundwater stocks.
- Rainwater and meltwater are stored in wetlands and released evenly over time. Peak runoff levels are diminished and flood damage mitigated. By reducing the energy of waves during storms, wetland vegetation helps to reduce erosion.
- Wetlands function as "sinks" for sediments and toxic substances. In zones where the water flows slowly, suspended matter is able to settle. Pollutants, such as pesticides, which adhere to particles are extracted in this way from the water.
- Wetland vegetation may retain nutrients and remove nitrate from the water (denitrification), thus helping to improve the quality of neighboring waterbodies (Mitsch, 1994). These characteristics are used in many countries as a means of treating wastewater in man-made wetlands (Kadlec, 1994; Brix, 1994). Worldwide, wetlands store many times more carbon than the total CO_2 in the atmosphere (Section D 4.4).

Moreover, wetlands are used by humankind in many ways:

Table 1.2-1
Global occurrence of wetlands.
Source: WCMC, 1992

	Former USSR	Europe	Southeast Asia	Africa	Canada	USA and Alaska	South America
Area (in 1,000 km^2)	1,512	154	241	355	1,268	553	1,524
Percentage of total area	28.3	2.5	3.9	5.8	20.8	12.2	25.0
Major landscape-types	boreal moors	-	-	Rift Valley lakes	lake areas	boreal moors	swamps
Major rivers	Ob, Yenisey, Kolyma	Danube, Rhone	Mekong	Congo, Nile	-	Mississippi	Amazon, Orinoco

Other areas (in 1,000 km^2): Middle and Far East (19), China (32), Australia and New Zealand (15), and Central America (18).

- They are used to harvest raw materials such as fuelwood, timber and bark, resins and other non-wood products, e.g. for medical purposes.
- Highly productive wetlands are often rich fishing grounds. In Africa and many other tropical regions, fish are the most important source of animal protein. Many other animal products, such as skins and furs, eggs and honey, come from wetlands. Rice cultivation and aquaculture are the most important source of nutrition for large sections of the population in Southeast Asia, South America and Africa.
- Wetlands are used for recreation and leisure pursuits (hunting, sport fishing, bathing, sailing, contact with the natural world, eco-tourism).

1.2.5 Biodiversity of limnic ecosystems

Biodiversity refers to the richness of an environment in plant and animal species, genetic variety and ecological processes, i.e. the diversity of ecological functions and linkages within and between biological communities (Heywood and Watson, 1995). The biodiversity of many aquatic (and terrestrial) habitats is altered by progressive degradation (Section D 4.4). Little is known at present about the general role of biodiversity in the functioning of ecosystems, but there is no doubt that the preservation of biodiversity is essential in order to maintain the services and utilization functions of ecosystems for people; for it is the mix of microorganisms, fauna and flora that make up biological communities, and the interactions between them, that constitutes and ensures a healthy functioning ecosystem. The elasticity (resilience) of an ecosystem determines the response to external influences, including human impacts. An important function of biodiversity is its role in preserving this elasticity (Perrings et al., 1995).

The ecosystems of inland waterbodies contain more than 1,000 species of vascular plants and about 8,400 species of fish – or 40% of all fish species worldwide (WCMC, 1992). The mean number of species in freshwater habitats is far higher than the average number of species across all other habitats. A small number of very large and ancient habitats contain enormous numbers of species (Fig. D 1.2-2).

It is surmised that there are about 10,000 different species of phytoplankton, only about 10% of which have been described so far (WCMC, 1992). Freshwater ecosystems can also exhibit a high degree of species diversity, but, just as in other ecosystems, there is a paucity of research on the biodiversity of many microorganisms, animal and plant phyla. The number of species in the simply structured pelagial of lakes is surprisingly great and has been termed the "paradox of the plankton" (Hutchinson, 1961). Whereas previously it was assumed that species diversity increases in line with the "maturation" of ecosystems (Odum, 1971), it is now assumed that the variability of environmental factors over time continuously produces new conditions, thus permitting numerous different species to coexist (Sommer, 1985). The limnological analysis of lacustrine ecosystems has produced some basic and seminal findings on the functioning of eco-

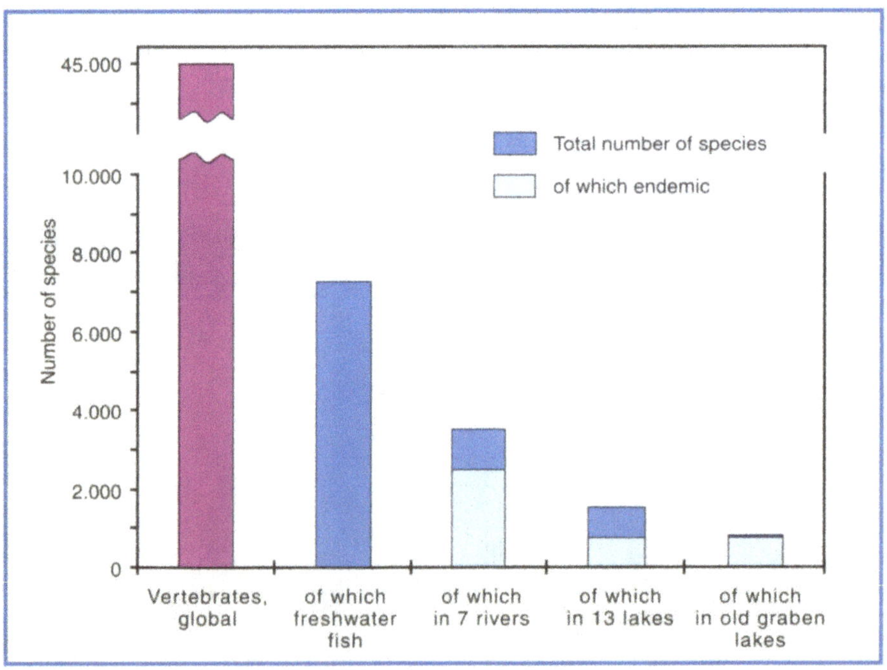

Figure D 1.2-2
Number of species of terrestrial and aquatic vertebrates.
Source: WCMC, 1992

systems in general (Mooney et al., 1996). The evolution of new species is a function of time. This is a major reason why geologically ancient ecosystems often have great species richness and a large number of endemic species (i.e. species which only exist within that particular ecosystem, e.g. Lake Tanganyika and Lake Baikal; see Box D 1.2-1 and Fig. D 1.2-2).

Communities with little biodiversity (e.g. monocultures) are often highly sensitive to perturbations (e.g. pests). On the other hand, habitats with high biodiversity (e.g. tropical rainforests, coral reefs) have proved to be vulnerable as well. There is obviously no simple relationship between biodiversity and susceptibility to perturbations. However, it is true that loss of biodiversity reduces the range of response options to environmental changes (Mooney et al., 1996). Impacts on particular keystone species may have drastic consequences for the functions of an ecosystem, while the disappearance of other species may go unnoticed. Island ecosystems and ecosystems in arid regions are often particularly sensitive to human interference. Individual species may often be the only representatives of certain functional groups, so that their role cannot be taken over by organisms with similar ecological demands and functions (Schindler, 1990; Frost et al., 1994). Loss of biodiversity, if manifested in shortened food chains, may have considerable impacts on nutrient cycles. This is shown by examples from the USA, where degraded aquatic habitats can no longer convert into biomass the nitrogen emissions from road traffic, which are now released directly to the groundwater (Koppes, 1990; Carpenter et al., 1996).

Freshwater ecosystems and their biodiversity are exposed to threats from a large number of anthropogenic influences (Section D 4.4). When damaging factors cease to operate, recovery is astonishingly fast in some cases, particularly in biological communities of running waters. This capacity is attributable to the fact that organisms in rivers and streams are adapted to a high variability of environmental conditions (Section D 1.5), and because recolonization is possible from upstream reaches if these are not similarly impaired. In other cases, regeneration takes much longer and often fails to reestablish the original state. For example, the reduction of phosphate loads to Lake Constance to about 20% of the previous maximum has led to considerable improvement in water quality, but phytoplankton biomass and primary production have only declined by about 30% so far (Tilzer et al., 1991). About 66% of the continental species of flora and fauna that have recently become extinct were freshwater organisms (Denny, 1994). One fifth of all extinct species worldwide are freshwater fish. The most common factors contributing to North American freshwater fish extinctions were habitat modification or loss (73%) and introductions of exotic species (68%) (Fig. D 1.2-3). The percentage of species threatened with extinction is much higher in the industrialized countries than in other countries of the world. Amphibian and mussel populations are especially threatened by the drainage of wetlands. About 43% of all freshwater mussel species and subspecies in North America is under severe threat of extinction, or has already disappeared (WRI, 1994).

1.2.6 Recommended action and research

Research recommendations

- Data collection and compilation of existing data in a global database; the latter should contain information on the characteristics, extent (water volume, surface area and mean depth) and location of freshwater habitats, be updated on a regular basis, and be made accessible to a broad range of users.
- Studies on structural and functional interrelationships of communities in freshwater ecosystems and adjacent habitats are imperative for assessing reactions to anthropogenic disturbances and the susceptibility of ecosystems to external disturbances.
- Studies on the potential impacts of introductions (e.g. of fish) and invasions (e.g. through the ballast

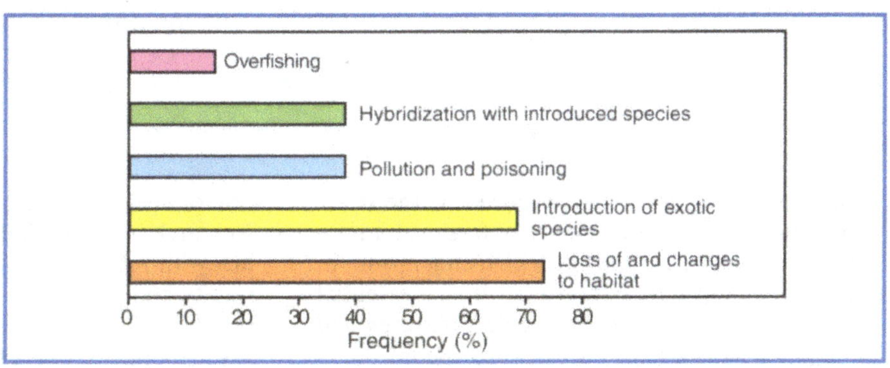

Figure D 1.2-3
Factors contributing to North American freshwater fish extinction (1900–1984).
Source: Miller et al., 1989

BOX D 1.2-1

Lake Baikal: One of the most important natural laboratories of evolution

Lake Baikal lies in the Siberian part of Russia and has a water volume of 22,995 km³, which is 18% of the global stock of freshwater stored in freshwater lakes and equal to the total volume of water in all the Great Lakes (Canada/USA). Lake Baikal was formed approx. 35 million years ago, making it the oldest inland lake in the world – (Tilzer and Serruya, 1990). Its catchment area extends over approx. 500,000 km², an area inhabited by 1.2 million people. The zone near the lakeshores is sparsely populated; the dominant vegetation is pine, fir and maple forests, and dry steppe.

Lake Baikal is worthy of protection for the following reasons: (1) it is the largest reservoir in the world of clear, uncontaminated freshwater. (2) It is also the oldest lake in the world, and its geographical isolation led to the evolution of many new species of flora and fauna. As a result, Lake Baikal contains the largest pool of endemic freshwater species. Of the 2,400 species identified so far in the lake, 84% are endemic, i.e. they are found only here (Kozhov, 1963). Particularly noteworthy is the Baikal seal (*Phoca sibirica*). The deep water in the lake is always well oxygenated, thus creating a rich habitat at great depths. Fish of the *Comephoriidae* family and certain amphipods are particularly worthy of mention. (3) Because of these unique features, the lake represents an irreplaceable ecosystem and a subject for many areas of research. (4) The lake and its surroundings could serve as a valuable recreation area (national park).

The threat to Lake Baikal: heavy metals, sulfide and eutrophying nutrients, e.g. from household sewage, are discharged into the lake from various point sources. It was only recently that the cellulose factory in Baikalsk on the southern end of the lake was finally equipped with wastewater treatment facilities. Sediment, fertilizer and pesticide loads from diffuse sources must be expected as a result of soil erosion in forestry areas and farming methods. Acid rain produced by emissions from power stations is deposited in the lake from the atmosphere. Severe impairment of water quality would result not only in the destruction of the ecosystem, but also in the extinction of endemic fauna and flora, and thus the loss of a rich gene pool. The greatest danger to this ecosystem, however, is the introduction of exotic species, because these may displace endemic species from their ecological niches and destroy their populations as a consequence.

Lake Baikal and the region surrounding it were recently put on UNESCO's World Heritage List. This step will make it much easier to push through the urgently needed protective measures at national and regional level and to canvas the massive international support required to achieve this.

Recommended action to protect the lake: (1) Conservation of water quality by installing sewage treatment facilities and air filter equipment in factories situated upwind from the lake basin; monitoring water quality. (2) Protecting stocks of endemic species by preventing the introduction of exotic species. (3) Preservation of the landscape in the immediate vicinity of the lake. Germany could assist in implementing the requisite measures by transferring scientific and technical know-how and by providing financial support.

By creating an infrastructure for eco-tourism and ensuring compliance with stringent environmental protection regulations, the economic situation of the region could be improved, and its dependence on further industrialization reduced. The Council recommends the implementation of a joint German-Russian research program on Lake Baikal, closely coordinated with projects already organized with other nations (especially the USA and Japan). There is enormous interest on the Russian side in intensifying scientific cooperation with Germany with regard to research on Lake Baikal. Research establishments (Limnological Institute Siberian Branch of the Russian Academy of Sciences, Irkutsk) and suitable research ships are already present.

water of ships) of non-native species on the structure, function and performance of freshwater habitats, especially in systems with high endemism. An understanding of the system as a whole is essential both for assessing anthropogenic pressures and interference, as well as for assessing the introduction of new organisms.

- Research into the biodiversity of freshwater habitats with reference to genetic diversity, species diversity and ecological diversity provides the basis on which to determine the importance of biodiversity in biotic responses to anthropogenic influences.

Recommended action

The Council recommends action to:

- support existing efforts and collaborate in the creation of a global database for recording and classifying the global stock of freshwater ecosystems (with priority given to those in special need of protection).
- exert pressure to ensure that the banning of major environmental damage to waterbodies and adjacent ecosystems is given priority over the utilization rights of riparians in the negotiations on the planned United Nations Convention on Non-Navigational Uses of International Watercourses.
- promote the designation and protection of other unique freshwater ecosystems as part of the World Heritage List.
- prevent the further loss of wetland areas and prioritize the restoration of previous wetlands on account of their multiplicity of functions.

1.3 The hydrological cycle

Basis for analyzing the climate system – Importance for energy balance – Diversity of interactions with the atmosphere – Impacts of vegetation on global water balance – Vegetation protects water quality – Climate simulation: more precipitation on land – Regions which gain or lose from climate change

Quantitative analyses of the global climate system (atmosphere, biosphere, geosphere, hydrosphere and cryosphere; see also WBGU, 1997 on the coupled system) are based on what are called balances. These enable the mass and energy budgets of components within the system to be described and characterized as changes in reservoirs, fluxes, sinks and sources. Where these fluxes link several components with each other, cycles of measurable intensity can arise. In the analysis of global environmental changes, these budget analyses deliver the following information: (1) The perturbation of a natural cycle can be quantified by a change in flux against the background of the unperturbed, natural flux. (2) Since the balance equations have to result in a balanced overall budget when combined, a quantitative understanding of the system as a whole can be obtained by estimating the various fluxes.

The quantitative analysis of changes in state within the climate system, and especially of their causes, is based on our understanding of the processes and interactions that characterize climate dynamics – be they isolated within one sphere or in exchange with one or several other components. Specific chains of cause and effect can be quantified in terms of their climate sensitivity.

1.3.1 Water balance

The variability of the atmosphere involves very different scales of space and time. Its spatial elements span about 12 orders of magnitude – from cloud droplets to planetary waves, from the microclimate of a single leaf to the climate of the planet. The total variance of dynamics is determined by 12-hour and diurnal cycles, short-term weather disturbances and weather anomalies lasting several days to months, by semi-annual and annual cycles, by quasi biannual oscillation and other processes. Although the ultimate driving force for all atmospheric processes is solar radiation, water vapor plays a major role, as a greenhouse gas and through condensation and evaporation, in the radiative and energy balance of the planet, as well as in climate dynamics.

In the long-term mean, gains in the global atmospheric water budget through evaporation are balanced out by equally large losses through precipitation. The mean volume of water stored in the atmosphere is estimated at approx. 2.5 g m^{-2} (or 13,000 km^3); the predominant form is water vapor. Global precipitation amounts to about 110 g cm^{-2} $year^{-1}$ (or 550,000 km^3 $year^{-1}$; Fig. D 1.3-1). More precipitation falls on landmasses (111,100 km^3 $year^{-1}$) than is evaporated from them (71,400 km^3 $year^{-1}$; Table D 1.3-1); the difference is runoff. These variables are determined by the mean atmospheric residence time of water vapor, which is only about 10 days. This means that water has a short "memory" with regard to changes. Although water vapor in the atmosphere corresponds to a liquid water column of only 2.5 cm,

Table D 1.3-1
Continental water balances.
Source: Baumgartner and Reichel, 1975

	Precipitation	Evaporation	Runoff
	(1,000 km^3 $year^{-1}$)		
Africa	20.7	17.3	3.4
Antarctica	2.4	0.4	2.0
Asia	30.7	18.5	12.2
Australia	3.4	3.2	0.2
Europe	6.6	3.8	2.8
North America[a]	15.6	9.7	5.9
South America	28.0	16.9	11.1
Total	111.1	71.4	39.7
Oceans	385.0	427.7	-39.7

[a] incl. Greenland

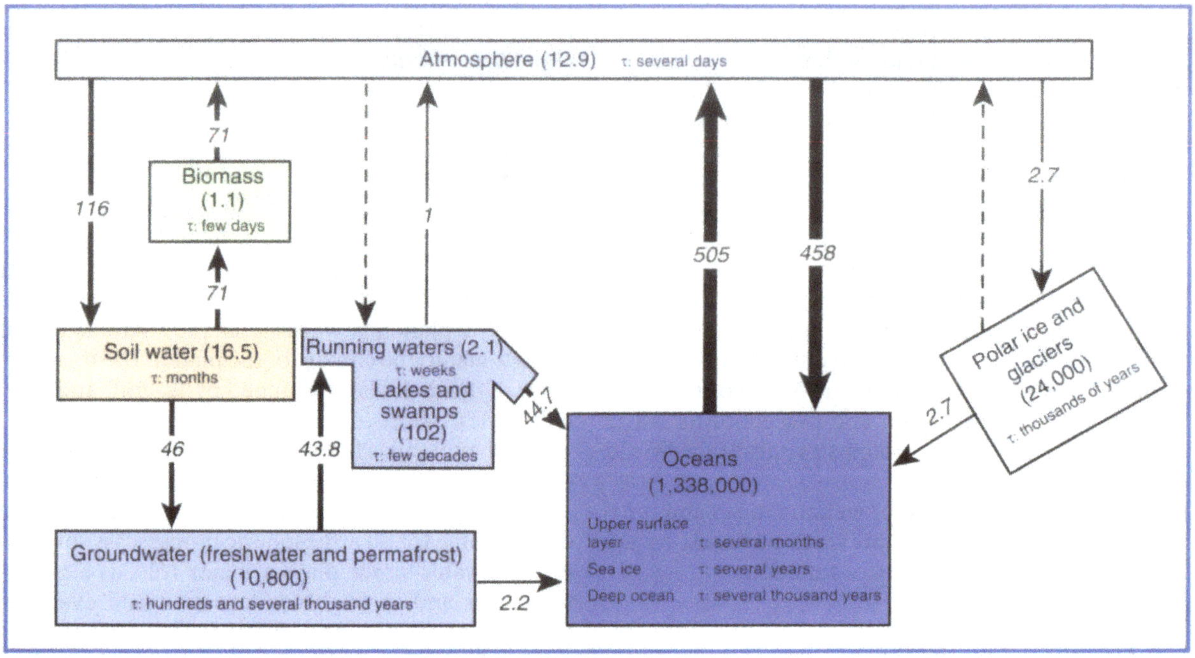

Figure D 1.3-1
Global hydrological cycle: reservoirs (in 1,000 km³), fluxes (in 1,000 km³ year⁻¹, italics) and typical residence times τ.
Sources: Shiklomanov and Sokolov, 1985; Baumgartner and Liebscher, 1990

its importance for atmospheric circulation cannot be overestimated due to its rapid conversion in the atmosphere and its interaction with other spheres. The uncertainties in balancing the global hydrological cycle are illustrated by the discrepancies between different estimates, each of which is based on observations (Fig. D 1.3-2 and Table D 1.3-1).

When analyzing the regional water budget of delineated areas, it is possible to neglect the volumes stored in the atmosphere and in soil for the climate mean. The atmospheric water vapor which has condensed over the catchments and has not evaporated flows off in rivers as well as underground. The catchments of rivers and the continents provide a suitable scale for water balances (Table D 1.3-1). Their mean climate and climate variability show a more or less pronounced interannual and seasonal variability of runoff, whereby dry and wet periods may be of extreme intensity and duration (Box D 1.3-1). This natural characteristic of the climate system is mainly determined by the dynamics of atmosphere-ocean interactions.

The total volume of water withdrawn by humans for domestic, industrial or agricultural use is estimated at 3,500–5,000 km^3 year^{-1} (WRI, 1990; UNDP, 1994). Although this amounts to only about 1% of annual global precipitation (Fig. D 1.3-1), the proportion rises to 5% of precipitation over landmasses (111,000 km^3 year^{-1}) and to 10% in relation to the total runoff of continental riverine systems, which is estimated at about 30,000–50,000 km^3 year^{-1} (Table D 1.3-1). Given the substantial interannual and seasonal variability of precipitation, water resources may become scarce in densely populated regions.

It would appear that humankind interferes only slightly with the global hydrological cycle, withdrawing only 1% of global annual precipitation from surface and groundwater, provided that analysis is confined to the water mass budget. This is a low figure in comparison with other anthropogenically influenced element cycles, such as the carbon cycle (about 5%), the nitrogen cycle and the sulfur cycle (each about 50%; Beran, 1995).

1.3.2
The hydrological cycle in the atmospheric energy balance

Water, in the form of vapor, clouds, snow and ice, as well as through its specific phase transitions, plays a cardinal role in the warming and cooling of the Earth. The driving force for this perpetual cycle of evaporation, transport, condensation and precipitation is the energy provided by the sun. The global energy balance results from the radiative balance and the thermal balance of the Earth's atmosphere and surface.

Mean incoming solar radiation to the Earth's atmosphere is 342 Wm^{-2} (100% in our context; Fig.

BOX D 1.3-1

Runoff variability of selected African rivers

The annual runoff curves result from the annual course of climate in the river basin, and the range of fluctuation from the interannual climate variability. Extreme variance in climate may result in a severe scarcity of usable water that is not sufficiently compensated during dry periods. This is particularly the case when there is a low natural variance in runoff, as is clearly demonstrated by the Senegal River (Fig. D 1.3-2).

Senegal: the Senegal River basin is situated in the sub-humid region of northwest Africa, and to a large extent in the Sahel. At Matam, the river has a catchment area of 230,000 km², and the mean annual runoff is 24 km³, or 2 km³ per month. The annual cycle of runoff is characterized by seasonal precipitation and a protracted dry period, as well as strong interannual variability. The water surface shows little seasonal variation. Only one sixth more land is flooded during annual peaks than in the annual mean. A larger floodplain would certainly have a stabilizing effect on the streamflow regime, in particular by enhancing the baseflow and by shortening the duration of runoff minima during the dry period.

Congo: the River Congo has the highest streamflow of all African rivers, at 1,269 km³ year⁻¹, or 106 km³ month⁻¹. At Kinshasa, the Congo has a catchment area of 3,475,000 km². Due to the low seasonal variability of precipitation in the humid tropics, the Congo manifests a seasonally balanced streamflow regime.

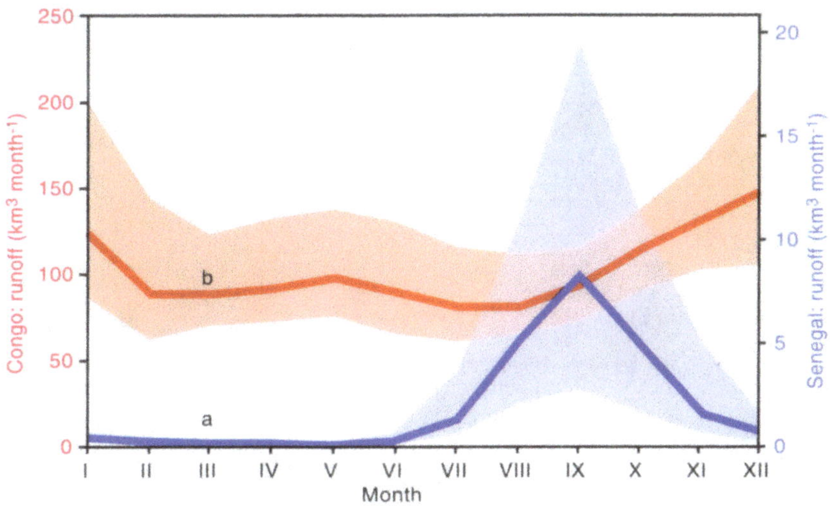

Figure D 1.3-2
Seasonal runoff and interannual variance (minimum and maximum) a) of the Senegal at Matam, 1903-1973, and b) of the Congo at Kinshasa, 1912-1983.
Source: Global Runoff Data Centre of the Federal Institute of Hydrology, Koblenz

D 1.3-3). About 22% of this incoming radiation is reflected back into space by aerosols and clouds. Another 20% is absorbed by water vapor and ozone. The direct insolation which remains, including the back radiation from the atmosphere, reaches the Earth's surface as the "solar constant". Another 9% of total incoming solar radiation is reflected by the surface, which leaves about 49% (or 168 Wm^{-2}) on average to warm the Earth's surface (Fig. D 1.3-3).

If all the solar radiation absorbed by the Earth's surface was reflected, the Earth's mean surface temperature would only be about -18 °C (Mitchell, 1989). The prevailing mean temperature of around +15 °C is achieved because the atmosphere absorbs a large proportion of the long-wave surface radiation and reflects the bulk of this to the Earth's surface as back radiation. This process, known as the natural greenhouse effect, is mainly attributable to the absorptive capacity of water vapor, carbon dioxide and clouds; in a narrow spectral range of around 10 µm wavelength, only about 12% of the solar radiation absorbed by the Earth's surface is directly emitted to space in the form of long-wave thermal radiation. Some trace gases absorb part of the thermal radiation leaving the Earth's surface, acting as a blanket that prevents energy from being emitted directly to space through the atmospheric "window".

Although the fluxes of radiation at the top of the Earth's atmosphere are balanced on a global average (short-wave insolation equal to long-wave emissions), this is not the case at lower layers: the radiation fluxes on the Earth's surface are mainly balanced by thermal flows, namely the sensible heat flux and the latent heat flux, that draw energy from the

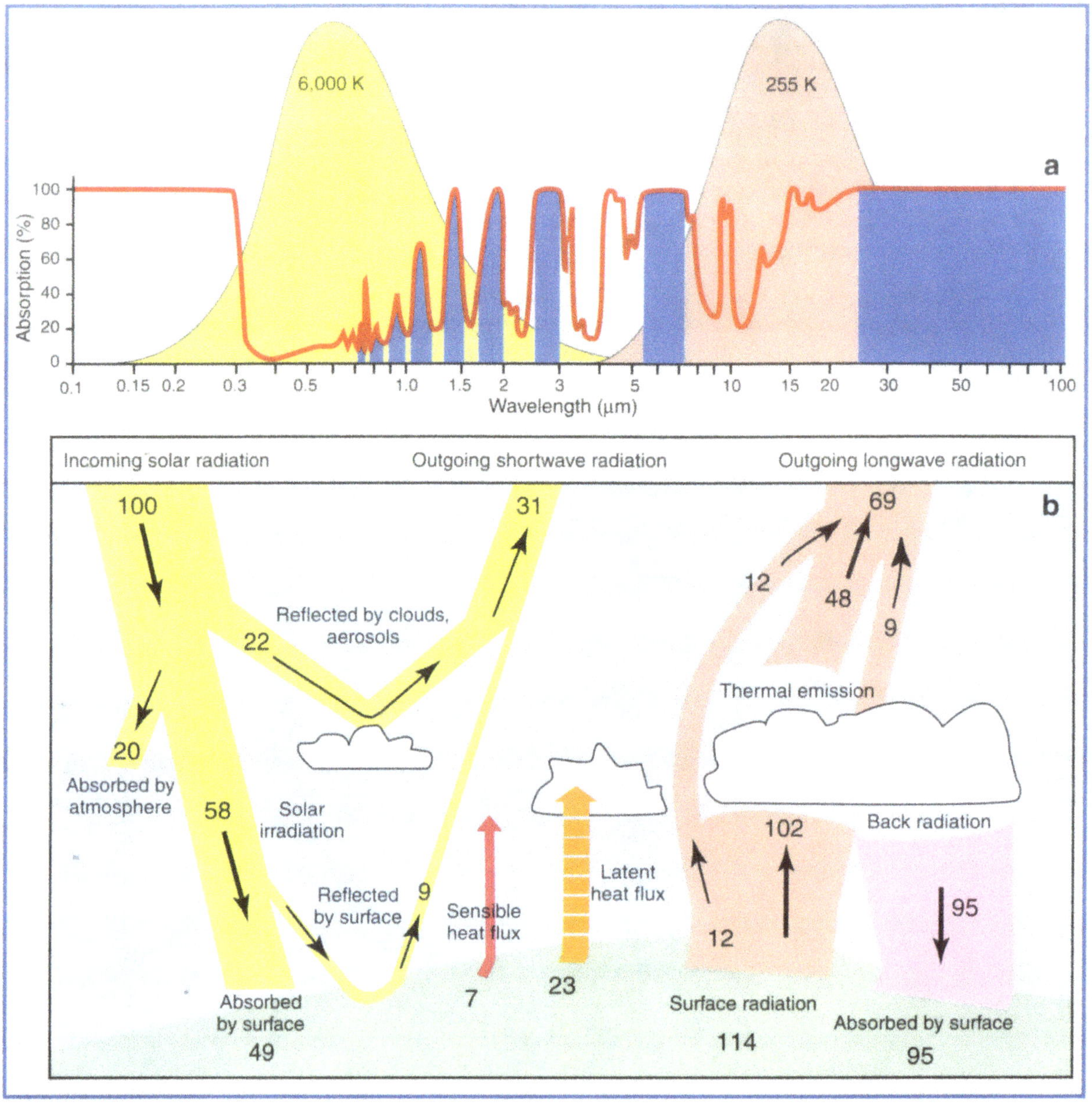

Figure D 1.3-3
a) Percentage absorption in the atmosphere (blue = H_2O-bands, absorption through water vapor) and spectral distribution of emitted radiation of black bodies at temperatures equivalent to the temperature of the sun (6,000 K) and the Earth's temperature (255 K). b) Global radiative and energy balance. Figures in percent of mean insolation (342 Wm^{-2} = 100%).
Sources: adapted from Mitchell, 1989; IPCC, 1996a

Earth's surface through the evaporation of water, only to release it again through condensation. If one were to take only radiation into account, then the atmosphere would cool down by about 1 °C per day. Sensible and latent heat fluxes warm it by about 0.3 °C to 0.7 °C per day. These figures highlight the role of the hydrological cycle for the conversion and transport of energy in the climate system.

1.3.3 Interactions with the atmosphere

The atmospheric part of the hydrological cycle links all the components of the climate system and their respective processes, which in turn impact on the hydrological cycle. These diverse and complex interactions are of great significance for the stability of the system as a whole. The response of the climate system (e.g. warming) to an external influence (e.g.

higher atmospheric concentrations of greenhouse gases) can be enhanced through self-reinforcing effects (positive feedbacks), thus destabilizing the system. Conversely, attenuating effects (negative feedbacks) operate to stabilize it, in that they are able to balance out external impacts. Examples of the role of the hydrological cycle in positive and negative feedbacks (in the context of global warming) are explained in the following.

1.3.3.1
Radiation, water vapor and clouds

Water affects the *radiative balance* of the climate system in many different ways. While the planet's ice caps reflect a high percentage of sunlight (reflectivity as high as 85%) and the oceans very little (reflectivity 5–10%), the impact of water in the atmosphere is much more complex. On the one hand, water vapor exerts a positive feedback on the greenhouse effect; on the other, the reflectivity of clouds reduces warming.

Water vapor absorbs short-wave solar and long-wave thermal radiation. It is the most important greenhouse gas, accounting at present for about 65% of the greenhouse effect; CO_2, by comparison, accounts for only one third. Warming of near-surface air increases evaporation, thus raising the concentration of water vapor in the atmosphere. Being a greenhouse gas, higher concentrations will enhance warming (positive feedback of water vapor on the greenhouse effect). A declining variation in daily temperature is another effect of increasing water vapor concentration, and has already been observed (IPCC, 1996b).

Clouds contribute to the greenhouse effect and hence to global warming through their high back radiation of sunlight, but they also help to cool the atmosphere. Their radiative characteristics are dependent on their size, height and temperature, the liquid water they contain (the volume of which can vary by more than one order of magnitude in clouds of different types), and the aggregate state and size of droplets and ice crystals. Other important variables are the presence of additional light-absorbing substances in the drops or ice particles (especially soot), and the concentration of water vapor above the cloud layer. This means that clouds are highly sensitive components of the climate system. They can respond with positive or negative feedback to a global temperature change. A change in cloud cover of only 1% can induce a change in radiative balance in the same order of magnitude as doubling CO_2 concentration. The net effect is to make the planet colder at present than it would be in the hypothetical case of atmospheric water being found exclusively in its gaseous state (by approx. 12 °C, according to Peixoto, 1995). To what extent this net cooling effect will continue to operate in a changed climate depends on the new distribution of lower and higher clouds, the former inducing cooling, the latter warming.

1.3.3.2
Atmospheric chemistry and aerosols

The chemical transformation of water in the atmosphere forges close links between the hydrological cycle and chemical processes. Chemical processes in the troposphere are influenced by water, even though water is a relatively unreactive compound. Ozone and water vapor are the most important sources for the hydroxyl radical (OH), a strong oxidant which operates as a kind of atmospheric detergent. It destroys the greenhouse gases methane and ozone, so that an increase in water vapor concentration and hence the concentration of hydroxyl radicals will slow down the increase of the greenhouse gases methane and ozone, thus reducing warming (negative feedback). At the same time, however, the decrease in hydroxyl radicals (which are themselves broken down during the destruction of methane and ozone) is slowed, so the oxidation capacity of the troposphere will tend to rise again, thus reducing the atmospheric lifetime of many trace substances (Fuglestvedt et al., 1995). Hydroxyl concentrations have declined since pre-industrial times, albeit with regional variability, but there has been no significant change over the last 15 years (Hauglustaine et al., 1994; Prinn et al., 1995).

Conversely, the chemical processes in the troposphere influence the atmospheric part of the hydrological cycle through the action of aerosols: cloud droplets and ice particles form only on certain aerosol particles, known as cloud condensation nuclei and ice nuclei. Formed of trace substances emitted biogenically and, increasingly, anthropogenically (Andreae, 1995; Schwartz and Slingo, 1995), these nuclei are sufficiently available over landmasses, but limit the formation of ice clouds over large segments of the upper troposphere, over oceans especially. There are also impacts which attenuate the current trend towards global warming: it is now being discussed whether there are negative feedbacks between the marine biosphere and its physical environment as a result of additional cloud condensation nuclei originating from the enhanced metabolism of marine phytoplankton (dimethyl sulfide emissions) (Charlson et al., 1987; IPCC, 1996b). The backscattering capacity of clouds consisting of many small drops is higher than clouds with the same liquid content, but with larger drops.

A slowing of global warming can be expected, in large regions of the northern hemisphere at least, due to a human-induced increase in cloud condensation nuclei. However, this can only be assessed to an imprecise degree: current estimates for so-called *indirect radiative forcing by aerosols* range from 0 to -1.5 Wm^{-2} (Schwartz and Slingo, 1995; IPCC, 1996b). The cumulative radiative forcing by anthropogenic greenhouse gases is an almost certain 2.45 Wm^{-2}. Closely related to this, but not yet predictable, is the impact of additional cloud nuclei on precipitation patterns. In those regions where cloud formation is limited by the availability of ice nuclei, increasing cloud formation can lead to increased precipitation. Outside clouds, increased backscatter due to more particles is over-compensated by a similarly enhanced absorption of thermal radiation. This causes reduced warming (*direct forcing by aerosols*, radiative forcing from 0 to -1.5 Wm^{-2}; IPCC, 1996b). Direct and indirect radiative forcing by aerosols are not cumulative, like greenhouse gas emissions – due to their rapid removal from the atmosphere (in days to weeks), they are dependent on the trace compound fluxes at the respective time.

1.3.3.3 Cryosphere and ocean

The cryosphere and the ocean are important subsystems of the climate system. The positive feedbacks on global warming exerted by these two subsystems are of major significance.

Snow, sea ice and continental ice sheets and glaciers cover about 16% of the Earth's surface, whereby fluctuations can be substantial. Without this cover, which reflects a high percentage of incoming solar radiation, surface temperatures would be higher and local-scale intra-annual variation much greater (about 2–3 °C, according to an energy balance model; Oerlemans and Bintanja, 1995). The positive ice-temperature feedback loop contributes significantly to climate dynamics: an increase in near-surface air temperature causes a decrease in sea-ice surface cover and reduces snowfall. As a result, the reflection of short-wave radiation by the Earth's surface is reduced and the absorbed proportion increased. Only the latter is able to enhance the greenhouse effect in a direct way, as thermal radiation. In fact, continental snow cover has declined in recent decades (by 10% in the northern hemisphere over the last 21 years). Moreover, observations of mountain glaciers (covering 0.3% of land surface) have revealed significant shrinkage over the last 100 years (IPCC, 1996b).

The influx of freshwater may have considerable effects on ocean circulation: the salinity of sea-surface water would decrease under the impact of the enhanced runoffs expected for the high latitudes. This could reduce convection and weaken the thermohaline circulation in the North Atlantic, which could lead in turn to less warming at high latitudes and more at lower latitudes. A general instability of ocean circulation would have unforeseeable impacts. Causal links of this kind seem probable, but definitive predictions cannot yet be made (IPCC, 1996b).

1.3.3.4 Vegetation in arid and semi-arid regions

When climate variations cause changes in vegetation cover, subsequent changes in the hydrological cycle exert feedback on climate. Extreme dry spells cannot be buffered by the elasticity (resilience) of the respective ecosystems, especially in the arid and semi-arid climate zones, with the result that vegetation cover responds more sensitively there than in other climate zones. In fact, the interannual and seasonal variability of precipitation is higher there than in humid climates, so that changes in precipitation and temperature due to nonlinear impacts on evapotranspiration and soil water have very strong impacts on runoff.

For example, if forested savannah is replaced as a result of drought by shrub with only partial vegetation cover, this weakens the hydrological cycle (reduced reservoir of soil water, less transpiration due to reduced rooting depth). The change in vegetation exerts feedback on atmosphere dynamics as well, due to the modified surface roughness. Increased reflectivity results in local cooling, leading to descending air in the atmosphere over such regions, and hence the downwelling of drier air from higher altitudes. These processes were at work in the late 1960s when a series of very dry years in the Sahel due to natural variability of the Earth's climate was followed by a more protracted drought period than in other regions of Africa and Asia. This was clearly a result of the region's geographical situation, topography and other features (Shukla, 1995). Human responses to these climate impacts tend to produce additional positive feedbacks: overgrazing and excessive logging can be expected during droughts, and were prevalent in the Sahel during the 1970s.

1.3.4
Interactions with vegetation

Water is an essential, life-giving and non-substitutable resource for all living beings. Water availability exerts strong selection pressure on the continued existence and development of ecosystems and their biological communities. In the course of evolution, innumerable adaptations and survival strategies have developed which enable life under the most diverse water regimes (e.g. extreme aridity, flooding, tidal zones). The changes in water budget associated with climate change can have severe impacts on the biosphere (e.g. shifts in vegetation zones, changes in species composition, decline of certain ecosystems) (Kirschbaum et al., 1996). Conversely, the biosphere for its part modifies the hydrological cycle and forms an integral component of the climate system (Melillo et al., 1996). Due to the poor data on the biosphere as a whole, the following description covers only the interactions between the hydrological cycle and vegetation.

1.3.4.1
Impacts on water balance

Vegetation and land use exert a critical influence on the distribution of precipitation, evapotranspiration and runoff. The impact of evapotranspiration on the water balance is highly dependent on regional climate.

INFLUENCE OF PLANT PHYSIOLOGICAL RESPONSES

The assimilation of carbon dioxide by plants (photosynthesis) involves the loss of water due to the coupling of the diffusion pathways of CO_2 and H_2O. In physiological terms, plants have to accept a compromise between the risk of desiccation during photosynthesis and of starvation when protecting against water losses. This problem is exacerbated by the fact that the relationship between evaporation and CO_2 assimilation is proportional to the ratio of water vapor and CO_2 concentrations in the air. Since the concentration of water vapor in ground level air is always about 2–3 orders of magnitude greater than the concentration of CO_2, the amount of water used by photosynthesis is 100–1,000 times greater than the expected level of CO_2 assimilation. If differences in available radiation are also taken into account, then the cultivation of crops in arid areas involves the consumption of twice to three times as much water per unit of biomass produced than crops growing in temperate zones.

The problem of efficient water use is not quite the same for animals. The oxygen they need for metabolic processes is available in air at concentrations 10–50 times greater than that of water vapor. The efficiency with which animals use oxygen, relative to water vapor, is always greater than 1, while the efficiency of CO_2 use in terrestrial plants relative to water vapor is always less than 1. This means that in arid areas, grazing and meat production make more sense than growing crops, which depends under such conditions as irrigation. Conversely, in temperate zones, crop growing is more efficient than meat production as far as water use is concerned.

The aperture size of plant pores (stomata – see Box D 1.3-2) is mainly controlled by the plant water status, which is dependent, in turn, not only on water loss through transpiration, but also on water uptake (Schulze, 1994). Rooting depth plays a crucial role here (Schulze et al., 1994; Kleidon and Heimann, 1996). In general, woody plants have a greater rooting depth than herbaceous plants, while natural vegetation has a greater root depth than crop plants selected by humans (Jackson et al., 1996). The greatest known root depths of trees are around 100 m. Crops mostly have a root depth of less than 1 m, and are therefore more vulnerable to aridity than natural vegetation in all climatic zones. Rooting depth also plays a major role in the salt balance of soil. A eucalyptus forest, for example, is able to keep its groundwater level through its high water consumption and by striking roots to depths of approx. 10 m below the surface. The horizon in which salts can dissolve and concentrate in soil water is therefore far below the soil surface. If the same site is converted to a wheat field, the soil layer where water uptake occurs moves to higher soil layers and with it the layer of salt accumulation. The salinization of wheat-growing regions in Western Australia is one example of this process (Barrow, 1994). The process is purely a result of different rooting depths, and does not depend on land-use intensity.

Stomatal apertures are also subjected to physiological control through nutrition. Through fertilization and nitrogen emissions, humans affect this regulation process and cause severe interference to the water balance. The size of the stomatal aperture is independent of the type of plant, and is directly proportional to the concentration of nutrients in the plant (Schulze et al., 1994). This means that a well-fed plant closes its stomata later than a badly nourished plant when exposed to aridity stress. Not only do badly nourished plants face a greater risk of desiccation at low precipitation levels, they also consume soil water at a faster rate. What is more, enhanced nitrogen concentrations promote the growth of shoots more than that of roots (Stitt and Schulze, 1994), causing

BOX D 1.3-2

The stomata of plants

Stomata are microscopic pores in the epidermal tissue of plants which open and close to enable water and carbon dioxide to diffuse through the otherwise impermeable plant surface (Fig. D 1.3-4). They are mainly located on the undersurface of leaves and provide the link between ambient air and the intercellular spaces inside the leaf. It is through the stomata that plant uptake of CO_2 for photosynthesis and the diffusion of water vapor occur. The pores are bounded by two guard cells, epidermal cells with a specialized structure. Changes in stomatal aperture are controlled by several different regulation mechanisms and interacting factors. Plant water status and CO_2 concentration inside the plant as a result of photosynthesis perform a highly important regulatory function. A good nutrient supply causes a reduction in inner-leaf CO_2 concentration.

A distinction is made between evaporation, the loss of water from surfaces, and transpiration, which refers to the loss of water vapor from the inside spaces of plants. While the first process is purely physical, the second is subjected to physiological control. The sum of the two processes is referred to as evapotranspiration.

Evaporation from the land surface depends on:

- the stomatal resistance of leaves (r_s), which is controlled by physiological responses of the plant,
- the available energy (R_n) (available radiation),
- the temperature-dependent air saturation deficit (D),
- the aerodynamic resistance (roughness) of the surface (r_a), and
- the extent of plant cover (E_{soil}/E_{plant}).

In a diffusion model, evaporation (latent heat E) depends on the above parameters as follows:

$$E_{plant} = \frac{s \cdot R_n + \rho \cdot c_p \cdot D/r_a}{s + \gamma \cdot (1 + r_s/r_a)} \text{ and } E_{soil} = f(R_n)$$

with the constants:

s: slope of the saturation vapor pressure curve
ρ: density of air
c_p: specific heat of air
γ: psychrometric constant

This model gives rise to two interesting borderline cases. If the aerodynamic resistance is much greater than the stomatal resistance ($r_a >> r_s$), it follows that evapotranspiration (E) is directly dependent on the available energy (R_n) (e.g. grasslands). If, on the other hand, the stomatal resistance is much greater than the aerodynamic resistance ($r_s >> r_a$), then evapotranspiration is determined by the ratio between the air saturation deficit and the stomatal resistance (D/r_s) (e.g. forests).

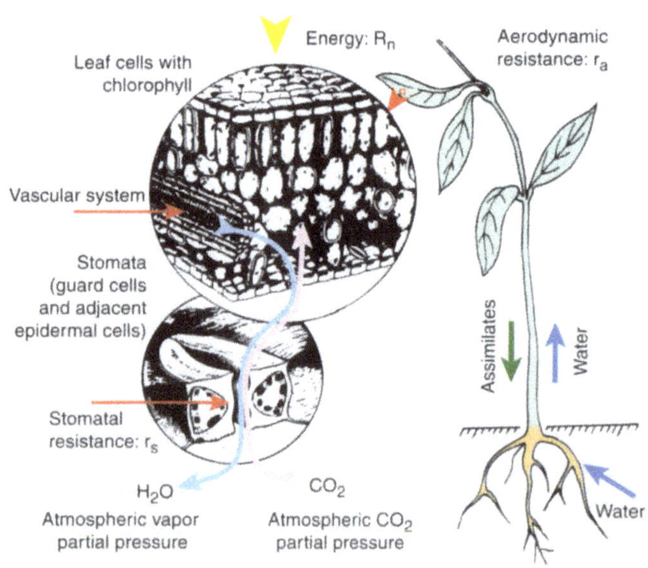

Figure D 1.3-4
Diagram of stomatal structure showing gas and water exchanges.
Source: adapted from Evenari et al., 1982

the rooting depth to shrink – with all the consequences this implies for plant growth in dry years (temperate zone) or dry climate (subtropics).

The meteorological consequence of stomatal regulation is that incoming solar radiation is largely converted into latent heat when stomata are opened, thus leading to temperature decline but also less groundwater formation. When stomata are closed, the rate of groundwater renewal is higher, but this is accompanied by an increase in sensible heat (e.g. the sweltering heat of the bush savannah).

Influence of plant structure on evapotranspiration

Within a specific climate zone and for a given wind speed, it is the roughness of the vegetated site, which depends primarily on the height of vegetation, that determines whether evaporation is driven more by the radiative balance than by the air saturation deficit (Box D 1.3-2). Under identical climatic conditions, transpiration is determined more by the air saturation deficit in the case of a forest stand, due to its high aerodynamic resistance, but more by the radiative balance in the case of grasslands or grain crops (Kelliher et al., 1993). The impacts on the hydrological cycle induced by a change in vegetation structure (e.g. conversion of forest to pasture) depend on climate conditions. In semi-arid and arid regions, for example, there is a very close relationship between land-use, evapotranspiration and precipitation (Savenije, 1996). The most important source of precipitation in these areas is moisture recycling, which means that a large proportion of the precipitation consists of evaporation from the region in question. Urbanization or conversion of forest to cropland would increase runoff, but precipitation would be reduced on the whole. In addition, the conversion of forest to cropland modifies the seasonality of water use. Transpiration increases during the vegetation period, but runoff increases outside the growing season (Schulze and Heimann, 1997). A decrease in evaporation and/or an increase in surface reflectivity (see below) as a consequence of large-scale clear-felling of tropical forests are likely to induce a decrease in atmospheric water transport in the inner-tropical convergence zone and in precipitation. As a result of feedback, the potential area for tropical rainforests and tropical seasonal forests would then decrease (Melillo et al., 1996).

Plant morphology influences the water balance not only through transpiration, but also through the interception of precipitation, fog and cloud droplets on the plant surface and subsequent evaporation (interception losses). This effect can have very different effects depending on the specific climate. The mist forest on Tenerife, for example, owes its existence to its capacity to extract sufficient water vapor from clouds. This source of water disappears when the forest is converted to cropland. The land is irreversibly drier and springs dry up. In humid climates, an increase in interception can also lead to negative changes in the water balance. The conversion of broad-leaved forest (low interceptive potential) into coniferous forest (high interceptive potential) in the low-lying parts of mountain ranges has resulted in a 10–20% reduction in spring discharge (Schulze, 1982).

Influence on radiative balance

Evapotranspiration increases with available radiation from the Arctic to the tropics, and is modified by the presence of clouds and aerosols (Section D 1.3.2). However, the radiative balance is also altered in complex ways by vegetation.

In arid regions, plants protect themselves against excessive solar radiation with waxy surfaces and/or cilia, which increase the radiation reflected from the plant surface. Vertical positioning of leaves has the same effect (e.g. the shadow-less eucalyptus forests in Australia). Thus, natural vegetation displays features that contrast sharply with those of cultivated plants, which are selected for their ability to absorb the maximum amount of sunlight. Compared to the natural vegetation of semi-arid climate zones, this results in much higher water demand by cultivated plants during the growing season. Hence, the major discrepancy in the water consumption of cultivated plants is due less to the intensity of agriculture or the growing of particular varieties, but is pre-programmed by human modifications to the natural plant cover. It is interesting in this respect that old agrarian cultures often used plant varieties that resembled natural vegetation in their radiative properties (e.g. the fruit of the date palm as food for humans, or atriplex as camel fodder).

In Arctic or alpine regions, the plant world tries to counteract the decline in temperature in the poleward direction or with increasing elevation by increasing absorption. High latitude warming as a result of climate change would cause the tree line to move northwards and hence to positive feedback (lower albedo, especially during the snow season, accompanied by increased transpiration). Over 50–150 years, this would increase the warming in northern mid- to high latitudes by more than 50% (Melillo et al., 1996).

Regulation of runoff

Vegetation influences the speed and volume of surface runoff. In addition to snowmelt intensity and heavy rains, the water storage capacity of ecosystems is essential for regulating water levels in surface waterbodies over time.

Certain types of vegetation are able to store large amounts of water and to release it to surface waterbodies at a gradual rate (Carter et al., 1979; Novitzki, 1979; Tayler et al., 1990; Lugo et al., 1990). Wetlands are a particularly important example (bogs with *Sphagnum* peat, lowland moors with *Carex* peat, swamps, ponds and lakes), some of which can be subject to substantial fluctuations in water level and thus equalize the large-scale variations in precipitation. If one compares forests, pastures and cropland as forms of land use, a forest is found to have the greatest proportion of large pores (channels left by old roots), which are able to siphon large amounts of water into the substrate (infiltration). The humus layer of the forest acts like a sponge, enabling water to be stored temporarily. Grasslands have a lower proportion of large pores than forests, and they do not have a top layer of humus. Nevertheless, the infiltration capacity of grasslands is greater than that of cropland, where the so-called sheet-flood effect can occur (fine particles in the topsoil swell up through precipitation and seal the soil against further infiltration of precipitation). A large proportion of the precipitation over croplands becomes surface runoff that flows at intense rates into the receiving water. This is all the greater for sealed surfaces.

The interceptive potential of vegetation, which increases with density of stand, delays and reduces the amount of water that infiltrates into the soil. Especially after heavy rains, this can lower surface runoff and erosion. However, this also means less water stocks in soils, springs and rivers.

1.3.4.2 Impacts on water quality

By virtue of their internal matter cycles, ecosystems are able to compensate anthropogenic influences within certain limits (e.g. substance loads through fertilization or immissions). Generally speaking, these are non-linear relationships that shape a saturation function or which are linked to certain physiological properties of plants, respectively (Marschner, 1990).

The non-linear relationship between fertilizer input and crop yield means that disproportionate amounts of fertilizer must be applied when the supply of nitrogen is already high in order to achieve an increase in yield. In practice, the surplus nitrogen resulting excessive fertilizer use enters groundwater as nitrate. Eutrophication of springs in rural parts of central Europe, and the need to build long-distance water supply networks not only for urban agglomerations but also for agricultural areas, are consequences of such farming practices (Mohr and Lehn, 1994).

There are clear indications (Tilman and Downing, 1994) that the quality of infiltrate increases with species richness. In a system with high species diversity, there is sufficient differentiation in the use of nitrogen (use of ammonium as opposed to nitrate or amino acids) and rooting depth, so that the nitrogen can be used up completely. A reduction of species diversity caused by eutrophication has a self-reinforcing effect on nitrogen loading to groundwater. Planting monocultures in forests and in agriculture operates in the same way.

We refer to Section D 1.2 ("Water as habitat") regarding the water purifying function of aquatic organisms. Terrestrial vegetation performs a key function both in the large-scale purification of precipitation as well as in the landscape-related transport of pollutants in water. The very fact that water has such a long residence time in wetlands gives the latter a special function in the purification of water. The lack of oxygen in wetlands causes denitrification of oxidized nitrogen, i.e. nitrate is converted back to molecular, atmospheric nitrogen (N_2) and N_2O (nitrous oxide) (Pinay et al., 1994; Weller et al., 1994). Studies in central Europe show that no nitrate is emitted from a boggy forest spring, while the nitrate load from a freely running forest spring planted to the edge with spruce can be substantial (Durka, 1994).

The filter function of vegetation is shown particularly clearly in riparian zones. Riparian forests lead to the breakdown of many pollutants that are laterally emitted from agricultural land along with infiltrate into the receiving waters, and thus ensure better water quality of surface waters. This is the case not only in Europe, but in all agricultural regions in the temperate zone (Cooper, 1990; Howarth et al., 1996; Peterjohn and Correll, 1984; Pinay et al., 1995).

The protective effect of vegetation with regard to soil erosion was described in Section D 1.3.4.1. The soil removed by runoff and deposited as sedimentation in rivers and lakes impairs aquatic ecosystems (species diversity) and water quality (eutrophication, self-purification potential). The soil particles finely distributed in the water cause problems for water purification (Pereira, 1974). To summarize:

- The structure of vegetation determines how much water is lost through evapotranspiration, infiltrates soil or which enters streamflow immediately as surface runoff. It therefore affects not only the availability of water but mediates as a component of the climate system the exchange of water and energy between the land surface and the atmosphere.
- As a capacitative variable, vegetation is able to

compensate for large fluctuations in precipitation with respect to runoff.

- The soil horizon from which water is withdrawn is dependent on the rooting depth of the plants, and this has implications on the salt budget in arid areas.
- Vegetation also has a filter effect, the effectiveness of which is dependent on biodiversity.
- Special types of vegetation have a filtering effect that is disproportionate in relation to the amount of cover. In temperate zones, these include the wetlands and riparian forests.

These interactions mean that vegetation performs numerous different functions of relevance to water resources, functions which need to be protected on a large scale:

- Protection against eutrophication of surface waterbodies.
- Assurance of groundwater quality and quantity.
- Protection against salinization of soils and waterbodies, thus ensuring their potential for sustainable use.
- Water storage as protection against floods and low streamflows.
- Preservation of species diversity and genetic resources.

1.3.5 Model: hydrological cycle in the present and future

How are key elements of the hydrological cycle changing as a result of climate change? Recent decades have seen a decline in precipitation at low latitudes and in southern Europe, against a weakly positive global trend (+1% since 1900) that is dominated by increasing precipitation at mid and high latitudes (Bradley et al., 1987; IPCC, 1996b). These trends are statistically significant for the mid latitudes of the northern hemisphere and for large areas of the subtropics (particularly in North Africa). However, it is not yet clear whether these trends are causally related to a warmer climate (Henderson-Sellers and Hansen, 1995), because climate models still display major shortcomings in describing the hydrological cycle. Evapotranspiration and cloud formation, two processes of immense importance for the hydrological cycle, are described in very rudimentary form. However, all climate models are consistent in projecting more precipitation, especially at mid and high latitudes. As one survey of coupled atmosphere-ocean general circulation models has shown (Lau et al., 1996), the ECHAM climate model displayed the smallest deviation from observed data. However, as far as the distribution patterns of precipitation are concerned, there is still considerable uncertainty in the projection of specific regional effects (IPCC, 1996a).

For the following analysis, characteristic features of the hydrological cycle in today's and in a projected climate were taken from an ECHAM4-OPYC coupled atmosphere-ocean model run with anthropogenic forcing (greenhouse gases with transient change in equivalent CO_2 concentration since 1860, but without consideration of anthropogenic aerosols) (German Climate Computing Center and Max Planck Institute for Meteorology; Oberhuber, 1993; Roeckner et al., 1996) and compared with each other using monthly mean figures. The 1980–1990 decade is compared with the decade of double today's CO_2 concentration (2070–2080), based on monthly averages. The simulated increase in global mean temperature (air temperature, 2 m above the ground) between these two decades is 2.6 °C. CO_2 equivalent concentration refers to the concentration of CO_2 that would ensue if other greenhouse gases are also taken into account (by converting them into the equivalent CO_2 concentrations for the radiative balance). The baseline emission profile for the future is similar to that obtained by projecting current greenhouse gas emissions (IPCC's IS92a scenario, 1992). The selected period of only 10 years implies that discrepancies cannot be attributed solely to the increasing greenhouse signal, but rather to the combined effect of anthropogenic climate change and the fluctuations arising from natural climate variability. The spatial resolution of the simulation corresponds to approx. 300 km near the Equator. Realistic simulation of the processes emulated by the model can only be expect-

Table D 1.3-2
Continental intercomparison of observed and modeled annual precipitation. The observed time periods are not perfectly congruent. Figures in 10^3 km^3 $year^{-1}$. Sources: see table

	Baumgartner and Reichel, 1975	Legates and Willmott, 1990	Cramer and Leemans, 1992	ECHAM4-OPYC-Model
Africa	20.7	21.9	19.1	24.5
Asia	30.7	28.7	23.9	28.2
Australia	3.4	4.1	3.0	4.1
Europe	6.6	7.1	5.5	6.7
North America	15.2	13.8	11.0	17.2
South America	28.0	29.7	26.1	27.5

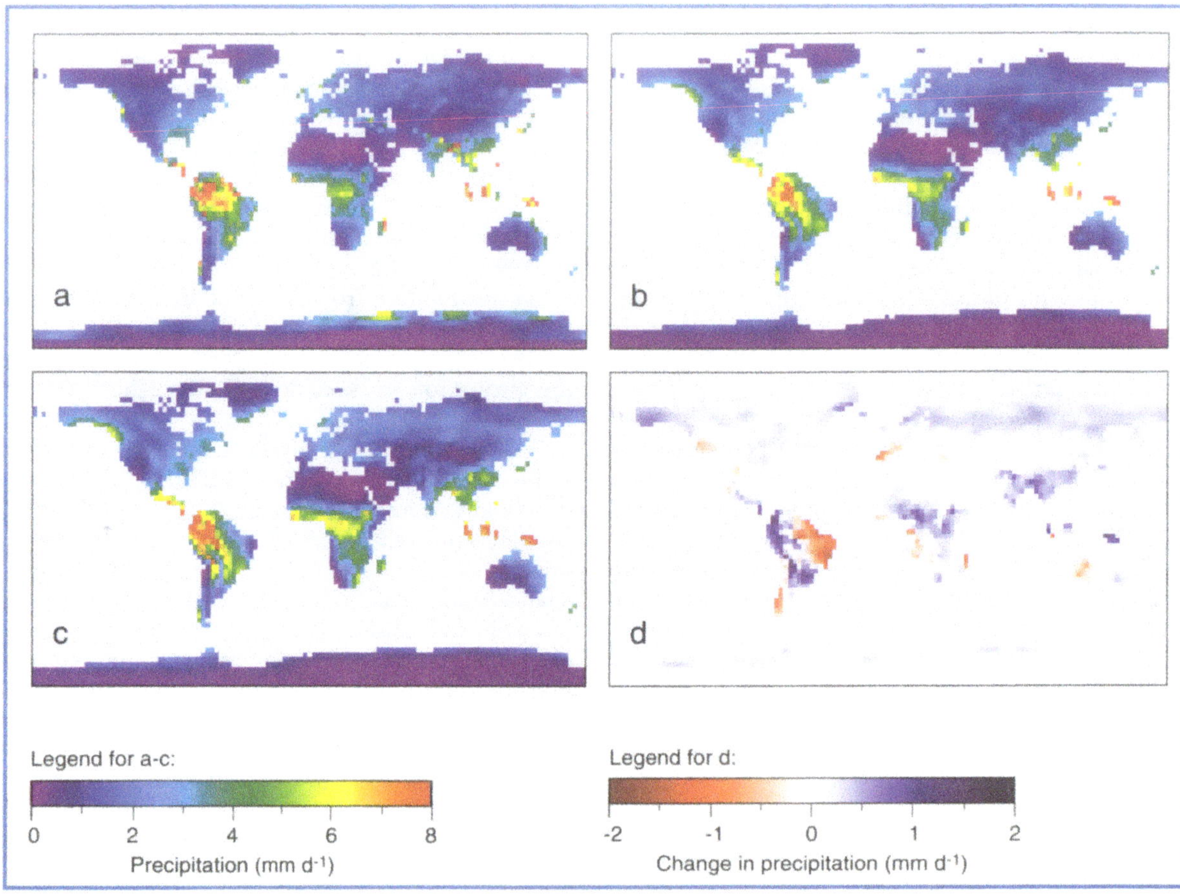

Figure D 1.3-5
Global distribution of annual precipitation. a) Observed data according to Legates and Willmott (1990) in the resolution of the model runs. b) Simulation of today's climate in the ECHAM4-OPYC model run. c) Simulation of a climate with double CO_2 equivalent. d) Difference between model simulations of the future (double CO_2 equivalent) and present day climates.
Source: Max Planck Institute for Meteorology and WBGU

ed, however, for structures at least several times greater in magnitude. Geographically smaller entities cannot be adequately resolved. South America, for example, is the continent with the highest precipitation and runoffs per unit area, but it is not possible with the model to simulate the strong precipitation gradients in the Andes region.

1.3.5.1
Comparison between observations and simulations of present climate

Table D 1.3-2 and Fig. D 1.3-5a and 5b show the observed precipitation data compared with the results obtained from the model. The climate model describes continental precipitations well on the whole, while those at middle and high latitudes and in Australia are very good. The model's weaknesses are the excessive winter precipitations in the prairie and Rocky Mountain regions of the North American continent, in northwest Canada and in Alaska. The South Asian summer monsoon is too weak in the model, and does not extend sufficiently to Hindustan, the Ganges valley and the Western Ghats. In West Africa, the boundary of the Sahel area is too far north, and the summer rains in South Africa are exaggerated.

1.3.5.2
Simulated changes in the hydrological cycle under CO_2 doubling

Simulation with the coupled model shows that in a warmer climate more precipitation falls on land, especially at high latitudes and in parts of the tropics and subtropics. In other regions, precipitation is then reduced. The regions most affected are large parts of Brazil, southwest Africa as well as Western and northern Australia (Fig. D 1.3-5c and 5d).

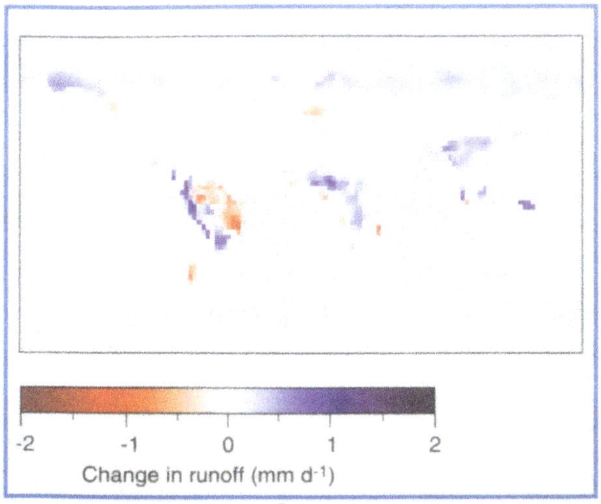

Figure D 1.3-6
Global distribution of annual runoff. Difference between model simulations of the future (double CO_2 equivalent) and present day climates.
Source: Max Planck Institute for Meteorology and WBGU

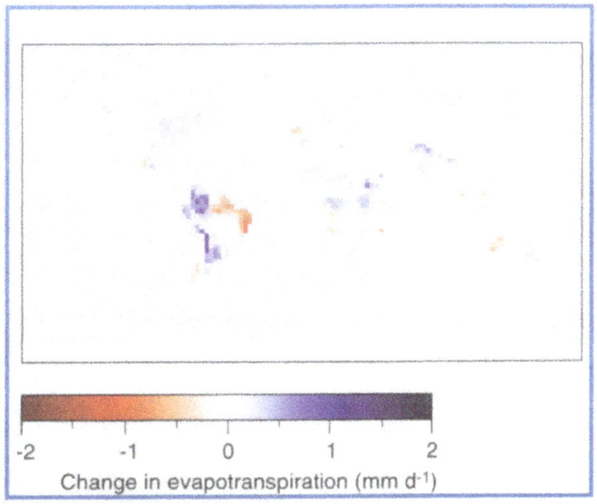

Figure D 1.3-7
Global distribution of annual evapotranspiration. Difference between model simulations of the future (double CO_2 equivalent) and present day climates.
Source: Max Planck Institute for Meteorology and WBGU

Today's seasonal variations at high latitudes are amplified: in the annual mean, drier summers are overcompensated by much wetter winters, especially for Canada, Alaska, Siberia and northwest Europe. Parts of western Europe experience both drier summers and winters. Much less precipitation falls in southwest Africa and, most drastically, in northeast Brazil (wet season during the southern hemisphere summer months). In equatorial Africa, precipitation increases during the June to August period. Suitable observation data are not available for a global analysis of evapotranspiration and runoff, since more densely located point measurements would be needed for interpolation to area data. The model reveals that their geographical patterns and their changes in projected climate generally resemble those of precipitation (Figs. D 1.3-6 and 7). The importance of evaporation declines naturally towards the high latitudes. This implies that, if precipitation reductions are evenly distributed, water scarcity is more likely to occur in tropical and subtropical regions, where losses through evaporation are higher. Lower precipitation in Columbia and Venezuela in June to August reduces runoff levels, although the simulation also

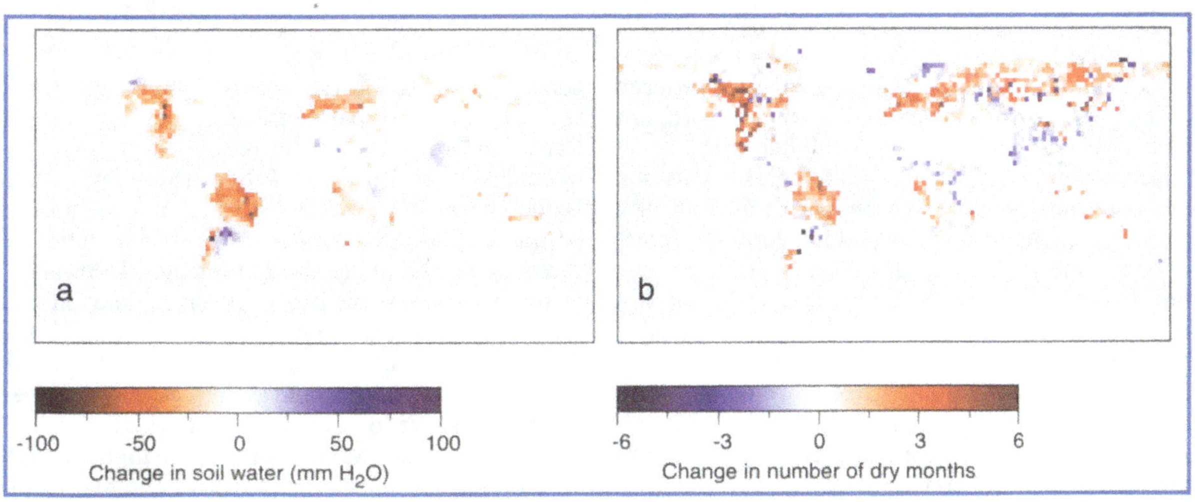

Figure D 1.3-8
Difference between the model simulations of today's climate and climate with double CO_2 equivalent. a) annual soil water totals. b) Number of months of aridity stress in which soil water concentration falls below a critical threshold.
Source: Max Planck Institute for Meteorology and WBGU

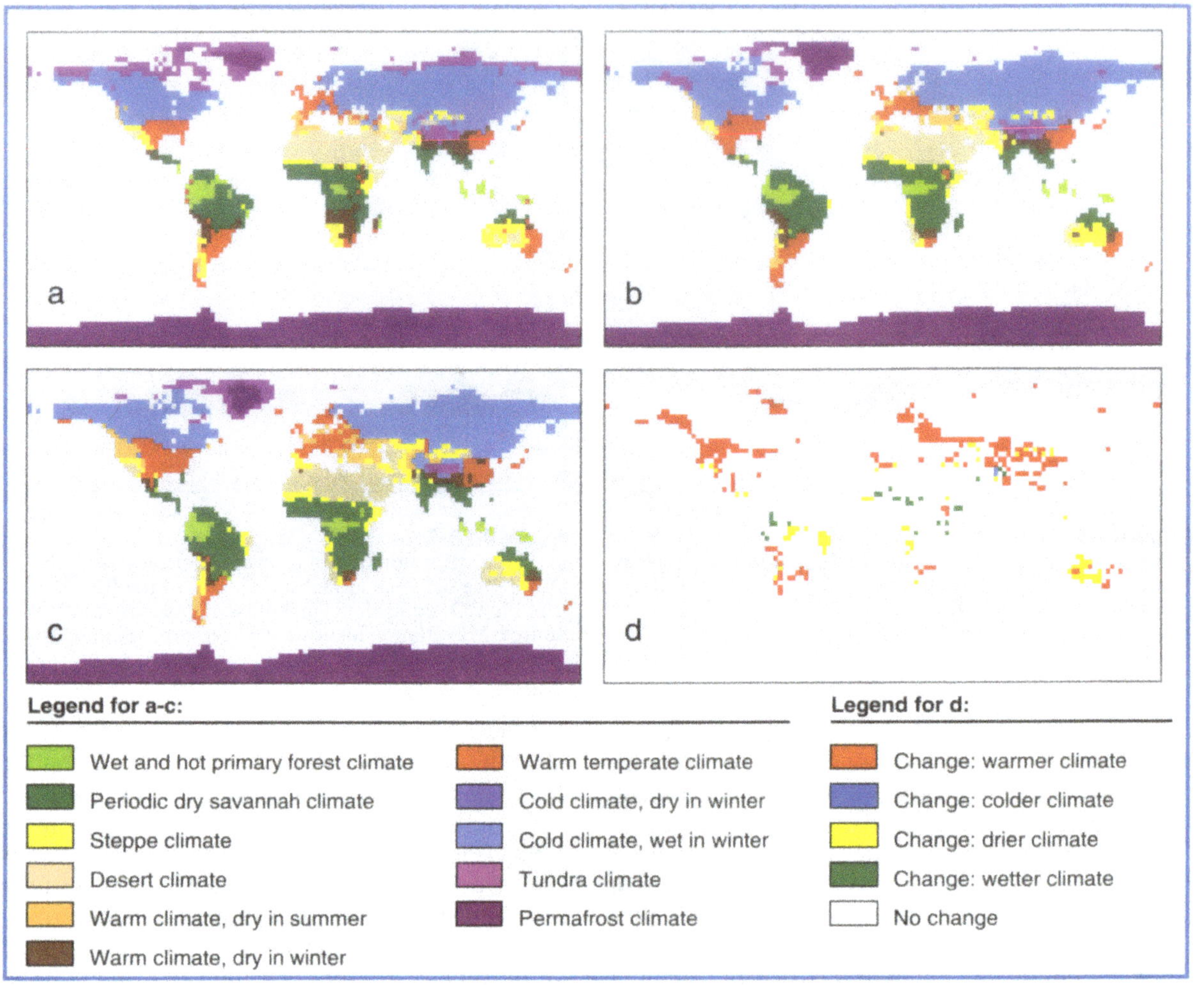

Figure D 1.3-9
Global distribution of climate zones according to Köppen. a) Observations according to Cramer and Leemans (1992). b) Simulation of today's climate in the ECHAM4-OPYC model run. c) Simulation of a climate with double CO_2 equivalent. d) Difference between model simulations of the future (double CO_2 equivalent) and present day climates. Source: Max Planck Institute for Meteorology and WBGU

shows that evaporation in fact increases. Reduced precipitation in sub-Saharan Africa and in northeast Brazil has a strong impact on runoff due to low or reduced evaporation. Fig. D 1.3-8a shows that soil moisture increases in some regions as a result of a warmer climate, both at low and mid latitudes (parts of North and East Africa, as well as South America), but declines markedly in others (e.g. large parts of Europe, Brazil, and the former prairie regions of North America). In these regions, in almost all of Siberia and in some other areas, more months with aridity stress occur in the annual cycle (Fig. D 1.3-8b). Aridity stress is inferred as the ratio of soil water to the total water retention capacity.

The transition to a climate influenced by a doubled CO_2 equivalent involves changes not only in the hydrological cycle but also in temperatures. What this might mean for the physical distribution of climate zones is shown in Fig. D 1.3-9: warmer climates come to predominate. In parts of Alaska and neighboring Canadian provinces, tundra recedes to be replaced by cold climates. In parts of Europe (Scandinavia, the Baltic and northwest Russia) and America (former prairie regions of North America), cold climates are succeeded by humid temperate climates. Tropical climates expand in the south of Brazil and on the African continent to the Abyssinian highlands. While drier climates recede in parts of the Sahel and the Horn of Africa, they expand in the western provinces of China and are formed anew in northeast Brazil.

In summarizing, we must emphasize that although anthropogenic climate changes force the hydrological cycle, according to the best scientific knowledge available today, this will be bound up with major disparities – in other words there will be "winning" and "losing" regions.

1.4 Current and future water withdrawals by agriculture, industry and for domestic use

Major growth in water withdrawals – Increase in developing countries, stagnation in industrialized countries – Low proportion of domestic use – water withdrawals by industry on the rise

1.4.1 Definitions and data situation

Statistical surveys on national water use usually distinguish between agricultural, domestic and industrial withdrawals (WRI, 1996). Although this classification scheme is commonly used, many statistics are not comparable with each other because, on the one hand, they relate to different kinds of use and, on the other, they are based on different categorizations regarding the quality of the water used.

The various types of water use can be broadly divided into *withdrawal* (off-stream) use and *in-situ* (in-stream) use (Kulshreshtha, 1993; Young and Haveman, 1985; PAI, 1993). In the former case, water is taken from the original source. Part of this water is consumed (water consumption), the remainder being returned to the source either polluted or warmed. In the latter case, water is used without being withdrawn from the original source. Examples include the transportation function of water, or its use in power generation.

In 1991, total water withdrawals in Germany amounted to 47.8 billion m³, the greatest proportion of which (29.1 billion m³) was used as a coolant in power stations (BMU, 1994). However, in the data collected by the World Resources Institute, using water as a coolant is not always treated as water withdrawal in the sense defined above, which gives rise to considerable distortions in the relevant statistics (WRI, 1996).

The terms water needs, water consumption, water demand and water use refer, quantitatively speaking, more to water withdrawals than to water consumption (Barney, 1991; Engelman and Leroy, 1995; Alcamo et al., 1997). Useful though the different terms may be, they tend to obscure other important analytical distinctions – the minimum requirement of water that an individual needs to survive is not necessarily a statistical or econometric variable, even though it is frequently (and wrongly) equated with demand or consumption (BMU, 1994).

Between 1940 and 1996, the world population grew by a factor of 2.5 from 2.3 billion to 5.8 billion, while annual global water withdrawals rose five-fold over the same period (Engelman and Leroy, 1995). It would be wrong to conclude from these figures that per capita water needs (in the sense of a minimum requirement) doubled. Patterns of consumption and behavior have indeed changed, however.

A lot of statistics relate, as far as quality is concerned, to water that is fit for human consumption. However, there are different notions of "fitness for human consumption". The WHO defines water as "safe" if it is either obtained from the public water supply system (piped water, public standpipe), or treated surface water, or untreated water from protected sources (e.g. springs and sanitary wells) (WRI, 1992; UNFPA, 1995; UNDP, 1995). Unfortunately, not all surveys are based on this definition. Moreover, the spatial and temporal variation in the quality of water used is so great that cross-regional forecasts and comparisons are virtually impossible to make (Nash, 1993). Box D 1.4-1 and Table D 1.4-1 contain information on the use of fossil waters.

The following description of global water withdrawals today and of the 1987 figures for agriculture (2,235 km³, or 69% of total withdrawals), industry (745 km³, or 23%) and domestic use (259 km³, or 8%) is not immune to the problems just mentioned, because a range of different surveys and literature sources is used. As a basic principle, however, the overview is based on the definition of water withdrawal made above.

1.4.2 Present rates of water withdrawal

WITHDRAWALS BY AGRICULTURE

In 1993, approx. 1.45 billion hectares of the world's cropland, or 3% of the Earth's surface, was used by agriculture, which was responsible for 65–70% of all water withdrawals (WRI, 1996; WWI, 1996). Water is a key resource for agriculture (Section D 4.3). Irrigation became especially important after the introduction of high-yielding varieties (Wolff, 1994; Ghassemi et al., 1995; Barsch and Bürger, 1996) (see also Section D 3.3). Between 1965 and 1985, more than 50% of the increased yield in global food production was achieved by irrigated agriculture (WRI, 1996). Irrigated crops account for 40% of global food production. The total area of irrigated land has risen five-fold in the last 100 years, from approx. 50 million hectares in 1900 to 95 million hectares in 1950, to approx. 250 million hectares in 1994. Over the same period, water withdrawals for irrigating land increased even faster, by a factor of six (Table D 1.4-2).

Agriculture based on irrigation is mainly found in Asia, which has a 64% share of all irrigated land in the world, far ahead of North America (9%), Europe

BOX D 1.4-1

Fossil water resources

The Earth's groundwater stocks (to depths of 2,000 meters) are estimated at 23.4 million km^3, and the proportion of freshwater at approx. 45%. In many parts of the world, water tables are falling up to several meters per year, because water withdrawals for irrigation (USA, China, India, Arabian peninsula) and tourism (many island states, especially in the Caribbean) exceed the rate of recharge.

Fossil water stocks are a special type of groundwater stocks that are renewed very slowly, if at all. They were mostly formed during other climatic regimes or when the ice caps melted. This means that they cannot be used sustainably, or at best at extremely low withdrawal rates. In most cases, the extent and renewal rate of these resources are poorly known; a global survey dating from 1990 provides estimates only (Table D 1.4-1). Major stocks are found in North America and North Africa especially.

The groundwater reserves underneath the Sahara (Nubian sandstone, continental intercalary aquifers) were formed over a period of about 140,000 years during the late Pleistocene period, with the greater part accumulating 18,000–40,000 years ago. According to recent studies, total stocks in the Eastern Sahara alone amount to 150,000 km^3 (Klitzsch, 1991; for comparison, this is 2,000 times the annual runoff from the River Nile). About 0.7–2 km^3 per year is renewed in the eastern Sahara and approx. 2 km^3 per year in the western Sahara (Margat, 1990). There are no indications yet of any depletion of stocks in the Libyan and Nubian desert (Klitzsch, 1991), particularly since the high rate at which some water tables are sinking means that withdrawal costs are becoming a limiting factor. In contrast, stocks on the Arabian peninsula are being rapidly depleted – all states in the region, with the exception of Oman, are withdrawing more than the rechargeable supply. Eighty percent of Saudi Arabia's water supply is taken from non-renewable stocks. The deep aquifers that are now being exploited were created more than 10,000 years ago. Between 1985 and 2010, these stocks will probably be halved (Gleick, 1993).

Table D 1.4-1
The world's major aquifers.
Source: Margat, 1990

Country	Name	Area (km^2)	Volume (km^3)	Average recharge (km^3 $year^{-1}$)	Recharge time (years)
Australia	Australian lowlands	1,700,000	20,000	1.1	20,000
Egypt, Libya, Sudan, Chad	Nubian sandstone	2,000,000	75,000	~1.0	75,000
Saudi Arabia	Aquifers of the sediment basin	~1,000,000	35,000	~1.05	33,000
Algeria, Tunisia	Continental aquifer	780,000	60,000	0.85	70,000
Niger, Mali, Nigeria	Continental aquifer	~500,000	10,000–15,000	~0.8	10,000–20,000
USA	Ogallala aquifer	450,000	~15,000	6–8	2,000
USA	Central Valley (California)	80,000	130	~7	160
Brazil	Sediment basin of Maranho	700,000	80,000	4	20,000
China	Hebei plain	136,000	5,000–10,000	35	150–300
Russia	Donez basin	250,000	175,000	5	35,000

Table D 1.4-2
Agriculture's share of global water withdrawals, 1900–1995.
Source: compiled from Clarke, 1993 (a); WRI, 1994 and 1996 (b); Alcamo et al., 1997 (c)

Year	Withdrawals by agriculture (km^3)	Total withdrawals (km^3)	Proportion (percent)
1900[a]	525	578	91
1940[a]	893	1,057	84
1950[a]	1,130	1,367	83
1960[a]	1,550	1,985	78
1970[a]	1,850	2,586	72
1980[b]	2,090	3,020	69
1987[b]	2,235	3,240	69
1995[c]	3,106	4,145	75

(7%) and Africa (5%). A total of 730,000 km^2 of land has come under irrigation worldwide since 1970 (FAO, 1996c), with Asia (58%) and the former USSR (13%) showing the largest growth rates. However, this expansion of irrigation is facing its environmental and economic limits, in that marginal lands now have to be farmed to an increasing extent. This demands greater inputs of labor and capital, yet involves greater uncertainty with regard to yields. The slowdown in the growth of irrigation that was clearly manifested in the ten years between 1980 and 1990 compared to 1970–1980 is attributable to the limited availability of cultivated lands and water resources, the high costs of developing and maintaining irrigation systems, and the increasing damage caused by salinization.

There are great regional variations, in terms of both absolute withdrawals and per capita consumption, in the amount of water withdrawn by agriculture (Table D 1.4-3, Fig. D 1.4-1). Continents where more than 80% of all water withdrawals are for agriculture, such as Africa and Asia, contrast with Europe and North America, where agriculture accounts for only 39% and 49% of total withdrawals respectively. In the arid and semi-arid developing countries, more than 90% of all water withdrawals is for agricultural purposes, whereas in humid regions the figure is only 30%.

In Africa, the continent with the highest percentage of agricultural water use, sub-Saharan countries show withdrawal levels that are low in both absolute and per capita terms. In Asia, agricultural water withdrawals are the highest in the world, firstly due to the dominance of irrigated cropland and secondly because more than 60% of the world population lives there (Fig. D 1.4-1a). Although less than 5% of the world population lives in North America, water withdrawals by agriculture are very high. This high absolute level of withdrawals is reflected in a similarly high level of per capita withdrawals. In the USA (1,117 m^3), per capita withdrawals run at twice the figure for China (550 m^3) or India (550 m^3), and are higher than the levels in arid countries such as Egypt or Libya, where per capita withdrawals are 900 m^3 and 1,050 m^3 respectively. The world's highest per capita water withdrawals for agriculture, at 2,000 m^3, are found in semi-arid to arid continental regions (Iraq, Turkmenistan or Azerbaijan).

Withdrawals by industry

The main determinants of industrial water demand in a given economy are population size, productive output, the sectoral structure of production (including the type of energy production) and the efficiency of the technologies used (Klemmer et al., 1994; Stanners and Bourdeau, 1995). In 1987, water withdrawals by industry in selected regions of the world reached the percentage figures of total annual withdrawals shown in Table D 1.4-4.

There are major differences among states. Within Europe, industrial use accounts for as much as 80–85% of total withdrawals, for example in Finland, Germany and Belgium, whereas in Greece, Portugal and Spain the figure is less than 30% (Stanners and Bourdeau, 1995). The proportion in developing countries shows considerable variation depending on the level of development, the economic structure and the technologies in use. Industrial use is between 10% and 30% (BMZ, 1995). In industrialized countries there are growing pressures and incentives to man-

Table D 1.4-3
Annual, continental water withdrawals by agriculture.
Source: WRI, 1996

	Year	Total withdrawal (km^3)	Per capita withdrawal by agriculture (m^3)	Withdrawal by agriculture (percentage of total withdrawals)
Africa	1995	145	175	88
Asia	1987	1.633	460	85
Europe	1995	455	244	31
North America	1995	608	711	49
South America	1995	106	196	59
Oceania	1995	17	199	34

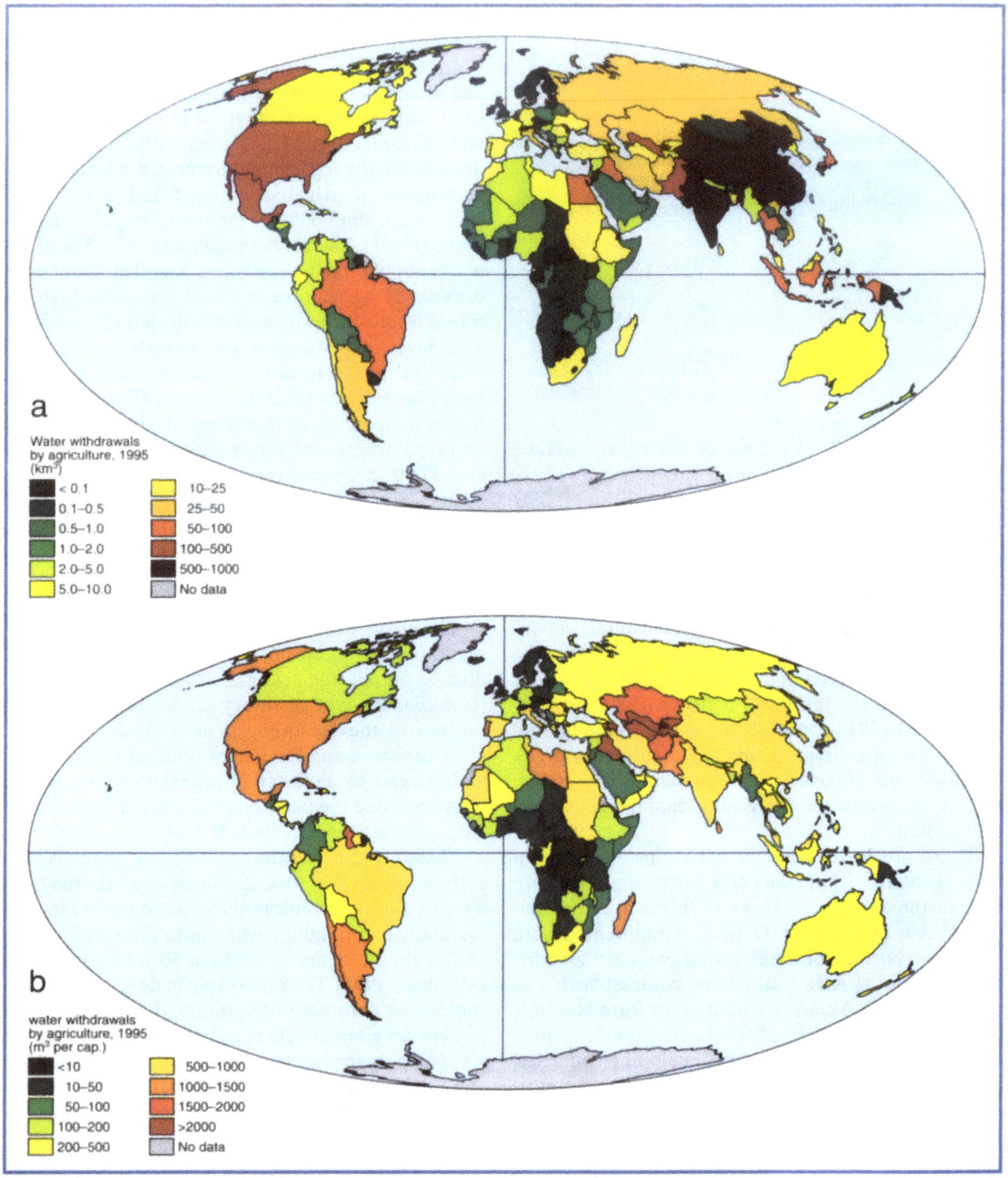

Figure D 1.4-1
a) Water withdrawals by agriculture, 1995. b) Per capita water withdrawals by agriculture, 1995.
Source: WBGU, using Alcamo et al., 1997

age and recycle water resources within a closed cycle, whereas in developing countries the technically feasible potential for multiple use in cooling and in industrial processes is hardly tapped at all (BMZ, 1995). Water consumption per unit of industrial production has declined considerably in the industrialized countries over the last twenty years (WRI, 1996).

In Germany, 62% of water withdrawals are made by thermoelectric power stations, 26% by manufacturing industry and 12% by the public water supply system (BMZ, 1995). As shown by the example of the

Table D 1.4-4
Water withdrawals by industry.
Source: WRI, 1996

	Industrial use (percent)
Africa	5
Asia	9
Europe	55
North and Central America	42
Oceania	2
South America	23
World	23

USA since 1950, these ratios are shifting towards higher demand by power stations (Table D 1.4-5).

Water demand by the energy production industry is determined by the amount of power generated, the specific types of power station (technologies deployed, type of energy carrier) and their capacity utilization. In the energy production sector, recycling can result in decoupling of water needs and water demand (Klemmer et al., 1994), meaning that the volume of water used may be much less than the volume of water actually consumed. In Germany, a ratio of 2:1 is generally assumed.

Specific water consumption in manufacturing industry depends on the product mix and the type of products being made. Relatively large amounts of water used in manufacturing are needed for cooling, whereas only about 20% is needed for production itself. Water-intensive industries, i.e. those consuming more than 15% of their water withdrawals, include the cellulose and paper industries, the cement industry and oil refineries (Stanners and Bourdeau, 1995). The utilization factor in German manufacturing industry is much higher than the 2:1 ratio in public-sector power stations. It can be as high as 3.9:1, because there are many ways in which water can be used repeatedly (Klemmer et al., 1994). The water coefficient (ratio of water consumption to value added) has been substantially reduced in the past. The principal reason for such reductions is the recycling of water and hence the enhancement of the utilization factor. There is also a trend towards a decoupling of total water consumption from growth in production output. The downward trend in specific water consumption and the more efficient use of water is also evident at European level (Stanners and Bourdeau, 1995). Developments in water reutilization rates in selected sectors of US industry are shown in Table D 1.4-6.

Industry consumes relatively small amounts of water in the sense of removing it permanently from the hydrological cycle (Shiklomanov, 1993). However, some of the water returned to the hydrological cycle is seriously polluted or warmed (BMZ, 1995). In addition to passing on the costs of supplying water, major incentives for reducing water consumption could be generated by passing on the costs for wastewater treatment as well – a rare occurrence at present (WRI, 1996). Global water withdrawals by industry are shown in Fig. D 1.4-2.

WITHDRAWALS FOR DOMESTIC USE

Domestic water withdrawals usually include drinking water, use by public-sector facilities and local authorities, as well as by companies and private households (World Bank, 1993). In practical terms, domestic use refers primarily to the public supply of drinking water to the population, partly including the water supply to small businesses (BMU, 1996). Terminological irregularities and lack of standardization in the collection of data are common problems in this area, generating considerable uncertainty in any assessment of domestic demand, such as the following.

In 1991, the public water supply system in Germany delivered about 6.5 billion m^3 to the population – equivalent to 13.5% of total withdrawals (BMU, 1996). However, only 4.1 billion m^3 was actually supplied to households and small businesses (i.e. 8.3% of total withdrawals). The difference is made up by public-sector water use, but the survey data do not take account of this factor in a consistent way. The World Resources Institute (WRI, 1996) states "domestic use" in Germany to be 11% of total water withdrawals in 1991, without indicating whether this is simply the average of the two figures above.

Global water withdrawals from the natural cycle in 1987 amounted to 3,240 km^3, of which a mere 8% was for domestic use (WRI, 1996). In Germany, per capita consumption of drinking water was 132 liters a day in 1995 (BMU, 1996) (Table D 1.4-7). With domestic use accounting for 8.3% of all water with-

Table D 1.4-5
Growth of water consumption by industry in the USA (km^3 year^{-1}).
Source: Gleick, 1993

Year	1950	1955	1960	1965	1970	1975	1980	1985	1990
Power generation	55	100	140	180	240	280	290	257	269
other industries	51	54	53	64	65	62	62	42.1	41.3

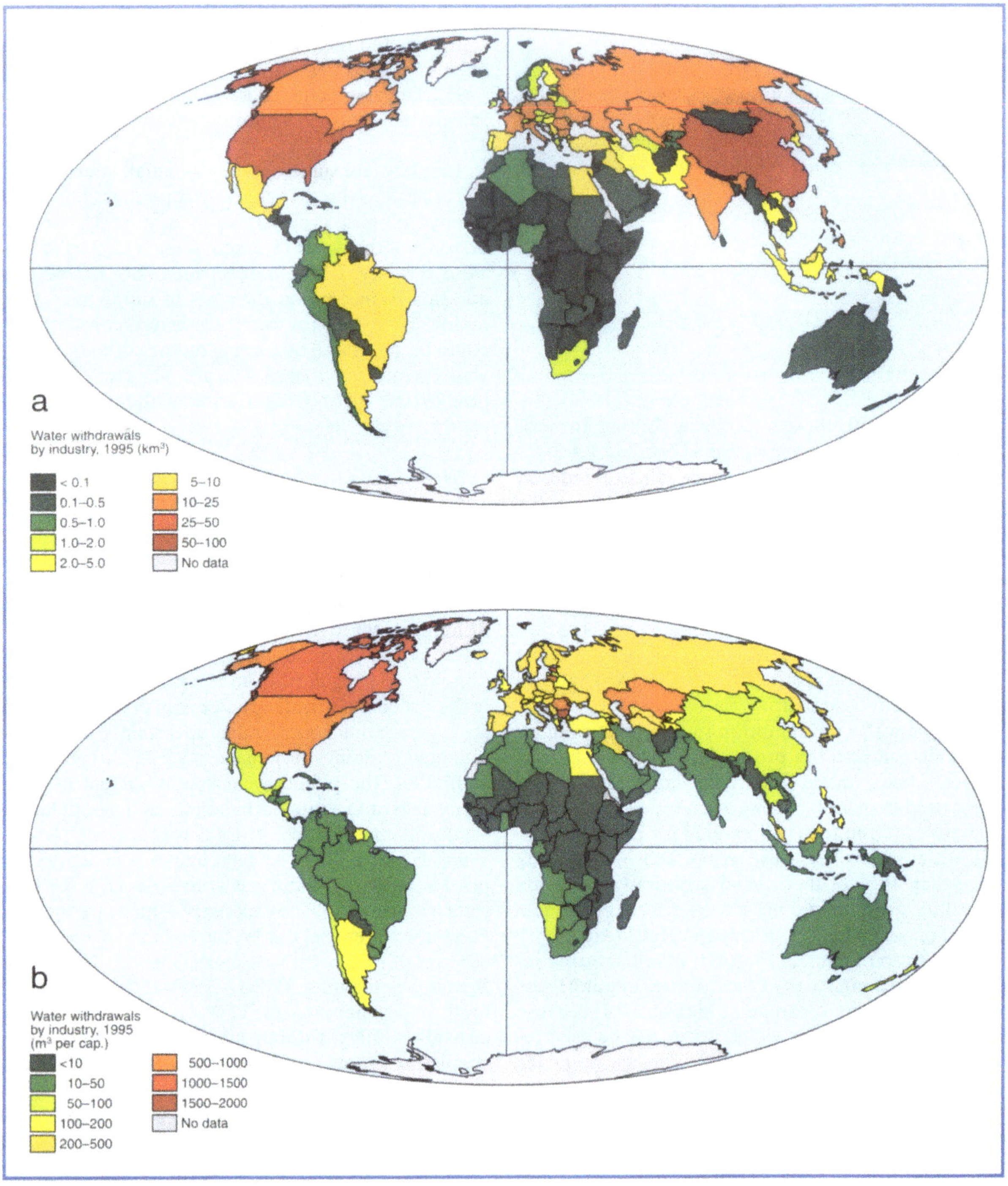

Figure D 1.4-2
a) Water withdrawals by industry, 1995. b) Per capita water withdrawals by industry, 1995.
Source: WBGU, using Alcamo et al., 1997

drawals, Germany is well below the European average (Table D 1.4-8).

In Germany, water withdrawals for domestic use have declined by approx. 8% from 1990 to today. Possible reasons for this trend are improved environmental awareness, use of water-saving techniques and substantial price increases in some parts of the country. In a global comparison, individual and total absolute withdrawals are close to the average (Fig. D 1.4-3).

Table D 1.4-6
Water recycling rate (number of times a given volume of water is used).
Source: Gleick, 1993

Year	Paper industry	Chemicals industry	Oil and coal industry	Heavy industry	Manufacturing industry
1954	2.4	1.6	3.3	1.3	1.8
1959	3.1	1.6	4.4	1.5	2.2
1964	2.7	2.0	4.4	1.5	2.1
1968	2.9	2.1	5.1	1.6	2.3
1973	3.4	2.7	6.4	1.8	2.9
1978	5.3	2.9	7.0	1.9	3.4
1985	6.6	13.2	18.3	6.0	8.6
2000	11.8	28.0	32.7	12.3	17.1

1.4.3 Future withdrawal trends

Forecasts of water withdrawals are very difficult to make. Growth in industrial water use, for example, depends on many factors such as changing preferences, technical advances, policy decisions, incomes, production level and the structure of industry (Kulshreshtha, 1993). This uncertainty is compounded by the very limited availability and comparability of water data (Stanners and Bourdeau, 1995; WRI, 1996).

Projecting agricultural water withdrawals is hampered by similar problems, as seen in the significant deviations between the estimates produced by various institutions (FAO, WRI, WWI). The uncertainties are even greater when assessing the amount of potentially irrigable land. A joint study carried out in 1990 by the World Bank and UNDP refers to a potential 110 million hectares in the principal irrigation regions in Asia, Africa and South America. If growth rates akin to those between 1980 and 1990 were to develop, this potential would be exhausted by the year 2025 (World Bank and UNDP, 1990). If one assumes, to simplify matters, that cropping practices based on irrigation will not change in the period up to 2025, then water withdrawals by agriculture will run to 3,697 km^3 in 2025. This would signify a 35% increase relative to 1990, or a doubling of water withdrawals since 1970.

Table D 1.4-7
Drinking water consumption per person and day in Germany, 1995.
Source: Absolute consumption figures from BMU, 1996; relative proportions from UBA, 1991

Type of use	(Liters)	(Percent)
Drinking and cooking	4	3
Bodily hygiene (bath, shower, wash basin)	48	36
Toilet flushing	42	32
Textile cleaning	18	14
Dishwashing	8	6
Household cleaning	4	3
Other activities (washing car, gardening)	8	6
Total	132	100

The FAO (Kendall and Pimentel, 1994) forecasts an annual rate of increase of 0.8% in the developing countries (excluding China). About two thirds of the additional area is expected to be in Asia. Further expansion potential for irrigated farming is seen in Africa and South America especially. Growth will be concentrated in Bangladesh, Brazil, China, India, Nigeria and Turkey (Postel, 1989; FAO, 1996c; UNDP, 1992). Uncertainties regarding increases in irrigated area are compounded by the enormous variability in the water-use efficiency of irrigation systems. Gravity irrigation systems with efficiencies between 30% and 60% account for 90% of the total, which means on average that only 40% of the water actually reaches the crops (BMZ, 1995). It is difficult to assess the respective water-saving potentials that each country might achieve in the future through technological improvements, better adaptation of irrigation systems, or the selection of plant varieties.

Population trends in individual states can be predicted with relative certainty. However, it is only one factor determining future demand for water. Assuming, for the sake of simplicity, that per capita water consumption in 2025 will be the same as in 1995, then basic trends can be identified by combining this assumption with projected population growth (Fig. D 1.4-4).

Water withdrawals will increase most in Africa and parts of Asia, in line with regional population growth. Demand will more than double in some parts of these regions. Growth rates of up to 40% are expected in North America, Oceania and China. Forecast growth in South America is as much as 80%. In Europe, by contrast, water withdrawals are expected to stagnate and even decline slightly in some areas. This simple model ignores increases in per capita income, and hence stronger growth in national income relative to population trends, as well as changes in the composition of industries and more efficient water use.

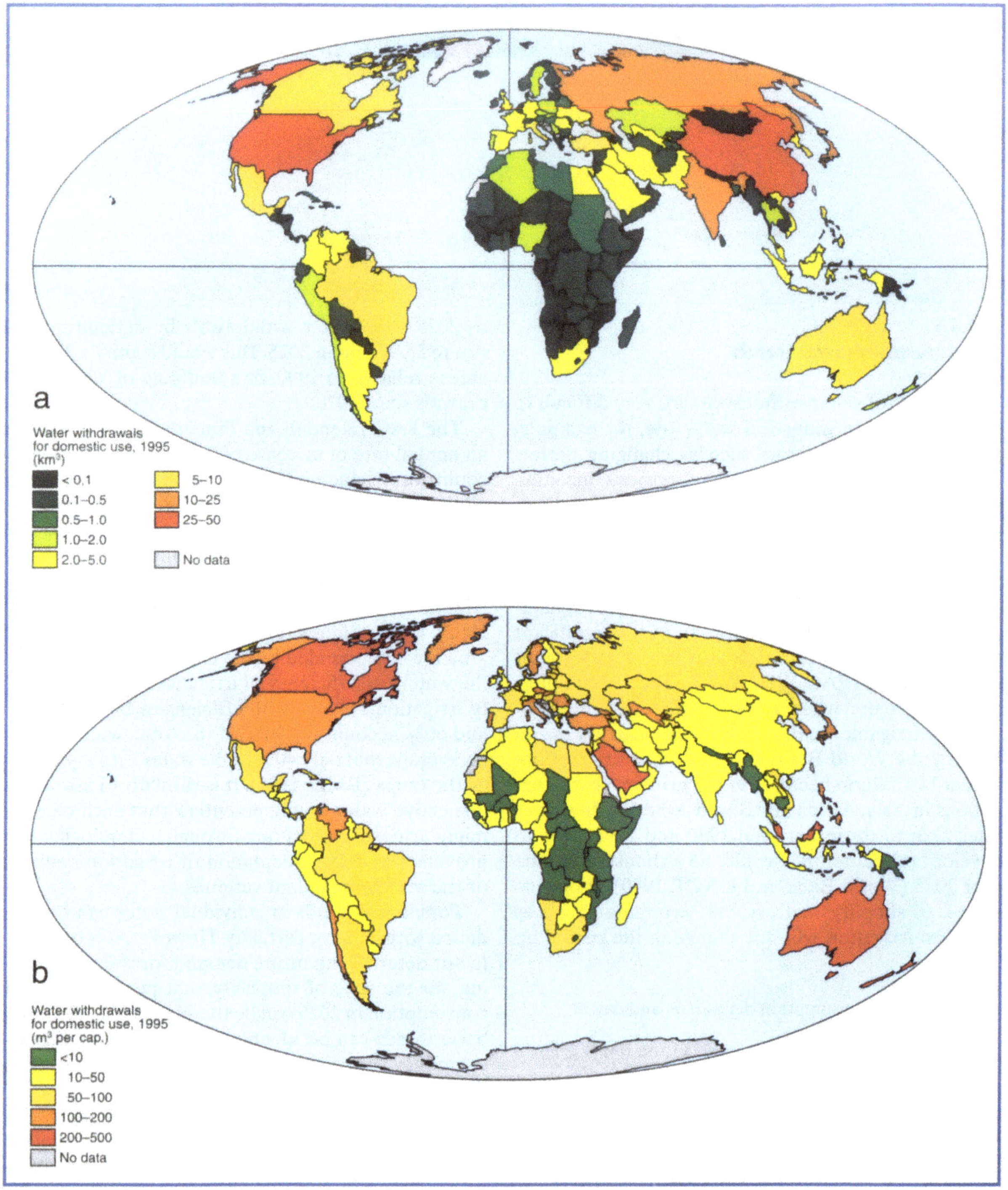

Figure D 1.4-3
a) Water withdrawals for domestic use, 1995. b) Per capita water withdrawals for domestic use, 1995.
Source: WBGU, using Alcamo et al., 1997

WaterGAP (Water – Global Assessment and Prognosis: Alcamo et al., 1997) is a recently developed predictive model for water use that takes into account not only world population growth, but also its spatial distribution, economic trends, and potential for more efficient management of water resources. With the help of WaterGAP, three scenarios have been generated for water use in the agricultural, industrial and domestic sectors in 2025 and 2075, based on the assumptions shown in Table D 1.4-9.

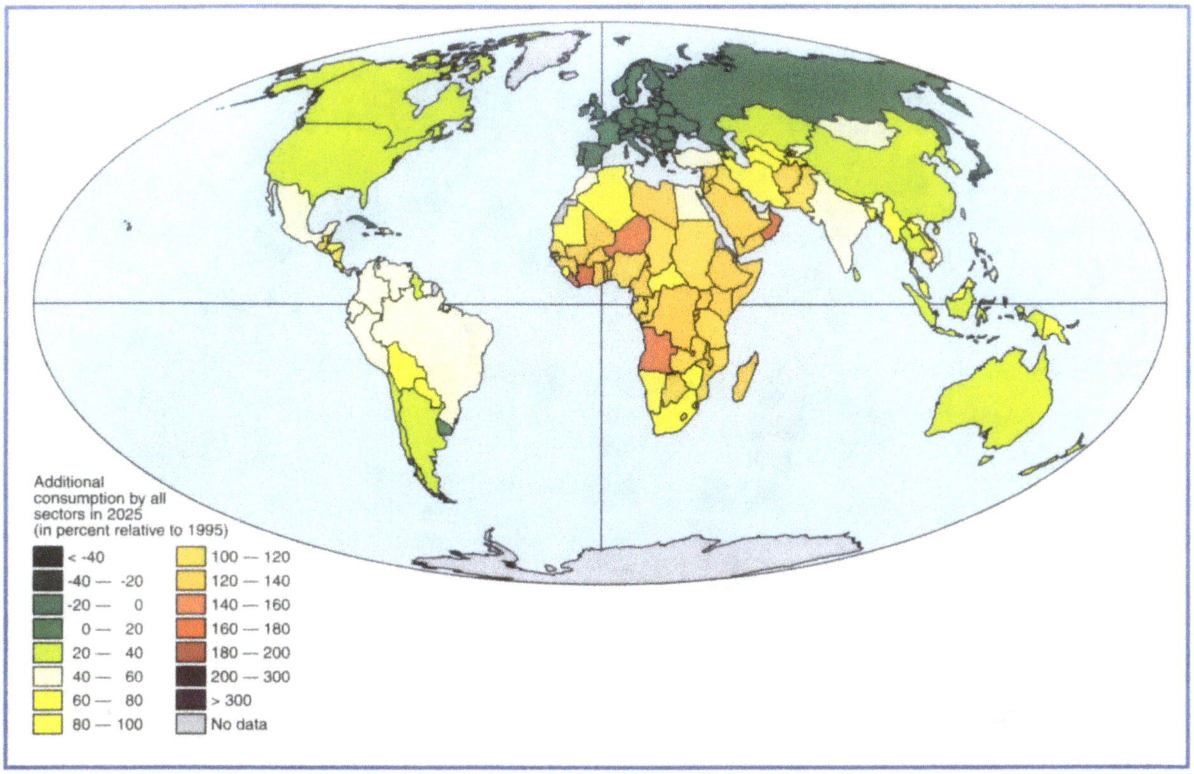

Figure D 1.4-4
Relative growth and decline in total water withdrawals as a result of population growth, 1995-2025.
Source: WBGU, using Alcamo et al., 1997

Changes in agricultural water withdrawals depend on an increase in irrigated land, improved efficiency of irrigation systems, and intensified cropping.

The increase in irrigated land is based on the assumptions of Alexandratos (1995) and was correlated with the growth rates in food production (cereal production) in the IMAGE 2.0 model (Alcamo, 1994). Global growth rates were obtained by dividing the world into 13 regions according to development type (Alcamo, 1994) and assessing each one separately. Stagnation in the industrialized countries contrasts, for example, with potential growth of 114% in Africa (not counting the North African states) (scenarios M and L). For efficiency improvements in irrigation systems, a linear increase of 0.5% per year until 2025 was assumed. Borrowing from Alexandratos (1995), an increase in cropping intensity in the 13 regions was taken into consideration. By the year 2025, five regions (Latin America, Middle East, India, South Asia, East Asia) are expected to show an increase in this regard – 20% in the case of India and South Asia, and 10% for the other regions.

In order to estimate industrial water demand, a distinction was made between countries with high and others with low per capita water stocks, with the dividing line set at 1,000 m³ per capita per annum. It

Table D 1.4-8
Annual domestic water withdrawals, by continent.
Source: WRI, 1996; BMZ, 1995

	Year	Total withdrawals (km³)	Per capita withdrawals (m³)	Withdrawal by households (percent)
Africa	1995	10	199	7
Asia	1987	98	542	6
Europe	1995	455	626	14
North America	1995	608	1,451	9
Oceania	1995	17	586	19
South America	1995	106	332	18
World	1987	3,240	645	8

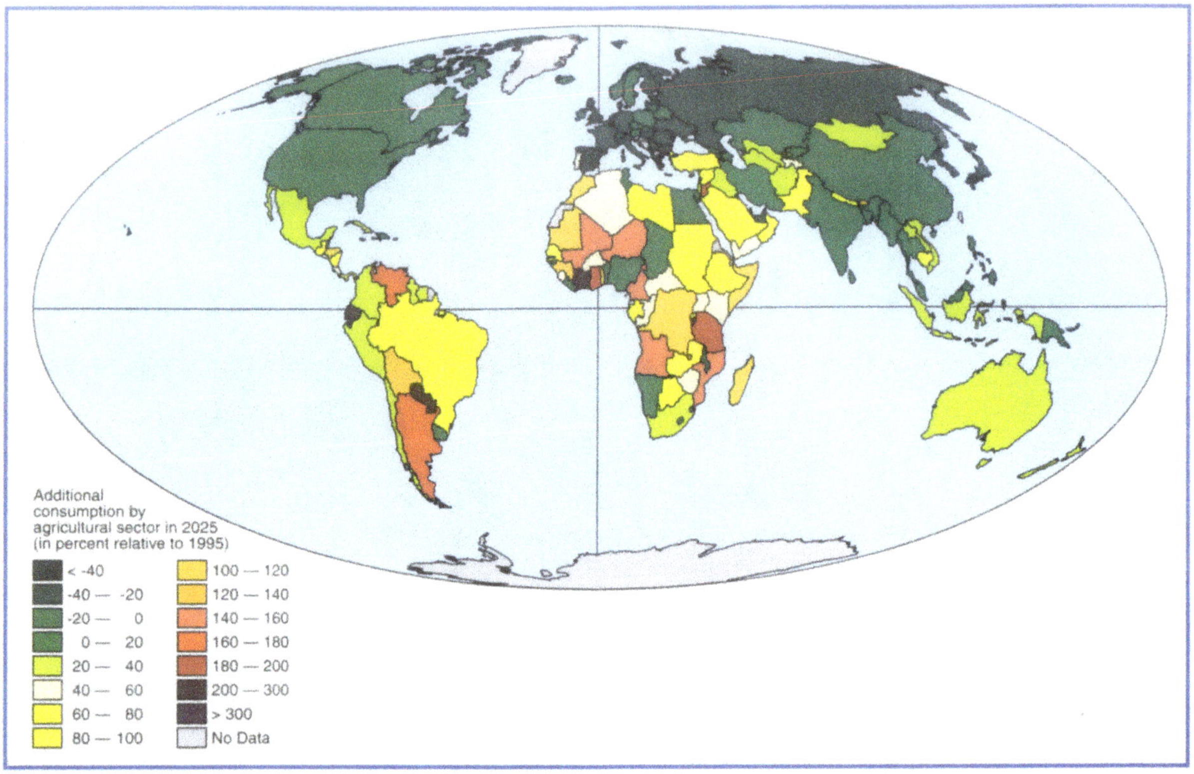

Figure D 1.4-5
Relative growth and decline in water withdrawals by agriculture, 1995-2025.
Source: WBGU, using Alcamo et al., 1997

was assumed for countries with high water availability that current water intensity (water withdrawals per unit of industrial GDP) will decline to 50% within a short space of time for per capita incomes exceeding US$ per annum (Scenario M). For countries with low water availability, this assumption was made where per capita incomes are US$ 5,000 per annum or more – due to the greater pressure to conserve water engendered by low availability.

As far as domestic use is concerned, Scenario M ("best guess") assumes a continuous increase in water intensity for annual per capita incomes up to US$ 15,000. After reaching an average income of US$ 15,000, the figure for water withdrawals declines to 50%, where it remains until further growth in incomes. Figure D 1.4-5 shows the estimated water withdrawals for agricultural, industrial and domestic use in 2025 obtained by applying the assumptions of the "best guess" scenario.

FUTURE WITHDRAWALS BY AGRICULTURE

Running the WaterGAP 1.0 model produces a figure of 3,655 km³ in total agricultural water use. This is equivalent to an increase of 550 km³, or about 18% since 1995. Despite this increase, agricultural use declines as a proportion of global water withdrawals to 56%, 19% less than the 1995 figure.

With the exception of Australia, the industrialized countries manage to reduce their water withdrawals. Factors contributing to this decline are stagnating population growth and the anticipated efficiency improvements in irrigated farming. One striking aspect is that even China, with a forecast population growth of more than 300 million people by the year 2025, is calculated to have 4% less water use than in 1995. Growth rates in China are expected to be lower than the predicted rates of efficiency enhancement because there is little potential for expanding the area under irrigation, so water withdrawals will fall slightly. The situation in Africa and South America is totally different. Projected growth in irrigated area of 114% in Africa (excluding North Africa) is reflected in much higher water withdrawals. Countries like Tanzania, Ghana or the Côte d'Ivoire reach growth rates of more than 180%. The picture is similar in South America, although there the enhanced level of withdrawals is lower than in Africa. Argentina, Venezuela, Paraguay and Ecuador are particularly conspicuous with at least 180% more water withdrawals.

Comparing these outputs with the purely demographic projection for the year 2025 confirms the disparities between the industrialized countries, where withdrawals decline, and continents like Africa or

Table D 1.4-9
Basic assumptions about water use in the agricultural, industrial and domestic sectors for the L, M and H scenarios of the WaterGAP model.
Source: Alcamo et al., 1997

Scenario	Private households	Industry	Agriculture
Optimistic Scenario L	Water withdrawal increases in line with rising incomes and peaks when per capita income reaches US$ 15,000 per annum. Withdrawal declines rapidly to 40% of the maximum for incomes up to US$ 20,000, approaching the 5% level asymptotically at higher incomes.	Depending on the availability of water, withdrawals remain at a constantly high level for incomes from US$ 5,000 (low availability) to US$ 15,000 (high availability). This is followed in both cases by rapid decline to 50% of the initial level, approaching the 5% level asymptotically at higher incomes.	Irrigated area remains constant. The efficiency of irrigation increases.
Best guess Scenario M	Water withdrawal increases in line with higher incomes, peaking when per capita income reaches US$ 15,000. This is followed by a rapid decline to 50% of the peak value. Withdrawal remains at the 50% level as incomes increase.	Depending on the availability of water, withdrawals remain at a constantly high level for incomes from US$ 5,000 (low availability) to US$ 15,000 (high availability). This is followed in both cases by rapid decline to 50% of the initial level.	Additional areas under irrigation in most developing countries. Efficiency of irrigation increases.
Pessimistic Scenario H	Water withdrawal increases in line with higher incomes, peaking when per capita income reaches US$ 15,000. Withdrawal remains at this high level as incomes increase.	Water withdrawal remains at the high initial level.	Additional areas under irrigation in most developing countries. No increase in irrigation efficiency.

South America, where there are major increases in water withdrawals. In the scenario generated by the WaterGAP model, these contrasts are amplified even further, and differences within the African and South American continents are greater. Withdrawal levels in Africa do not diverge from the demographic projection for 2025 as much as they do in South America – an indication of demographic inducing of additional water withdrawals in African agriculture.

In the purely demographic projection, total water withdrawals by agriculture in 2025 will amount to 4.466 km³, which is around 900 km³ more than the figure predicted by the WaterGAP model. The markedly lower values predicted by the model illustrate the impact of technological development and efficiency improvements in irrigation systems (Fig. D 1.4-5).

Future withdrawals by industry

Table D 1.4-10 shows the forecast population of individual states in 2025 and projected figures for water withdrawals by industry. These figures are obtained, on the one hand, from projections of current per capita water consumption and the expected population growth rate (column 5), on the other from the work of Alcamo et al. (1997) (column 6).

As the estimate in column 5 shows, industrial water use will increase as a result of population growth alone. Although the world population will grow by a factor of 1.45 over the 1995–2025 period, water withdrawals by industry will rise by only 1.22-fold. The difference arises from the fact that population growth will be relatively fast in those countries where industrial consumption of water is presently low.

Alcamo et al. (1997) conclude that the volume of water withdrawn by industry will increase by a factor of 2.98, which is significantly more than population growth (Fig. D 1.4-6). This higher increase is caused by the industrial growth forecast in the IPCC IS92a scenario. It should be emphasized here that, mainly as a result of industrialization in the developing world, the proportion of total water withdrawals accounted for by industry in 2025 may reach the same level as agriculture, which has by far the largest share of water consumption in the world today.

These calculations are based on ad-hoc assumptions about water intensity and consumption trends, without explicit consideration of determining factors such as water prices, cultural influences and other, institutional conditions for water withdrawal.

Country	Population in 1995 (millions)	Population in 2025 (millions)	Withdrawals in 1995 (millions m³)	Withdrawals in 2025 (due to demographic changes) (millions m³)	Withdrawals in 2025 (WaterGAP) (millions m³)
Afghanistan	20	45	60	135	189
Egypt	63	97	7,182	11,058	24,551
China	1,221	1,526	64,957	81,183	403,474
Germany	82	76	32,792	30,392	41,853
India	936	1,392	43,618	64,867	246,106
Indonesia	198	276	2,772	3,864	8,887
Iran	67	123	2,204	4,047	3,788
Italy	57	52	13,053	11,908	14,149
Japan	125	121	35,638	34,497	74,730
Mexico	94	137	7,407	10,796	17,166
Pakistan	141	285	4,174	8,436	31,977
Rumania	23	22	15,974	15,203	46,590
Spain	40	38	9,243	8,776	8,039
Thailand	59	74	3,806	4,773	10,974
Turkey	62	91	7,198	10,565	14,687
USA	263	331	179,997	226,536	357,083
World	5,692	8,261	738,231	897,421	2,202,617

Table D 1.4-10
Growth of water withdrawals by industry in selected countries.
Source: compiled from WRI, 1996 and Alcamo et al., 1997

FUTURE WITHDRAWALS FOR DOMESTIC USE

Domestic water use is dependent on many factors. To be able to predict withdrawal levels with any degree of precision, it is necessary to have data not only on demographic trends, but also on the extent to which consumption patterns alter in relation to economic trends, water-sustainable technologies are developed and deployed, and infrastructural measures are implemented in order to enable access to potentially available water resources.

According to WHO estimates, 40–60% of all water supply facilities in the rural areas of the developing world is not operational (BMZ, 1995). Potential is left unexploited, or considerable quantities of drinking water are wasted because even the most simple technical equipment for controlling water withdrawals is not available. In many cases, only a fraction of water withdrawals is actually put to any use.

It is difficult to assess how the factors influencing water withdrawals will develop in the future. The World Resources Institute assumes, however, that domestic water use will increase by 2–3% relative to agricultural and industrial use, with major differences between continents (WRI, 1996). In total, water withdrawals by private consumers have risen substantially since the 1980s (Table D 1.4-11).

Forecasts show that, presuming a comparatively low baseline value, there will be enormous growth in water withdrawals, especially in Africa and Asia. In Europe and South America, by contrast, a downward trend is discernible. There is a close connection between rising living standards and the demand for water availability and water quality. Even today, it is apparent that the level of per capita water withdrawals in the developing countries approaches that of the industrialized countries as soon as houses have direct access (BMZ, 1995). Comparing the increase in water

	Percentage of total withdrawals by households (1980s)	Percentage of total withdrawals by households (2000)	Percentage growth in water withdrawals (1980s to 2000)
Africa	7	13–14	200
Asia	6	10	127
Australia and Oceania	19	19	34
Europe	14	12–15	17
Former USSR	6	7	52
North America	10	11	36
South America	19	17	67
World	8	10–11	74

Table D 1.4-11
Growth of domestic water withdrawals (1980s to 2000).
Source: WRI, 1990

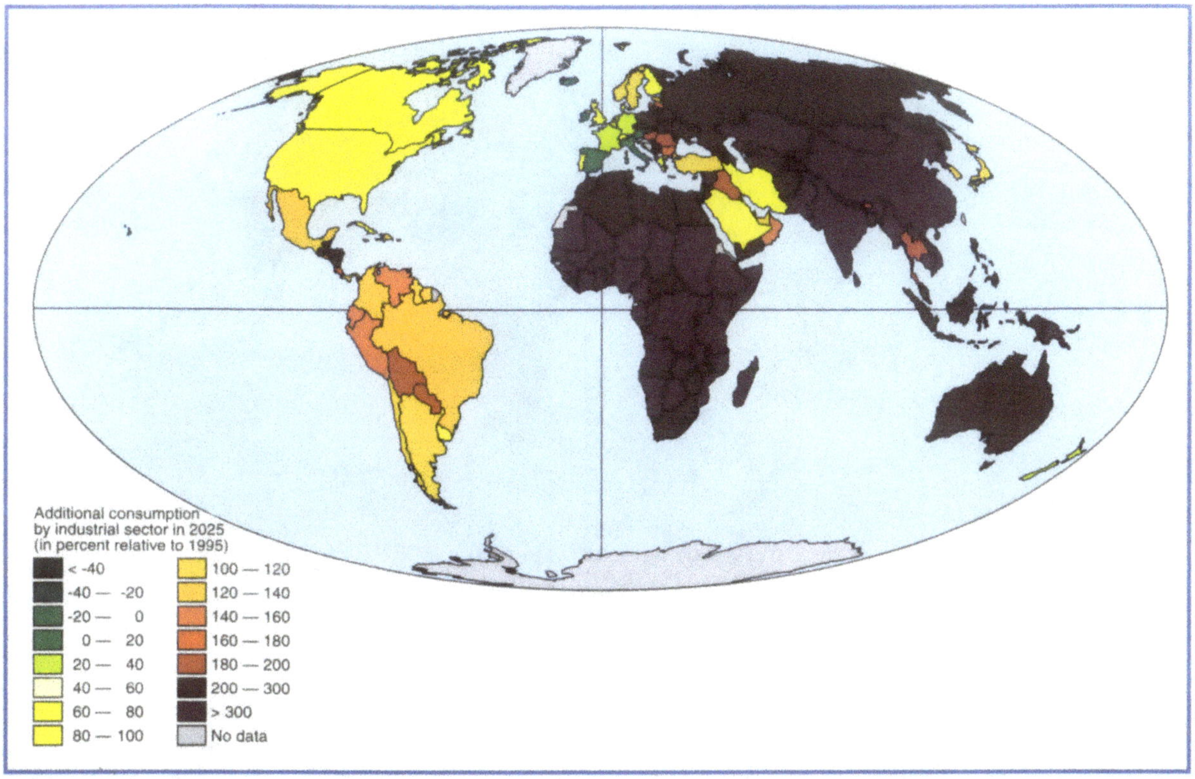

Figure D 1.4-6
Relative growth and decline in water withdrawals by industry, 1995-2025.
Source: WBGU, using Alcamo et al., 1997

withdrawals over the 1980–2000 period (Table D 1.4-11) and Scenario M for the period 1995–2025 reveals a stabilizing trend (Fig. D 1.4-7). The highest growth rates occur in Africa and Asia (as high as 300%), followed by South America and Oceania. However, in contrast to the growth rates for Europe (17%) and North America (36%), as calculated on the basis of projected population figures, Alcamo et al. (1997) expect stagnating or even declining withdrawals. According to the latter analysis, the highest savings will be achieved in western Europe.

1.5 Water quality

Imbalances in global data acquisition – Precipitation transfer problems – Quality of surface water – Delayed appearance of pollutants in groundwater – Drinking water specifications – Water as a means of agricultural production

Water quality – a normative term – is assessed using a broad array of parameters. Different variables are of importance when examining the various natural and usage-related functions of water (see Section D 1.1). The answer to the question of which values are considered acceptable or desirable depends on the way in which the respective waterbody is to be utilized. The assessment of water quality is also influenced by sociocultural values. For example, the use function of "safe drinking water for human use" is linked to a definition of water quality that varies from country to country (different variables and quality criteria). By contrast, water must meet only a few broadly defined requirements in order to perform its transport function. When assessing the habitat function of freshwater ecosystems, on the other hand, a crucial factor in most cases is the existence of a pristine condition, i.e. a state characterized by minimal human influence.

Although "water quality" has different meanings according to freshwater's many natural functions and types of use, it is possible to define water quality with a set of identical parameters that fall into the following categories:

- Physical properties: temperature, amount of suspended matter, color
- Chemical properties: gases, organic and inorganic compounds dissolved in water
- Biological properties: populations of organisms (indicator species, diversity indices). In connection with human consumption of water, the biological properties also include hygienic characteristics,

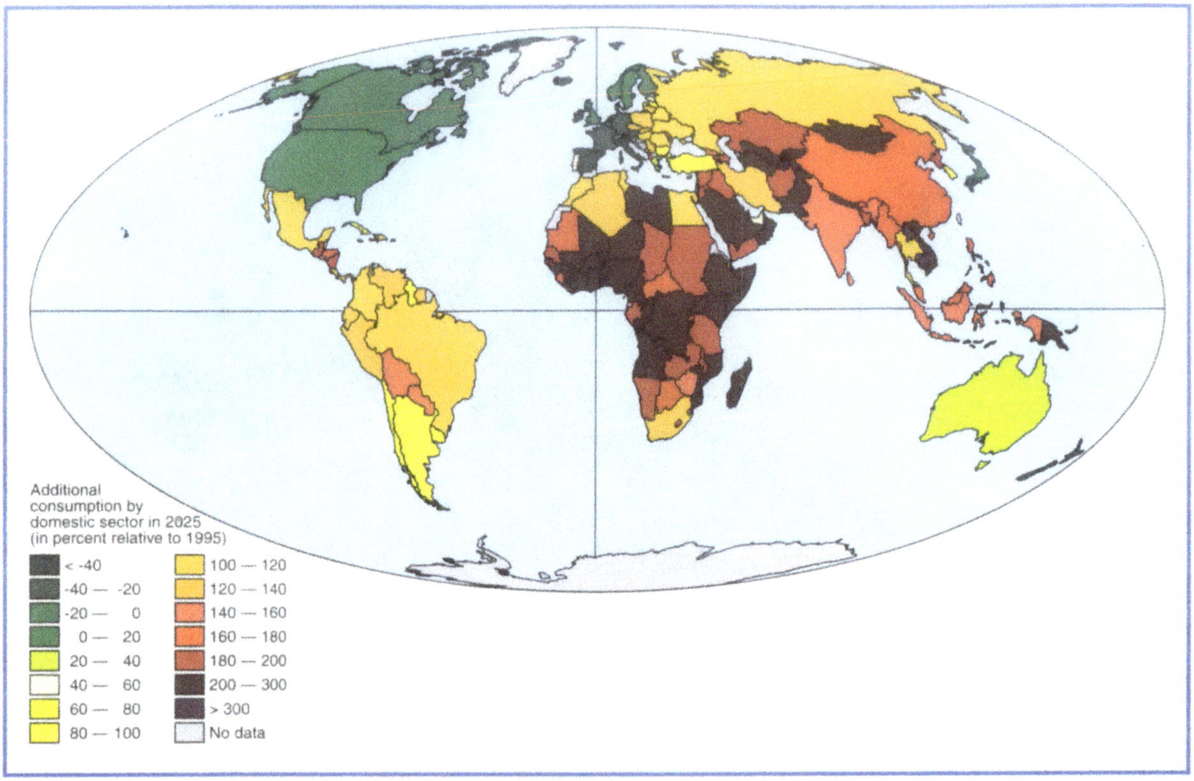

Figure D 1.4-7
Relative growth and decline in domestic water use, 1995-2025.
Source: WBGU, using Alcamo et al., 1997

Table D 1.5-1
Factors affecting water quality.
Source: Chapman, 1992

	Natural function or cultural function						
	Drinking water	Irri-gation	Aquatic wildlife, fisheries	Indus-trial uses	Power and cooling	Recre-ation	Trans-port
Contamination from excrements	2	1	0	2	na	2	na
Suspended solids	2	1	2	1	1	2	2
Organic matter	2	+	1	2	1	2	na
Eutrophication	1	+	1	2	1	2	1
Nitrate	2	+	1	2	na	na	na
Salinization	2	2	2	2	na	na	na
Metals	2	1	2	1	na	1	na
Organic micropollutants	2	1	2	?	na	1	na
Acidification	1	?	2	1	1	1	na

+ = beneficial, 0 = no impairment, 1 = minor impairment, 2 = marked impairment causing major treatment or excluding the desired use, na = not applicable.

such as bacterial counts, viruses and animal pathogens (see Section D 4.2).

Water quality can be influenced by natural factors, and especially by human use, to such an extent that its natural and cultural functions are altered or impaired (Table D 1.5-1).

Different quality requirements for the various types of usage can lead to conflicts over use, especially in connection with natural waters. Utility companies supplying drinking water, for example, are interested in water with a low nutrient content, because this minimizes the amount of treatment that is necessary. Fisheries, by contrast, are interested in high

Figure D 1.5-1
The sequence of water quality issues arising in industrialized countries, 1850 - present.
Source: Chapman, 1992

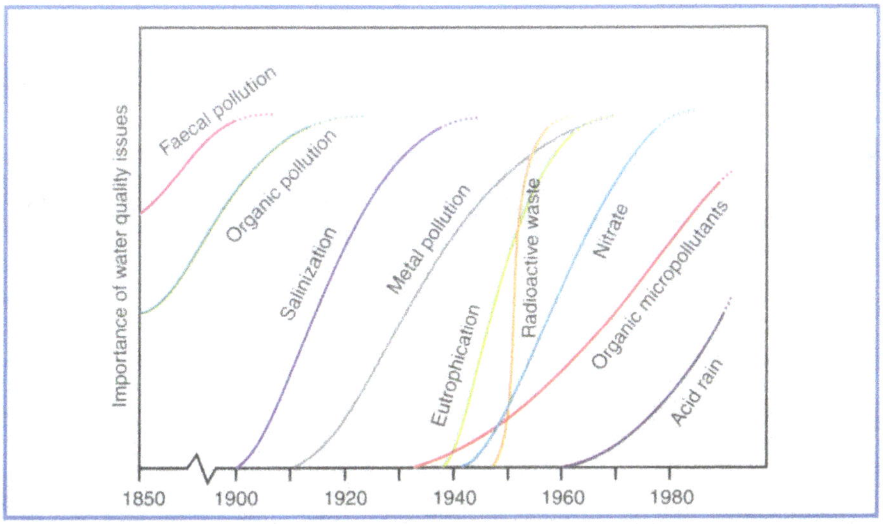

yields, which cannot be achieved from nutrient-poor waters. Conflicts may also arise over the discharge of wastewater into receiving waters, necessitating a cost-benefit analysis of the expenditure on wastewater treatment versus the possible consequences of reduced quality of the receiving waters.

Problems with water quality have been known to humankind since the Roman Empire and the Middle Ages. Since the onset of industrialization, many new stress factors have been added. The historical sequence of problems experienced by the industrialized countries (Fig. D 1.5-1) is being repeated in the developing nations in a similar fashion, though concentrated within a considerably shorter period of time. In many developing countries, water quality problems typical for industrialized countries make themselves felt even before the more "traditional" problems can be adequately managed.

1.5.1 Inventorying of water quality

Since the beginning of this century, an enormous wealth of data has accumulated worldwide as single measurements or in results of regular monitoring programs. In many ways, the reasons for collecting data vary as much as the measured variables themselves and the survey methods employed. Since 1974, coordination and collection of comparable data on water quality have been advanced on a global scale through GEMS (Global Environmental Monitoring System), a program jointly operated by WHO, UNESCO, WMO and UNEP. Since the program began, data have been collected from an increasing percentage of the 1,200 sampling stations in rivers, lakes and groundwater that according to GEMS are needed for sufficiently representative global coverage. The geographic imbalance in the collection of water-relevant data corresponds with the distribution of sampling stations used by the World Meteorological Organization (WMO). In 1989 there were 48,000 such stations, only 20% of which were located in developing countries. In Africa there were only 361. The number of these sampling stations doubled worldwide between 1977 and 1989. However, there was a downward trend in the Asian and Pacific region over the same period.

Waterbodies are linked to each other through the global hydrological cycle and are basically exposed to the same natural and anthropogenic influences. The quality-related problems that occur in waterbodies and are described in the following are basically similar, but will vary depending on local conditions and hydrological factors (renewal rate and mixing).

1.5.1.1 Precipitation

Atmospheric water is circulated about 40 times a year, with an average residence time of nine days (Bliefert, 1994; Häckel, 1990). This means that substances released to the atmosphere can be washed out 40 times a year on average and then deposited on the Earth's surface. Precipitated water free of anthropogenic influences contains a range of dissolved ions whose composition and concentration are closely related to their respective origin (marine or terrestrial) (Table D 1.5-2). As a result of rainout within clouds and washout during precipitation, gases and particles are removed from the atmosphere. Due to this cleansing effect, the chemical composition of precipitation changes, and the concentration of the compounds contained in the water declines in the course of the respective precipitation event. In gen-

Table D 1.5-2
Typical concentrations of the major ions in continental and oceanic precipitation. Continental precipitation refers to areas remote from anthropogenic sources.
Source: Berner and Berner, 1996

Ion	Continental precipitation (mg l^{-1})	Oceanic precipitation (mg l^{-1})
Na^+	0.2–1	1–5
Mg^{2+}	0.05–0.5	0.4–1.5
K^+	0.1–0.3	0.2–0.6
Ca^{2+}	0.1–3.0	0.2–1.5
NH_4^+	0.1–0.5	0.01–0.05
H^+	pH = 4–6	pH = 5–6
Cl^-	0.2–2	1–10
SO_4^{2-}	1–3	1–3
NO_3^-	0.4–1.3	0.1–0.5

eral, concentrations of ions and acids in clouds and fog are higher than in precipitation (rain, snow and dew). High concentrations are detected in rime.

ACIDIFICATION

Increasing emissions resulting from the combustion of fossil fuels lead to higher concentrations of acid-forming ions in precipitation water and, as acid rain, has become a major environmental problem in the industrialized regions of North America, Europe and East Asia. This trend will very likely spread to other regions (Southeast Asia, South America, Fig. D 1.5-2). Emissions of sulfur dioxide (SO_2), mainly from power stations without desulfurization facilities, are by far the most important driving factor behind the increasing acidification of rainwater, followed by nitrogen oxides (NO_x), which are predominantly generated by road traffic.

The distance traveled by the 150 million tons of sulfur in sulfur dioxide and 49 million tons of nitrogen in nitrogen oxides that are emitted annually may range from several hundreds to thousands of kilometers. Emissions from central and western Europe, for example, caused acidification of surface waters in Scandinavia (see Section D 4.4). In Germany, the steps taken to desulfurize power stations since the 1980s have already reduced sulfur deposition by 40–60%. In spite of this enormous reduction, the pH value of precipitation did not rise significantly because of the simultaneous decline in neutralizing, alkaline cations in precipitation water (Hedin et al., 1994).

The expansion of intensive agriculture leads to large-scale ammonia emissions amounting to a total of 54 million tons of nitrogen per year worldwide. Ammonia in the atmosphere is converted to ammonium, which causes eutrophication when deposited in precipitation water, as well as acidification of surface waters after nitrification. With an average 0.65 mg of ammonium per liter of precipitation, Europe has the highest figures for atmospheric deposition in the

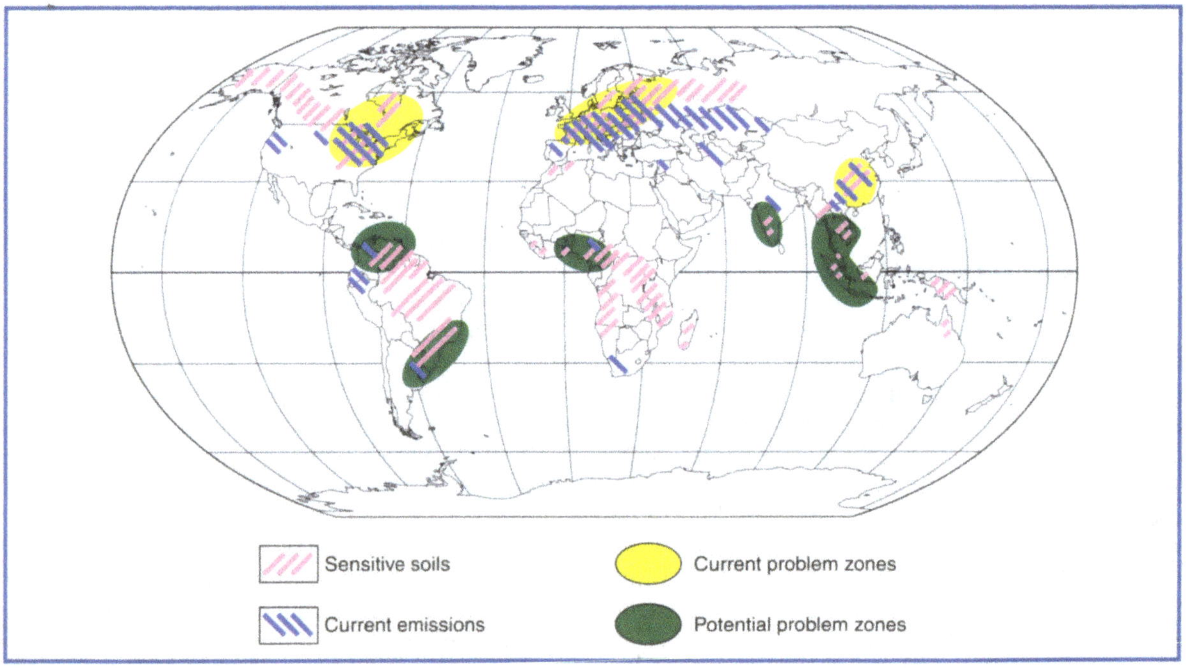

Figure D 1.5-2
Threat of waterbody acidification.
Source: adapted from Rhode, 1989 in GEMS/Water database, 1997a

world. Given the regional distribution of sources responsible for atmospheric loading and the short atmospheric residence time, the spatial and temporal distribution of ammonium is highly heterogeneous. In predominantly agricultural areas of the USA, ammonia concentrations in precipitation are 0.2–0.3 mg l^{-1}, while values in similar regions in The Netherlands are 0.5–1 mg of NH_3 l^{-1} (Lovett et al., 1992; Vermetten et al., 1992; Berner and Berner, 1996).

HEAVY METALS AND ORGANIC POLLUTANTS

Compared to most inorganic acids, organic acids are relatively weak. Nevertheless, they can become major acidifying agents in precipitation if the inorganic precursors are present only in very low concentrations relative to the organic precursors. Such conditions prevail in tropical forest areas, for example, due to the oxidation of isoprene (C_5H_8), resulting in precipitation with pH values below 5. Organic micropollutants are released into the environment primarily through the use of agricultural chemicals or in connection with other, unclosed fluxes (e.g. hexachlorocyclohexane, polychlorinated biphenyls). The industrialized countries have succeeded in reversing the growth of metal emissions, at least for lead, cadmium and zinc. Heavy-metal emissions in the developing countries, on the other hand, are continuing to increase (Nriagu, 1992).

1.5.1.2 Surface waters

As befits the role played by surface waters as the most important source of freshwater for human use (Meybeck et al., 1992), and the availability of international data, special attention is dedicated here to the quality of running waters. The natural properties of flowing waters are governed by the type of soil, the vegetation cover and the type of surficial rock. Because of their zonation (see Section D 1.2) and interspersed lakes and wetlands along their course, running waters display marked horizontal gradients. Smaller rivers, for example, show strong seasonal fluctuations in water level, whereas conditions in large rivers are less variable over time. Anthropogenic impacts also change along the course of a river. In upper reaches, the most significant factors are material deposition as a result of rainfall, increased erosion due to felling operations and damming of rivers to supply power and water. Contamination from sewage, pollutant loads from agriculture and industry, as well as regulatory modifications to the river bed become increasingly important factors further downstream.

Temporal fluctuations in running waters are usually extremely high compared with other types of waterbody. Changes may result in a matter of minutes (storms, chemical spills), in the course of a day (light- and heat-dependent processes of synthesis and degradation), in the course of a year (climatic conditions) or over even longer periods of time (industrialization, land-use changes, climate variability). The enormous spatial and temporal variability of the relevant variables makes it difficult to identify global trends and to distinguish natural variability from human-induced changes. What is required, therefore, is a solid basis of data with an adequate temporal and geographic resolution on reference waters subject to little human influence and on waters of various types and in different locations.

PARTICULATES

The amount of suspended matter transported by running waters is extremely variable and ranges from a few mg l^{-1} to 30 g l^{-1} in some Asian rivers. Particulate substances reduce the depth to which light can penetrate, thus affecting aquatic fauna and flora. Pollutants and nutrients (heavy metals, organic pollutants, pathogens, phosphate) are adsorbed on the surface of suspended particles, with the result that sediments of deposited particles are sometimes extremely polluted (such as in the Elbe, Lower Rhine and Neckar; e.g. Lozán and Kausch, 1996). Suspended matter is undesirable in drinking water because of its absorptive properties; filtering it out is a costly process, especially when particles are fine.

Particulates are the nutrient base for innumerable organisms (shellfish, worms and small crustaceans). Toxic substances bound to particulate matter are transmitted via food chains and enriched (biomagnification), and are a threat to end consumers (predatory fish, seals, sea birds, humans). Large sediment loads may destroy valuable habitats, spawning grounds and coastal ecosystems (coral reefs) (see Section D 3.4.2.3). Through the adsorption of pollutants, sediments play a role in the self-purification of waterbodies, but they can also release pollutants as a result of resuspension and changing environmental conditions (e.g. concentrations of salts, oxygen and acid) (Müller, 1996). Many of the chemical transformations occurring in sediments have not been sufficiently studied.

The total amount of particulate matter is closely linked to conditions in the catchment and is heavily dependent on anthropogenic influences (land use, changes in stream bed). The fact that two thirds of the total particulate load reaching the oceans come from waterbodies in Southeast Asia is due mainly to the combination of heavy rainfall, a topology with steep slopes and easily eroded topsoils, as found, for exam-

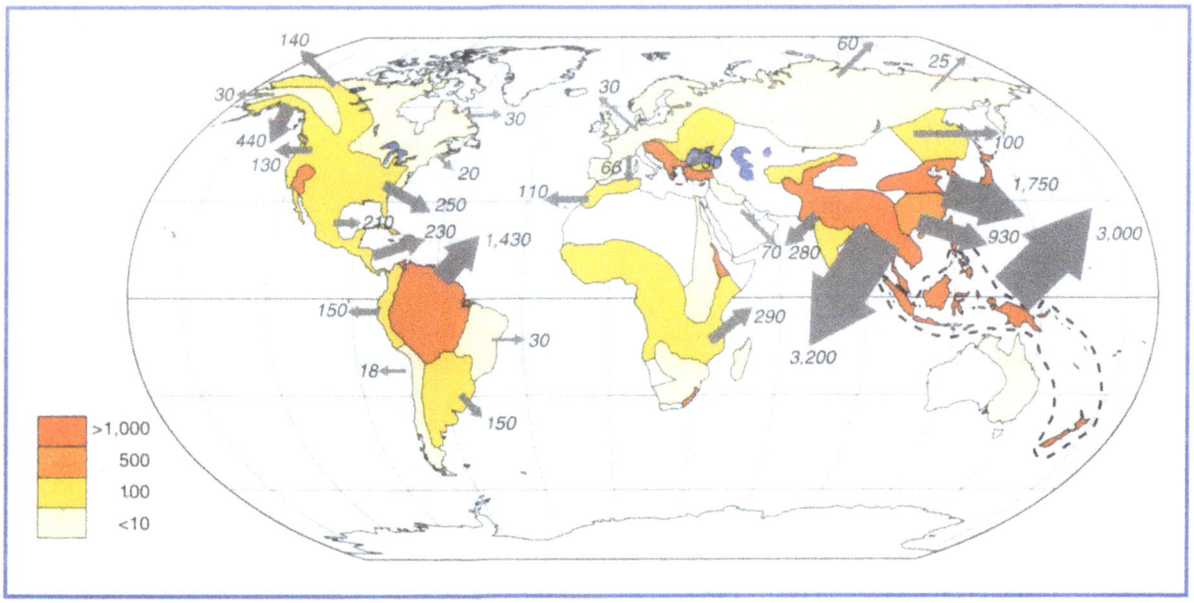

Figure D 1.5-3
Sediment transported by rivers. Absolute sediment yield (million t year^{-1}) and relative sediment yield per unit area of catchment area (colored areas, in t km^{-2} year^{-1}).
Source: adapted from Milliman and Meade, 1983 in GEMS/Water database, 1997b. © 1983 by The University of Chicago Press

ple, in the northern Chinese loess belt. The annual sediment load of a flowing water system, calculated in terms of the size of the catchment area, is a key indicator of *soil erosion* (Fig. D 1.5-3). In many rivers, the amount of suspended matter rises with river discharge. Peak values many times higher than the annual mean occur during a few days each year. Therefore, calculations of sediment load (and related pollution) are reliable only when sampling is carried out with a fine temporal resolution.

Oxygen concentration

Oxygen concentration is of paramount importance for the water quality of surface waters in general. In running waters, the primary source for dissolved oxygen is the atmosphere. The concentration of organic matter plays a crucial role, in addition to temperature, because oxygen is consumed when organic matter is decomposed. There is a significant correlation between the population density of water catchment areas and organic pollution. The biochemical oxygen demand (BOD_5) is a measure of the level of biodegradable organic pollution. The global mean BOD_5 in running waters is around 2 mg l^{-1}. Substantially higher values are reached in receiving waters for untreated sewage, wastewater from agriculture and food production (e.g. palm oil), as well as industrial effluent from certain sectors such as pulp and paper production. However, a large proportion of organic substances, particularly in industrial effluent, is degraded extremely slowly or is completely resistant to degradation. The chemical oxygen demand (COD) gives an indication of the proportion of such materials (e.g. organochlorine compounds, tensides, etc.). In the extremely polluted Espierre River, which flows through France and Belgium, the mean BOD_5, for example, is around 179 mg O_2 l^{-1}. A much higher COD value and a COD:BOD_5 ratio of about 20 is indicative of the degree of pollution from industrial effluent. A low COD:BOD_5 ratio is typical for running waters polluted with sewage in the rapidly growing megacities of Asia and Latin America. The Banjir Canal in Jakarta has a mean BOD_5 of 10.4 mg l^{-1}; the COD:BOD_5 ratio here is around 1.3.

Overfertilization of waterbodies with plant nutrients such as nitrate and phosphate produces a high biochemical oxygen demand (eutrophication; see Section D 4.4).

The decomposition of organic matter by microorganisms is part of a self-purification process in which aquatic and terrestrial plants adjacent to the river are also involved. The self-purification capacity of waterbodies and the extreme dilution of anthropogenic loads by large quantities of water at low levels of pollution are often sufficient to keep the waterbody in an acceptable condition. However, in many regions pollution is far in excess of the maximum tolerable level (see Section D 4.4). The self-purification capacity of waterbodies should not conceal the fact that only easily degradable organic matter is removed.

Table D 1.5-3
Biodegradable organic substance (measured as biological oxygen demand – BOD) and non-biodegradable substance (measured as chemical oxygen demand – COD) and oxygen concentration in European rivers and worldwide.
– = no data available.
Source: Meybeck et al., 1989; Stanners and Bourdeau, 1995

	Number of sampling stations	Proportion of sampling stations where BOD, COD and oxygen content do not exceed the specified values (mg l^{-1})				
		10%	25%	50%	75%	90%
European rivers						
BOD	645	1.4	1.9	2.8	4.7	7.9
COD	470	4.5	7.8	15.0	25.0	36.6
Oxygen	620	6.4	8.4	9.7	10.7	11.6
Virtually natural European rivers						
BOD	11	–	1.2	1.6	2.7	–
COD	23	–	5.1	13.3	29.9	–
Oxygen	8	–	10.2	10.6	11.1	–
Rivers worldwide						
BOD	190	1.6	–	3.0	–	6.5
COD	127	6.0	–	18	–	44

Substances of low or zero degradability remain in the water, accumulate in the food chain, or are deposited in sediment (Table D 1.5-3).

Water quality classification

Various indices based on selected species in running waters have been developed for assessing water quality. In Germany the saprobic index pursuant to the DIN 38410 standard is one of the most important. It is calculated from the frequency of a number of heterotrophic macroorganisms and microorganisms, weighted according to their ecological requirements. Although the saprobic index is suitable, at best, for identifying the impacts on biological communities in flowing waters caused by biodegradable organic pollutants (Simmann, 1994), it is the basis for the quality map of flowing waters in Germany published by the *Länder* Working Group for Water (*Länderarbeitsgemeinschaft Wasser*). Although commonly used, the saprobic assessment of running waters would require, strictly speaking, a wide geographical distribution of the indicator species, which is not the case even within Germany (Zauke and Meurs, 1996). Other major points of criticism regarding the saprobic system include the lack of consideration given to autotrophic organisms (e.g. diatoms), the historical development of biological communities, and the fact that the distribution of indicator species is dependent on much more than one single isolated factor.

Salt concentration

Dissolved salts in surface waters are primarily composed of positive sodium, potassium, calcium and magnesium ions as well as negative chloride, sulfate and bicarbonate ions. The natural salt concentration of inland waterbodies depends on the rock types found in the catchment area and on the hydrological regime. In running waters the salt concentration may vary by more than four orders of magnitude (Fig. D 1.5-4). Mining, oil production, changes in the natural water balance and irrigation measures are some of the major anthropogenic influences on salt concentration.

Enhanced concentrations of calcium and magnesium are found particularly in regions where salt and potassium are mined (in Europe, along the Upper Rhine and Werra rivers). High sodium chloride concentrations are generally found in arid or semi-arid regions. Anthropogenic factors have caused the concentration of sodium chloride in many rivers of the world to rise to 10 to 20 times above the natural values. In the industrialized countries, using salt to de-ice roads is a cause of salinization.

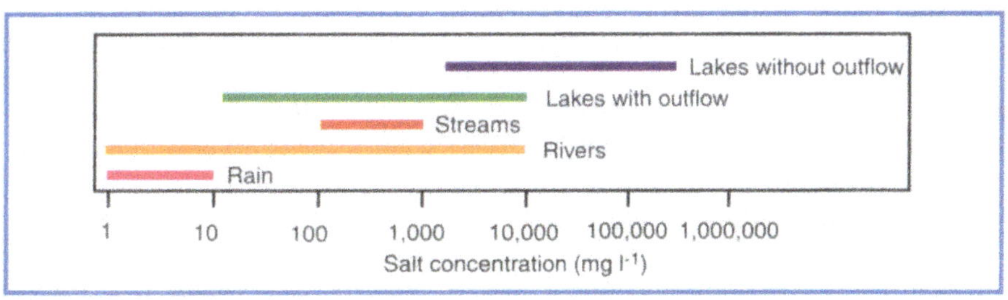

Figure D 1.5-4
Salt concentrations in different freshwater bodies worldwide.
Source: adapted from GEMS/Water database, 1997c

	Number of sampling stations	10% (mg l⁻¹)	50% (mg l⁻¹)	Maximum (mg l⁻¹)	Waterbodies with maximum concentrations
Arsenic	38	<0.001	0.0025	0.03	Klang, Malaysia
Cadmium	56	<0.001	0.001	0.312	Missouri, USA
Chromium	58	0.003	0.01	1.675	Espierre, Belgium
Copper	66	0.004	0.01	0.08	Espierre, Belgium
Lead	64	0.002	0.006	0.05	Ohio, USA
Mercury	59	<0.00002	0.0001	0.0005	nine Japanese rivers
Manganese	61	0.01	0.05	1.350	Rio Rimao, Peru
Zinc	51	0.005	0.02	0.4	Missouri, USA

Table D 1.5-4
Metal concentrations in the world's waterbodies. Total concentrations not exceeded by 10% and 50% of the measured values, and maximum values. Source: Meybeck et al., 1989

The concentration of bicarbonate ions is highly important for the acid buffer capacity of waterbodies and soils. The risk of acidification is much higher in waterbodies with low bicarbonate concentration. In the future, large areas in humid parts of South America, Africa and Asia will be affected by acidification because of their low buffer capacity (Fig. D 1.5-2).

Metals

Various metals are naturally present in significant amounts wherever rock formations containing metal are found. Some metals, like copper, manganese, zinc and iron, take part in the physiological processes of organisms and are therefore essential for life in aquatic habitats, although they may have toxic effects at increased concentrations. The toxicity of metals depends on the chemical form in which they occur. Little is known about the impact of some highly toxic metals (beryllium, thallium, vanadium, antimony, molybdenum) on humans and the natural biota in ecosystems.

The assessment of metal concentrations in water is expensive, technically complicated and subject to high errors. Significant contamination may occur during sampling and storage of samples. Many data sets have been based on different methods of sample pretreatment, which makes them very difficult to compare. At present there is a lack of reliable data on the natural background concentration of various metals. Measurements of metal concentrations are available for only 35% of the GEMS/Water stations, above all from North America, Europe and Japan. At half of all GEMS/Water sampling stations, metal concentrations fall short of the WHO guidelines for drinking water quality and of EU standards (Table D 1.5-4). At 10% of the sampling stations, the accepted maximum values for arsenic, cadmium, chromium, manganese and zinc are reached or exceeded. In some cases, the highest observed values lie significantly above these standards (cadmium in the Missouri, chromium in the Espierre).

In addition to the release of metals from the smelting and metals industry, the direct use of metals (chromic salts in tanneries, copper as a plant protection agent and lead in fuels) is (or was) a major anthropogenic source of contamination. Serious pollution originates from municipal landfills and mining tips around the world, especially when metals come into contact with acidic solutions. Acidification of waterbodies and soil, the consequences of eutrophication (oxygen scarcity) and the deposition of organic chelating agents (phosphate substitutes) make it likely that metal pollution will continue to rise in the future. Knowledge about the availability and mobility of metals in the environment is still very patchy.

Organic trace substances

Many thousands of different organic micropollutants enter the aquatic environment as a result of human activities. Little or nothing is known about their distribution and behavior in the environment. Chemical analyses require sophisticated technical methods and expertise. Thus, it is almost impossible to summarize the global situation.

Worldwide, crude oil and its derivatives rank among the most frequent pollutants in waterbodies. The 800 or more compounds identified in mineral oils can be found in water in the form of surface films, emulsions or deposits adsorbed onto suspended particles. The different solubilities, boiling points, surface tensions and other properties of these substances have an impact also on biochemical, photochemical and microbial processes in water. Given the high risk of environmental damage and toxic effects on humans, one of the most important objectives for the international community is to monitor and control possible sources of emissions.

Approximately 10,000 different pesticides are used worldwide in agriculture and industry (e.g. in timber processing) as insecticides, herbicides and fungicides, as defoliants and to control aquatic plants and disease-transmitting organisms. The most important groups include organochlorine pesticides, organophosphorous pesticides, carbamate pesticides and triazine herbicides. Whereas organophosphorous compounds often decompose in the environment

Table D 1.5-5
Classification of surface waterbodies according to trophic level. Unless otherwise stated, these are limit values. Source: OECD, 1982 in Meybeck et al., 1989

Degree of trophication	Total phosphate ($mg\ m^{-3}$)	Chlorophyll annual mean ($mg\ m^{-3}$)	Chlorophyll maximum ($mg\ m^{-3}$)	Mean visible depth according to Secchi (m)	Visible depth minimum (m)	Oxygen saturation in deep water (percent)
Ultra-oligotrophic	4	1	2.5	12	6	90
Oligotrophic	10	2.5	8	6	3	80
Mesotrophic	10–35	2.5–8	8–25	6–3	3–1.5	40–89
Eutrophic	35–100	8–25	25–75	3–1.5	1.5–0.7	40–0
Hypertrophic	100	25	75	<1.5	<0.7	10–0

within a few weeks, organochlorine pesticides tend to remain stable for long periods of time. Although DDT has been subject to legal restrictions in many industrialized countries for several years now, it continues to be used in developing countries. Traces and decomposition products can even be detected in animal tissues in the most remote regions of the world (Arctic and Antarctic).

In contrast to the risks posed by pesticides, public awareness of the significance of volatile organic compounds such as chloroform and benzene is slow to develop. Chlorinated or brominated substances are created, for example, by reactions with organic substances contained in water. This is particularly the case when chlorine is used to disinfect drinking water and swimming pools. In countries where the water contains large quantities of organic matter, the use of chlorine for drinking water disinfection is not recommended for reasons of health and environmental protection.

Tensides from detergents and cleaning agents in industrial effluent and municipal sewage rarely reach toxic concentrations. The environmental impacts of modified behavior on riverine organisms such as small crustaceans have been difficult to assess. As a result of the foam generated by tensides, gas exchange and self-purification processes in running waters are reduced. Pollutants and pathogens may become concentrated in such foam. In recent years there has been a growing tendency to use easily degradable tensides that decompose within a few days, consuming oxygen in the process. This property can only bring about a significant reduction of water pollution if decomposition occurs in wastewater treatment plants.

Especially between 1960 and 1970, polychlorinated biphenyls (PCB) were produced on an industrial scale, for example as flame retardants and plasticizers. PCBs have proven to be extremely persistent. They accumulate in the fatty tissue of aquatic and terrestrial organisms at the end of the food chain. Extremely high values were measured in seagull eggs in Baltic Sea states during the mid-1970s. Despite their current decline, PCBs will remain a problem in aquatic habitats for a long time to come.

Eutrophication

Besides contamination by trace elements, metals and organic trace substances, eutrophication is globally the most significant cause of damage to the water quality of standing waters. In most inland waterbodies, phosphate is the most important plant nutrient limiting algal growth and hence the principal cause of eutrophication (Section D 4.4).

The maximum concentration of total phosphorus in winter is closely related to the degree of eutrophication of a waterbody (Table D 1.5-5), which, under certain regional conditions, may be high even in the

Table D 1.5-6
Trophic levels in lakes and reservoirs worldwide. Source: Meybeck et al., 1989

	Percent oligotrophic	Percent mesotrophic	Percent eutrophic and hypertrophic	Number of waterbodies studied
Canada	73	15	12	129
USA	7	23	70	493
Italy	29	28	43	65
Germany	8	38	54	72
Baltic States	15	38	31	130
South and Central America	24	20	56	25
South African reservoirs	31	41	28	32
18 OECD countries	18	17	65	101

Table D 1.5-7
Possible levels of complexity for a general monitoring of running waters program.
Source: Chapman, 1992

	Sampling points	Sampling frequency (per year)	Water analysis	Sediment study	Biological variables	Necessary resources
Simple monitoring	10	6	°C, pH, conductivity, oxygen, particle content, main ions, nitrate			small team, standard analysis as in agricultural and medical laboratories
Intermediate-level monitoring	100	6–12	as above, additionally phosphate, ammonium, nitrite BOD, COD	inorganic trace substances (metals)	species, diversity, saprobic indices, etc.	specialized lab, with hydrobiologists
Advanced-level monitoring	100 –1,000	more than 12	as above, additionally dissolved organic substances, dissolved and particulate organic carbon, chlorophyll, inorganic trace substances	as above, additionally organic trace substances	as above, additionally chemical study on bio-accumulators	specially equipped laboratory with specialists (e.g. national research institute)

absence of human influences. 30 to 40% of all lakes and reservoirs in the world are affected by eutrophication (Table D 1.5-6). Contamination of standing waters with inorganic and organic trace substances, acidification and salinization is as serious as it is in running waters.

1.5.1.3 Groundwater

Groundwater quality is currently monitored within the GEMS/Water program at only 61 stations worldwide. The major factors impacting on groundwater quality today are nitrate pollution, salinization and contamination with inorganic and organic trace substances. Rising nitrate concentrations in surface waters almost all over the world is now spreading to groundwater as well. Apart from natural sources, nitrate enters groundwater mainly through agriculture and sewage, as well as locally from cesspools, municipal and industrial landfills. As early as the late 1970s, the growing use of nitrogen fertilizers as a result of intensified agriculture led to rising nitrate concentrations in groundwater resources in the USA and Europe. A similar picture is now emerging in countries with a rapidly developing agricultural sector, such as India. Nitrate applied over a large area enters the groundwater after a certain time delay. For this reason, one can expect a further rise in nitrate values over the next 10–20 years even in the event that land-use patterns are changed drastically. In addition to diffuse loads from agriculture, point-source loads from pit latrines and septic tanks play a major role, particularly in developing countries (Lewis et al., 1982). Problems for drinking water supply frequently arise in medium and smaller sized towns, where water supply and water disposal systems may be closely linked in spatial terms. In many parts of India, Botswana and Nigeria, contaminants in drinking water have already exceeded the maximum levels established by WHO.

The natural salt concentration of many groundwater systems is generally high. Irrigation causes salinization of water and soils in many arid and semi-arid regions. In coastal regions, excessive depletion of groundwater stocks enables the intrusion of saltier water from the adjacent sea. Lowering of water tables in coastal mining regions can also lead to an influx of sea water (e.g. North Carolina). Disposal of the saltwater that accumulates in the process of oil production has led to the salinization of many aquifers in the USA and the former USSR.

Metals such as iron and manganese are usually contained in groundwater at higher concentrations than in other waterbodies, because the combination of a long residence time and low oxygen concentration releases extremely high concentrations. Heavy metals and organic trace substances can enter groundwater directly via the atmosphere, from wastewater or from leachates. Of the 100,000 landfills in the USA, only about 10% can be considered safe for groundwater resources. In Germany, an estimated 240,000 sites are suspected of being seriously polluted (UBA, 1995). The risk of groundwater pollution in the newly industrializing and developing countries is assessed as high.

1.5.1.4
Monitoring water quality

MONITORING

Monitoring of aquatic habitats can be carried out with different objectives in mind. Background monitoring provides the basis for defining natural water quality. The purpose of trend monitoring is to study medium- to long-term trends. Early warning systems may be necessary to protect drinking water supplies or in the case of multiple uses (e.g. the Rhine). General monitoring is carried out to ascertain the suitability of water resources for single or multiple use. Isolated events (chemical spills) are monitored with respect to their temporal and spatial impacts. When designing a monitoring program, the selection of variables, time frame and duration should be decided on the basis of the objective and the characteristic properties of the waterbody to be monitored. In practice, the human resources, technical equipment and logistical capacities frequently act as limiting factors on the planning and implementation of monitoring programs. Wherever feasible, the highest possible stage of complexity should be striven for (Table D 1.5-7).

Continuous sampling methods are primarily used in running waters, which are subject to high temporal variability. Continuous sampling methods are used as early warning systems against serious risks, or when potential contamination would involve severe consequences (drinking water supply). There are limits to the number of variables that can be recorded with continuous methods. To reduce the risk of undetected contamination, many early warning systems monitor the overall impact on living organisms rather than measure single variables.

TOXICITY TESTS

In dynamic toxicity tests, a continuous flow of water from the waterbody under investigation is put through a special vessel containing fish, small crustaceans or protozoa. If their normal behavior changes due to acute toxicity (e.g. fish stop swimming), an appropriate warning signal can be given. Such methods are not suitable for assessing pollution levels that remain below the critical threshold or which have long-term impacts.

In the industrialized countries, highly advanced test procedures exist for examining chemicals with respect to their possible acute and chronic toxicity, prior to their approval for use in the environment. Standardized tests are mostly carried out with animals in free water, rarely on plants and, despite the importance of sediments, seldom on benthic organisms. Because of the specialized facilities and expertise required, tests are too infrequently conducted with respect to possible impacts on higher organizational levels (populations, ecosystems), on bioaccumulation (except in Japan), under natural conditions. The combined effects of substances acting simultaneously are seldom examined.

It has been shown that risk assessments of individual systems have no global relevance. Under tropical conditions, for example, microbial and photochemical decomposition may be faster and the accumulation of toxic substances concomitantly lower than in temperate regions. However, the toxicity of many substances is greater at higher temperatures, and the sensitivity of tropical ecosystems is generally rated higher because of their higher species numbers and small population sizes of individual species (Römbke and Moltmann, 1996).

1.5.2
Water quality standards

Water for human use (drinking, preparing food and other household purposes) must meet the highest quality standards. Microbiological variables are of paramount importance in this context (Section D 4.2). Irrigated agriculture has the largest share of global water consumption (see Section D 1.4). The concentration of certain ions (sodium, chlorine, boron) is critical in this connection. Water is used by industry mainly for cooling, transport, cleaning and power generation. The quality standards required for industrial water use vary accordingly, although they are rather low in most cases. For some types of use, such as irrigation, fisheries or in industry, quality criteria have been defined only roughly.

The adoption of water quality standards is focused worldwide on health protection for the population and on protecting the environment (Stabel, 1997). Regulations concerning water sports and agriculture have played a lesser role. The USA, Canada and Great Britain operate water quality standards for livestock watering. In some nations guidelines exist for irrigation water. Some standards are aimed at preventing economic damage (e.g. corrosion by tap water).

Within current European Union law, there are no fewer than 19 directives, follow-up directives and drafts dealing with the protection of water resources. Some relate to public health (directives on surface waters, bathing water, shellfish waters and drinking water), others to the environment, e.g. to groundwater, urban wastewater and the discharge of dangerous substances into waters (Lehn et al., 1996). In the view of experts, the EU Bathing Water Directive has contributed greatly to the protection of waters by virtue of its popularity.

1.5.2.1 Drinking water

In the 1993 WHO guidelines, standard values were laid down for drinking water quality. They include aesthetic variables, microbiological criteria, indicators of organic pollution, the concentration of particulate matter, components containing nitrogen, salts, organic trace substances such as pesticides, and inorganic trace elements, e.g. heavy metals and radioactive substances (Gleick, 1993). The WHO guidelines on drinking water quality are not legally binding. They can serve only as minimum requirements when developing national standards. The same applies to the EU directives. EU legislation requires that the directives are implemented as national law within two years.

Chemicals in drinking water can be classified into the following groups according to their significance for health.

- Type 1:
  Substances of acute or chronic toxicity. The toxicity rises with increasing concentration. Below a threshold value, however, no impairment to health or long-term damage is detected (various metals, nitrate and cyanide).
- Type 2:
  Carcinogenic, mutagenic and teratogenic substances without threshold value. Even minimal quantities increase the likelihood of impairment [organochlorine compounds (pesticides), polychlorinated phenyls (PCB) and arsenic].
- Type 3:
  Essential substances required by the human organism and harmful in high concentrations (fluoride, iodide, sodium chloride).

Membership of one of these categories is a decisive criterion when assessing the quality of drinking water.

The drinking water standards currently in force in North America, Europe and the former USSR vary in some cases by several orders of magnitude, depending on the specific variable (Table D 1.5-8, data only partly valid at present). Even within the specific countries the standards need not apply equally to all drinking water (e.g. exceptions in the new states in eastern Germany). There are several reasons for these differences. A reliable threshold value could not be determined for many toxic substances because, for instance, animal test data and empirical long-term studies were either inadequate or non-existent. For substances that have no toxicological threshold value, it is only possible to define standards involving a hypothetical "residual risk". Some aspects related to environmental safety and health standards vary considerably between countries, such as how much residual risk is accepted, which mathematical models are to be used as the basis for calculations, and how high additional safety standards must be. Sociocultural factors, which are located in the context of specific historical developments and events of major importance, are another reason why the perception and acceptance of potential risks differ between countries.

The heated debate on water quality standards and our growing knowledge generate more and more new recommendations and amendments to existing standards. Permissible threshold values can provide only a relative degree of protection against health hazards. The reasons for this include the different sensitivity of people depending on age, their previous exposure to contamination, their state of health and nutrition as well as the synergistic impacts of concurrent substances. The very imponderability of these factors gives rise to demands for zero concentrations of many substances in water, in particular carcinogenic, mutagenic and teratogenic substances. Theshold values, or the lack of same, are often a reflection of economic considerations or technical feasibility, rather than protection against health risks (Gleick, 1993).

Unless supported by further regulations, imposing permissible thresholds for drinking water may actually result in them being exploited, i.e. that actual concentrations increase to the permissible maxima without any attempts to keep concentrations well below these limits over the long term (Lehn et al., 1996). The *Länder* Working Group for Water (LAWA) has developed a concept for protecting waterbodies (LAWA 2000) which takes into account the hydrological cycle, i.e. the fact that the various forms in which water occurs on Earth are in a process of continuous exchange (Section D 1.3). The quality targets defined for specific environmental resources, e.g. aquatic biota, drinking water and the marine environment, vary according to the resource in question and are designed as orientation values. The LAWA recommendations are currently being implemented in a voluntary trial phase for which there is no legally binding commitment.

Because of its generally good quality, groundwater is the most suitable source of drinking water that can be used without costly treatment. In Europe, groundwater has traditionally formed a large percentage of the drinking water supply, amounting to over 80% in

Table D 1.5-8
Comparison of selected parameters for drinking water standards.
Source: modified from Chapman, 1992

	Variable	WHO	EU	Canada	USA	USSR
Physical	Color (color units)	15	20	15	15	
	Turbidity (turbidity index)	5	4			
Chemical	pH value	6.5–8.5	6.5–8.5	6.5–8.5	6.5–8.5	
	Oxygen (mg l^{-1})					4
	Dissolved salts (mg l^{-1})	1.000		500	500	
	Hardness (mg l^{-1} $CaCO_3$)	500				
	Ammonium (mg l^{-1})		0.5			2
	Nitrate-nitrogen (mg l^{-1})	10		10	10	
	Nitrate (mg l^{-1})		50		10	
	Nitrite-nitrogen (mg l^{-1})		1			
	Nitrite (mg l^{-1})		0.1			1.0
	Phosphorus (mg l^{-1})		5			
	Sodium (mg l^{-1})	200	150–175			
	Chloride (mg l^{-1})	250	25	250	250	350
	Sulfate (mg l^{-1})	400	25	500	250	500
	Fluoride (mg l^{-1})	1.5	1.5–0.7	1.5	2	1.5
	Cyanide (mg l^{-1})	0.1		0.2		0.1
	Arsenic (mg l^{-1})	0.05	0.05	0.05	0.05	
	Lead (mg l^{-1})	0.05	0.05	0.05	0.05	0.03
	Cadmium (mg l^{-1})	0.005	0.005	0.005	0.01	0.001
	Chromium (mg l^{-1})	0.05	0.005	0.05	0.05	0.1–0.5
	Iron (mg l^{-1})	0.3	0.3	0.3	0.3	0.5
	Copper (mg l^{-1})	1	0.1	1	1	1
	Nickel (mg l^{-1})		0.05			
	Mercury (mg l^{-1})	0.001	0.001	0.001	0.002	0.0005
	Crude oil (mg l^{-1})		0.01			0.3
	Sum of pesticides (µg l^{-1})		0.5			
	Individual pesticides (µg l^{-1})		0.1			
	Aldrin, Dieldrin (µg l^{-1})	0.03		0.7		
	DDT (µg l^{-1})	1		30		
	Lindan (µg l^{-1})	3		4	0.4	
	Benzene (µg l^{-1})	10			5	
	Hexachlorobenzene (µg l^{-1})	0.01				
	Pentachlorophenol (µg l^{-1})	10				
	Phenols (µg l^{-1})		0.5	2		1
	Detergents (mg l^{-1})		0.2		0.5	0.5
Biological	BOD (mg O_2 l^{-1})					3
	Faecal coliforms (in 100 ml)	0	0	0		
	Coliforms (in 100 ml)	0–3		10	1	

no data: variable not specified

Denmark, Portugal, Germany and Italy. In the USA, around 40% of the total public water supply, and in rural areas as much as 96% of domestic water is supplied from groundwater. Water demand in Latin America, and hence in some of the largest cities in the world, e.g. Mexico City, Lima and Buenos Aires, is met by groundwater resources. In rural areas of Africa, untreated groundwater will continue to be the main source for years to come.

Rivers, lakes and reservoirs are also used as sources of drinking water. Most cities in Africa and Asia are supplied with surface water. However, treatment of surface water for drinking water supply is much more expensive because of naturally occurring substances and contamination by wastewater. Surface water is much more likely to be contaminated with germs and is usually rich in particulate matter. Prime importance is attached to groundwater protection, in addition to the prevention of wastewater loads to surface waters.

1.5.2.2
Water in agricultural production

IRRIGATION

The suitability of water for irrigation purposes cannot be assessed on the basis of a common standard, but instead must be subjected to a differentiated analysis of the specific climate, soil properties, crop plants, irrigation methods and irrigation management, as well as the available water supply. Irrigation water is taken from surface waters (rivers, lakes, reservoirs) or groundwater (wells, springs, qanats). In addition, wastewater or brackish water with a salt concentration of up to 2,000 mg l^{-1} and more is used.

Physical properties

Water containing coarse suspended particles such as gravel or sand is unsuitable for irrigation. Negative consequences can be avoided by means of simple collecting and filtering devices. Suspended matter has a potentially fertilizing effect on the soils. The sealing effect of small suspended particles in irrigation channels may also be desirable, although the related siltation of reservoirs can represent an additional cost factor.

If the temperature of the irrigation water is too low (< 15 °C), slower plant growth and reduced yield may result. A water temperature of approx. 25 °C is regarded as optimal for most crops. Temperature-sensitive crops, such as beans, melons and rice, produce lower yields even when water temperatures fall to below 20 °C (Achtnich, 1980).

Chemical properties

The concentration of dissolved salts is one of the main properties determining the quality of irrigation water, and is a crucial factor determining the suitability of the water for the respective crops. Irrigation water already contains dissolved salts as a result of weathering processes and the percolation of water through rock and soils. Natural salt concentrations vary considerably around the world, depending on the geology of the catchment area and the local climate (Fig. D 1.5-4). Due to the osmotic pressure, enhanced salinity makes it more difficult for plants to take up water from the rooting zone. Salt concentration in irrigation water is measured in terms of electrical conductivity (in µS cm^{-1}). Yields of salt-sensitive crops, such as beans or apricots, decline by 25% at salt concentrations with a conductivity of over 2,000 µS cm^{-1} (or approx. 1 g of dissolved salts per liter of water) (Ghassemi et al., 1995). In recent decades, various standards that take account of the salt tolerance of the plants and the soil conditions have been developed for salt concentrations in irrigation water.

Another irrigation problem concerns the clogging and compaction of soils by calcium leaching when there is an overabundance of sodium. This, too, can cause reduced yields due to shortages of air, water and nutrients, and may greatly impair tillage operations. The sodium adsorption ratio (SAR) of the water is used to assess this risk (Table D 1.5-9). With the help of this key empirical index, it is possible to forecast the extent to which sodium ions are exchanged in the soil for calcium and magnesium as a result of the water input. When the sodium adsorption ratio rises, so does the risk of cation exchange and the resultant damage on soil structure.

The toxicity of irrigation water is primarily determined by the concentration of boron, chloride and sodium. An excessive concentration of nitrogen (as nitrate or ammonium) enhances vegetative growth, promotes blade storage and delays maturity. Compared to the WHO guidelines for drinking water, the maximum acceptable concentrations of metals in irrigation water recommended by the FAO (Table D 1.5-10) are twice as high for some metals (e.g. ar-

Irrigation problems	Damage to crops		
	none	increasing	severe
Salinity EL (mS cm^{-1})*	<0.75	0.75–3.0	3.0
Sodium adsorption value**	low	medium	high
Boron (mg l^{-1})	<0.75	0.75–2.0	2.0
Chloride (mmol l^{-1})	<4	4-10	<10
Sodium	low	medium	high
Nitrate or ammonium nitrogen (mg l^{-1})	<5	5–30	<30
Bicarbonate (mmol l^{-1})	<1.5	1.5–8.5	8.5
unusual pH value		(normal range 6.5–8.5)	

*Electrical conductivity
**Measured as $c_{Na}/c_{Ca}+c_{Mg}$

Table D 1.5-9
Water quality assessment for irrigation.
Source: Bretschneider et al., 1993

Table D 1.5-10
Recommended maximum concentrations of toxic substances in irrigation water for continuous irrigation systems.
Source: FAO, 1976

Substance	Concentration ($mg\ l^{-1}$)	Substance	Concentration ($mg\ l^{-1}$)
Aluminum	5	Lead	5
Arsenic	0.1	Lithium	2.5
Boron	0.75	Manganese	0.2
Cadmium	0.01	Molybdenum	0.01
Chromium	0.1	Nickel	0.2
Copper	0.2	Selenium	0.02
Iron	5	Zinc	2

senic, cadmium, chromium), but may be 10 times (lead, iron) or up to 25 times higher (aluminum).

Biological properties

The hygienic condition (germ count) of the water is of key importance for irrigation. Because of the risk of spreading pathogens capable of reproduction, wastewater polluted with human or animal excrement may be used only after mechanical and biological purification for the irrigation of certain crops, such as fodder turnips, sugar beet or oil seeds.

LIVESTOCK FARMING

The water qualities required in livestock farming depend on the water turnover of the respective species and the weight of the individual animal. As a rule, very high salt concentrations in livestock water may induce physiological stress or even the death of an animal. Guidelines issued by the National Academy of Sciences in the USA permit a basic classification of salt tolerances for water given to livestock and poultry (Table D 1.5-11).

Table D 1.5-11
Suitability of saline water for cattle.
Source: FAO, 1976

Salt concentration ($mg\ l^{-1}$)	Suitability/problems
<1,000	suitable without restriction for livestock and poultry
1,000–3,000	well-suited for livestock and poultry after adaptation
3,000–5,000	still suitable for livestock needs after long-term adaptation, less suitable for poultry, high mortality, low growth
5,000–7,000	restricted suitability for some animal species such as sheep and pigs
7,000	not suitable

Table D 1.5-12
Maximum values recommended by the National Academy of Sciences, USA, for metals and salts in water drunk by cattle.
Source: FAO, 1976

Substance	Concentration ($mg\ l^{-1}$)	Substance	Concentration ($mg\ l^{-1}$)
Aluminum	5	Lead	0.1
Arsenic	0.2	Mercury	0.001
Boron	5	Nitrate	90
Cadmium	0.005	Nitrite	10
Chromium	0.05	Selenium	0.05
Cobalt	1.0	Vanadium	0.10

A guideline for the maximum level of toxic substances in livestock water has been drawn up by the same institution (Table D 1.5-12). This is partly based on the WHO drinking water guidelines (Table D 1.5-8) (e.g. cadmium, chromium), but allows much higher values for other substances (arsenic, selenium, aluminum). In contrast to the drinking water standards, there are no guidelines for irrigation or livestock water that include contaminant groups such as organic trace substances and pesticides.

1.5.3 Recommended research and action

RESEARCH RECOMMENDATIONS

- Investigation of the fixation and accumulation of pollutants in aquatic habitats by means of physicochemical and biotic processes, and of the transformation and decomposition of pollutants in waterbodies, soils and neighboring habitats, with particular reference to their significance for self-purification capacities and for strategies aimed at protecting and restoring such environments.
- Research on pollutant groups inadequately studied to date, with respect to their formation, turnover and impacts (including chelating agents, volatile organic compounds, hormonal substances, artificial fragrances, persistent organic pollutants).
- Intensification of research on the impacts of anthropogenic pressures in climate zones that have received little attention until now (e.g. the tropics), also with regard to possible climate changes. Given the critical importance of anthropogenic factors in the assessment of water quality, it is essential to investigate physiological processes (cellular and subcellular systems), relevant groups of organisms (particularly in sediments) and the use of ecosystem models. Hitherto underrepresented fields (tropical organisms, ecotoxicology) merit

special attention.

- Development of assessment criteria (indicators and combined variables) for aquatic habitats that can be applied, regardless of edaphic conditions and biogeographical region.
- Definition of water quality standards for the various uses of water, such as for agriculture and industry, as a basis for freshwater resource management and with special reference to public health.
- Increased water monitoring in regions (Asia, South America, Africa) and waterbodies (wetlands, groundwater, lakes) for which limited data are available, in order to provide the basis for a global database. Extending monitoring to include parameters that have not been adequately measured worldwide (e.g. metals, pesticides, organic trace substances).
- Further development of simple and reproducible measuring techniques in order to facilitate monitoring and to allow data comparison. Expansion of international collaboration in the use of sophisticated analytical methods. Acquisition of reference data on background pollution and research into the natural variability of relevant parameters (e.g. lake sediments).

RECOMMENDED ACTION

The Council recommends

- that maximum pollution levels be defined for the water quality of surface waters, groundwater and neighboring habitats, and to ensure that pollution levels are kept below these limits, paying particular attention to self-purification capacities. In the case of substance loads, it is imperative to focus not only on point sources but also on diffuse sources.
- that efforts be made to promote the collection of data on quality-related properties of surface waters, groundwater and freshwater habitats in a global database, together with existing data, as a basis for developing regional and global strategies of action.

1.6 Water and disasters

Floods – Causes of floods – Higher probability due to climate change – The Rhine floods – Landslides – Flood waves – Imprecise models – Protecting against remaining risk – Mapping floodplains – Classifying damage – Planning process – Factors influencing risk awareness – Recommendations

1.6.1 Introduction

Extreme conditions in water management regimes result from the variability of meteorological and hydrological conditions. Floods and droughts have always been the natural disasters causing the greatest economic damage and loss of human life. Whereas droughts are brought about by unusually severe and long-lasting water scarcity in agriculture, floods involve extremely large amounts of runoff. Both are extreme situations that may lead to disasters unless sufficient protective action and countermeasures are taken by the society concerned. This section examines the mechanisms by which floods originate, the influence of global changes on flood events and possible ways of mitigating the risk of floods. Drought disasters are an equally important complex requiring a separate analysis that cannot be provided in the present context.

Water's life-giving and life-sustaining force turns into the opposite when too much water, in the form of floods, endangers human life and the environment. People tend to build their largest settlements near rivers because the water they hold can be used as drinking or irrigation water, because floodplain soils are usually fertile and well-suited to growing crops, and because bulk transport of goods is relatively easy on waterways. Since time immemorial, humans have accepted the irregular occurrence of floods as the price to be paid, or have tried to protect themselves against floods by technical means.

The enormous financial damages that are caused by floods worldwide are not wholly attributable to environmental factors such as meteorological conditions or local possibilities for runoff. Aside from direct human-induced causes, such as hydro-engineering schemes or surface sealing, the extent of damage incurred is largely determined (as is the case with other natural disasters) by the behavior of the people involved, both during and after flood events (Ketterer and Spada, 1993). For example, people settled in flood-prone areas fail to take precautionary measures, resist evacuation despite acute danger, etc. Therefore, precautions against flood events must take into account not only natural causes, but also the changes in the cause-effect chain that are brought about by human activities as well as people's anticipated reactions.

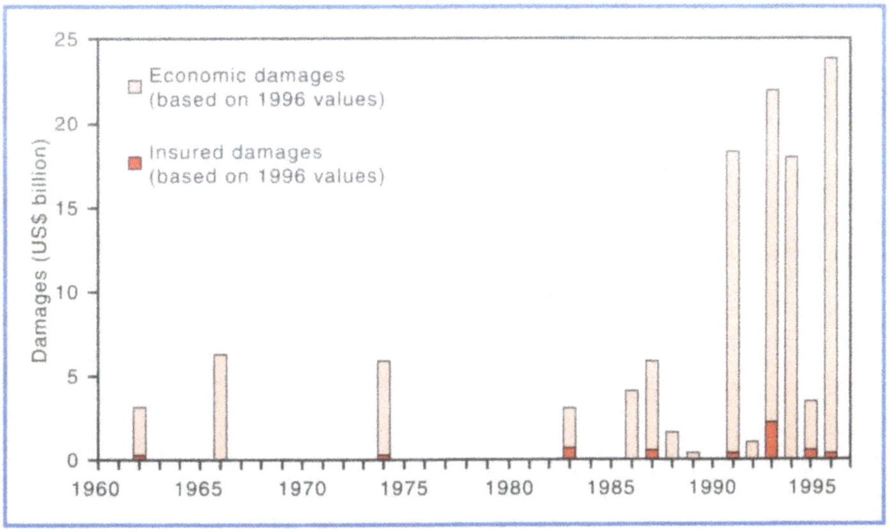

Figure D 1.6-1
Flood damage: costs to the economy and insured damages caused by major floods in the 1960–1996 period involving more than 100 deaths and/or US$ 100 million in damages.
Source: Münchner Rückversicherung, 1997

1.6.1.1
Flood damage trends

Although humans have learned to cope with floods since the first settlements, they have repeatedly fallen victim to devastating floods – such as the "Great Mississippi Flood 1993", which caused damage totaling more than US$ 12 billion. Each year, the *Münchner Rückversicherung* reinsurance company publishes damages statistics which indicate that in most years flood damage is responsible for the worst economic losses worldwide. The flood damage statistics for the 1960–1996 period are shown in chart form in Fig. D 1.6-1.

Flood damage has risen substantially in recent years. One of the main reasons is the increase in precipitation. Studies at many sites in Europe indicate a rise in precipitation in recent decades (e.g. Scotland, Mansell 1997). For Norway, Hanssen-Bauer (1997) calculated an increase of 8–14% over the last 90 years. In the Rhine area, a general rise in discharge during the winter months has been measured (Engel, 1997). These increases in precipitation have been attributed in some cases to climate variability. However, the mounting damage potential has a more severe impact. In the industrialized countries, the number of facilities and assets that are affected by floods is increasing.

Table D 1.6-1
Number of people (in thousands) affected by natural disasters, 1991–1995.
Source: WMO, 1997

	Floods and landslides	Strong winds	Earthquakes	Volcanoes
Africa	1,674	6	63	1
sub-Saharan Africa	1,503	6	50	1
northern Africa	171		13	
America	2,407	3,344	303	449
Central America	395	262	208	376
Caribbean	896	1,797	7	6
South America	1,057	642	88	67
North America	64	644		
Asia	775,245	72,578	2,272	894
East Asia	482,274	49,225	1,485	65
South Asia	274,532	1,070	204	
Southeast Asia	18,421	22,233	366	829
West Asia	18	50	217	
Europe	2,784	3	550	7
EU Member States	380	2	15	7
Non-EU Member States	2,404	1	535	
Oceania	119	2,872	15	106
World	782,230	78,804	3,203	1,457

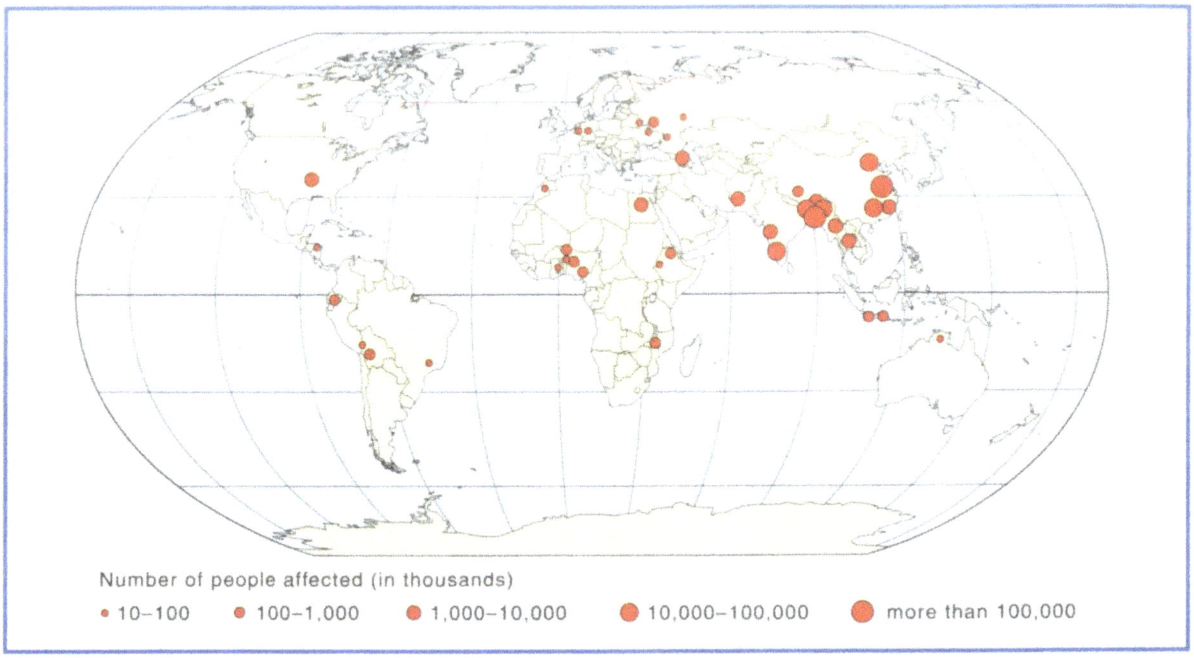

Figure D 1.6-2
Global distribution of flood risk.
Source: adapted from Münchner Rückversicherung, 1997; WMO, 1997

Statistics on the number of people affected by natural disasters are of greater relevance than damage statistics. Table D 1.6-1 shows figures for a five-year period (1991–1995) which highlight the tremendous number of people suffering damages through floods compared to other natural events. These figures are on the rise, too, because not only does vulnerability grow, but also the number of people threatened.

The present world population is about 6 billion people, and will rise to more than 10 billion by the middle of the next century. This means that more and more people will be threatened by floods because flood-prone areas will have to be settled (Fig. D. 1.6-2). Since people in a socially weaker position tend more than others to settle in threatened areas, one can expect that the increased risk of flooding will affect the poor and the vulnerable in particular. This is exacerbated by a possible increase in the frequency and severity of extreme events as a result of climate change. Preventing the resultant disasters is a major challenge to which humanity must respond. The United Nations has recognized this challenge and declared 1990–1999 as the International Decade for Natural Disaster Reduction (IDNDR).

1.6.1.2
From heavy rains to flood damage

Any study of the causes of flood events must include separate analysis and assessment of meteorological, hydrological and hydraulic factors, as well as of economic conditions, settlements development and population trends.

If specific conditions are unfavorable, any one of these factors may contribute to an increased risk of flood damage. If all factors reinforce each other in an adverse constellation, flood events and severe damages occur. The particular conditions and combinations of factors that may occur are extremely wide-ranging and to a certain extent randomized. The absence of only one adverse factor (within a generally unfavorable constellation) may make the difference between a major flood disaster and runoff conditions involving no significant damage. Each of these factors is obviously influenced by human activities, though to differing degrees.

1. Meteorological factors
   In most cases, heavy precipitation is necessary for a flood event to occur. Critical factors are the duration, intensity and distribution of rainfall. Rapid melting of snow (often combined with rainfall) increases the quantity of water that appears as runoff. Other flood events that are not caused directly by heavy precipitation are rare and mostly occur after seaquakes, dam breaches, mountain or

glacier slides, or as a consequence of ice jam.

2. Conditions in the catchment basin
Heavy precipitation causes extreme discharge if a significant portion of the precipitation does not infiltrate and flows instead into stream channels as surface runoff. The infiltration capacity of the soil is most commonly exceeded as a result of natural factors. On the one hand, very intensive rainfall may exceed the maximum infiltration rate of the soil surface at that particular time. On the other, the water table may be raised over a wide area by persistent prior rainfall, causing expansion of the saturated areas near waterbodies and generally leading to a higher degree of soil saturation, with the result that little or no free pores are available to absorb the precipitation. Moreover, the water retention capacity of the catchment area may be reduced by crusting, compaction and sealing processes on the soil surface. Other significant factors include the state of vegetation cover as well as the regional topography and microtopography.
3. State of the hydrological regime
The hydrological conditions in the streamflow system determine how much and how fast water flowing into a surface waterbody can be transported in the river bed or in riparian flood basins and what proportion is kept in additional retention areas – whether desired or not. One of the basic problems here is that channel conditions which reduce the likelihood of flooding in the respective area due to high discharge capacities in a section of river make the situation worse for downstream areas because of the unattenuated inflow of water.
4. Potential damage
Damage caused by floods is not perceived as such until people or things that they value are adversely affected. The more densely populated the flood areas, the more intensively they are used, the higher the material assets are and the fewer the precautions taken against flood risks, the greater the potential damage. It is quite obvious, therefore, that population growth and an increase in the value of assets in the riparian habitats are sufficient to increase the damage potential. This situation is particularly critical in regions where population pressure leads to areas being settled that were previously left aside as buffers against river flooding.

1.6.2 Classification of different flood types

Flood control measures must be designed to match local conditions and the type of potential flood event. An extreme precipitation event alone does not trigger flooding; other factors must also be operating.

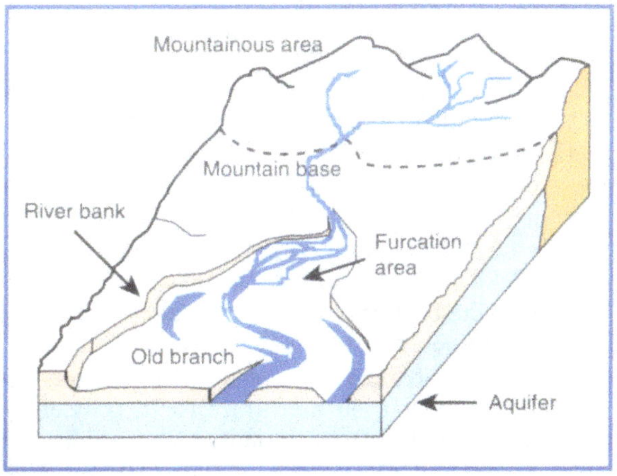

Figure D 1.6-3
Parts of a watershed.
Source: Petts and Amoros, 1996

It is useful to distinguish between these causative factors according to the type of precipitation and the specific catchment.

Floods and flood control measures in mountainous areas

The different types of catchment area are shown diagrammatically in Fig. D 1.6-3. In the mountainous area of the upper reaches, the terrain has a pronounced structure and the valleys are cut deeply into the landscape – more deeply in geologically young mountains, such as the Alps, than in old mountains. Heavy local precipitation combined with snowmelt produces flood situations in spring, especially when the ground is still frozen. Severe flooding in mountainous regions can have particularly devastating impacts. Because of the steep gradient, the water flows at high speed and sweeps everything along with it – destroying bridges and roads, digging deep erosion channels in the flood areas and leaving behind a landscape strewn with debris.

Due to the steepness of slope in mountainous areas, floods may result from other natural causes that are not directly linked with precipitation, such as closure of a valley by a landslide, behind which the water that is prevented from flowing out accumulates until it flows over the natural dam and destroys it through its erosional force. A similar phenomenon is known from the tributaries of the large rivers that drain the Himalayas, namely the "lakes of sorrow", which in the course of time break through the barriers that restrain them, causing huge flood waves that destroy everything in their path. In India, for example, there are reports of a flood wave that was still 40 m high even 80 km below such a lake. It occurred in August 1893 when water broke through the lake in the upper reaches of the Ganges in the Gona Valley

(Gupta, 1974). Similar disasters engendered by landslides are also known in the Alps. They usually occur in connection with long periods of precipitation, such as the dramatic Valpola landslide in northern Italy in 1987, when the eastern flank of Mt. Zandila slid into the Adda River (which flows into Lake Como), destroying several towns and claiming the lives of 27 people (Azzoni et al., 1992). Similar impacts result when a lake is formed behind the ice barrier of a glacier. When the barrier ice thaws, the lake water is released and flood disasters may be the consequence.

Landslides are likely to occur with greater frequency and severity in the future for two essential reasons. Firstly, forests in mountainous regions are being cleared in many parts of the world, without consideration for the inevitable consequences. These include a reduction in the natural consolidation of soil by tree roots, the destabilization of slopes, while heavy rainfall invariably produces further erosion that then leads to increased concentrations of suspended solids in rivers.

Secondly, the gradual warming of air that has been observed in conjunction with global climate change is a contributory factor in the destabilization of slopes that were not previously threatened. This results primarily from the shift in the permafrost line to higher elevations brought about by warming. Melting of permafrost ice at high mountain locations lowers the cohesive force of the ice in slope material. As water declines in volume during the phase transition from ice to liquid, cavities are formed into which the slope water penetrates, creating excess pressure in the interstitial water and thus reducing the supportive effect of the slope elements. The overall impact is a decrease in the angle of internal friction of the slope material, combined with a greater risk of rockslides and accumulation of old geological talus material. This effect is also supposed to have been one of the factors responsible for the Valpola landslide.

A similar human-induced flood situation occurs when flood waves are released by dams. These discharges are necessary, for example, to drain a reservoir in anticipation of heavy precipitation, or to provide water for downstream use. A particularly devastating type of human-induced flood wave occurs when a dam breaks. The failure of some reservoirs abroad (e.g. Teton Dam, 1976) as well as some small dams (retention basins) in Germany are a clear demonstration that dams are not absolutely safe and that every dam harbors the risk of failure (see Section D 3.4).

Floods in hilly regions

Severe floods occur in the small valleys of hilly regions as a result of thunderstorms in summer, locally reinforced by orographic effects, or following heavy spring rains in conjunction with snowmelts and frozen soil. Extreme local events that are difficult to predict and cause great damage in the respective area are repeatedly triggered in this way. Larger valleys, by contrast, are more vulnerable during persistent cyclonal weather. Flood control in such regions must therefore involve two types of operation – heavy local precipitation must be absorbed and downstream reaches protected. A possible approach to flood control in the face of these twin problems is to set up flood retention basins in the upper valleys.

Human activities have particularly severe impacts on the structure of small catchments. The worst effects are caused by construction work and soil compaction. Under normal circumstances, soil acts like a sponge that can absorb and temporarily store large amounts of water. When land is built on, the porous space in the ground is "sealed". In city centers, for example, precipitation can no longer be absorbed and stored in the soil, so the entire volume of water runs off directly and rapidly to the river. Under normal conditions (natural soils densely covered with vegetation), less than 10–20% of precipitation becomes surface runoff. This percentage is called the runoff coefficient. It depends very heavily on the extent of building development and may be as high as 90–95% in urban areas. Building development also causes runoff to be accelerated. Peak runoff occurs earlier in urban areas than in rural areas, and is reinforced by the fact that the runoff is channeled through sewers and discharged into waterbodies as rainwater. However, this need not necessarily lead to a higher river level. In fact, there is frequent evidence that cities located along the lower reaches of smaller rivers can relieve the pressure on stretches further downstream. Because the flood wave has already flowed away from the urban region before the flood peak coming from rural areas arrives, the peak discharge (the key variable for dike design) is reduced.

The runoff coefficient is dependent not only on land use, but also on the amount and type of natural vegetation cover. Forests generally have the lowest runoff coefficient. Geological features are equally important: extremely permeable layers have a low runoff coefficient, loess areas a higher one. The third factor is the previous state of the soil. If it has rained for a long time prior to the extreme precipitation event, the soil is saturated down to the impermeable bedrock and can absorb only a small portion of the precipitation. The same applies to frozen ground in which the pores are closed due to the expansion of the water during ice formation. It is therefore unlikely that building development has a major influence on runoff outside urban areas during extreme floods. Detailed studies on the relationship between runoff, different types of land use and tillage methods may provide new findings (Mendel et al., 1996). Some research projects are studying the influence of non-traditional agricultural techniques mainly aimed at preventing erosion and at the same time enhancing the capacity of small-scale artificial retention basins, such as furrows.

FLOODING OF PLAINS

The streamflow characteristics of flat-land rivers differ significantly from those in other types of terrain. Under natural conditions, river flooding on plains causes inundation of large tracts of land. The river may change its course during the flood event and will tend to meander or form numerous branches, depending on the grain size and composition of the soil particles in the riparian area, and on the slope of the region. Whenever a river comes from the steeper mountainous areas and flows into a plain, matter suspended in the water is deposited. The river silts up in the course of time and the water level rises, thus increasing the risk of flooding – or it may even become completely clogged with debris and effectively dam itself, as has happened in many cases on Indian and Chinese rivers with high levels of debris flow. Since the water depth is small relative to the width of the river, and the velocity comparatively slow due to the small gradient, such rivers need to increase their width in order to discharge fully. In the process, they interact with the groundwater, which is elevated in the floodplain areas and may cause reduced discharge levels, depending on the previous water level.

TIDAL FLOODS

Enormous damage may be wreaked by floods in the estuarine regions and deltas of large rivers all over the world, because the large ports with their concentrations of assets and high populations are located in these exposed coastal areas. The number of potential victims is accordingly high. These regions are threatened from the sea by large tidal floods (an issue not dealt with in this report) and by tidal river flooding when extremely high tides are coupled with strong inflow from upper reaches of the river into the estuary. The combination of an imminent tidal flood from the sea and extreme discharges from upstream reaches is especially dangerous. Such situations have occurred on repeated occasions worldwide, resulting in large-scale flooding, devastation and loss of life. Examples in Europe include the Hamburg flood disasters of 1962 and 1976, as well as the floods that hit London and the entire Rhine delta in 1953. From the global perspective, some of the worst floods are those which regularly occur in the Ganges-Brahmaputra delta in Bangladesh.

1.6.3 Effects of climate change on floods

There has been a great deal of discussion, in connection with the extreme flood events that have occurred with increased frequency in recent years, as to the extent to which climate change, river regulation or landscape changes have increased the force or probability of flood events. Only some of the cause-effect relationships are understood and scientifically verifiable. The linkages between causes, effects and subsequent impacts are summarized in Fig. D 1.6-4 in the form of a "cascade of flood risks". It is based on those aspects of global change of relevance to floods:

- climate changes,
- changes in land use and landscape,
- changes in waterbody systems,
- increasing human settlement.

In a simplified analysis, the risk of floods is a combination of natural factors influencing the frequency of floods and human-related factors that influence flood impacts.

Of the flood-inducing factors just mentioned, meteorological conditions are of greatest significance as far as possible climate changes are concerned. The state of vegetation and soils in the catchment area may also be impaired by climate changes, with concomitant feedback on flood occurrence. Generalized statements about changes in flood activity can be made to a very limited extent only, since we are dealing with an extremely non-linear system that is dependent on naturally high spatial and temporal variabilities of meteorological and topographical factors, soil, vegetation, climate, groundwater and waterbodies. The impacts of climate change on flood peaks, in particular, are extremely difficult to quantify with any degree of uncertainty. However, it is possible to make the following three generalizations regarding the impacts of climate change as they pertain to floods.

1. Sea-level rise enhances the risk of coastal storm floods and tidal river floods
   The rise in sea level induced by global warming increases the probability of tidal river floods, because an increase in mean sea level produces a rise in extreme water levels by the same amount. This fact now plays a role in establishing the design criteria for flood control structures in the estuary regions of larger rivers, as in the case of the flood protection gate in Rotterdam that was put into operation in 1997 and where the design water level was set 50 cm higher to take into account the expected sea level rise.
2. Elevated temperatures force the global hydrological cycle

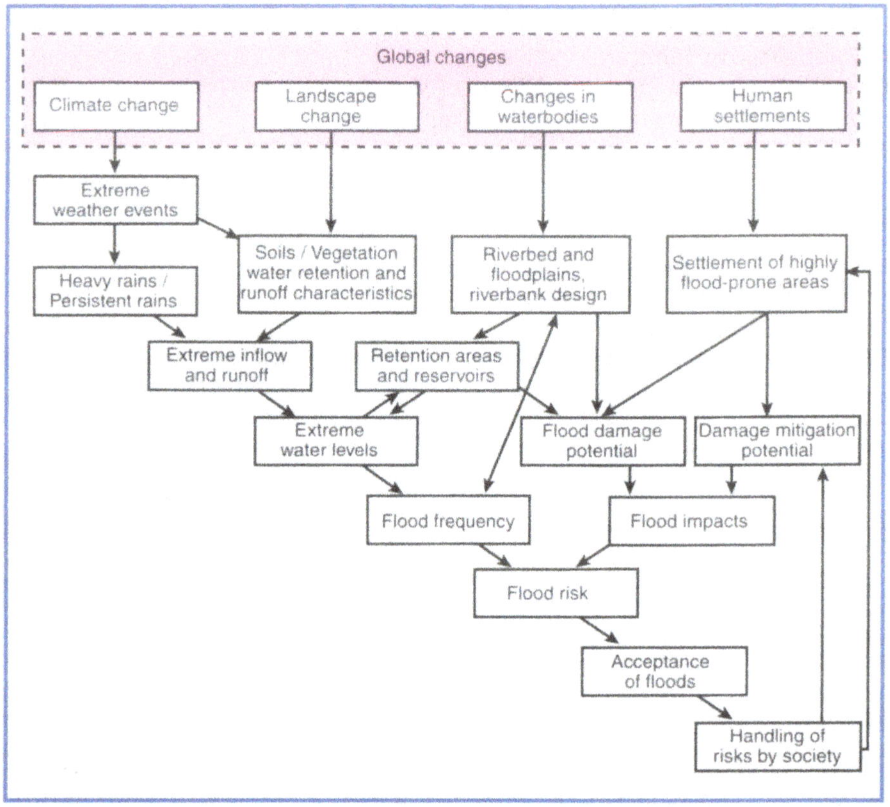

Figure D 1.6-4
Cascade of flood risk. Source: adapted from Plate, 1997

The energy and hydrological cycles are closely linked systems that mutually influence each other. At the global level and from a thermodynamic analysis, it can be assumed that a temperature rise will lead to a general intensification of the hydrological cycle. For assumed global temperature rises of between 2.8 °C and 5.2 °C, Mitchell (1989) estimates an increased global rate of evaporation and precipitation ranging from 7% to 15% (see Section D 1.3). In combination with expected higher climate variability (see below), this results in a significant increase in the amount of precipitation, and probably in heavy precipitation as well, in various regions of the globe.

3. Climate changes increase the frequency of extreme weather events
Slight changes in the mean climate or mean climate variability can bring about relatively large changes in the frequency of extreme events. A small change in variability has greater impacts than a small change in mean behavior (IPCC, 1996a). Flohn (1988 and 1989) concludes that the frequency of extreme large-scale weather anomalies will presumably increase at all latitudes as a consequence of climate change. An increased occurrence of weather anomalies would mean a higher frequency of those weather patterns that cause flooding (and droughts).

1.6.3.1 Observed precipitation and runoff trends

PRECIPITATION TRENDS IN GERMANY AND EUROPE

With the present level of scientific understanding, a rise in mean precipitation at global level can only be derived and calculated on the basis of theoretical considerations (see Section D 1.3). On a regional scale, however, certain trends over the last 100 years or so have been statistically verified and measured. Engel (1995), for example, reports on annual precipitation since 1890 in the Rhine area as far as Cologne, and identifies a rising overall tendency with pronounced periodic fluctuations.

In addition to an increase in the total annual precipitation figures, observations indicate a tendency towards a seasonal shift in precipitation distribution from summer to winter. If the two trends are summarized, the result is a more or less constant mean level of precipitation in the summer months (June–October) and a markedly rising mean winter precipitation (November–May) over the last 100 years. The statistical analyses of precipitation changes in Europe from 1891 to 1990 carried out by Rapp and Schönwiese (1995) show a distinct increase in winter precipitation in central and northern Europe as well as a decline in southern and southeastern Europe. These observations are consistent with the trends measured in the Rhine region. If these precipitation trends continue, the risk of extreme flooding of central European rivers may increase (further).

Changing weather pattern frequencies in Europe

Bárdossy and Caspary (1990) investigated the weather pattern frequency for Europe on the basis of the daily time series of atmospheric circulation patterns over Europe and the North Atlantic since 1891. A statistically significant rise in the frequency of westerly patterns was verified for this time series, with the breaking point in the time series being in the 1970s. Statistically speaking, the rise in zonal patterns is highly significant. Since these weather patterns are considered typical for long-lasting, large-scale precipitation events in central Europe, high correlations between the increases in these weather patterns and frequent flood events can be calculated for some catchment areas. Further detailed studies are necessary to elucidate the interrelationship between this weather pattern and specific flood events. Due to the lack of recorded weather map data going further back in time, it is not possible to specify the extent to which the observed trend of the 115-year time series is part of more long-term natural fluctuations.

1.6.3.2 Other possible changes to flood hydrology due to climate change

The Rhine floods of recent years

Grünewald (1996) reports that the cause of the spectacular Rhine floods in 1993/94 and 1995 was to be found not in the Alps, but in the low absorption capacities of soils in the hilly regions of relevance to flood events. The low infiltration rates were attributable, in particular, to the periods of widespread heavy rains prior to the actual flood period (e.g. December 7–18, 1993) or to snowmelt and frozen soil (in January 1995).

These phenomena illustrate how important snowmelt conditions can be for the occurrence of floods. Changes in the scope, frequency and time of occurrence of snowmelt events may also have an impact on the hydrological regime of a river and thus on its flood behavior. Engel (1997), for example, has pointed out that a shift in the typical flood behavior of the Alps region from snowmelt floods (May to July) to heavy-rain floods (usually in winter) as a result of warming may severely enhance the flood risk potential in the Rhine region – with possibly disastrous consequences – because then the peak discharges from the Alps would coincide with peak discharges from hilly stretches, something which has never happened until now.

Landslides and mudrock flows

If the permafrost line in high mountainous regions climbs as a consequence of long-term rises in mean temperature, thawing and thus destabilization of moraines may be the result. This can lead over a transitional period lasting several decades at least to a more frequent occurrence of landslides and mudrock flows, with all the devastation this implies. It has yet to be clarified whether the greater frequency of mudslides in Switzerland in recent years is related to the warming in the Alps region that has taken place since 1850 (Vischer, 1996).

Flash floods

In the long term, a change in climate can also induce a change in natural vegetation cover. As a general rule, a combination of soil degradation, less vegetative growth, a reduced proportion of macropores and an increasing tendency towards crusting and siltation will cause a reduction in infiltration capacity, thus increasing the proportion of direct runoff in the case of heavy precipitation.

A reduced density of vegetation has an impact on the occurrence of floods due to reduced interception, on the one hand, as well as the reduced protection of the soil surface against the precipitation energy. The latter can result in sealing and thus reduced infiltration, and may be the factor triggering flood peaks in the first place, especially after very heavy rainfall. Problems of this kind are particularly frequent in semi-arid regions (for example, the Mediterranean area), but also occur in agricultural regions in Germany. However, changes in the composition of natural vegetation and in soil structure due to different climate conditions can only take place over long periods of time and for that reason can only have a significant impact on flood regimes over the long term (i.e. not until later in the future).

Flood mitigating factors

Burkhardt (1995) reports on a decline in extreme flooding in the Elbe basin over the last 200 years. This is presumably a consequence of the reduction in flood events brought about by ice jams and ice barrier breaks. The reduction in ice jams is due to regional warming after the end of the last short ice age after 1800 (as well as, to a lesser extent, to the increase in salt concentration and water temperature resulting from cooling water discharges).

1.6.3.3 Modeling

METEOROLOGICAL AND HYDROLOGICAL CALCULATIONS

Predicting the size, location and timing of flood events necessitates complex and detailed mathematical models. In the ideal case, many of the above described observations could be integrated into such a system. Ultimately, this would enable an assessment as to which of the observed phenomena are due to changing climate conditions or other changes in the period under study, and which are accounted for by the natural variability of meteorological and hydrological systems. The modeling procedure requires three basic components before a flood can be predicted:

1. Meteorological model (weather forecasting):
   The quality of weather forecasting models has improved substantially in recent years. This makes it possible today to calculate information on the relevant values, such as temperature, wind and precipitation, up to a week in advance. The output from such modeling is a significant advance towards comprehensive flood prediction.
   However, the predictions of the place and amount of precipitation do not yet meet the requirements for calculating the amount and location of flood discharges. In particular, the data on the location and intensity of local convective heavy rainfall (thunderstorms) are still inadequate for direct prediction of flash floods.
2. Hydrological model (conversion of precipitation to runoff):
   Recent years have also seen considerable improvements in the hydrological models used to convert precipitation to runoff. However, their usefulness for predicting runoff prior to the occurrence of floods has been very limited until now. This is due to the fact that – particularly for conditions during extremely heavy rainfall – there are still basic shortcomings in the hydrological models (for example, in calculating the percentage of precipitation that becomes surface runoff, and the groundwater response to heavy precipitation). In addition, there are a number of scaling issues in the computation of hydrological processes that have yet to be clarified. Another problem is that the data base essential for operating these models is inadequate in many cases.
   It can be expected, however, that the gap between the specific precipitation forecast and the formation of runoff in the catchment area will gradually be closed and a basis established for predicting the occurrence of floods as these models are further improved. For large regions, this will require much more in the way of human and financial resources than is presently available.
3. Hydraulic model (flood discharge in running waters):
   Hydraulic models calculate the discharge of flood waters within the waterbody system. They are not focused on the question of how much water flows into the rivers, but on how the water that has already entered the waterbody system is then discharged. They require water levels and/or discharge rates at the start of the system (usually certain levels in upper reaches of the river or in tributaries) and calculate the water level, discharge rate and in some cases the size of flood areas for the waterbody covered by the model.
   These models have been developed to a high level of sophistication for large rivers in western Europe and have provided water-level forecasts in connection with floods for years. Since they require the water flowing into the system as an input, they can only be operated with relatively short warning times (equal to the flow time between the site for input levels and the site at which calculations are made). The warning time for the River Rhine around Cologne, for example, is little more than two days. For smaller rivers, the flow times are so short that hydraulic models are of little use for real-time flood prediction.

An important task for the future is to continue improving the modeling systems mentioned above and to couple them with each other. This could enable warning times to be achieved that correspond to the reliability of weather forecasts. With such a coupled system it would also be possible to study the longer-term impacts of climate change in scenarios. This means calculating the response of a given catchment in terms of flood occurrence to typical states of a future climate, then deriving typical flood characteristics for future climate conditions. These results can then be compared with present or historical conditions in order to produce qualitatively reliable conclusions for the region being analyzed.

RELEVANCE OF CLIMATE MODELS FOR FLOOD-RELATED ISSUES

Currently the most efficient climate models are coupled global circulation models (GCMs), which compute equations for the transportation of thermal energy, pulse, atmospheric water vapor and salt concentration (in the ocean) for the entire globe on a three-dimensional basis. Such models have various levels of horizontal resolution. Horizontal grids of approx. 500 km and 250 km are used routinely at present. The spatial resolution of global circulation models (i.e. their horizontal grid size) is far too rough to produce useful output for the analysis of flood events. Moreover, the scale on which global models deliver realistic results is always several times greater than the grid size used when running the model. Even with a spatial resolution of 250 km (i.e. a unit area of 62,500 km^2), Germany, for example, would be covered by only five values. These five values provide virtually no useful information on the specific precipitation events that might trigger floods.

Nevertheless, various regionalization methods have been developed over the last few years that enable regional climate change scenarios to be produced. One type of regionalization involves the use of regional climate models. In contrast to GCMs, these models look at only one section of the Earth and model this section at a higher spatial resolution. For regional climate models, grid sizes of approx. 50 km and less are used. The output from the GCM defines the climatic conditions at the margins of the regional section. This spatial resolution is much better for the study of floods. However, there are a number of problems in this area as well, which means that the calculation or even prediction of precipitation cannot be done with the accuracy needed for flood management.

- Quite often, a systematic error in the atmospheric dynamics within the region is imported with the boundary conditions of the regional model that are taken from the GCM. Thus, errors in the GCM impose restrictions on the performance of the regional climate model.
- The parameterization of major diabatic processes, such as cloud formation, soil or land surface processes, has not been resolved yet to the extent necessary to simulate natural variability or to detect a possible climate change signal for each weather pattern.
- The resolution of regional climate models is certainly detailed enough to represent large-scale precipitation events. Small-scale, convective formation of precipitation (e.g. thunderstorms), by contrast, cannot be integrated with adequate detail into such models as yet. Although subgrid-scale processes can be parameterized by splitting grid cells into a cloudy and a non-cloudy portion, it is still not possible to localize several convective systems within a grid cell. With increasing resolution, more and more parameterizations would have to be replaced by explicit modeling, which is not yet possible.

Furthermore, the serious gaps that still exist in the scientific understanding of certain processes mean that the strength of predictive models is limited (see Section D 1.3.5).

Regional climate models can thus provide valuable information necessary to calculate heavy precipitations of relevance to flooding. This applies above all to weather conditions involving precipitation over large areas. However, the output from these models is still too imprecise and spatially undifferentiated to generate information, with the degree of accuracy necessary for computing flood discharge, on the location, quantity and intensity of precipitation as well as on changes in precipitation due to global climate change.

In addition, running fine-resolution models requires a great deal of computation time, so the extended simulations needed to assess climate change are still very rare.

For this reason, several alternative regionalization methods have been developed. By classifying weather patterns, for example, the computational requirements for the regional models can be reduced significantly. As an alternative to using regional models for statistical regionalization, empirical interrelationships between observed data on the large and regional scale can be used to produce statistical models. One advantage of these statistical methods is the minimal amount of computer capacity needed. On the other hand, it is essential that long time series of observation data are available.

Much research needs to be done before any of these regionalization methods can generate scenarios of regional climate that are sufficiently realistic to assess flood risks in a changing climate.

1.6.4
Management and control of flood risks

As with most natural and technical risks, one characteristic of flood risks is that the potential damage is inversely correlated to the probability of occurrence. The greater the damage, the lower the probability that this damage will actually occur. Frequent minor flooding poses a minimal threat because people can take precautionary steps and be prepared for any damage caused. But if a rare flood event occurs and the water level exceeds the level for which protection has been provided in the form of technical or non-technical measures, the impacts can be catastrophic. Protective measures that help to prevent large-scale damage in the event of small or medium-sized floods may prove counterproductive in the case of rare extreme events. Smaller dams or levees hold back the waters initially, but release them suddenly and all at once when there is a breach or overflow.

It is not possible to protect oneself against every conceivable flood event since the maximum height of a flood remains undefined. People can, of course, avoid areas with a high flood risk, but as already mentioned, precisely such regions attract human settlement for a number of reasons. In addition to the

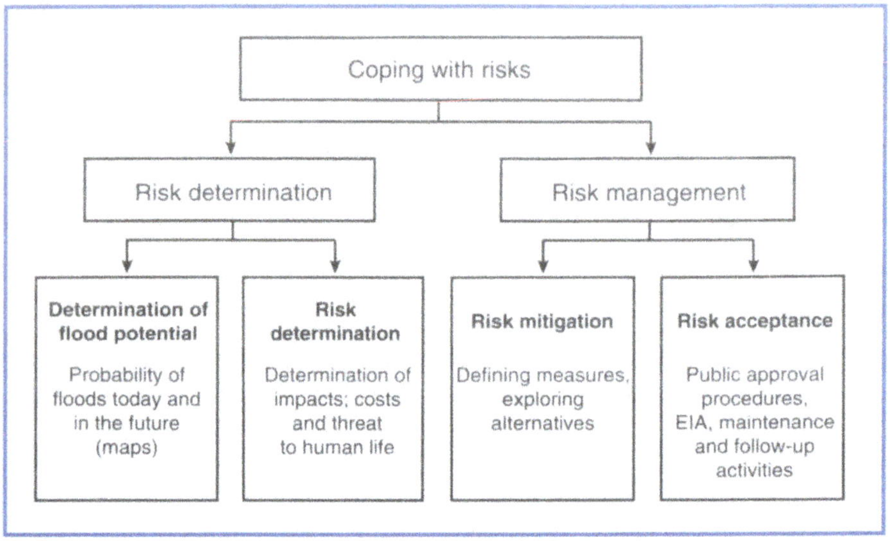

Figure D 1.6-5
A risk management system.
Source: adapted from Plate, 1997

physical constraints on complete flood protection, there are also economic limitations. It makes little economic sense to develop flood protection by means of very cost-intensive measures to such an extent that people living on rivers even protect themselves against extreme rare events that are expected only once in several hundred or even a thousand years. Therefore, no matter what technical precautions are taken, there is still a residual risk that can be regarded as barely acceptable. To safeguard against floods, protective or mitigating measures will be initiated and implemented as long as the local residents deem it acceptable after weighing up the costs and the residual risk. Thus, the objective is not to prevent flooding at all cost, but to reduce the risk to an acceptable level and minimize the damage in the event that a severe flood and the concomitant damage occurs after all. Since individual persons have limited capacities in this respect, flood control is the responsibility of the state or of specialized organizations established for this purpose. In Germany the responsibilities are stipulated in the Federal Water Act of November 12, 1996, which is supplemented by water laws at the *Länder* level.

The ways in which risks can be limited and managed are shown in Fig. D 1.6-5 on risk management aspects. It consists of two separate parts: risk determination, in which the analytical basis for decision-making is defined, and their implementation as practically effective measures to minimize risks, designated here as risk management.

Risk determination includes determining the flood potential and the quantitative extent of risk, i.e. calculating a function linking probability of occurrence and amount of damage. The former is referred to as risk identification, the latter as risk assessment or risk modeling. Risk management is similarly split into two subdomains. Risk minimization or mitigation encompasses all measures aimed at limiting either the probability of occurrence or the damage potential. The degree of limitation here is a policy issue. Risk acceptance refers to all processes and measures for ascertaining the level of acceptable residual risk, as well as communication with those concerned. Since the issue of "risk" will be covered in detail by the Council in one of its next Annual Reports, only the basic problems and problem-solving paths taken in connection with risk determination and management in flood control will be discussed here.

1.6.4.1
Determination of flood risks

The first step in dealing with risks is to identify the risk and to assess it in numerical terms (risk assessment). The basis for this is the determination of the potential damage associated with the intensity of flooding, and calculation of the likelihood of occurrence. The level or intensity of the flood itself is merely an intermediate step in determining the risk, the purpose of which is to establish the damage potential in terms of the human lives, material assets and environmental assets at stake. It is only when the flood has impacts on people or on facilities built by people that flood becomes a risk. Therefore, the second major step in determining the flood risk is to assess the negative consequences, i.e. the damage that may be incurred due to the flood.

Determination of flood potential

Identification of the flood potential is essential when calculating the damage potential. This is a task for hydrologists, who have to assess flood events ex-

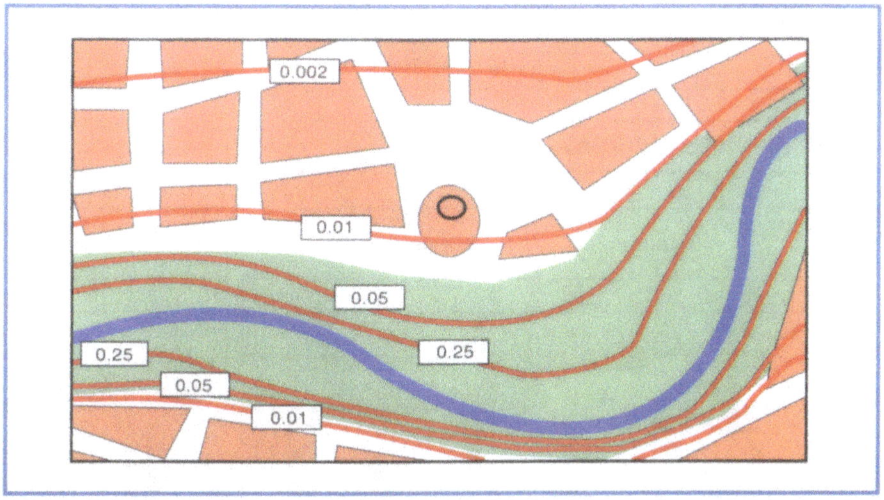

Figure D 1.6-6
Flood risk exposure, shown by the isolines for water levels during extreme flood events with different return periods. Source: Plate, 1997

pected in the future in terms of the likelihood of their causing damage. Within flood control systems, the job of hydrologists is to provide reliable forecasts. Hydrologists distinguish between two types of predictions here: the so-called forecast and the actual prediction of a specific flood event. The (real-time) prediction is part of the flood warning procedure (see Section D 4.4.1). The forecast, on the other hand, consists of abstract determination of flood events as a function of the probability of their occurrence. It forms the basis for ascertaining the flood potential and thus for designing flood control systems.

Depending on the time frame, hydrologists can define the maximum flood event to be expected. A base-10 scale is generally used. What flood events can be expected every 10 years, every 100 years, every 1,000 years, etc.? On the basis of these calculations, a political decision has to be made regarding the maximum event for which flood control is to be designed. In the industrialized countries, protective measures are usually geared to a flood event that can be expected once every 100 years.

Expected values for flood potentials and probabilities are calculated on the basis of statistical analysis of past observations, or using model calculations. Expected values therefore represent approximate figures only, i.e. higher water levels may also occur within shorter intervals, as the Rhine example showed. In December 1993, the peak water level in the lower reaches downstream from Koblenz was close to that of the past 100 years. This was followed in February 1995 by an almost equally high flood peak.

Determination of risk exposure: determination of flood areas

To derive damage potential from flood events, the flood areas for each event must be determined after identification and definition of the flood events. The respective water levels must therefore be ascertained. This can be done by means of measured discharge rates, which can only be determined at a few locations, as well as by using hydraulic calculations that also enable flood risk areas to be mapped (see schematic example in Fig. D 1.6-6). The obvious problem with such maps is that they influence property prices and land-use concepts. However, since the data base is not precise and the hydraulic calculations for converting water quantities into water levels are complex, reliable flood risk maps cannot be expected at present – unless there are sufficient records on historical flood events from which a statistical distribution of frequency can be derived. The unreliability of most flood risk maps is in stark contrast to the economic and political impacts they may trigger.

In the communication of risk-related information, it is therefore essential to draw attention to the uncertainty of the input data. Nevertheless, satellite pictures and other remote sensing techniques offer ways of identifying the areas that will be affected by a flood event. Geographic information systems (GIS) are useful in this context.

Risk determination: flood impact assessment

The next step in risk management is to ascertain the flood damage and other impacts. They are expressed qualitatively in the basic principles of providing flood protection (e.g. LAWA, 1995). These are taken into account in quantitative terms by numerically linking the extent of damage and the probability of occurrence. A typical assessment quantifies the threat to human life as the probability that a person will lose his or her life when a critical flood event occurs (an event for which the protective measures do not suffice). In this case, risk means the average number of flood deaths expected every year.

The diversity of potential damage makes it necessary to have a comparative scale for the different cat-

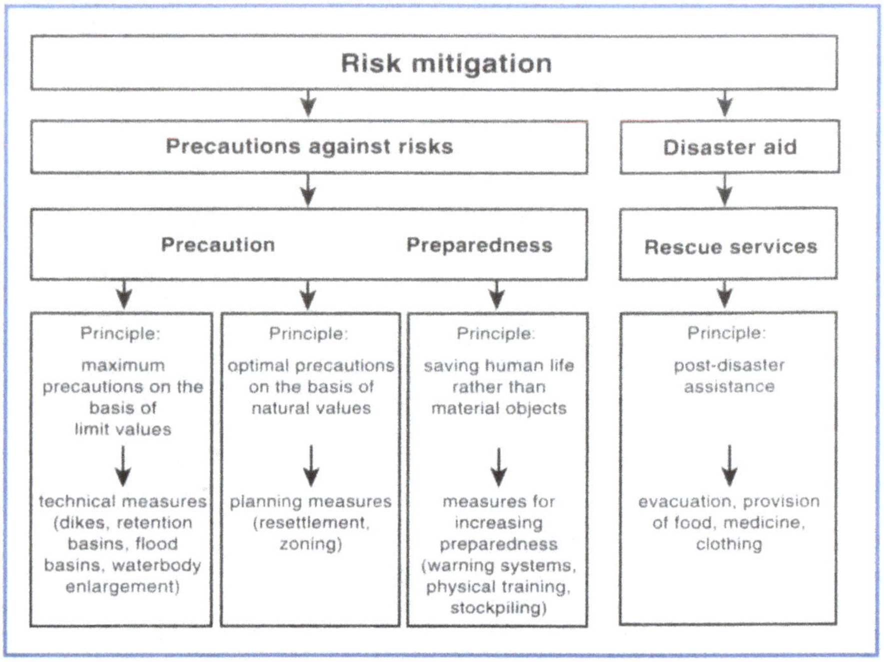

Figure D 1.6-7
Elements of a flood control and protection system.
Source: adapted from UNDRO, 1991

egories of damage. As a rule, the potential damage is specified separately for the categories of loss of human life, loss of assets and threats to ecotopes (Buck and Pflügner, 1991). In the recently revised emergency protection plans in Switzerland, as many as seven different categories of damage are defined, although these may also be aggregated, depending on the specific case. For the purposes of risk management, however, decision-makers tend to favor separate classification.

1.6.4.2
Managing flood risks

Determination of risk provides the essential basis for the decision-making process, in which the necessary level of protection and the related precautionary measures are decided upon. The aim here is to reduce the risk, by means of technical or non-technical measures, to an acceptable level, i.e. to the residual risk regarded as acceptable by society for the protection of those concerned. The risk can be minimized in two basic ways: by reducing the probability of occurrence or by reducing the range of damage. The methods available for this purpose are summarized in Fig. D 1.6-7. The question of which to apply depends on a number of factors that are addressed below.

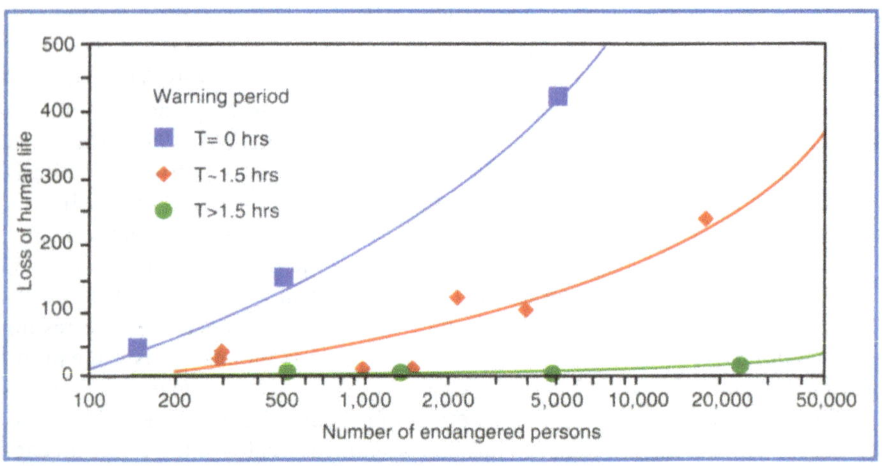

Figure D 1.6-8
Benefits of advance warning periods in the case of dam breaks.
Source: von Thun, 1984

BOX D 1.6-1

Non-technological methods of flood prevention and control

1. Methods for reducing vulnerability to damage
   - 1.1 Acquisition of land
     - a. complete
     - b. partial
   - 1.2 Evacuation
     - a. permanent (resettlement)
     - b. temporarily (in serious cases)
   - 1.3 Protection of property
     - a. compulsory, possibly subsidized
     - b. initiated through information
   - 1.4 Planning methods
     - a. land-use plans designating areas at risk, including identification of wetlands, restrictions on use
     - b. urban development plans involving urban renewal
     - c. building regulations
2. Methods for mitigating flood impacts
   - 2.1 Flood defense, deployment plans for local personnel
   - 2.2 Deployment of emergency and disaster control personnel (technical relief organizations, etc.)
3. Methods of damage compensation
   - 3.1 Flood insurance
     - a. subsidized
     - b. compulsory elementary damage insurance for all homeowners
   - 3.2 Government assistance in the form of
     - a. grants or cheap loans
     - b. tax reductions

Source: Buck and Lee, 1980

Prevention

Risk reduction using non-technological measures

In general, reducing the damage potential or the vulnerability of people, property or natural assets threatened is the obvious solution if the existing threat is considered unacceptable. There are several options to choose from in this connection (see Box D 1.6-1). First of all, one can reduce the number of threatened elements, e.g. by resettling the people affected or relocating industrial facilities outside the flood areas in order to minimize the damage potential. Even temporary resettlement is an option when a good early warning and well-developed general warning system exists.

Real-time flood prediction, i.e. the most accurate possible calculation in advance of the temporal and spatial course of an expected flood event, is the basis for any operational warning system. American studies on cases of dam failure and extreme floods show that the effectiveness of risk reduction by means of predictions and subsequent evacuation is primarily dependent on the amount of advance warning. With a minimum warning time of 1.5 hours, fatalities during local flood events in smaller regions can be almost wholly prevented (von Thun, 1984). Fig. D 1.6-8 shows the relationship between risk reduction and advance warning time.

Another major factor in connection with possible risk minimization, besides the amount of advance warning, is the capacity of the population to respond. After the warning, the protective measures must be implemented quickly and effectively. In extreme cases, evacuation must be completed within a few hours. It is therefore of vital importance that decisions concerning protective action are rapidly and efficiently implemented. One aspect of damage minimization is the avoidance of false alarms. Because predictions made long in advance are frequently very imprecise, the activities themselves, i.e. the specific protective measures, must be carried out in phases that are staggered in such a way that the respective steps are not taken until an adequate degree of certainty has been achieved regarding the intensity of the extreme event.

The effectiveness of protective action depends to a crucial extent on the readiness of those concerned to respond appropriately to the warning. In contrast to common preconceptions, American studies demonstrate that in emergency situations the population reacts without panicking and with great composure. Three main conditions are essential, however: (1) the people concerned must have confidence in the warning signals and the information issued by the public safety authorities; (2) they have to know exactly what kind of behavior is expected of them in the situation; and (3) the infrastructure must be so designed that appropriate behavior in the emergency is also objectively possible (otherwise panic will indeed result).

Many authorities are faced with the fundamental dilemma that their credibility suffers if they issue a precautionary warning and the event in question does not occur, but they are held politically responsible if they fail to issue an alarm in a situation where the likelihood of critical flood levels being exceeded appears too low, and then the improbable occurrence of a flood disaster occurs nevertheless. This dilemma is one of the challenges that risk communication has to tackle and it requires an intensive exchange between the authorities and the local population even before a flood event takes place.

If these aspects are also taken into consideration, a flood warning system will consist of the following components (see also Homagk, 1996):

1. a measuring system that supplies the data for early warning,
2. a communication system that transmits the information from the gauging station to the control station,
3. a model for forecasting the event, including its intensity and temporal profile,
4. a communication system for passing on the forecast to the decision-maker,
5. an evaluation system that enables assessment of the expected impacts and translates them into a warning,
6. a communication system for transmitting the warning to those people who take action to implement the warning and who activate the deployment plan.

Research is still needed on all these steps, particularly with respect to the design of the communication system to take account of the degree of uncertainty of the message and the binding nature of the action thus triggered.

Risk reduction using technological measures

The technically feasible measures for risk reduction depend on the specific features of the region being protected and the type of flood.

Floods in high mountainous areas

Technical measures for protection against floods and mudslides at torrents and mountain rivers and, in some cases, the landslides they induce, tend to focus on confining the river or streambed by reinforcing the banks and occasionally by means of large-scale lining (outer concrete walls) for direct protection of roads and houses. Every type of confinement or dam involves a residual risk of overflow or breach. Since such breaches are usually signaled beforehand, there are calls in many countries, e.g. in Switzerland, for dam operators to draw up disaster control plans designed to ensure the well-organized evacuation of people from the threatened area prior to an imminent breach in a dam.

Floods in hilly regions

The small valleys of hilly regions face a dual problem with floods in many cases. On the one hand, the severest floods occur as a result of summer thunderstorms locally reinforced by orographic effects, or due to heavy spring rains combined with snowmelt and frozen soil. Larger valleys, on the other hand, are threatened more by long-term cyclonal weather patterns. One possible approach to flood control in such regions is to set up flood retention basins in the upper valleys that serve to protect the residents immediately below them, and, together with a system of retention basins, can also protect the entire catchment area against large-scale flooding. Suitable calculation methods are available for this purpose (e.g. Ihringer, 1996). Flood control does not stop at retention basins nowadays. Efforts are also made to regulate rivers at major danger points, e.g. in towns, such that no narrow sections exist where the river can flow out of its bed.

In small catchments, human interference in the structure of the catchments is the factor having the greatest impact on the flood regime. The building development structure of the catchment area can increase both the runoff coefficient and the runoff concentration (flow velocity). Desealing and artificial infiltration measures have been considered for some time now, and indeed carried out in some places as partial compensation for these effects. Such measures can mitigate to some extent the occurrence of floods in densely built-up areas. However, this mitigation has less of an impact the heavier the flood-inducing precipitation and the greater the soil moisture in upstream parts of the region. Precisely those floods that are caused by extreme precipitation are less intensified by building development than are the more frequent flood events against which people are usually protected anyway. Accordingly, the flood-mitigating effects of desealing measures carry less weight in connection with extreme precipitation.

Flooding of lowland areas

Under natural conditions, peak discharges on plains produces large-scale flooding. The obvious solutions for flood control on plains are dikes that fix the river bed and prevent inundation of the floodplains. In addition, the slope is increased by cutting off river loops. However, the disadvantage of such methods is the severe interference with the river regime they involve, which is difficult to compensate for, not to mention the environmental consequences and landscape changes that can be expected.

In addition to dikes and flood basins, flood control of lowland rivers can take the form of flood channels (levees) parallel to the existing riverbeds, or shortened routes from tributaries to the main receiving waters via branch canals. Two rows of dikes, a common practice in the past, also deserves mention: a so-called winter dike situated a substantial distance away from the river, forming a protective wall against severe floods, and a summer dike, which is built closer to the river and offers protection against smaller floods in summer. The land between the dikes can be used for agricultural purposes.

Tidal floods

Technical precautionary measures against flooding in the estuary and delta regions of large rivers include the creation (and maintenance) of river dike systems as well as the construction of large, controllable barriers at river deltas. The purpose of these barriers is to stop the flow of seawater into the river system at extremely high tidal water levels and thus provide sufficient space for temporary retention of inflow from the upper course of the river. When the high tide level in the sea has dropped, the barrier is opened and the river water can flow into the sea. To ensure that the temporary storage of the river water does not cause any flooding, the river dikes must have a certain height even when barriers are present. Such barriers are impressive structures. They have been built to protect such cities and ports as London, Hamburg and Rotterdam. Even when dike systems and barriers exist, technical failure or floods exceeding the design assumptions mean that flood disasters in these regions cannot be ruled out.

Decision-making in response to risks

In order to compare the costs of hydraulic and organizational measures with the benefits gained in terms of reduced risk, it is essential to have a meticulous planning process in which the need for safety and the costs of safety are weighed up and assessed, taking secondary impacts into consideration as well. The latter may include impairment of alternative uses of the waterbodies and riparian areas due to the measures implemented, as well as changes in the appearance of the landscape and unwanted environmental impacts.

In Germany, the priority of protective measures considered should be based on those developed by the Americans after the Great Mississippi Flood of 1993 and codified in the form of development principles (Rasmussen, 1994):

Principle 1:

Prevention of flood risk by means of resettlement and property protection.

Principle 2:

Where this is not possible, mitigation of flood impacts to the greatest possible extent.

Principle 3:

Mitigation of flood damage impacts through insurance and local self-help.

Principle 4:

Implementation of principles 1–3 in such a way that the natural environment of riparian areas is protected and its benefits enhanced.

In this context it is planned to transfer the responsibility and costs to local bodies in order to ensure an efficient form of precautionary disaster control, while government agencies provide for overall coordination, advisory services and fundamental research.

After the event

The simplest way of coping with risks is to do nothing until the disaster occurs and then ensure that its effects involve the least possible damage. Focusing on action after the event may be cheaper than taking precautionary measures in some cases, especially if human life or significant assets are not at stake. In Germany, for example, it is customary today to allow flooding of agricultural areas, whereby the assurance is given that if damage due to flooding occurs, the farmers affected will receive appropriate compensation.

Risk acceptance

The decision on the level of risk a society considers tolerable, i.e. the residual risk it is prepared to accept, is based on criteria that vary from country to country. There are also different preferences for or aversions to specific protective measures. Whereas evacuations are a normal part of disaster control in the United States, they are regarded as a last resort in Germany. Cultural and social factors are also key factors influencing the acceptance of measures.

The ways in which people deal with the threat of floods, whether and how they take precautionary measures against possible damage or react to damage caused by floods depend to a decisive extent on the perception and assessment of these risks. Research conducted by the social and behavioral sciences on perception of the risk of natural disasters has identified a whole range of factors that may influence this risk awareness, e.g.:

- features of the natural event or situation (in the case of floods: lack of controllability, suddenness, chronological sequence, intensity, recurrence),
- problems relating to the ways that humans process information when coping with uncertain events or probabilities (resulting in incorrect assessment of the risk),
- cognitive and/or emotional strategies for dealing

with the threat (e.g. denying or playing it down),
- personal experience of flood risk, of one's own or communicated by others (familiarity, amount of damage in the past),
- attitudes, values, personality characteristics of those concerned as well as motivational aspects (convictions regarding control, voluntariness of risk assumption, interests),
- social norms (group pressure),
- sociocultural influences (including economic, political, legal and technological frameworks).

The above factors have a bearing on any risk assessment, regardless of whether it involves natural disasters such as floods, technical risks (e.g. nuclear, genetic or chemical engineering) or environmental risks (e.g. ozone hole, greenhouse effect) (Jungermann and Slovic, 1993).

Although human activities are often contributory causes of floods, the latter are primarily regarded and experienced as natural calamities. Floods are characterized by an inherent lack of controllability as well as by the fact that third parties (other people, the government) cannot be made responsible for them. Moreover, they take place relatively rarely and the group of persons affected by a flood event can be specified (Karger, 1996). To that extent, flood disasters are widely regarded as "twists of fate" that must be accepted and which are beyond human control.

The actual or apparent naturalness of these risks is also the reason why the dangers posed by natural disasters are deemed by most people to be far lower than those caused by human agency, despite the amount of potential devastation that may be involved (Brun, 1992). Although the extent of damage caused by floods (expressed as the expected number of deaths, for example) is usually overestimated (Lichtenstein et al., 1978), people seem to focus more on the likelihood of damage than on the potential scale of damage when assessing the risks (Slovic et al., 1978). If the probability of a flood disaster occurring is underestimated in such circumstances, the associated risk will be assessed as minimal – with obvious consequences for behavior. For example, people live in flood-prone areas without even taking out insurance (much less changing their place of residence) and frequently believe after the occurrence of a flood disaster that it will not be repeated (Burton et al., 1978; Kunreuther, 1978).

This underestimating the risk of a natural disaster such as a flood, and the resulting behavioral responses, can be attributed to the following:
- The tendency, essential for survival, to ignore risks having a low probability of occurrence (Slovic et al., 1978).
- The function of denying or playing down uncontrollable threats as a basically "healthy" mechanism for coping with emotional stress (Lazarus and Folkman, 1984; Evans and Cohen, 1987).
- The incidence of cognitive "errors" (heuristics) when evaluating uncertain events or when dealing with probabilities (Gardner and Stern, 1996).

If behavioral patterns based on an underestimation of flood risk are to be influenced in the context of comprehensive risk management, merely informing the population about the "real" threat is a pointless approach. Instead, an appropriate strategy of risk communication must target different groups, and cover all variables of relevance to risk assessment, including the knowledge, experience, values, attitudes and ways of processing information among those concerned.

The acceptance of measures is governed by a number of a factors. A protective strategy must therefore include greater efforts to communicate risks and to integrate the population into the precautionary planning process. Particularly in the USA, relatively good results have been achieved with Citizen Advisory Panels, which are involved in the preparation of disaster control plans and act as communicators vis-à-vis other local residents. Furthermore, they have played a contributory role in reviewing the technical measures recommended by the authorities with an eye to weak points and unsubstantiated assumptions about behavior. This practice should also be implemented in the developing countries especially, because the authorities there are often confronted with a loss of credibility and are unable to operate effective organizational measures, such as comprehensive warning systems.

Public participation is essential nowadays when implementing large-scale flood control projects. The issue of risk acceptance as it pertains to flood control has always been a question of the solidarity of those people who are not affected with those who are. Such solidarity is facilitated when the benefits produced are not confined to the local residents only. This is the case if other aims besides flood control can be achieved. Integrating environmental aspects into the "Integrated Rhine Program" enabled a decision based on solidarity that was politically accepted in all the riparian nations. However, the political decision to develop a flood control system does not always meet with the approval of local residents, because the construction of higher dikes or flood basins often requires private land and therefore curtails other uses. If those affected are themselves the beneficiaries of a protective measure, the acceptance is much greater, of course, than in the case of an upstream resident who has to create retention basins solely for the benefit of downstream residents.

1.6.5 Research recommendations

In recent years, there have only been isolated research projects on questions concerning the occurrence of floods, possible exacerbation and more effective flood management. The following research topics are therefore important:

- Integrated analysis and modeling of the entire causal chain, from precipitation, formation and concentration of runoff, the course of flood events (also in flooded areas), to damage assessment.
- In the meteorological field especially, precipitation forecasts for regional or local catchments are important for practical activities. Downscaling methods must be further developed and validated, particularly with respect to the determination of precipitation frequencies in space and time. This applies to the linking of measured values with precipitation data, as well as weather patterns with precipitation events.
- Derivation of scenarios for extreme weather situations, both on a regional and local scale, on the basis of global warming scenarios, global and regional climate models and analysis of the influence of cyclonal weather patterns on precipitation in Germany.
- Development of methods for evaluating the stress-bearing capacity of our rivers. The conflicting interests associated with the navigability of waterways, land use, environmental functions and flood control must be taken into account. How do floods affect the relationship between these aspects and how does the damage potential change indirectly as a result?
- The damage potential of floods is a consequence of human settlements in flood-prone areas. The interrelationships between the perception and handling of risks and the consequences of floods must be studied in more detail, because the responses to risk also determine the potential for damage reduction.
- Interest should focus on the general social processes of perception, communication and response in connection with flood risks, compared to other individual and civilizational risks. What role do limit values play in the acceptance of risks?

2 Water in the Global Network of Interrelations – the causal web

Water-centered Global Network of Interrelations – key trends in the hydrosphere

2.1 Trends in the hydrosphere

In its various Annual Reports to date, the Council has already specified a number of trends that describe the major changes occurring in the hydrosphere. Within an analytical description of global change, these trends are seen not simply as a result of human activities, but also as active elements within a network of mutually impacting trends in human society and the environment.

The interactions between these trends may be represented by diagrams showing the reinforcing and attentuating effects that can appear. The type of interaction thus shown is not mechanical in any sense (as in the type of model used in physics), but symbolizes instead the qualitative, phenomenological consistency in the occurrence of the two trends in question.

These elements can be used to construct networks of interrelations representing a simple form of qualitative expert system: in standardized form, they summarize existing expert knowledge concerning the constitutive elements of global change. The diagrams – as a formal instrument for qualitative environmental system analysis – are not only able to illustrate the high degree of connectivity between the various aspects, but also convey an idea about how specific problematic trends are embedded within the overall phenomenon of global change (for a detailed description see WBGU, 1997). This instrument can help answer the questions: What trends within global change exert a significant influence on the freshwater crisis? Where are the critical feedbacks?

In a diagram illustrating the hydrosphere-centered Global Network of Interrelations, the impacts and effects of the various trends on the hydrosphere must be shown separately due to the complexity of the interrelations (Figs. D 2-1 and D 2-2). In the following, the most important relationships within the water-centered Global Network of Interrelations are described, with special reference being made to the different regional manifestations, which are not visible in the charts.

One key problem within the hydrosphere is increasing water scarcity. The intensity of this trend varies considerably between one region and another. Africa and parts of West Asia seem to be especially vulnerable to this environmental problem. Northwest China, western and southern India, large parts of Pakistan and Mexico, the west coast of the USA and northeast Brazil are seriously affected as well. The main dimension of the trend is the high increase in withdrawals at local level. The following examples describe some of the more serious developments:

- Excessive groundwater extractions: depletion of the Ogallala aquifer in the USA and of the Disi aquifer in Jordan and Saudi Arabia, extraction of fossil groundwater stocks in Libya and Yemen, as well as in India and Southeast Asia.
- Diverting water for irrigation projects: desiccation of the Aral Sea (see Section D 3.4.3.1) or of the Everglades in Florida.
- Intrusion of salt water into deltas and coastal aquifers as a consequence of excessive use of groundwater and surface waters: e.g. in Israel, China, Vietnam, the Gulf of California or the Gaza Strip.
- Losses from supply pipes, in irrigation (evaporation) and in production are also important factors causing water scarcity.

Within the hydrosphere, the main reinforcing factor affecting the amount of potable freshwater ("freshwater scarcity") results from "declining water quality". The causes of this decline can be pollutant loads from air (acid rain, dust, substances washed out by precipitation), from processes using water (industry, households, sewage disposal) or from soils (agriculture and waste dumps). A number of problematic substance groups are relevant here: heavy metals, salts, acids, persistent organic pollutants (e.g. from chlorine chemistry), nutrients (feces, particulate matter from eroded soils) and pathogens. "Declining water quality" – one of the main reasons for the decrease in usable freshwater ("freshwater scarcity") –

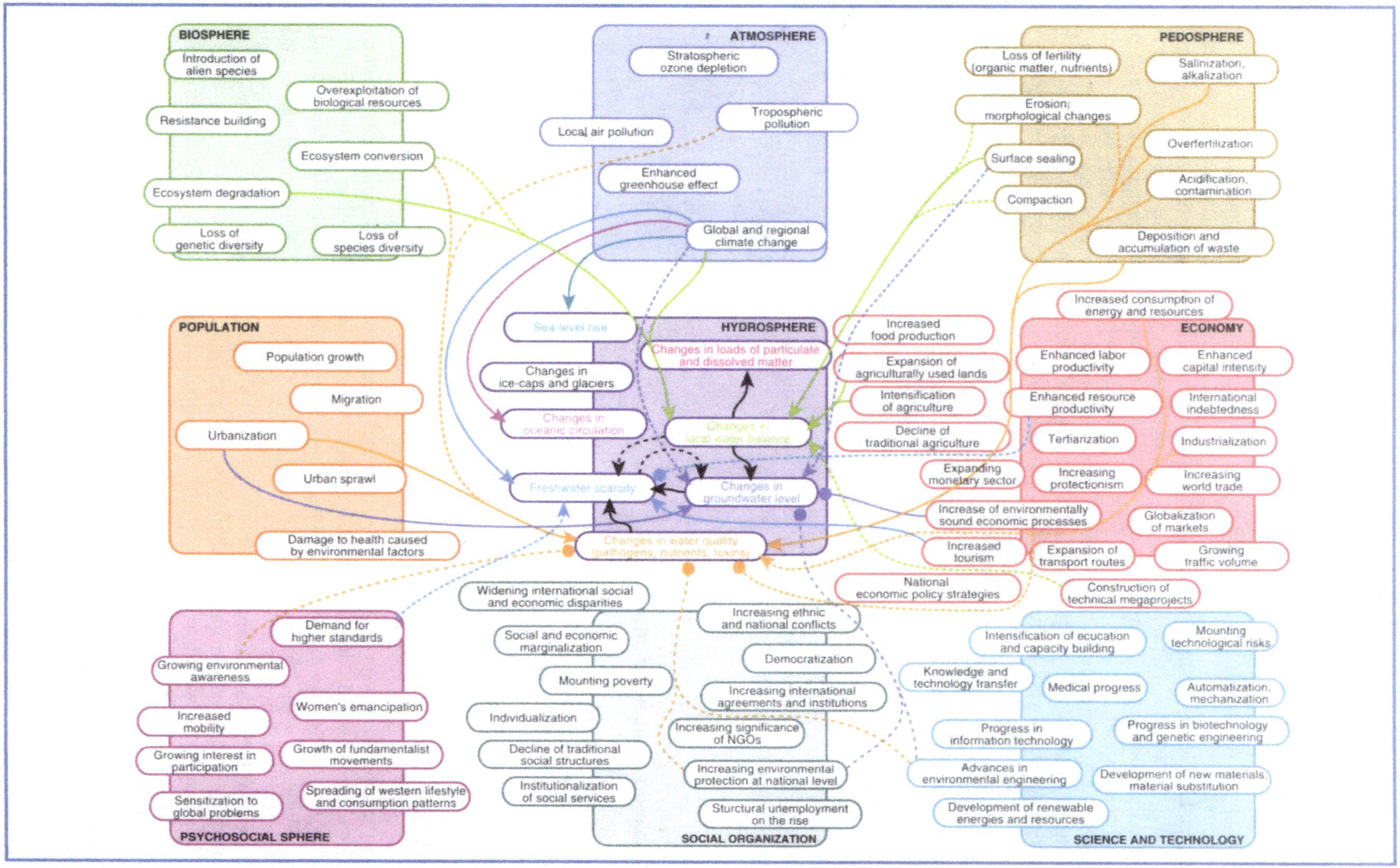

Figure D 2-1
The water-centered Global Network of Interrelations. The solid lines denote a permanent causal link.
Source: BMBF "Syndrome Dynamics" project, PIK Core Project QUESTIONS and WBGU

is manifested in different forms depending on the specific region:

- Africa: high salt concentrations, large sediment loads in rivers at the beginning of the rainy season.
- North America and Europe: pollutant loads from industry (toxins, acidification), nitrates and biocides from intensive agriculture (eutrophication of surface waters).
- South America: high level of bacteriological and organic pollution (pathogens and nutrients);
- Middle East: salinization due to intensive agriculture, intrusion of salt water.
- East and Southeast Asia: organic pollution, pesticides and eutrophication.

In the water-centered network of interrelationships, the trend labeled "intensification of agriculture" is characterized primarily by "expansion of irrigation", which is a major factor in the global increase in water consumption (see Section D 1.4). Irrigation always has impacts on "changes in the local water balance" because it affects both runoff and evapotranspiration. If a reinforcing effect is not present in every case, this information is represented by broken lines (Figs. D 2-1 and D 2-2). This is done to show that the phenomenological correlations between these trends of global change may involve regional variations.

2.2 Global mechanisms of the water crisis

Freshwater scarcity – Declining water quality – Groundwater depletion – Changes in the local water balance – Changes in loads of particulate and dissolved substances

2.2.1 Impacts on hydrosphere trends

Impacts on freshwater scarcity

"Demand for higher standards" in the psychosocial sphere leads to an increase in water consumption by households, on the one hand, as well as by trade and industry as a result of increased production, on the other. The water withdrawals involved are equivalent to roughly 11% of the world's total renewable water stocks. This percentage will increase still further as a consequence of the "spreading of western lifestyles and consumption patterns". Per capita water consumption is currently high, especially in the industrialized countries, even though individuals' direct consumption of water as a foodstuff is quite low. Much more drinking water is consumed for cleaning or sanitation.

Rising water prices have established the economic conditions for a reversal of these trends. "Enhanced resource productivity" is gaining in importance, above all in the industrialized countries: water withdrawals by industry and for domestic use (e.g. washing machines and dishwashers) have decreased substantially, and multiple use now plays a greater role. In contrast to the agricultural sector, where water as an environmental asset is often less relevant as a cost factor due to the subsidies for free access to surface water or groundwater, the manufacturing and production sectors generally rely on the supply of drinking water from waterworks. These withdrawals are therefore a visible cost factor, so reductions in consumption are desirable for users. One example is the increase in productivity per unit of industrial water use in Japan, which has more than trebled over the last 20 years or so.

Another trend of global change that can exert a very strong influence on freshwater scarcity at regional level is "growth in tourism". Local water withdrawals can increase heavily in popular tourist regions, which in many cases are water-poor and arid. Mass tourism, through which the creation of tourist infrastructure becomes profitable in the first place, causes a severe drain on local water resources. Swimming pools, activity zones, parks and gardens requiring constant watering, in addition to a high level of individual consumption, can result in the depletion of renewable drinking water resources at local level.

Impacts on "declining water quality"

Worldwide, the quality of surface waters and hence the quality and amount of directly accessible freshwater stocks are undergoing changes. The severity of impacts differs considerably from one region to another, and is linked to several trends. The "tropospheric pollution" trend plays a special role in this context on account of the eutrophication and acidification of waterbodies. In northern Europe, the acidification of waterbodies has declined in severity, mainly because of flue-gas desulfurization, whereas the trend has intensified in other regions (East and Southeast Asia). In recent years there has been an increase in nutrient loads – especially nitrogen – via the atmosphere to waterbodies, which in turn can reduce the quality of surface waters through increased biomass production.

This phenomenon results to a much greater extent from the trend towards "intensification of agriculture" and the "overfertilization" this often involves – eutrophication is induced here primarily by the enhanced loads of phosphorus and nitrogen into waterbodies (see Section D 4.4). The situation is exacerbated by pollution through biocides in many developing countries, where older, more persistent pesticides

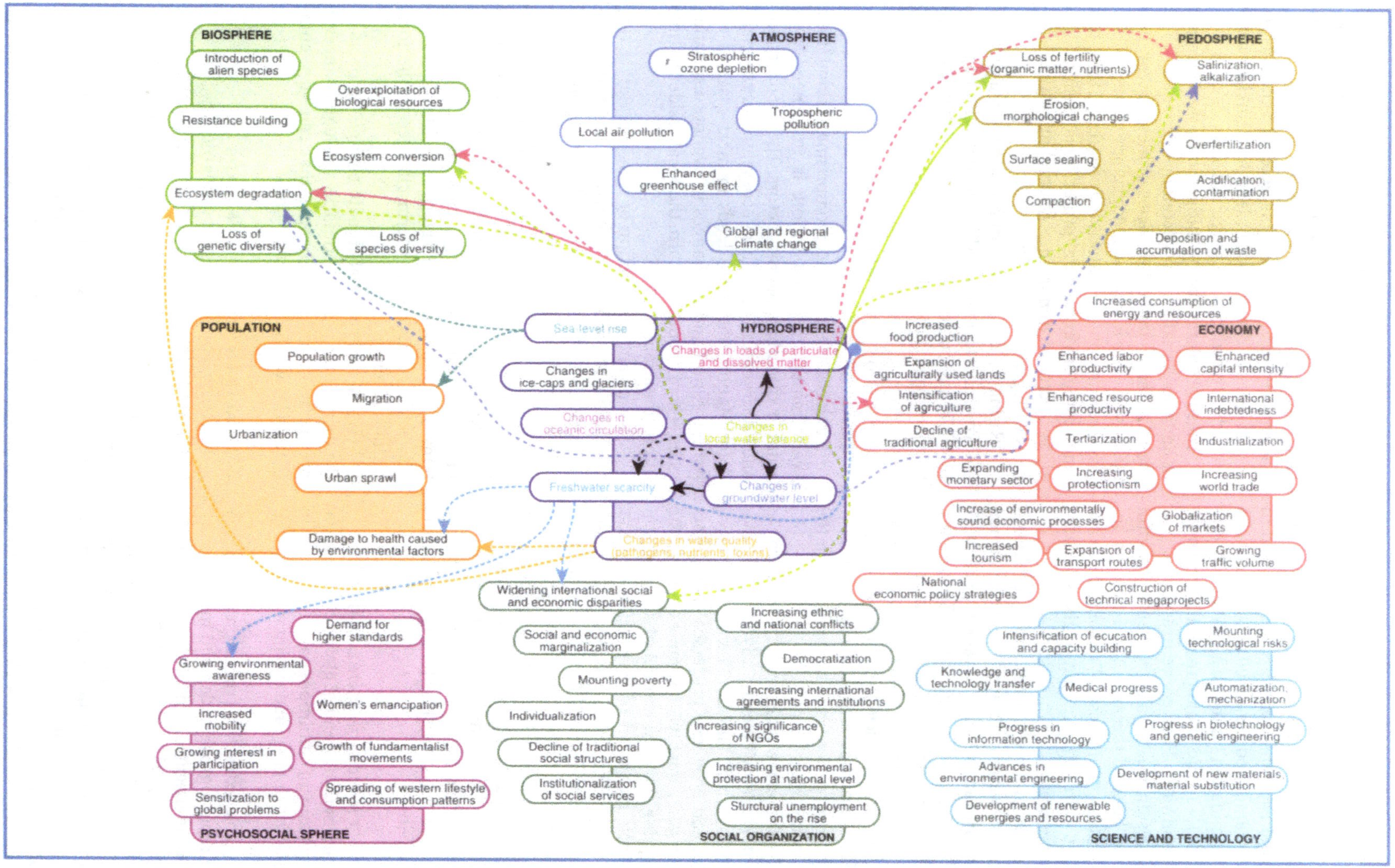

Figure D 2-2
The water-centered Global Network of Interrelations: Impacts. The broken lines denote regionally varying causal links.
Source: BMBF "Syndrome Dynamics" project, PIK Core Project QUESTIONS and WBGU

and herbicides are still frequently used (see Section D 3.3 on the Green Revolution Syndrome).

Complementary trends causing deterioration of water quality are manifested in other regions – the "Deposition and accumulation of waste" on and in soils (e.g. at contaminated industrial sites) can lead – as in the "contamination" trend (e.g. leakage of pollutants, oil spills) – to the contamination of riparian zones and groundwater. High concentrations of polluting substances in waste (also in sealed landfill sites) are "time bombs" threatening to unleash severe water pollution. "Soil erosion" can also have adverse effects on water quality due to the sediment loads that are generated.

Cities are some of the worst hot spots: "urbanization" exacerbates "water pollution" most of all when the urban infrastructure fails to keep pace with growth (see Section D 3.5 on the Favela Syndrome). The "increasing consumption of energy and raw materials" trend may also have certain impacts, especially when obsolete and inefficient technologies are in use. This problem is particularly prevalent in the newly industrializing countries (see Section D 3.2 on the Asian Tigers Syndrome).

Figure D 2-1 shows few tendencies that reduce the "declining water quality" trend: "advances in environmental engineering" and the "increase of environmentally sound economic processes" counteract the "decline in water quality". Examples include the setting, monitoring and enforcement of environmental quality targets in respect of production and use, for example compliance with limits on contaminant concentration and the use of efficient technologies for wastewater purification. The driving forces behind these trends are the "increasing environmental protection at national level" and "growing environmental awareness". These are essential if improvements in water quality are to be achieved by means of technical processes – as well as through changes in consumer behavior.

Impacts on "changes in water tables"

A particularly serious trend of global change is the worldwide problem of anthropogenic alteration of water tables. Both "sinking" and "rising water tables" can lead to damage.

Non-sustainable extractions occur predominantly as a result of rapidly growing demand in cities ("urbanization") or in irrigated agriculture ("intensification of agriculture"). However, enhanced demand due to "growth in tourism" can lead to the depletion of groundwater reserves if withdrawals from surface waters are not an option, or if the latter resources require costly treatment before use.

In addition to these trends, which are associated with excessive withdrawals, the regional potential for groundwater replenishment is changing. "Surface sealing" caused by "urbanization" and "overdevelopment", as well as the "expansion of transport routes" and "changes in the local water balance" (e.g. through land-use changes or as a consequence of "regional climate change"), can lead to a sinking water table due to reduced recharge rates.

These problems can be mitigated by the "increase of environmentally sound economic processes". The use of water-saving technologies or methods for multiple use lead to lower water consumption and thus to less pressure on groundwater resources. Here, too, "increasing environmental protection at national level" is of crucial importance in combating the problems with groundwater resources.

"Rising water tables" can result from inappropriate irrigation techniques ("intensification of agriculture") in association with poor runoff, which can lead to waterlogging of soils.

Impacts on "changes in the local water balance"

The local water balance, comprising precipitation, evapotranspiration and runoff, is influenced by a number of anthropogenic factors. The "conversion" and "degradation of natural ecosystems" (especially deforestation) therefore leads in most cases to enhanced runoff and to changes in evapotranspiration rates, which reduces, in turn, the local residence time of precipitation.

"Sealing and compaction of soils" accelerate runoff, as does the shoring up of river banks and the canalization of watercourses. "Intensification of agriculture" plays a key role in this context. The "expansion of agriculturally used lands" and the intensification of land use by agriculture can lead to major changes in the runoff regime as a result of the draining of wetlands ("conversion of semi-natural ecosystems") and damming of riverside fields and floodplains, thus causing long-term changes in the local water balance.

"Global and regional climate change" impacts on the local water balance primarily through changes in precipitation patterns, changes that can vary enormously between different regions.

Impacts on "changes in loads of particulate and dissolved substances"

One important component of water quality is the amount of suspended sediment and dissolved salts in running waters. The sediment load transported by rivers is increased primarily by soil erosion, the effects of which are passed on through changes in the runoff regime ("changes in the local water balance"), and which is itself caused mainly by changes in land-use practices (WBGU, 1995). Water development projects, from changes to the riverbed to the construction of dams, are other key factors changing the transport of sediment (see Section D 3.4). Due to losses through evapotranspiration – especially in arid and semi-arid regions – large-scale water withdrawals for "irrigated farming" bring about the intense accumulation of salts in return water, which can then reinforce "salinization" of downstream soils (see Section D 3.3).

2.2.2 Effects of hydrosphere trends on other spheres

Effects of "freshwater scarcity"

"Freshwater scarcity" compels people to use water of lesser quality, exposing them to the risk of "increasing damage to health" through the spread of epidemics, for example, or poisoning (see Section D 4.2). In agriculture, water scarcity prevents "increased food production".

"Freshwater scarcity" may increase the potential for conflict between neighboring states dependent on the same surface waters. For example, a major conflict of interests is developing between Syria and Iraq, who have barely sufficient water resources to maintain their intensively irrigated agriculture, and Turkey, which is able to control inflows by means of dams (see Section D 4.1). Conflicts over resource distribution are also possible within countries, where they cause "social and economic disparities" to increase.

Effects of "declining water quality"

The main effect of water pollution concerns the quantity of water that can be used to meet human needs ("freshwater scarcity"). The increasingly poor quality of regional water reserves, primarily caused by the pollution entering surface waters and groundwater from industrial, agricultural and urban sources, can lead to "increasing damage to health" in urban agglomerations, especially in developing countries. Examples of epidemics induced by low-quality water supplies are typhoid, cholera and diarrhea (see Section D 4.2). However, gradual poisoning by heavy metals or other contaminants in water is also of significance. The enrichment of waterbodies with nutrients, and the eutrophication that ensues, is an example of the "degradation of natural ecosystems".

Effects of "changes in water table levels"

A "sinking water table" means a reduction in or scarcity of freshwater stocks. This causes an increase in the expense and effort needed to extract water from ever-deeper groundwater reservoirs. An additional threat can develop for areas further away, which are deprived of their groundwater resources by the long-distance supply networks for urban centers. Their natural ecosystems are degraded by a "sinking water table".

In coastal zones, saltwater intrusion can replace the depleted freshwater in a sinking water table. During the 1950s, the water table in the Tel Aviv area sank below sea level over an area of about 60 km^2 as a result of groundwater pumping. A large-scale system of basins for replenishing the aquifers with freshwater was needed to avert the threat of salinization from saltwater intrusion into the groundwater aquifers.

"Rising water tables" – as a consequence of poorly managed irrigation, for example – can lead to waterlogging of soils. A special problem arises when the capillary link between soil water and groundwater is broken, thus promoting the "salinization" of the soil. In the same vein, a "rising water table" due to "ecosystem conversion" (e.g. logging of eucalyptus forests in Australia and reduced evapotranspiration) can lead to the "salinization" of upper soil horizons through the mobilization of salt reserves in deeper horizons.

Effects of "changes in the local water balance"

Within certain limits, the local water balance represents an element of regional climate that can be actively influenced by vegetation cover. Changes in vegetation can induce changes in the regional climate on account of the different evapotranspiration rates ("degradation of natural ecosystems"). There are similar feedbacks between the water balance, as part of the conditional framework for habitats, and natural ecosystems. If ecosystem degradation is induced by changes in the runoff regime, for example, species diversity can undergo significant changes leading in extreme cases to the total transformation of the ecosystem in question (e.g. disappearance of wetlands, see Section D 4.4).

Humans, too, may be directly exposed to threats of various kinds: free waterbodies artificially produced by holding back runoff, as occurs in irrigation projects, for example, can foster the spread of waterborne diseases (e.g. schistosomiasis or malaria; see Sections D 3.4 and D 4.2). Shifts in the water balance due to inappropriate irrigation methods (enhanced evapotranspiration) often cause "salinization" of productive soils; this is one of the most important interactions behind the global problem of soil degradation (see WBGU, 1995).

EFFECTS OF "CHANGES IN LOADS OF PARTICULATE AND DISSOLVED SUBSTANCES"

This trend has major consequences on the pedosphere as well as indirect impacts on the biosphere. The removal of soil particles by water can lead not only to the loss of productive soils due to water erosion and eluviation ("loss of fertility"), but also to changes in the morphology of a landscape. Dams and the concomitant decreases in flow rate act to reduce the "sediment load" transported by rivers, which in turn causes a reduction in sediment deposits in lower reaches coupled with enhanced riverbed erosion, with all the implied consequences for coastal and estuarine ecosystems ("degradation" and "conversion of ecosystems"; see Section D 3.4).

Enhanced throughput of salts by irrigation systems can have similarly adverse impacts on ecosystems; the threat of "salinization of soils" is magnified downstream through renewed use of river water for agricultural purposes.

3.1 The criticality index as a measure of the regional importance of the water crisis

Complex indicator developed – Innovative modeling approach – Assessment of the global water crisis with regional resolution – Future trends concerning supply, demand and crisis-resolving capacity – Assessing the severity of water stress and water scarcity

In this section, the Advisory Council attempts a prospective assessment of the global water crisis. This criticality assessment is one of the few global models integrating hydrological, climatological, demographic and economic factors, and enables conclusions to be drawn about the future dimensions of the water crisis as manifested at the specific regional level.

Until now, leading analyses have typically been based on the evaluation of water consumption figures for the various sectors within a given country. In older surveys, the water problem tended to be viewed from a mainly technical perspective – the underlying goal was to provide, by means of water development projects and other technical measures, a water supply of adequate quality and quantity in response to a given level of demand that was assumed to rise inexorably. Inadequate consideration was given to the finite nature of the resource, and trends in demand were not questioned. More recent reports, on the other hand, have been focusing more and more on the total volume of water resources available at national level (see Section D 1.4). This is essential for any comparison of existing use patterns and potential demand in the future and thus for assessing the extent to which the consumption of this renewable resource is sustainable. However, the key indicator needed here – "renewable freshwater stocks" – has not been defined with sufficient precision as yet. There is general agreement regarding the amount of water available to a country in the form of precipitation. But there are major disparities regarding the inclusion of surface water and groundwater, which may occur as inflows from other countries and hence substantially increase the local supply of water.

Another major problem, in the eyes of the Council, centers on the current definition of the terms water stress and water scarcity: one speaks of regular or periodic water stress, which may involve only occasional or locally confined water problems, when the amount of renewable freshwater available to each person is between 1,000 and 1,666 m^3 a year. Chronic water scarcity is when less than 1,000 m^3 of freshwater is available, because damage to human health and economic development are the result. If less than 500 m^3 of freshwater can be used per person and year, one speaks of absolute water scarcity (Falkenmark and Lindh, 1993). The three terms as thus defined may be useful categories when evaluating the water crisis on the basis of water availability per inhabitant, but they provide nothing more than a basic orientation and cannot qualify as precise limits with global applicability. Gleick, for example, refers to the 1,000 m^3 limit acknowledged by the World Bank as an indicator of water scarcity as "a level some suggest is an approximate minimum necessary for an adequate quality of life in a moderately developed country" (Gleick, 1993). For Gleick, the minimum level for preventing regular or periodic water stress proposed by Falkenmark is useful merely as a "warning signal" for countries with growing populations, i.e. as long as population size does not stabilize, most countries classified as water stressed risk becoming water scarce (Engelman and LeRoy, 1995). However, even if these threshold values were to serve as a basic orientation for assessing a water crisis, their objective relevance is questionable: comparing and contrasting the amount of water resources available and trends in population growth may be informative with regard to potential demand levels, but say nothing about consumptive behavior, the extent to which minimum needs are met, or about the technical and financial resources available for compensating for shortages. For example, a relatively prosperous country such as Israel is able to manage with 461 m^3 of freshwater per person per year without having to fear severe water scarcity and the devastating consequences this would have for economic development and human health – although the situation is very fragile.

For a realistic and regionally differentiated assessment of the global water crisis, the Council therefore recommends an approach similar to the one described in its Annual Report on the threat to soils (WBGU, 1995). This approach involves assessing the water crisis with a complex indicator combining natural water stocks, the drain on water resources caused by humans, as well as society's problem-solving or response capacity. In countries where water is scarce, the pressure of demand high and the capacity to respond low, the global crisis is manifested in a particularly acute form; in contrast, no crisis is seen to exist in countries or regions with minimal demand levels, abundant water supplies and various options for solving its problems. Most of the world's nations are going to be somewhere between the two extremes. The criticality index that needs to be developed should serve primarily as a means for assessing water crises in the near future, i.e. as a kind of early warning system. It must therefore be a dynamic indicator that takes current trends into account.

With the help of this "local", composite indicator, K(r),

$$K(r) = \frac{\text{water withdrawals}}{\text{water availability} * \text{problem-solving capacity}},$$

the world's freshwater problems could be assessed from the regional scale upwards by means of a criticality index. Each of the variables depends on specific factors – water withdrawals are determined by the local population density, the specific economic system (especially with regard to its water efficiency and water-polluting potential), the environmental conditions as well as cultural factors. For water availability, the responsible factors are climate, vegetation, soil characteristics, hydrography and topography, climate variability as well as the water resource development projects in place. Problem-solving capacity, the "softest" of the three variables in the indicator, could be measured in terms of the economic strength of a location (per capita GNP), an indicator for water-related know-how, in terms of the quantity and quality of existing infrastructure for the supply, distribution and treatment of (waste)water, and with an indicator for the efficiency and stability of the relevant political institutions.

At present, however, only some of the requisite input variables are available at sufficient regional resolution. On the other hand, collection and processing of the basic data needed will soon be essential for a more detailed assessment of the regional significance of freshwater problems. The Council sees an urgent need for research in this area, alongside further investigation of the functional interrelationships between the various spheres.

3.1.1
Modeling withdrawal trends

The Council has developed an initial approximation for the proposed criticality index in order to point out the direction that might be taken by an integrated and transdisciplinary assessment of the global water crisis. The following regional criticality assessment KA(r) operates from the basic analysis of K(r) using currently available data. It permits an assessment of the global water crisis that is adequate at the regional scale, albeit with gaps in respect of the many variables that need to be considered. This assessment is also prospective, in that it tracks the three decisive criticality factors until 2025 on the basis of scenarios obtained from climate research. The Council draws here on work carried out by the Center for Environmental Systems Research at the University of Kassel (Alcamo et al., 1997) in connection with "water withdrawals" and "water availability", and by a workgroup at the Potsdam Institute for Climate Impact Research concerning "water-specific problem-solving capacity" and integrated modeling (Lüdeke et al., 1997).

The unit of measurement in the criticality assessment of global water resources is defined as people's water withdrawals. Water withdrawals are mainly made by:

- agriculture,
- industry,
- for domestic uses.

Available data on national withdrawals were plotted for 1995 (older values so far) on the basis of the current distribution of population density and other data (e.g. distribution of irrigated area) with a geographical resolution of 0.5° x 0.5°. The determinants of withdrawal levels include, *inter alia*, the population size, the per capita gross domestic product (GDP) and the intensity of water use in industrial production. To calculate future water withdrawal trends until the year 2025, three different scenarios were developed, namely for low, medium and high withdrawal levels. The criticality assessment presented in this Report is based on the medium-level scenario (M). It is derived, firstly, from the regionally differentiated assumptions concerning population and economic growth in IPCC's IS92 scenario (world population: 8,205 million people, global average GDP per capita: US$ 7,314), and secondly from a set of assumptions about water efficiency trends as they relate to economic development, differentiated according to sector and economic geography. Water prices and changes in same were not explicitly included in this assessment.

Working from time series for industrialized and developing countries, it was assumed for the domestic sector that per capita withdrawals increase in proportion to per capita income. The figure reaches a maximum at US$ 15,000, after which it declines relatively sharply to half the maximum value, remaining constant at incomes of US$ 20,000 and upwards. Withdrawals by industry were based on sectoral figures for gross value added. The assumptions of the model are that a consistently high level of water withdrawals decreases to 50% once industry's gross value added per capita reaches a certain value; this value is set at approx. US$ 5,000 for countries with low water availability, whereby the reduction in withdrawals does not occur until approx. US$ 15,000. The reason for this is the assumption that industrial production can be adapted better than domestic production to the existing water supply due to greater flexibility in the choice of products and production processes. Water withdrawals by agriculture, the largest category of withdrawals worldwide, was sub-divided into global withdrawals for livestock farming and withdrawals for irrigated farming. Based *inter alia* on the FAO study by Alexandratos (1995), Scenario M assumes for these latter components that the irrigated area will remain constant in the industrialized countries, but will increase in developing countries at regionally varying rates in line with the growing population and its demand for food. Irrigation efficiency increases in all areas by 0.5% per year, or 16% in total by the year 2025.

3.1.2 Modeling water availability

Water availability was calculated with complex models based on a subdivision of the Earth into 1,162 catchments (Alcamo et al., 1997). Modeling of water availability was done with a grid resolution of 0.5° x 0.5°, matching the resolution of global databases. By integrating the driving climatic variables such as soil and vegetation characteristics, slope inclination and the hydrological base of the respective soil type, the daily water balance was calculated for each grid cell, then compared and calibrated with the empirically determined annual runoff.

At the catchment level, the total water availability can be derived as the sum of the annual surface runoff and groundwater recharge in the area in question. Even though the temporal resolution of the model calculations is finer, the results can only be taken as annual means, since lateral flows were not explicitly included. Given the retention capacity of the runoff systems and soil, this is a quite acceptable approach. In the view of the Council, one of the strengths of modeling water availability at the catchment level is that it enables hydrological conditions to be described with much greater precision than would be the case for a country-scale analysis, as is carried out in many other models of the global water crisis.

However, it should be mentioned that the significance of the model's output is qualified by a number of aspects that have not been integrated sufficiently, if at all, in the model so far. First among these factors is water quality. An essential requirement for any realistic assessment of the global criticality of water resources would be to have aggregated water quality indicators enabling international comparisons. Since deterioration in quality (e.g. through the discharge of toxic effluent into receiving waters) can render enormous amounts of freshwater unusable as drinking water or even as process water, this aspect represents a major limitation also in connection with the quantitative dimension of the water crisis. Sediment loads resulting from runoff may also affect water quality, and hence the direct availability of water resources. Various other aspects are not covered by the scenarios, such as intra-annual climate variability, unpredictable land-use changes and changes in vegetation cover. At present, major uncertainties also surround the assessment of the future precipitation patterns brought about by climate change. The Council stresses the need for research in this field, too.

In order to assess the impact of population growth, economic changes and climate change on the water crisis, the water withdrawal scenarios were supplemented by a water availability scenario in which future temperature and precipitation values for the year 2025 were determined with the help of the climate scenario of a coupled global atmosphere-ocean circulation model (GCM) generated at the Max Planck Institute for Meteorology in Hamburg (see Section D 1.3). This GCM output is based on the projection of greenhouse gas emissions in the IS92a scenario (IPCC 1992), which does not include any emission reduction strategies. According to that scenario, the CO_2 equivalent concentration in the atmosphere will increase 50% by the year 2025 relative to 1980–1990 levels. Given the current uncertainties with regard to estimates of regional precipitation, modeling of water availability was carried out using both the IPCC climate scenario and the results from an additional GCM run (Geophysical Fluid Dynamics Laboratory). Only at a large scale of resolution are today's simulations of global climate able to model precipitation distribution with any degree of reliability. Elaborating regional climate scenarios is therefore an area where research needs to be stepped up.

There are still uncertainties with regard to the precise impacts that climate change will have on the global water balance. It can be assumed with some certainty that the anticipated 1.5–4.5 °C rise in mean global temperature will lead worldwide to a 3–15% increase in mean annual precipitation (IPCC, 1996a). While enhanced precipitation will improve water availability (e.g. for agriculture), the rise in temperature will produce the reverse effect (e.g. through higher evapotranspiration levels). Even assuming that climate change will have a positive net effect on water availability, there is considerable uncertainty in respect of the regional and temporal distribution of precipitation. This is another area where the Council sees a major gap in research, closure of which would significantly boost the predictive strength of the criticality index outlined above.

The mean data now available must be treated with caution in any criticality assessment of global water resources. The very fact that agriculture is responsible for the largest water withdrawals worldwide – which means that the core problem of "freshwater scarcity" is very closely linked to the core problem of "threats to world food security" – implies that a global, integrated index should be based on a cautious assessment of water availability that takes full account of natural variabilities. For this reason, the scenario in question is based not on average monthly precipitation figures, but on the precipitation falling in dry years (only 10% of all years are drier).

3.1.3 Water-specific problem-solving capacity

Humans not only cause water crisis, they are also able to take action in order to mitigate the resultant impacts. Social actors and social systems possess numerous options in this respect, including (more details in Section D 5):

- long-distance transmission of water supplies from regions with surplus;
- water resource development projects for equalizing seasonal fluctuations in supply;
- improving water quality by means of purification techniques and/or by reshaping production;
- substitution of traditional water sources (e.g. through desalination of sea water).

These and other measures depend on the society's level of development, on its water-specific problem-solving capacity. Their main thrust is to bring about an effective increase in water supply, and they are distinct from measures leading to enhanced efficiency of water consumption in the relevant withdrawal scenarios. Depending on the preferred concept of development being applied, opinions will vary as to how this capacity should be defined and measured. This touches on a fundamental issue in the analysis of society-nature relations: namely the question whether humankind is able, by virtue of its economic and technological capacities, to cope with or compensate for scarcities – regardless of their natural or anthropogenic origins – through the (additional) deployment of socioeconomic resources.

Projecting this issue into the context of the economy-ecology debate, one can identify four basic, ideal-typical positions regarding the degree to which the stock of natural capital can be substituted by socioeconomic capital (Table D 3.1-1).

The Council considers the two extreme positions in this ideal-typical scheme – total substitutability and total non-substitutability – to be unrealistic. The idea that there are no limits to what a society can achieve provided the requisite financial resources are available makes little sense given the non-substitutability of water for human consumption. Conversely, humankind's thousands of years' experience in the management of scarce water resources refutes any assumption that modern societies are totally inflexible in their exposure to regional scarcities. This reduces the range of plausible positions to the two in the middle, which the Council includes in its analysis. In producing the criticality assessment, the Council deliberately steered clear of adopting a single position on substitutability. The Council considers both a moderately substitutionalist and a moderately non-substitutionalist position (referred to for simplicity's sake as "environmentalist economics" and "ecological economics") to be acceptable, and would like to include both in order to arrive at two different assessments of society's capacity for curing the global water crisis.

The criticality index proposed by the Council would take into account a number of other dimen-

	Substitutability of natural by socioeconomic capital			
	low to zero	low	high	high to total
Economic ideal-types	radical ecologism	ecological economics	environmental economics	radical economism

Table D 3.1-1
Ideal-typical conceptions of the economy-ecology relationship.
Sources: Jansson et al., 1994; Pearce and Turner, 1990; Radke, 1996; Rennings and Wiggering, 1997

sions to solving the problems that exist. For pragmatic reasons, it is assumed for the criticality assessment developed here that a society's level of economic development – measured as per capita gross domestic product (GDP) – can operate as an initial approximation for the water-relevant curation capacity. It almost goes without saying that a rich economy will possess a greater capacity for coping with crisis than a poor one. That said, there are differences in the extent to which human capital is able to counteract natural (and anthropogenic) crises. This aspect is reflected in the two scenarios described below. Classification of gross domestic product was based on four categories, numerically analogous to the four categories of water scarcity and the World Bank's grouping of countries into high, upper-middle, lower-middle and low per capita income. Figures for per capita GDP are available for national entities only, but this detracts little from the strength of the model, since one can assume that national policymakers are the decisive actors in this context.

Regardless of how high or low the substitutability of natural resources is considered to be, however, two things must be considered in any case:

- *Efficacy:* the measures adopted to overcome a water crisis must achieve the intended goals.
- *Opportunity costs:* the means deployed in surmounting the crisis are not available for other, perhaps equally important tasks.

Put more simply: an abundance of money can be badly spent, and money well spent cannot be used a second time. The per capita GDP criterion on which the assessment of the global water crisis is based has not been examined for the extent to which it meets these two conditions. Both scenarios assume a kind of seepage effect: higher GDP will increase the likelihood that mitigating measures are instituted. A syndrome-specific analysis of the global water crisis is essential if full consideration is to be given to the two aspects. Another point is that any assumption that the severity of the water crisis can be weakened by greater capital investment, as is the case with Scenario I in particular, may collide with the per capita GDP forecasts of the IPCC. These forecasts do not normally consider the growth-reducing impacts of resource scarcity and hence are probably too optimistic, especially in Scenario I.

3.1.4
Formulation of a criticality assessment

On the basis of the ratio of water withdrawal to water availability, the criticality of water withdrawals can be assessed for each cell in the grid. This parameter varies from the relatively small values characterizing little-used catchments to values greater than one. The latter designate withdrawal areas where users consume this environmental resource at a rate exceeding the renewal rate – by pumping fossil groundwater stocks, operating seawater desalination plants, or by recycling used water, for example.

Table D 3.1-2
Definition of the vulnerability index. The matrix elements correspond to the following allocations:
1: freshwater surplus, 2: low vulnerability, 3: freshwater stress, 4: freshwater scarcity.
Source: Kulshreshtha, 1993

	Ratio (water withdrawal : water availability)			
Water availability per capita (m^3 per annum)	<0.4	0.4–0.6	0.6–0.8	0.8
<2,000	2	3	4	4
2,000–10,000	1	2	3	4
>10,000	1	1	2	4

By combining the water withdrawal / water availability ratio with per capita availability, the method adopted by Kulshreshtha (1993), a *water scarcity index* is obtained that gives special emphasis to two critical aspects: firstly, the almost total withdrawal of renewable water stocks (withdrawal-availability ratio) and secondly a low per capita level of total remaining water availability. This second aspect reflects whether an identical withdrawal rate leaves a relatively large or only a relatively small amount of leeway (Table D 3.1-2).

The results obtained from this classification give a first estimation of the regional importance of the freshwater crisis. An analysis of the results reveals that, despite the expected global increase in water supply due to climate change, the situation is likely to become even worse in precisely those regions already faced with water scarcity (Alcamo et al., 1997).

However, the calculations do not yet include the water-specific problem-solving capacity at regional level. This variable, which represents the socioeconomic capacity to reduce or cure water-related problems, can be estimated initially from the growth rate of per capita GDP. Depending on the substitutability of the stock of natural capital, the assessment uses specifically designed evaluation matrices (Fig. D 3.1-1).

The basic idea behind the two matrices is relatively simple – if one assumes a high substitutability of water by GDP per capita, then crises and scarcity (water scarcity categories 3 and 4) can be relatively well compensated, whereas this is much less the case if low substitutability is assumed. However, in the case of the lowest income category both positions are

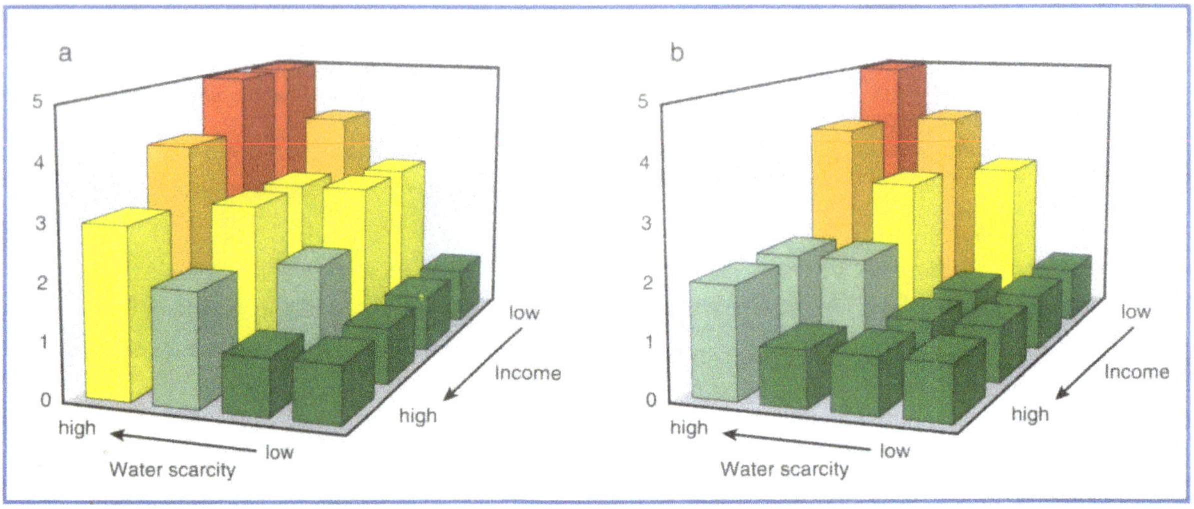

Figure D 3.1-1
Assessment matrices. a) Low substitutability (Scenario II). b) High substitutability (Scenario I) of natural by economic capital stock. The colors of the bars match the colors used in Figs. D 3.1-2a and 3a.
Source: BMBF "Syndrome Dynamics" project, PIK Core Project QUESTIONS and WBGU

agreed that the situation of water scarcity cannot be compensated. The criticality indices between these two extremities are obtained by interpolation.

If all the aspects discussed so far are integrated, then various scenarios are obtained for the global criticality of water resources between 1995 and 2025:

- Scenario I, which models current water scarcity and its future development under a changing climate until 2025, combined with a rather more optimistic assessment of society's response options due to the higher substitutability of natural capital.
- Scenario II, which models the same water scarcity until 2025, but with a more cautious assessment of the crisis due to a lower substitutability of water resources.

The results of the two scenarios are shown for 1995 in Figs. D 3.1-2a and 3a. A second step compares how each situation develops between 1995 and 2025 (Figs. D 3.1-2b and 3b). Special reference is made in this context to the number of people affected by very severe or severe scarcity (Table D 3.1-3).

Scenario I gives a regionally differentiated (catchment-based) picture of global water scarcity for 1995, with a high estimated capacity for curation (Fig. D 3.1-3a). In spite of this capacity and the relatively strong weighting attached to it, it is striking how many regions of the world suffer from a degree of water scarcity that cannot be (sufficiently) cured. These include parts of North, South and East Africa, poorer parts of the Arabian peninsula, extensive areas in the Near and Middle East, as well as in South and East Asia. Some regions like the southwest region of the USA, the northeast of France or large areas in Australia are designated as very water scarce – be it through heavy withdrawals from a large natural supply (France), or because their utilization of a low resource base is low in absolute terms, but large in relative terms (Australia). Yet they are able to compensate at regional level for this anthropogenic drain on their natural capital by mobilizing socioeconomic capital (e.g. in the form of long-distance transmission or groundwater pumping).

However, if one assumes, as in Scenario II for 1995, that the substitutability of the two types of capital is more likely to be *low* (Fig. D 3.1-2a), then this latter group of countries is also affected more adversely by the water crisis (e.g. USA and Australia). In this assessment, Saudi Arabia, Mexico and Libya are similarly unable to overcome the problems of scarcity using technical means. It must be stressed, however, that the results of both scenarios show the necessity of additional and/or specific aid. Regardless of whether the implications of high per capita GDP are assumed to be high or low, human survival and production in the respective regions will be adversely affected in ways that cannot be compensated. This result indicates the need for specially targeted counteractive measures and combinations of measures to combat the specific mechanisms behind the crisis. The type of analysis needed is therefore more than just regional in focus, but one that disaggregates the webs of cause and effect to a greater extent; examples of the latter are provided in the sections that follow (Sections D 3.2 to D 3.6).

Looking at the regional forecasts, shown as changes in criticality relative to 1995 (Figs. D 3.1-2b and 3b), one is struck in Scenario I by the worsening of

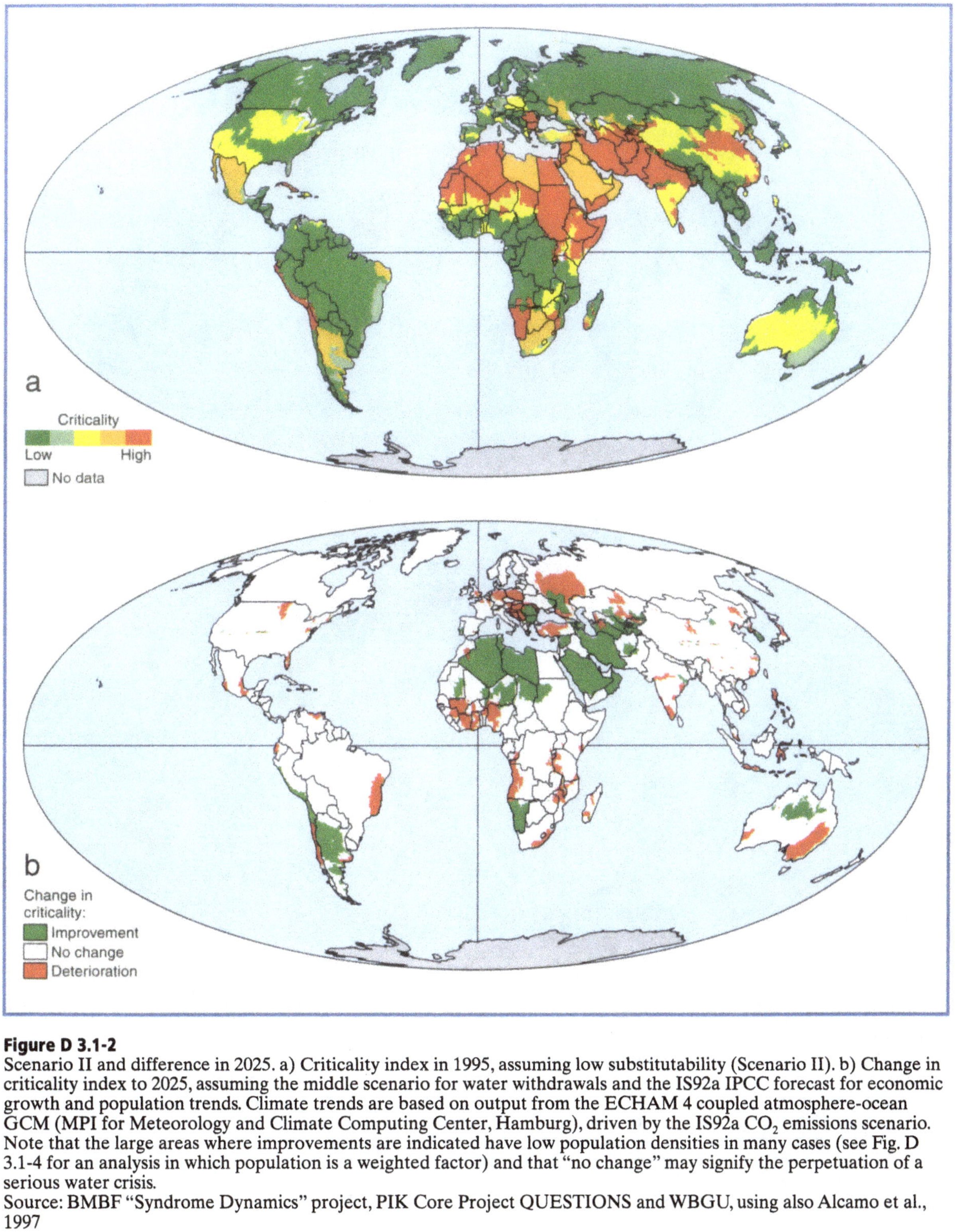

Figure D 3.1-2
Scenario II and difference in 2025. a) Criticality index in 1995, assuming low substitutability (Scenario II). b) Change in criticality index to 2025, assuming the middle scenario for water withdrawals and the IS92a IPCC forecast for economic growth and population trends. Climate trends are based on output from the ECHAM 4 coupled atmosphere-ocean GCM (MPI for Meteorology and Climate Computing Center, Hamburg), driven by the IS92a CO_2 emissions scenario. Note that the large areas where improvements are indicated have low population densities in many cases (see Fig. D 3.1-4 for an analysis in which population is a weighted factor) and that "no change" may signify the perpetuation of a serious water crisis.
Source: BMBF "Syndrome Dynamics" project, PIK Core Project QUESTIONS and WBGU, using also Alcamo et al., 1997

the water crisis in large parts of eastern Europe, while no change is computed for western Europe. Scenario II, in contrast, foresees deterioration of the water situation in certain parts of western Europe as well (e.g. the east German *Bundesländer* or the Greater London region). The scenarios differ also in their evaluation of trends in the densely populated countries of India and China. The advocates of high

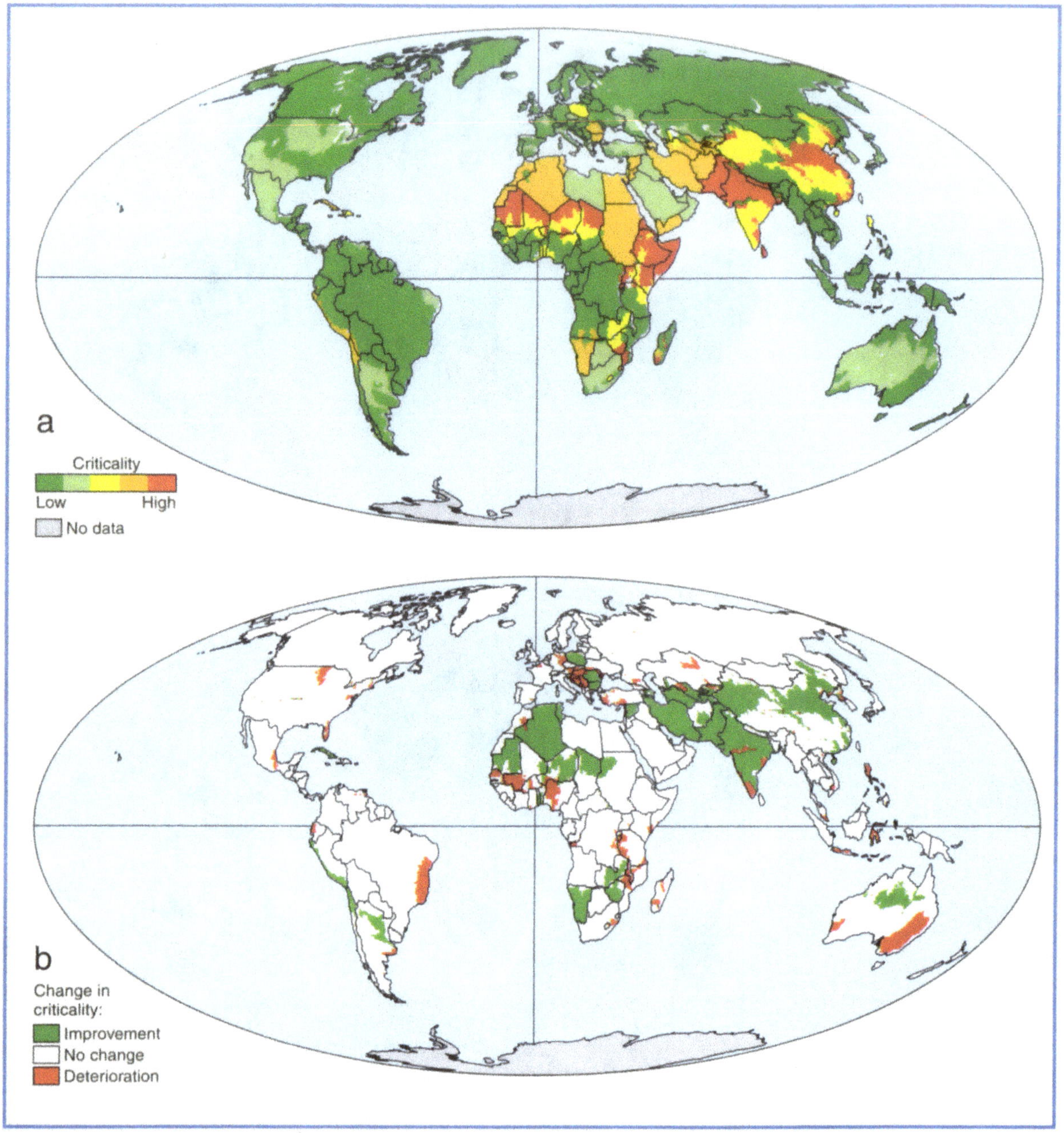

Figure D 3.1-3
Scenario I and difference in 2025. a) Criticality index in 1995, assuming low substitutability (Scenario II). b) Change in criticality index to 2025, assuming the middle scenario for water withdrawals and the IS92a IPCC forecast for economic growth and population trends. Climate trends are based on output from the ECHAM 4 coupled atmosphere-ocean GCM (MPI for Meteorology and Climate Computing Center, Hamburg), driven by the IS92a CO2 emissions scenario. Note that the large areas where improvements are indicated have low population densities in many cases (in parts of India, where the size of population is very large, KI falls from 5 to 4) and that "no change" may signify the perpetuation of a serious water crisis.
Source: BMBF "Syndrome Dynamics" project, PIK Core Project QUESTIONS and WBGU, using also Alcamo et al., 1997

substitutability (Fig. D 3.1-b) expect an improvement in the water situation in large parts of these countries – with the exception of some regions in the Indian states of Kerala, Tamil Nadu and Madhya Pradesh, where the situation is expected to worsen.

So far (Figs. D 3.1-2 and 3), regions have been identified that are or will be affected by the global water crisis. What is referred to in this non-specific manner is the entire natural and social resource base within a given geographical space. Yet a criticality assessment should also be capable of ascertaining the specific extent to which people are affected. Population density, for example, varies considerably around the world. Focusing purely on regions will not produce a sufficiently sensitive indicator for the criticality of water resources. After all, there is a big difference between two equal-sized territories classified as critical, if 10,000,000 people live in one and "only" 10,000 in the other. A regionally based index may suffice as far as ecosystem aspects are concerned. On the other hand, the potential for economic, social and political crises, as well as simple humanitarian considerations, make it necessary to design an index for the exposure of people to the global water crisis and its consequences.

The Council has chosen a relatively simple method to achieve this, namely by weighting the two most severe categories of water crisis (categories 4 and 5) with the population density today and as forecast for 2025. An increase in the number of people living in a critical region will exacerbate the crisis, whereas a decrease will reduce its scale. In other words, the global crisis can worsen not only through more regions having higher criticality values, but also by more people living in such regions. The situation in a region with a constantly high criticality level can thus be exacerbated through population growth alone. Conversely, the situation improves not only when regions move from critical to less critical on the scale, but also when the number of people affected by the water crisis declines – although this is a rare occurrence given the worldwide growth in population. In the two figures on the following pages, this interrelationship is illustrated cartographically (Fig. D 3.1-4a for Scenario II, Fig. D 3.1-4b for scenario I). In the regions colored yellow and red, the number of people affected (categories 4 and 5) is increasing, but decreasing in the areas marked green (usually because these regions experience a declining water crisis).

If the substitutability of water by human capital is assumed to be low (Scenario II), then the critical trends will mount in severity between 1995 and 2025 not only in extensive parts of Africa and Asia, but also in Mexiko and in northeast and east Brazil. The latter gives cause for serious concern, precisely in view of the rapid urbanization in these two countries. According to this scenario, some areas in Europe (Poland, Romania) will face increased water stress.

In terms of the number of people affected, the situation will not improve even if one assumes a relatively high level of substitutability (Scenario I). Although some countries will succeed, according to this scenario, in moving out of the critical categories (4 and 5) thanks to the growth in their average capital per capita, Poland and Romania, for example, will not be among them; a similar fate awaits Algeria, Iran, Namibia, Mexico and Brazil. In this scenario as well, total improvements will be exceeded by total deteriorations. The main reason is that the water crisis will become more severe – in terms of the number of people affected – in those Asian countries with the largest populations (India, China).

In view of these results, it should be noted that the scenario for economic and population growth, in which negative feedbacks due to resource scarcity are not included, may well be too optimistic.

If the total number of people affected by the global water crisis is taken as a criterion (Table D 3.1-3), the following picture emerges: for 1995, Scenario I calculates a figure of 1.9 billion people (34% of the world's population) affected by high or very high water stress, whereas the figure in Scenario II is 2.1 billion people (37%) for the same year. The relatively small difference is a clear indication that the global freshwater crisis is indeed a serious core problem

Table D 3.1-3
Number and percentage of people affected by the global water crisis.
Source: WBGU

Scale of the water crisis	Billions of people affected / Percentage of total population 1995	Billions of people affected / Percentage of total population 2025
	low substitutability (Scenario II)	
KI = 5 (very high)	1.8 / 32%	2.7 / 33%
KI = 4 (high)	0.3 / 5%	0.6 / 7%
Total	2.1 / 37%	3.3 / 40%
	high substitutability (Scenario I)	
KI = 5 (very high)	1.5 / 27%	0.3 / 4%
KI = 4 (high)	0.4 / 7%	2.4 / 29%
Total	1.9 / 34%	2.7 / 33%

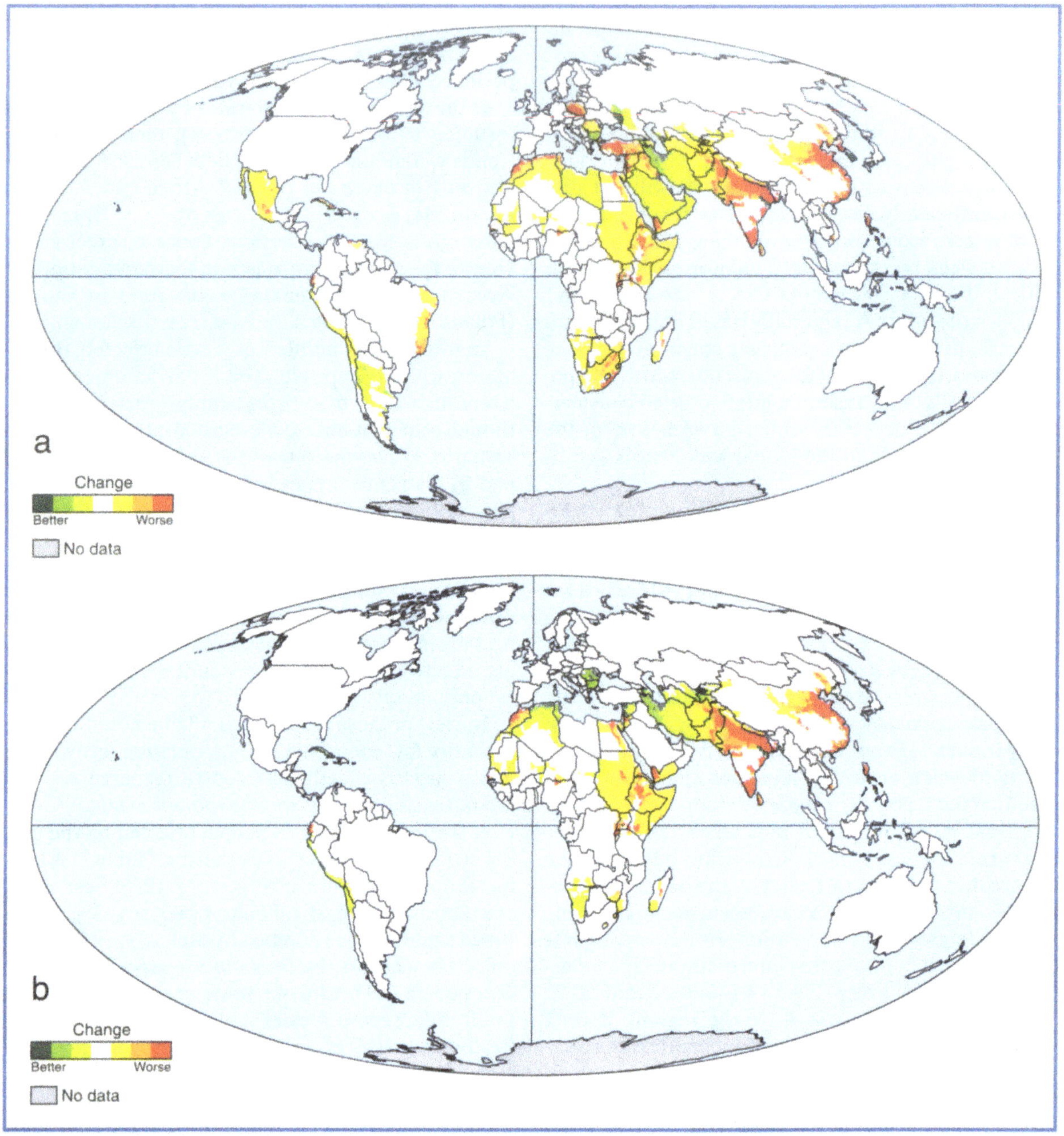

Figure D 3.1-4
Change in the number of people affected by severe or very severe water crisis (K = 4 or 5) between 1995 and 2025 (for total figures, see Table D 3.1-3). In the regions colored yellow and red, the number of people severely affected increases (by as many as +3,000 people per km² in the dark red areas). In the light- to dark-green regions, in constrast, the number of people severely affected decreases. a) Change in the number of people severely affected, assuming low substitutability (Scenario II). Net global effect by 2025: +1.2 billion people. b) Change in number of people severely affected, assuming high substitutability (Scenario I). Net global effect by 2025: +0.8 billion people.
Source: BMBF "Syndrome Dynamics" project, PIK Core Project QUESTIONS and WBGU, using also Alcamo et al., 1997

faced by humankind, irrespective of any differences in theoretical approach or methods of analysis.

For the year 2025, Scenario I indicates 2.7 billion affected people (33% of the world population), while the figure in Scenario II is 3.3 billion (40%). Focusing purely on the very high criticality category shows there are clear differences between the scenarios. However, since it transpired that this result is very sensitive to the way in which the evaluation matrix of Scenario I is chosen in detail for low incomes and high water stress, the result for the "high criticality" category was taken for the overall assessment in Table D 3.1-3 and in Fig. D 3.1-4. (This sensitivity is primarily due to the assessment of China's national income growth relative to its curation potential.) The absolute significance of water stress as a core problem will therefore mount in severity if current trends are projected, even assuming that problem-solving capacities will be high due to rising GDP, and even if global water availability increases as a result of climate change. The fact that water quality aspects were not explicitly included is something that needs re-emphasizing; were this factor to be integrated in the analysis, the outcome of the criticality assessment would probably be much worse. These findings clearly indicate that the global water crisis is a key dimension of global environmental change and that it will become increasingly evident in the future.

3.2 Syndromes as causal webs of relevance to the water crisis

Overview of syndromes – Water-related components – "Utilization" syndromes – "Development" syndromes – "Sink" syndromes

As a basic life-giving substance and key environmental medium, water is used and polluted by human societies in growing dimensions. Population growth, increasing consumer demands, increasing world trade, expansion of irrigation, and growing mountains of waste are some of the global trends contributing towards the global water crisis. The diverse ways in which water resources are utilized and put under strain, combined with the diversity of social stakeholders and systems that exploit the same resource for different purposes, make the global water crisis a multifaceted core problem of global change.

A geographical presentation of the typical manifestations of water crisis enables globally recognizable and functionally interrelated patterns to be found. The concept of syndromes of global change developed by the Council and the Potsdam Institute for Climate Impact Research (PIK) (see Box D 3.2-1; for more detailed descriptions see e.g. WBGU, 1995, 1996 and 1997; Schellnhuber et al., 1997) provides the analytical basis in this connection.

In Box D 3.2-1, the 16 syndromes of global change are briefly characterized with special reference to their specific water-relevant components. How do the active trends of a certain syndrome impact on the other symptoms? What are the impacts relating to the water problem? What feedbacks act on human beings and human societies?

As regards the consequences for water resources and their utilization for human ends, a distinction is made between primarily quantitative and primarily qualitative aspects of the water crisis (e.g. water scarcity and waterbody pollution) – in the awareness that both aspects are interrelated. A third aspect in this assessment concerns the impacts on human actors and social systems of those components of the respective syndromes that involve water problems. The key issue is this: how many people are affected and to what extent by the syndrome-specific water problems? From these three components – quality, quantity and people affected – we derive an overall evaluation of the global relevance of the 16 syndromes concerning water-related problems.

3.2.1 Relevance of individual syndromes for water resources

OVERCULTIVATION OF MARGINAL LAND: THE SAHEL SYNDROME

The Sahel Syndrome involves a complex web of factors causing environmental degradation when maximum sustainable crop yields are exceeded in regions where natural environmental conditions (climate, soil) restrict agricultural use (marginal locations) (WBGU, 1995). Key components of this pattern are soil degradation (erosion, loss of fertility, salinization), the spread of desert-like conditions (desertification), the conversion of semi-natural ecosystems (e.g. through deforestation), loss of biodiversity and regional climate changes. The Sahel Syndrome typically appears in subsistence economies where groups of rural poor and sections of the population threatened with marginalization are confronted with increasing degradation of their natural environment due to overexploitation of agricultural land (e.g. overgrazing, spread of farming to ecologically sensitive regions). The *syndrome-specific* problems of the population include mounting poverty, rural exodus, greater vulnerability to food crises as well as rising frequency of political and social conflicts over scarce resources. Besides the Sahel region itself, which manifests the arid form of the syndrome, other marginal locations in humid climate zones may be affected, such as tropical forests (marginality due to limited soil conditions) that are degraded by slash-and-burn agriculture (shifting cultivation).

As far as the water-related components of the syndrome are concerned, the central problems relate to:

- Overpumping of groundwater stocks from boreholes (arid variant).
- Changes in the regional water budget due to ecosystem conversion (arid and humid variants).

BOX D 3.2-1

Overview of Syndromes of Global Change

The syndrome concept breaks down the highly complex dynamics of the people-environment-relations within the Earth System into the basic underlying dynamic processes, the syndromes. The basic elements of the syndrome analysis are still the highly aggregated symptoms of global change listed by the Council in its previous Annual Reports (Figs. D 2-1 and D 2-2). This collection of symptoms represents an integrated, transdisciplinary summary of the key changes occurring worldwide and referred to collectively as global change. By analyzing the mutual interactions between these symptoms, it is possible to construct graphical and formal networks of interrelations which can then serve as an instrument for a qualitative environmental systems analysis capable of illustrating the highly complex and networked dynamics of these changes. However, the symptoms and interactions within the networks of interrelations can only be defined with relative precision in a regional analysis. This level of analysis, on the other hand, produces networks of interrelations in which specific interactions are shown, and globally typical patterns of people-environment interactions are identified. The basic evaluation of the adverse impacts of the syndromes listed below is not without its difficulties, however – the problem arises with syndromes that are not only negative in nature (e.g. Favela or Sahel Syndromes), but which came into being through an intentional and positively interpreted process: examples here include the Green Revolution, Asian Tigers or Aral Sea Syndromes. In this context, the Council prefers an evaluation philosophy based on "crash barriers" or "guard rails": as specified in the following for the Aral Sea Syndrome, maximum boundary conditions for tolerable pressures on the environment should be developed, specific to the country and/or situation at issue, with reference to the impacts of large-scale projects (e.g. a dam). For example, narrow limits would have to be drawn for irreversible damage to rare biotopes, whereas these limits could be wider for a small section of a widely distributed biotope. Any assessment of the benefits and adverse impacts of a dam project, as well as possible variants and alternatives (e.g. several smaller dams), must then be carried out with these limits in mind (for more detail, see Section D 3.4). In the following, the syndromes are briefly described in three groups: "utilization", "development" and "sink" syndromes.

"Utilization" Syndromes

1. Overcultivation of marginal land: Sahel Syndrome
2. Overexploitation of natural ecosystems: Overexploitation Syndrome
3. Environmental degradation through abandonment of traditional agricultural practices: Rural Exodus Syndrome
4. Non-sustainable agro-industrial use of soils and bodies of water: Dust Bowl Syndrome
5. Environmental degradation through depletion of non-renewable resources: Katanga Syndrome
6. Development and destruction of nature for recreational ends: Mass Tourism Syndrome
7. Environmental destruction through war and military action: Scorched Earth Syndrome

"Development" Syndromes

8. Environmental damage as a result of poorly managed or unsuccessful large-scale projects: Aral Sea Syndrome
9. Environmental and developmental problems caused by the transfer of locally inappropriate farming methods: Green Revolution Syndrome
10. Disregard for environmental standards in the course of rapid economic growth: Asian Tigers Syndrome
11. Environmental degradation through uncontrolled urban growth: Favela Syndrome
12. Destruction of landscapes through planned expansion of urban infrastructures: Urban Sprawl Syndrome
13. Singular anthropogenic environmental disasters with long-term impacts: Major Accident Syndrome

"Sink" Syndromes

14. Environmental degradation through large-scale diffusion of long-lived substances: Smokestack Syndrome
15. Environmental degradation through controlled and uncontrolled disposal of waste: Waste Dumping Syndrome
16. Local contamination of environmental assets at industrial locations: Contaminated Land Syndrome

- Changes in local (and possibly global) climate (e.g. precipitation patterns) (mainly humid variants).
- Exacerbation of water stress, with impacts on health (arid variants).

Overexploitation of natural ecosystems: the Overexploitation Syndrome

The Overexploitation Syndrome involves the conversion of natural ecosystems and the overexploitation of biological resources, and affects both terrestrial (forests, savannahs) and marine (overfishing) ecosystems. In both cases, ecosystems are overexploited without regard for their regenerative capacity, resulting in severe damage to the natural balance. Violation of the sustainability principle leads to degradation and even destruction of natural ecosystems, e.g. through outright clearance of forests, or overfishing. The immediate consequences are loss of habitat, the resultant loss of biological diversity, and erosion (particularly in mountainous regions). Emissions of CO_2 from biomass burning and soils contribute to the greenhouse effect. For the local population, the conversion of ecosystems means loss of livelihood, resulting, *inter alia*, in impoverishment and loss of cultural identity. One of the typical features of the Overexploitation Syndrome is the permitting of overexploitation for short-term gains.

Aspects of special relevance to water include:

- Lowering of water tables and perturbation of the water balance due to changes in surface runoff, leading in turn to increased variability of water levels in rivers.
- Enhanced deposition of sediments in rivers, inland waterbodies and coastal waters as a result of increased soil erosion (risk of floods and impairment of aquatic ecosystems).
- Regional climate change (including shifting precipitation patterns).

Environmental degradation through abandonment of traditional agricultural practices: the Rural Exodus Syndrome

The Rural Exodus Syndrome refers to environmental degradation caused by the abandonment of traditional land-use practices. In many cases, traditional farming methods can only be maintained with high inputs of manual labor. Labor-intensive methods for cultivating small plots of land, such as terraced slopes, small-scale irrigation systems, or protective measures against wind erosion, become increasingly unprofitable as socioeconomic conditions change. The reason is often the exodus of young males to urban centers in search of higher wages, better educational opportunities and a less "provincial" way of life (see Favela Syndrome, Asian Tigers Syndrome). Tilling the land is too labor-intensive for the women, children and elderly persons who are left behind. The lack of protective measures and proper care of the land causes erosion (often reinforced by excessive timber-felling on steep slopes, as in the Sahel Syndrome), mudrock flows and landslides.

The water-related component is not very distinct here:

- Abandonment of irrigation systems and increasing sediment loads in rivers.

Non-sustainable agroindustrial use of soils and bodies of water: the Dust Bowl Syndrome

The Dust Bowl Syndrome is a specific causal complex in which environmental destruction is caused by non-sustainable use of soils or bodies of water as biomass production factors, involving intensive deployment of energy, capital and technology.

Modern agriculture aims at achieving the highest possible labor productivity and profit by means of maximum yields per hectare, subordinating crucial long-term environmental aspects to this goal. High-yielding varieties, agrochemicals and mechanization form the basis of modern industrial biomass production. Farming enterprises in such agricultural systems are highly mechanized and automated (e.g. intensive livestock farming, modern irrigation systems, aquaculture, forest monocultures) and require only a small workforce. The conditional framework is determined by developed (international) markets as well as national and international agricultural policies (e.g. subsidies). Non-sustainable production methods can severely damage the environment (loss of ecosystem and species diversity, genetic erosion, CO_2 emissions, soil degradation).

The relatively large-scale presence of this syndrome has major quantitative and qualitative repercussions for water resources:

- Contamination of groundwater with pesticides and nutrients.
- Eutrophication of surface waters.
- Excessive extraction of fossil groundwater resources.
- Changes in runoff regimes and groundwater recharging.
- Acid rain.

Environmental degradation through depletion of non-renewable resources: the Katanga Syndrome

The Katanga Syndrome stands for the environmental damage caused by intensive mining of non-renewable resources above and below ground, with no consideration given to preservation of the natural environment. Although mining of non-renewable resources is usually effected over relatively short periods of time, it leaves behind permanent, sometimes irreversible environmental damage in many cases (toxic residues, changes in the morphology of landscapes). A typical feature of the syndrome is large-scale destruction of natural ecosystems and arable soils, whereby the latter problem is particularly prevalent in connection with open-cast mining activities in developing and newly industrializing countries.

The following mechanisms are relevant for water resources:

- Contamination of groundwater and surface water by (partly toxic) residual substances (crude oil, heavy metals).
- Regional water stress caused by water-intensive mining operations.
- Changes in runoff and greatly increased sediment loads (use of flotation, e.g. in Chilean copper mining industry).

Development and destruction of nature for recreational ends: the Mass Tourism Syndrome

The Mass Tourism Syndrome describes the environmental degradation caused by the unceasing growth of global tourism in recent decades. The driving forces behind the syndrome are rising incomes in the industrialized nations and falling costs of transportation, combined with a simultaneous reduction in working hours and fundamental changes in leisure behavior. Typical "hot spots" are coastal areas and mountainous regions (skiing, pony trekking, etc.). Mass tourism induces, for example, the destruction of semi-natural areas through the construction of tourist infrastructure (hotels, holiday homes, transport routes) and damage to or loss of sensitive mountain and coastal ecosystems (e.g. dune landscapes, salt marshes). The rapid growth of long-distance air travel in recent years causes pollution of the Earth's atmosphere.

The water-relevant symptoms are:

- Elevated demand for freshwater by tourists (pools, showers, etc.).
- Local groundwater depletion.
- Local but severe pollution by wastewater due to the concentration of tourist infrastructure.
- Changes in runoff from mountain slopes (including increased water erosion).
- Air traffic as a factor contributing to global climate change.

Environmental destruction through war and military action: the Scorched Earth Syndrome

Military conflicts and military infrastructure can have adverse impacts on the environment, both directly and indirectly as well as in times of peace (practice ranges, disposal areas, maneuvers). In the worst cases, the environment or its destruction is deliberately used as a military instrument (e.g. agent orange in the Vietnam War, the "flaring" of oilfields in the Second Gulf War). Contaminated military sites pose a long-term threat to humans and the natural environment.

Threats to freshwater resources emanate in particular from:

- Contamination of surface water and groundwater by toxic waste.
- Direct discharges of pollutants (oil, chemical weapons).
- Destruction of water-relevant infrastructure (dams, waterworks, power supplies, etc.).

Environmental damage as a result of poorly managed or unsuccessful large-scale projects: the Aral Sea Syndrome

The Aral Sea Syndrome refers to the problems induced by the bad management or failure of centrally planned, large-scale projects involving deliberate reshaping of the natural environment. As far as the hydrosphere is concerned, the main examples are irrigation and drainage schemes, the canalization and diversion of rivers, and dams. Such projects are essentially ambivalent: on the one hand, they provide the additional resources that are needed (water for food security, renewable energy), or they protect existing resources (flood control); on the other, they have severe impacts on the environment and society. The effects of these large-scale installations are rarely confined to the local or regional area, but can assume far-reaching and even international proportions. The syndrome is described in detail in Section D 3.4.

Key symptoms of significance for the hydrosphere are:

- Changes in the local water balance, especially changes in surface runoff as a result of water withdrawals from rivers, the canalization and diversion of rivers, and above all the construction of dams, in addition to what are sometimes severe losses through evaporation.
- Changes in loads of particulate and dissolved substances as a consequence, in particular, of changes in sedimentation dynamics.
- Deterioration of water quality.
- Greater exposure to health hazards in the form of water-based vectors (e.g. malaria, schistosomiasis).
- Raising and lowering of water tables as a result of interference with water resources.
- Risk of increasing international conflicts over water resources.

Environment and development problems caused by the introduction of locally inappropriate farming methods: the Green Revolution Syndrome

The Green Revolution Syndrome circumscribes the extensive, centrally planned and rapid modernization of agriculture with imported, non-adapted agricultural technology to ensure food security, whereby negative side-effects on geographical conditions of production and the social structure may occur, and indeed are tolerated (see Section D 3.3). The "evolution" of the Green Revolution Syndrome is characterized by a particular combination of geopolitical, biological, population and economic trends (the interplay of national interests, the "seed revolution" in agriculture, population growth and impoverishment respectively). The green revolution was always forced through "from above" within the framework of large-scale plans, on a global scale through the transfer of technology and know-how "from the rich to the poor" and on a national scale by elites ("opinion leader" strategy). The successes of the green revolution are mainly achieved in irrigated agriculture, leading to huge increases in irrigated areas and in the demand for water. In many cases, water-related problems are generated within the space of a few years. They include, in particular:

- Lowering of water tables due to rising demand for water on the part of agriculture.
- Salinization of soils due to the absence of specialized drainage systems.
- Contamination of groundwater and surface waters by pesticides and nutrients.
- Exposure of the rural population to health hazards in the form of toxic contaminants in water.
- Changes in the local water balance due to water withdrawals from rivers, the canalization and diversion of rivers, the construction of irrigation channels, and losses through evaporation.
- Increasing conflicts over the utilization of water resources.

Disregard for environmental standards in the course of rapid economic growth: the Asian Tigers Syndrome

Many regions in the newly industrializing countries are experiencing extremely rapid economic growth. The intensity and momentum of this structural transformation may have grave implications for human beings and the natural environment. Rapid, export-oriented industrialization is accompanied by a more or less deliberate disregard for environmental standards (policy failures). Rising industrial production and traffic volumes subject the affected regions to severe air and soil pollution.

Important water-related aspects of the Asian Tigers Syndrome are:

- Pollution of surface waters (by toxic substances in some cases) due to inadequate treatment of industrial effluents and domestic sewage (one of the consequences being major health risks).
- Scarcity of freshwater resources due to rapidly increasing consumption by the industrial and domestic sectors (growing competition over the use of resources with agriculture and with the needs of ecosystems).
- Acid rain.

Environmental degradation through uncontrolled urban growth: the Favela Syndrome

The Favela Syndrome refers to a process of unplanned, informal and thus environmentally harmful urbanization driven by rural-urban migration, induced by the "pull" effects of cities. Its features include various manifestations of poverty, such as the formation of slums and illegal shanty towns. These are accompanied by overloading, infrastructural and environmental problems, as well as by segregation of the population. As a result of increasing traffic and industrial emissions, and in the absence of adequate regulatory requirements and monitoring systems, air and noise pollution reach extremely high levels. Other aspects include the spread of surface sealing, uncontrolled accumulation of waste, and thus acute threats to the health of the population. Many Favela inhabitants settle on marginal sites (slopes, flood areas) and are therefore exposed to special risks. The urbanization trend manifested worldwide (megacities, but also the growth of smaller cities) means that this syndrome is affecting a large and constantly growing number of people. The syndrome is described in detail in this Annual Report (see Section D 3.5).

The main water-related aspects are:

- Contamination of surface water and groundwater by untreated municipal wastewater.
- Health hazards, sometimes of potentially epidemic proportions (outbreaks of cholera, typhoid and other diseases), due to inadequate water supplies and poor water quality.
- Changes in the water balance as a result of surface sealing and evaporation.
- Overexploitation of surface water and groundwater by long-distance transmission and boreholes.
- Water losses from defective or insufficiently maintained water distribution systems and as a result of illegal withdrawals.

Destruction of landscapes through planned expansion of urban infrastructures: the Urban Sprawl Syndrome

The Urban Sprawl Syndrome refers to urban expansion with far-reaching environmental impacts. The formation of urban agglomerations leads to entirely new spatial structures and a need for corresponding adaptation. The increasing spatial separation of living, shopping and working – bolstered by private cars, owner-occupied homes and the demand for higher standards in modern societies – lead to overdevelopment of large tracts of land and higher levels of traffic. As a consequence, anthropogenic stresses on soils and air (smog, ground level ozone, fragmentation of ecosystems, etc.) are worsening. In addition to the relatively large number of people affected by this syndrome (and who induce it in the first place), its global relevance derives primarily from the sheer intensity with which natural resources are consumed.

The water-related components include:

- High levels of per capita water consumption in modern agglomerations (demand for high standards).
- Pollution of water resources by what are sometimes toxic substances.
- Depletion of water stocks (including fossil reserves), e.g by groundwater pumping and long-distance distribution.
- Water losses in older water supply networks.
- High vulnerability and flood risk exposure due to settlement in flood-prone areas.

Singular anthropogenic environmental disasters with long-term impacts: the Major Accident Syndrome

The central feature of the Major Accident Syndrome is the mounting threat to the environment in the form of singular localized disasters caused by humans. Although the probability of these events is very low, they have grave and in many cases transboundary impacts. Major accidents seem to be gaining in significance within the global change context. The rise in global transport-sector activities and increasing local demand for energy and raw materials enhance the risk of tanker accidents or, more generally, the incidence of environmental disasters in connection with the transport of hazardous goods. Other major threats besides the latter are industrial accidents. They include the large number of old nuclear power plants, chemical and other industrial plants in newly industrializing countries, economies in transition and developing nations that do not deploy best available technology. Pollution of air and soils with high concentrations of what are mostly highly toxic substances, combined with long-lasting impacts on ecosystems and human health are the main features of this syndrome.

Symptoms of key importance for the hydrosphere are:

- Pollution by mostly toxic substances and/or radioactive contamination (often resulting in scarcities for short or long periods).
- Health hazards due to contaminated water.
- Dam breaks and the failure of water resource development projects.

Environmental degradation through large-scale diffusion of long-lived substances: the Smokestack Syndrome

The Smokestack Syndrome describes the remote effects of substance emissions following disposal in the environmental media water and air. The problem arises from the strategy of disposing of undesired substances simply and easily by distributing them as finely as possible in the environment, or by diluting them greatly in environmental media (water, air). However, high smokestacks do not eliminate air pollutants – they merely transplant the problem to other regions remote from industrial activity. A similar mechanism is at work in the "disposal" of hazardous industrial waste and other pollutants via the "wastewater pathway", especially through the emission of greenhouse gases. Depending on the emission patterns and the physical/chemical behavior of the substances in the various environmental media, the emitted waste may be distributed on a local, regional or even global scale. Long-distance transport occurs primarily via the atmosphere or through running waters. A distinction must be made, as far as environmental effects are concerned, between direct impacts on biological communities following distribution of the pollutant in the environment, and bioaccumulation of the substance in the system. Given that this pattern of people-nature interaction is found worldwide and in different environmental media simultaneously, its global relevance is comparatively high.

The following are the main problems as they afflict freshwater resources:

- Contamination of surface water and groundwater through the discharge of waste products and hazardous substances.
- Threats to aquatic organisms and ecosystems arising from direct discharges of pollutants into water resources and through the absorption of air pollutants by water.
- Health hazards from toxins and their accumulation in the food web.
- Eutrophication of surface waters.
- Acid rain.

- Global climate change and its impacts on the hydrological cycle.

Environmental degradation through controlled and uncontrolled disposal of waste: the Waste Dumping Syndrome

The Waste Dumping Syndrome refers to the disposal of residual and waste matter, and the consequences that ensue. In contrast to the Smokestack Syndrome, where the underlying intention is to avoid pollution by "diluting" it in the air or water, the characterizing feature of this syndrome is the "concentration" of waste through compaction and accumulation. Waste materials are collected in concentrated form at small-scale facilities and isolated from the environment to highly different degrees. In the last analysis, however, even with sophisticated landfill facilities, there can be no knowing how long uncombined pollutants can be kept separate from the surroundings. Major uncertainties exist with regard to the quality of seals and the decomposition processes at work. Depending on local environmental standards, contamination of soils and air may ensue; waste dumping also ties up financial and human resources for long periods of time, because remediation and restoration work must be carried out on a periodic basis. This syndrome is therefore of considerable global importance – at least as far as unclosed cycles are concerned – given the need to dispose of an ever-greater volume of waste worldwide. Waste exports from industrialized countries to developing nations and to countries undergoing the transition to the market economy are just one example of this fact.

The water-related aspects of the syndrome are:

- Contamination of groundwater by substances of low or zero degradability.
- Increasing scarcity of drinking water resources.
- Health risks (especially in developing countries).

Local contamination of environmental assets at industrial locations: the Contaminated Land Syndrome

The Contaminated Land Syndrome characterizes sites and regions with accumulated depositions of pollutants in soils or underground that are a hazard to human health and the environment. Contaminated sites are found at locations and in regions that used to feature intensive industrial, commercial or military activities, although they can also be found at abandoned and disused sites for storing solid municipal and industrial waste, or hazardous production residues. Severe pollution is inflicted on soils, in particular. The health risks to which people living in the vicinity are exposed can be alarming. Remediating and restoring such sites also involves enormous costs.

In this syndrome, water resources are affected in one key respect:

- Pollution of groundwater by toxic substances.

3.2.2 Systematic ranking of the syndromes

Now that the water-relevant aspects of the various syndromes have been described in qualitative terms, the next step is to evaluate them systematically with regard to their significance to the water crisis. Ranking is carried out using two main criteria: the syndrome-specific combination of water quality impairment and reductions in water quantity, and the severity of direct water-related impacts on the people affected. The latter aspect takes into account the prevalence of the syndrome, thus enabling an assessment of each syndrome's contribution to the current global water crisis.

Syndrome ranking is carried out on the basis of the following three aspects (Fig. D 3.2-1):

- Global significance of freshwater quality degradation.
- Reduction in the quantity of available freshwater, expressed as the global sum of regional scarcities.
- Number of people directly affected worldwide by the water-related aspects of the syndrome, weighted according to the severity of impact.

To enable an overall assessment of the relevance of a syndrome in respect of its water-related aspects, each syndrome was evaluated in terms of these aspects and plotted in a three-dimensional chart (see color coding in Fig. D 3.2-1: declining global relevance for water resources as color changes from red to green). The "people affected" aspect was weighted by a factor of two in order to keep this anthropospheric domain equal in value to the more ecospherical aspects of water quantity and water quality.

Some of the syndromes already subjected to extensive analysis (see Sections D 3.3 to D 3.5; WBGU, 1997; Schellnhuber et al., 1997) could be ranked using quantitative material from existing studies; other syndromes were ranked by evaluating a number of expert analyses.

From those syndromes in which the use of water resources by agriculture is a key element (the Dust Bowl, Green Revolution and Sahel Syndromes), it was found that the Dust Bowl Syndrome and the Green Revolution Syndrome are roughly equal as major causes of water scarcity – from the global perspective – whereas the Sahel Syndrome plays a lesser role in this regard. This assessment was derived from previously available material on syndrome intensity fields at global level, as well as national assessments of water scarcity attributable to irrigated agriculture

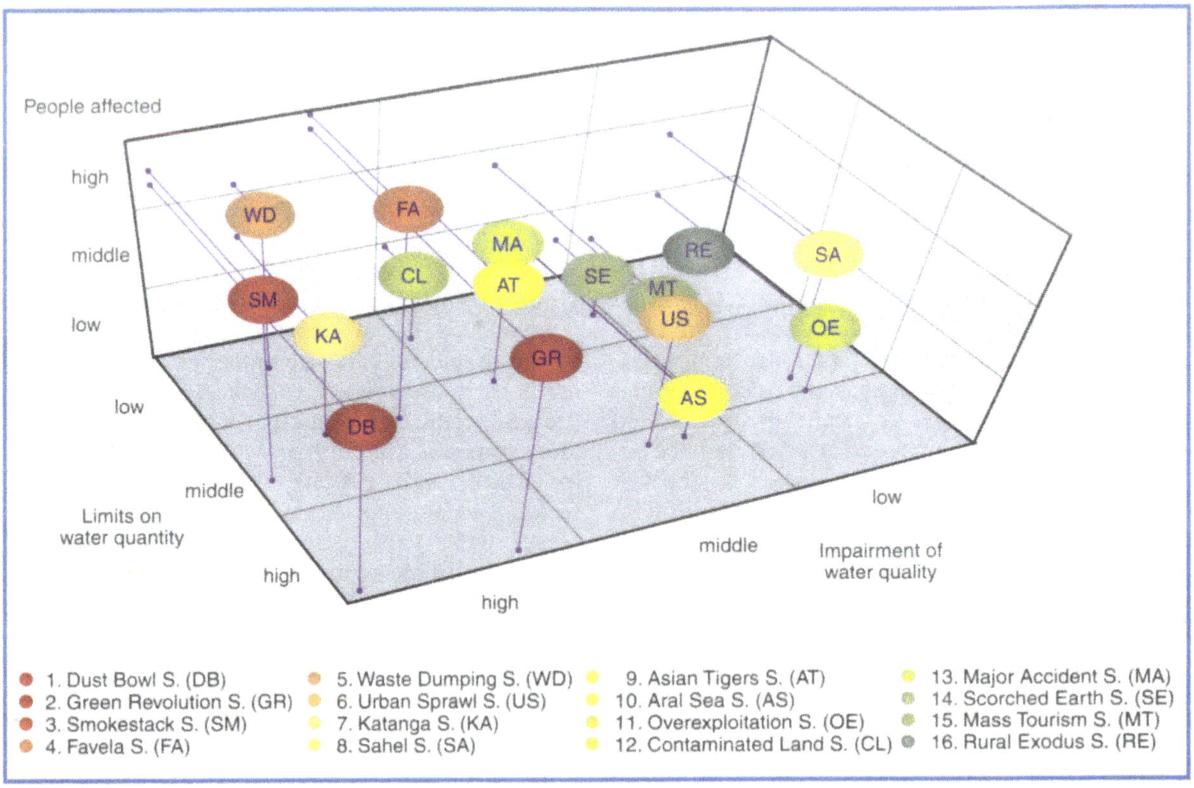

Figure D 3.2-1
Significance of the individual syndromes in terms of their contribution to the water crisis. The color codes from red to green designate declining relevance for global water resources, which is formed from a linear combination of the three axis categories. The "People affected" aspect was given double weighting in order to keep this anthropospheric aspect equal in value to the more ecospherical aspects of water quantity and water quality.
Source: WBGU

(proportion of renewable water supply withdrawn by agriculture). The agriculturally relevant syndromes just mentioned play a greater role as far as quantity-related aspects are concerned, firstly because the water withdrawals involved are so enormous, and secondly because they are mainly for irrigation purposes in arid areas and are lost to a major extent through evaporation (consumptive, one-off use). The high rate of evaporation underscores the eminent importance of the Aral Sea Syndrome on the quantity axis. Water withdrawals are less relevant in the other syndromes, and reductions in water quantity (primarily in terms of water availability) are due either to excessive groundwater pumping or a slow-down in groundwater recharging. Both aspects play a key role in the Urban Sprawl Syndrome, which makes a major contribution to the problem of global water scarcity on account of increasing surface sealing, primarily in industrialized and newly industrializing countries.

As far as the quality aspect is concerned, the agricultural syndromes are ranked as they are due to the total amount of pesticides and fertilizers used in the world, combined with the size of the regions affected, whereas the Smokestack and Waste Dumping Syndromes in the "sink" group involve point sources – albeit very frequently occurring ones globally speaking – that draw their relevance for the reduction of water quality from the high pollutant concentrations discharged.

The weight of the "people affected" criterion for the syndrome relevance is exemplified in the ranking of the Favela Syndrome, which does not play a leading role at global scale in respect of either quality impairment or quantity reductions. Although the syndrome plays only a middle-ranking role as far as the number of people affected by the syndrome is concerned, it acquires a high level of significance on account of the degree of impairment in the form of severe damage to health.

It is not possible in this Report to describe all the syndromes in detail, so three water-relevant syndromes from the "development" group are chosen as examples. The Green Revolution Syndrome is analyzed as the pattern of degradation involving agriculture (high relevance level for global water resources on all ranking scales and high topicality level in connection with the debate over the "second green revolution"). The problems associated with the mismanagement of urban development patterns are discussed with specific reference to the Favela Syndrome (highest ranking on the "people affected" scale). The Aral Sea Syndrome is examined, not least on account of its close ties with the Green Revolution Syndrome.

3.3
The Green Revolution Syndrome: Environmental degradation through the introduction of inappropriate farming methods

Increase in food cereal production through technology transfer – Mismatched technologies – History and stages of the syndrome – Impacts on humanity and nature – Syndrome indicator – Global occurrence – Syndrome coupling – Water requirements of new varieties – Salinization – Conflicts over use of irrigation water – Water-specific recommendations – The new green revolution in the light of syndrome analysis – Learning from the mistakes of the old green revolution – More variety in agriculture – Linkage with rural development – The example of India – Food security and international law – Participation in the planning process – Recommendations

3.3.1
Definition

The Green Revolution Syndrome circumscribes the extensive, centrally planned and rapid modernization of agriculture with imported, non-adapted agricultural technology, whereby adverse side-effects on ecospherical conditions of production and the social structure occur, and indeed are put up with.

3.3.1.1
Description

The term "green revolution" came into use at the end of the 1960s and refers to a set of major agricultural improvements that took hold in developing countries from 1965 onwards (Bohle, 1981). The green revolution was based on biological, technical and chemical innovations in the agricultural sector, most notably the successes achieved in the development of new, high-yielding food cereal varieties. Within the long history of agricultural intensification, the rapid changes induced by these improvements led from *green evolution* to the *green revolution*. The results were major boosts in food cereal production. From a technological perspective it looked as if the problem of food insecurity could finally be brought under control. From this point of view the syndrome has the character of a well-meant project with problematic consequences. One has to consider that in the face of the pressures existing, people could see no other options.

The greatest successes of the green revolution are achieved in irrigation farming, resulting in major expansion of irrigated areas and rising water demand. Within a few years, the emergence of water-relevant problems such as "freshwater scarcity", "lowering of water tables", "salinization" and "water pollution" is a common occurrence. There are several dimensions to the green revolution, as described below.

GEOPOLITICAL DIMENSION

The green revolution can be understood as the outcome of geopolitical interests. Colonial rule, for example, is a factor predisposing some countries. On the one hand, the infrastructures created under colonial rule enabled planned implementation of the green revolution. On the other, it frequently took place on soils that had already been subjected to intensive use by colonial governments. The uneven distribution of land tenure, often a feature of the precolonial period, became more pronounced during the colonial period. The green revolution reinforced this tendency still further.

In the postcolonial period, many developing countries were confronted with the problem of growing food demand. The deteriorating food situation and mass poverty were threatening to destabilize many countries. This challenge facing food and development policymakers coincided with strategic interests in the context of the Cold War. In addition, many countries had economic and political problems after the departure of the colonial powers. The green revolution was expected to foster peaceful nation building. The issue at international level – at a time of growing political tensions between East and West – was to keep countries away from the opponent's sphere of influence (with food imports) by making them dependent on their own side (by means of technology transfers).

TECHNICAL DIMENSION

The green revolution involved a major increase, triggered by technology transfers from the industrialized nations, in the yields per hectare that developing countries achieved in the production of staple cereals. This growth in production was made possible with a combination of high-yielding seed varieties, fertilizers, pesticides, irrigation and mechanization – "complementary inputs", so called. These changes were largely based on the seminal research on new cereal varieties carried out in the 1960s by the Centro Internacional de Mejoramiento de Maiz y Trigo (CIMMYT) in Mexico and the International Rice Research Institute (IRRI) in the Philippines (the "seed revolution"). Instrumental in achieving these successes, which related primarily to wheat, maize and rice, were the Rockefeller, Ford and Kellogg Foundations as well as the World Bank and the FAO, which put up the financial means (Barrow, 1995). The priority was to maximize soil productivity as rapidly as possible by increasing capital investments (e.g. energy, mineral fertilizers).

Development policy dimension

The green revolution should also be seen in the context of development policy goals. The intensification of agriculture also aimed at creating purchasing power, improving living conditions in rural areas and thus at helping to eliminate malnutrition. The focus of economic policy was on import substitution. This explains why the Green Revolution Syndrome is classified as a development syndrome. Technical change was seen as an alternative to political change (Bohle, 1981). Through the green revolution, countries with predominantly agricultural economies were supposed to become independent of cereal imports, although this meant they had to accept new import dependencies instead. Mineral fertilizers, machines and oil now had to be imported in return for scarce foreign exchange. The ultimate aim was to boost industrialization by modernizing agriculture.

Institutional dimension

The birth of the green revolution was accompanied by a worldwide network of specialized, newly established institutions. Included in this process were the Consultative Group on International Agricultural Research (CGIAR) and its sub-organizations, the establishment of national agrarian banks and the International Fund for Agricultural Development (IFAD), the setting up of distribution networks for agricultural inputs, extension services (know-how transfer) as well as the promotion of agrochemical and agrotechnological industries. In addition, new technologies and irrigation were extensively subsidized in order to ensure rapid dissemination. Rural elites were picked out for special support so that new techniques would trickle down to all groups in society.

3.3.1.2 Major features

The "evolution" of the Green Revolution Syndrome is characterized by a particular historical concurrence of geopolitical, biological, population and economic trends (the interplay of national interests, the "seed revolution" in agriculture, population growth and impoverishment, respectively). The green revolution was always introduced through large-scale planning from top to bottom, on a global scale from rich to poor (technology and know-how transfer), and on a national scale through national elites (opinion leader strategy). Other specific features of the syndrome are its rapid pace, its regionally non-adapted pattern, and the acceptance of the risks these implied in attaining increased soil productivity. From today's perspective, the green revolution was a very short-termist approach.

3.3.2 General description of the syndrome

3.3.2.1 Syndrome mechanism

The central trend of the syndrome is "increasing food production" (Figs. D 3.3-1 to 3). In the face of high "population growth", the threat of a production-based food crisis as well as widespread impoverishment, a national economic strategy is initiated in the countries concerned with the aim of "increasing food production" and rural development through the "intensification of agriculture", and in this way to counteract the trend towards "poverty" (hunger). The trends labeled "know-how and technology transfer", "industrialization" and "globalization of markets" act to further reinforce the "intensification of agriculture" trend.

Stages of the Green Revolution Syndrome

Stage I (ca. 1965 to the mid-1970s):
A successful start

The first stage of the green revolution was mainly characterized by positive trends; although adverse environmental impacts or impairment of social development were not apparent, their foundations were already laid. During this initial stage, only a few trends had impacts (Fig. D 3.3-1). The intensification of agriculture according to the green revolution pattern was highly successful in many countries at first, in terms of combating hunger and reducing the de-

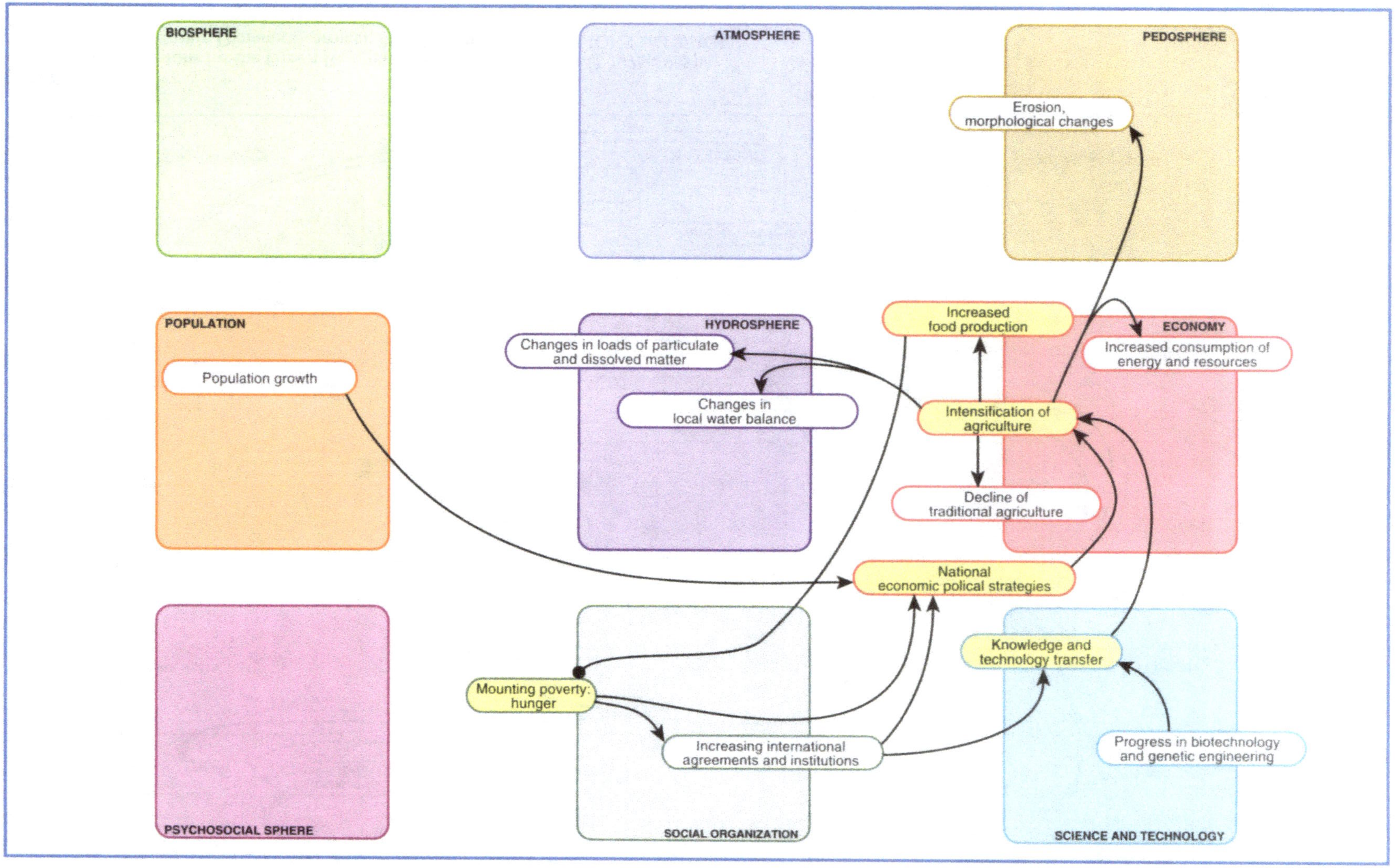

Figure D 3.3-1
Network of interrelations for the Green Revolution Syndrome, Stage I (ca. 1965–1975).
Source: BMBF "Syndrome Dynamics" project, PIK Core Project QUESTIONS and WBGU

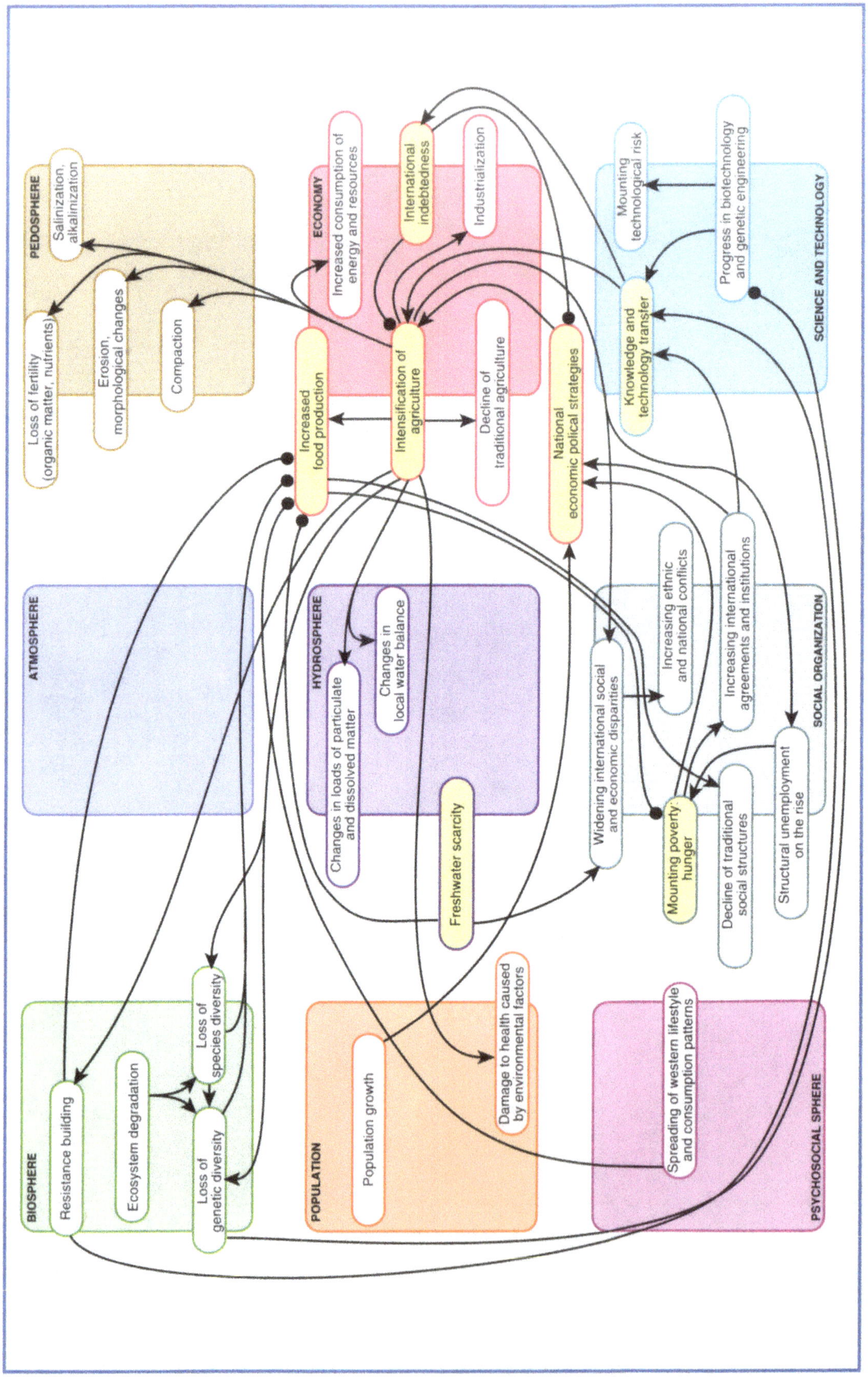

Figure D 3.3-2
Network of interrelations for the Green Revolution Syndrome, Stage II (ca. 1975–1985).
Source: BMBF "Syndrome Dynamics" project, PIK Core Project QUESTIONS and WBGU

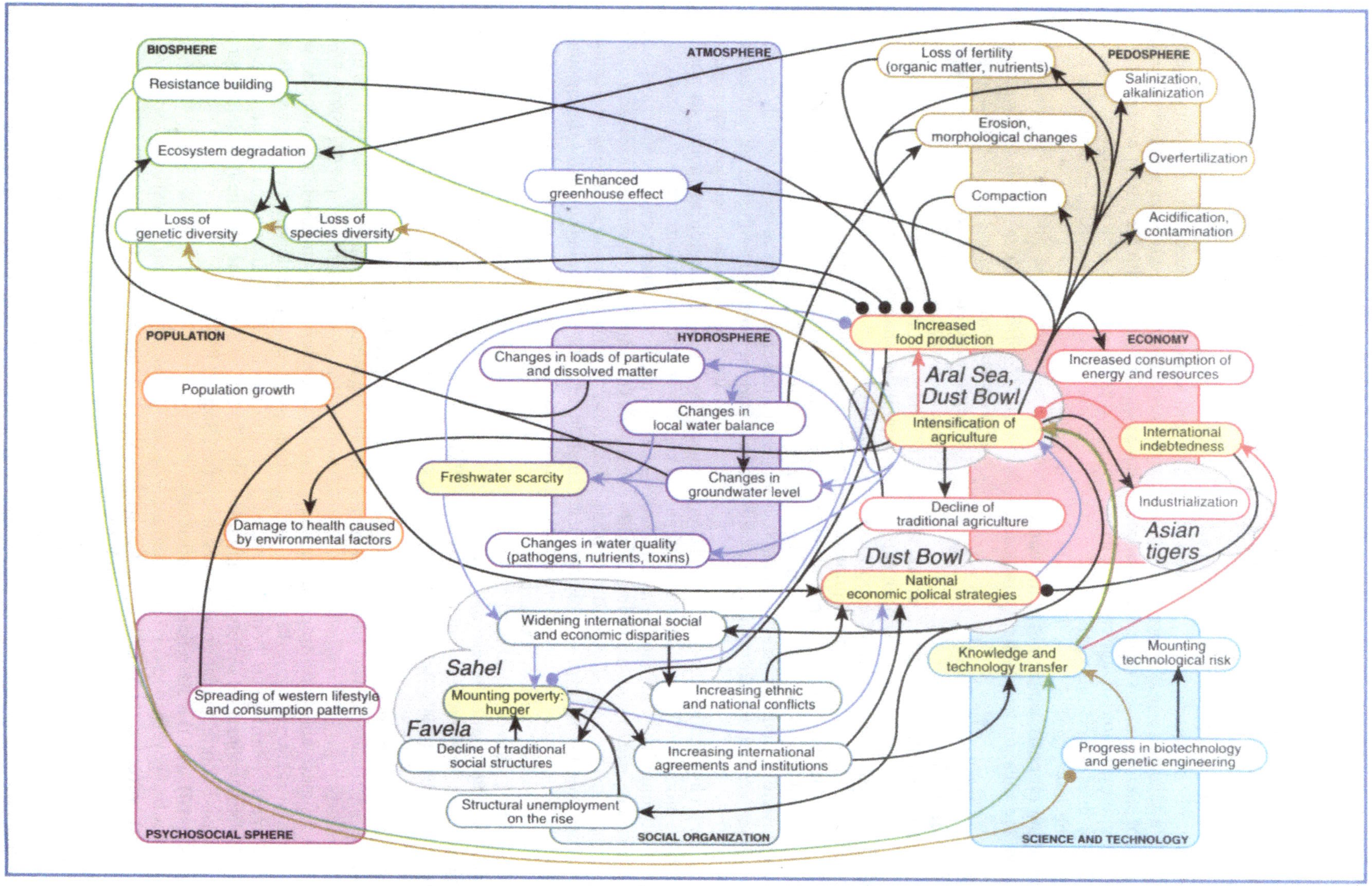

Figure D 3.3-3
Network of interrelations for the Green Revolution Syndrome, Stage III (ca. 1985–today). The colored arrows signify the syndrome-specific sub-networks.
Source: BMBF "Syndrome Dynamics" project, PIK Core Project QUESTIONS and WBGU

pendence on food imports. In some regions, new employment opportunities were created by the introduction of double cropping, although increasing mechanization gradually eliminated this initial benefit. The transfer of new biotechnology products was wholly welcomed under the acute pressure of imminent hunger crises. Little significance was attached to the "decline of traditional agriculture", a process that was welcomed instead as a sign of progress. The high "consumption of energy and raw materials" associated with the "intensification of agriculture" along green revolution lines played only a minor role in the 1960s due to the low prices at that time; with the emergence of the oil crisis, which marks the end of this first stage, the situation was to change radically.

Stage II (mid-1970s to mid-1980s): Negative impacts come to the fore

In many green revolution countries, cereals were grown on fields whose soil had already been exploited through the intensive cultivation of export crops under colonial rule. The combination of unfavorable soil and climate conditions, colonial exploitation as well as the structural and social framework in developing countries were mounting obstacles to the success of the green revolution in the years that followed.

Although the high-yielding varieties that had been introduced were producing high yields under optimal conditions, these conditions could not be maintained in tropical, humid or arid to semi-arid climates without considerable expense and effort, and to a large extent were only possible in favorable regions (e.g. Punjab and Haryana, India). Unfavorable edaphic conditions (sandy soils, laterites) exacerbated the lack of success. The great vulnerability of genetically uniform high-yielding varieties to pests or drought, as well as the lowering of water tables due to overexploitation by intensive irrigation, the emergence of salinized soils and groundwater resources in arid and semi-arid regions, and the risk of soil erosion were all factors that grew increasingly severe during this period (Fig. D 3.3-2).

The advancing mechanization of agriculture put more and more workers out of work, without adequate employment alternatives being provided. Structural problems within rural society (inequitable distribution of land tenure, feudal power structures, shortage of capital) and the state (e.g. insufficient capital resources, bad governance, corruption, failure of power elites) prevented the negative impacts on the ecosphere and the socioeconomic consequences from being equalized or at least cushioned. To secure food for a growing population by means of improved production techniques, despite the unsustainable developments that were emerging, intensification of agriculture was pushed forward along green revolution lines (e.g. higher pesticide doses, or a combination of different pesticides, deeper wells and more fertilizers). This self-reinforcing process was now a characteristic feature of the Green Revolution Syndrome, and the successes of the initial years receded into the background (Fig. D 3.3-2). Furthermore, the "know-how and technology transfer" that accompanied the green revolution proved to be a new form of dependence that, together with the rise in energy and raw material prices generated by the oil crisis, culminated in the international debt crisis that marks the end of this stage.

Stage III (mid-1980s to present): Structural adjustment modifies the green revolution, which is now compounded by far-reaching negative impacts with long lead times

Contrary to original expectations, the situation is not simplified thanks to the elimination of problems or problematic linkages; instead, interrelationships have become increasingly extensive and complex (Fig. D 3.3-3). Around the mid-1980s, the green revolution entered the age of globalization. The beginning of this stage is marked by the worldwide "debt crisis" and the structural adjustments that curtailed the very subsidies that had once been introduced to spread the green revolution as rapidly as possible. The network of institutions created as part of the green revolution, including extension services, agrarian banks and distribution systems for seed and fertilizer, also underwent a process of structural transformation. In many developing countries, the agroindustries and their distribution networks were privatized, or liquidized due to low profitability. Recent years have seen an emergent trend towards stagnation of yields per hectare, especially in Indonesia, Mexico, Pakistan and Tunisia; in some cases, yields have declined (Bangladesh and Sri Lanka). Even IRRI test fields have yielded diminishing returns, in spite of optimal inputs and improved varieties. Studies are being carried out to determine whether changes in soils may be a cause of this decline (WRI, 1994). Unsustainable developments and the interactions between them give rise to a whole series of other syndromes, such as the Sahel Syndrome, the Favela Syndrome, the Dust Bowl Syndrome and the Aral Sea Syndrome (see Section D 3.3.2.3).

Social, economic and cultural impacts

Without the green revolution, an "increase in food production" of this magnitude would not have been possible. The threat of production-related food crises was reduced but not entirely eliminated, because private entitlement rights play a key role in addition to national food supply. For example, after a good harvest India appears as a net exporter of grain, although about 220 million people in the country are chronically undernourished.

The "intensification of agriculture" along green revolution lines leads to "widening social and economic disparities" (FAO, 1996a) in many countries because of structural problems and conceptual weaknesses (top-down approach), a process described as growth without development (Bohle, 1989). The green revolution has shown its greatest success in irrigated regions, so that is where state efforts are concentrated – exacerbating regional disparities, which in turn may produce "increasing ethnic and national conflicts".

Socioeconomic disparities are intensified by the green revolution, because most small farmers do not participate to the extent expected and because alternative employment opportunities are not available for those people who are thrown out of work. In addition, agricultural producers become increasingly dependent on external inputs. For example, the yield of high-yielding hybrid varieties declines with each daughter generation, so seeds have to be purchased again and again. This dependence becomes problematic when the coping capacity of small farmers is low, and an unfavorable and relatively rapid change in the economic or infrastructural framework takes place due to exogenous factors (price fluctuations, supply bottlenecks, etc.). In states prone to crisis, the supply of these inputs may fluctuate considerably (Pilardeaux, 1995).

In many cases, the green revolution is incompatible with the social and cultural structures in the regions to which it has spread. One example concerns the problems introducing formal credit systems where there is no land register documenting people's land holdings, the basis for determining creditworthiness. The need for additional knowledge is another important factor in this context. The spread of the green revolution is also bound up with a "decline in traditional social structures".

As food production rises, dependence on cereal imports declines; this is linked within the green revolution to a growing dependence on imported agricultural technology, agrochemicals and seeds, at least temporarily ("know-how and technology transfer"). This trend, alongside many others, reinforces "international indebtedness". To combat the debt crisis, structural adjustment measures have been implemented in many countries since the late 1980s. The ensuing deregulation of national economic policy strategies profoundly changes the network of institutions and services created by the government in the course of the green revolution.

IMPACTS ON THE ENVIRONMENT

A whole range of negative trends in the ecosphere are reinforced by the "intensification of agriculture" according to the green revolution pattern (Fig. D 3.3-3). In the pedosphere, they include "overfertilization", "acidification", "loss of fertility" and "erosion and morphological changes to the soil". The latter refers, in particular, to the compaction of soils by agricultural machinery. Nutrient losses in agriculture, acidification and salinization are particularly prevalent in northern India, Bangladesh, Indonesia, northern China and North Korea (WRI, 1994). The inadequate protection afforded to soil by monocropping promotes soil erosion, which in turn impairs the water quality of surface waters ("changes in loads of particulate and dissolved substances") (see Section D 1.5). Intensive and frequently incorrect use of agrochemicals is a threat to the environment (soil flora and fauna, waterbodies, neighboring ecosystems, etc.) and to human health (see Section D 4.2).

Around 20% of the pesticides applied worldwide each year are consumed by developing countries. No comprehensive data are available on the extent to which they are used in food cereal production. Because of the large areas taken up by cereal crops, however, the quantity applied must be substantial. Although use of pesticides is relatively low in scale compared to the industrialized countries, it is growing rapidly in some developing countries, in contrast to most industrialized countries, where use is restricted or more effective, and less persistent products are applied. In India, for example, use of pesticides is rising at an annual rate of around 12%. Some developing countries already use pesticides more intensively than is the case in Germany. The negative impacts of such intensive use are particularly severe in the developing countries, where most pesticide-related deaths occur (House of Commons Agriculture Committee 1987, according to Pimentel, 1996). The main reasons are lack of information, inappropriate handling, storing and application of pesticides, inadequate labeling as well as a lack of protective clothing and cleaning facilities for rural workers. Non-degradable insecticides with toxic effects on both humans and the environment, such as Lindan and DDT, are frequently used in developing countries. In many industrialized countries, by contrast, use of such extremely harmful plant protection agents is now prohibited. In India, Lindan and DDT account for about 70% of all pesticides used. Many governments promote pesticide consumption by providing subsidies. Intensive pesticide use is often due to misconceptions on the part of farmers that this practice is extremely progressive and modern. This notion has been nurtured by years of "consulting" by represen-

BOX D 3.3-1

The green revolution in India: Water problems

India has been plagued by famines for thousands of years. The last great famine was in Bengal in 1943 and claimed 3.5 million lives within a year. Overall, the food situation on the Indian subcontinent worsened considerably during the first half of the 20th century. Cereal production was increasingly unable to match population growth, so per capita cereal production declined continuously. To make matters worse, the partition of the subcontinent after India's independence meant that Pakistan took over 30% of the particularly high-yielding irrigation areas but only 18% of the population.

After the failed attempts in the 1950s and early 1960s to raise yields by intensive irrigation alone, stagnation in production levels for many years and two years of drought (1965-1967), the Indian government made fundamental changes in its agricultural development strategy. Under pressure from the USA, which threatened to discontinue grain supplies, and at the insistence of its own large farmers, it switched over to the green revolution (Bohle, 1981). 1966/67 saw the introduction of the first high-yielding rice varieties, which can produce up to 5,000 kg per hectare in contrast to an average yield of traditional rice varieties of around 860 kg per hectare. Between 1960 and 1978, food cereal production rose by a total of 60%. Much of this growth was accounted for by substantial increases in wheat production, which rose by 157% between 1960 and 1975. Rice production increased by 43% over the same period. The successes of the green revolution in India were confined to wheat and rice crops.

These successes would have been inconceivable without changes in irrigation farming. Already established during the colonial period (construction of channels for watering cotton crops), irrigation underwent a development unparalleled in Indian history. Between 10% and 20% of the national budget was expended on irrigation development in the late 1960s and early 1970s. The area under irrigation rose subsequently by over 50%, from 20.9 million hectares in 1950 to 32 million hectares in 1976. Nevertheless, irrigated farming accounts for only about a fourth of India's total agricultural area. This means that the successes of the green revolution were geographically so concentrated that economic disparities between regions grew. India, Pakistan and China account for 45% of the world's irrigated areas.

Social disparities widened also, because the large number of small farmers did not get involved to the extent expected, but went into long-term debt instead, while the medium-sized and large landowners achieved disproportionate increases in profit. Those farmers who also produced for the market were guaranteed fixed and relatively high selling prices, so improved incomes were mainly restricted to this group.

The consumption of mineral fertilizer, most of which had to be imported for scarce foreign exchange, rose tenfold between 1960 and 1977. Imports of oil and mineral fertilizer accounted for almost 30% of India's total imports in 1975/76, consuming around 40% of total export earnings due to the demand for foreign exchange.

Application of mineral fertilizers necessitates increased water input because fertilizers must be dissolved before they can be absorbed by plants. As irrigation grew in importance, groundwater pumping was used to an increasing extent. Groundwater extraction facilities are particularly efficient when they supplement the water supply from existing irrigation systems, as in the alluvial plains of northwest India (channel irrigation) and the delta areas of southeast India. The use of groundwater pumps made it possible not only to achieve higher yields per unit area, but also to transform single- to double-cropping regions. Double cropping was additionally favored by the shorter ripening times of high-yielding varieties.

Water pumps were relatively expensive, costing ten times the annual income of a rural worker, and therefore could be afforded by wealthy farmers only. Extraction with deep tubewell boreholes increased to such an extent that the water table in the state of Tamil Nadu dropped by 25–30 m within a decade (FAO, 1996b). In the state of Punjab, India's grain belt, groundwater levels are falling by 20 cm annually on two thirds of the area, and in Gujarat declining groundwater levels were observed in 90% of the wells in the 1980s (Postel, 1996). Access to groundwater is becoming increasingly difficult for farmers with traditional wells, because these are unable to reach deeper water-bearing layers. Between 1970 and 1982, the proportion of total irrigated land in India watered from boreholes rose from 14.3% to 26.2% (Bohle, 1989). In the west Indian state of Gujarat, groundwater overdrafting by irrigators in the coastal region has led to the intrusion of saltwater

►

into the aquifer, contaminating village drinking water reserves.

Improving the efficiency of water use (e.g. rehabilitation of existing facilities) can increase yields, the potential of which is far from exhausted. In India alone, the irrigated area affected by the aforementioned problems is estimated at 10–13 million hectares (Postel, 1993).

tatives of the agrochemical industry and by government advisory services for farmers (WRI, 1994).

Extensive, frequently uncontrolled and in some cases preventive use of pesticides promotes "resistance building" among pests. Widescale monocropping and years of growing the same crop create ideal conditions for the growth of pest populations. Moreover, only a small percentage of the pesticides applied actually reach the target organism – less than 0.1% in the case of many insecticides. The harmful effects on natural predators and parasites, combined with the favorable living conditions provided by monocropping, allow small populations of pests to grow rapidly until they emerge as new plagues (secondary plagues).

The rapid growth of insecticide use in Indonesian rice farming between 1980 and 1985 reduced the populations of the natural enemies of the brown planthopper (a cicada species), resulting in a sharp rise in the planthopper population. Yield losses were so severe that Indonesia had to import rice for the first time in many years. The response of farmers to such secondary plagues – applying pesticides more frequently and in higher doses – accelerates the "resistance building" trend and increases the cost burden for more or new pesticides as crop yields fall. More than 500 insects and mite species are now resistant to one or more insecticides, and over 273 species of weeds have developed the capacity to neutralize one or more herbicides. Around 150 plant pathogens, such as fungi and bacteria, are now resistant to fungicides. Intensive nitrogen fertilization worsens infestation by lice (aphids), because the high level of nitrogen provided by the host plant promotes their reproduction. In Asia, the susceptibility of rice plants to certain diseases has been found to increase as a result of intensive nitrogen fertilization (WRI, 1994).

Monocropping and intensive application of pesticides lead to a loss of "genetic diversity" and "species diversity". This loss intensifies the vulnerability of the agricultural system to pests, diseases and climate fluctuations (FAO, 1996a). "Loss of genetic diversity and species diversity" has an attenuating effect on the trends "progress in biotechnology and genetic engineering" and "increasing food production", especially with regard to sustainability. One of the problems here is that traditional, highly diversified and locally adapted crop varieties have been eliminated by the widespread use of high-yielding varieties (Harlan, 1975; Heywood and Watson, 1995). Many green revolution countries (India, China and Mexico, for example) are important genetic pools for crop plants, but this richness in genetic resources is acutely threatened by the predominance of a few high-yielding varieties of limited genetic diversity. India, for example, is a major source of genetic diversity in respect of field crops, such as rice, sugar cane, mango, cucumber, avocado, medicinal herbs and spice plants. Losing the genetic diversity of cultivated crop varieties means losing an irreplaceable genetic resource for research. The research community is faced with a constantly shrinking genetic resource base, which thus inhibits the new and further development of modern agricultural varieties ("progress in biotechnology and genetic engineering"). The ever-changing situation as far as plant pests are concerned (due, among other things, to "resistance building") necessitates the continuous improvement of plant varieties by crossing in new characteristics from cultivated and wild varieties. The green revolution reduces the genetic resource base on which future "intensification of agriculture" relies.

Moreover, the confinement to a few cereal varieties has led to reduced cultivation of protein-rich legumes. In India, for example, 20% of the growth in food cereal production was achieved at the expense of legume crops. Between 1967/68 and 1979/80, cultivation of legumes in India declined at an annual rate of 1.1% (Pierre, 1987). The supply of protein for the population deteriorated accordingly. Confining production to a few cereal varieties means failing to meet subsistence demand for fuelwood, livestock fodder, etc. (see also Box D 3.3-1).

Another consequence is an *"enhanced greenhouse effect"* due to increasing methane emissions in rice farming and energy-intensive production equipment.

The impacts on the hydrosphere, which include "freshwater scarcity", "salinization", "eutrophication" and contamination with pesticides, are dealt with in Section D 3.3.3.2.

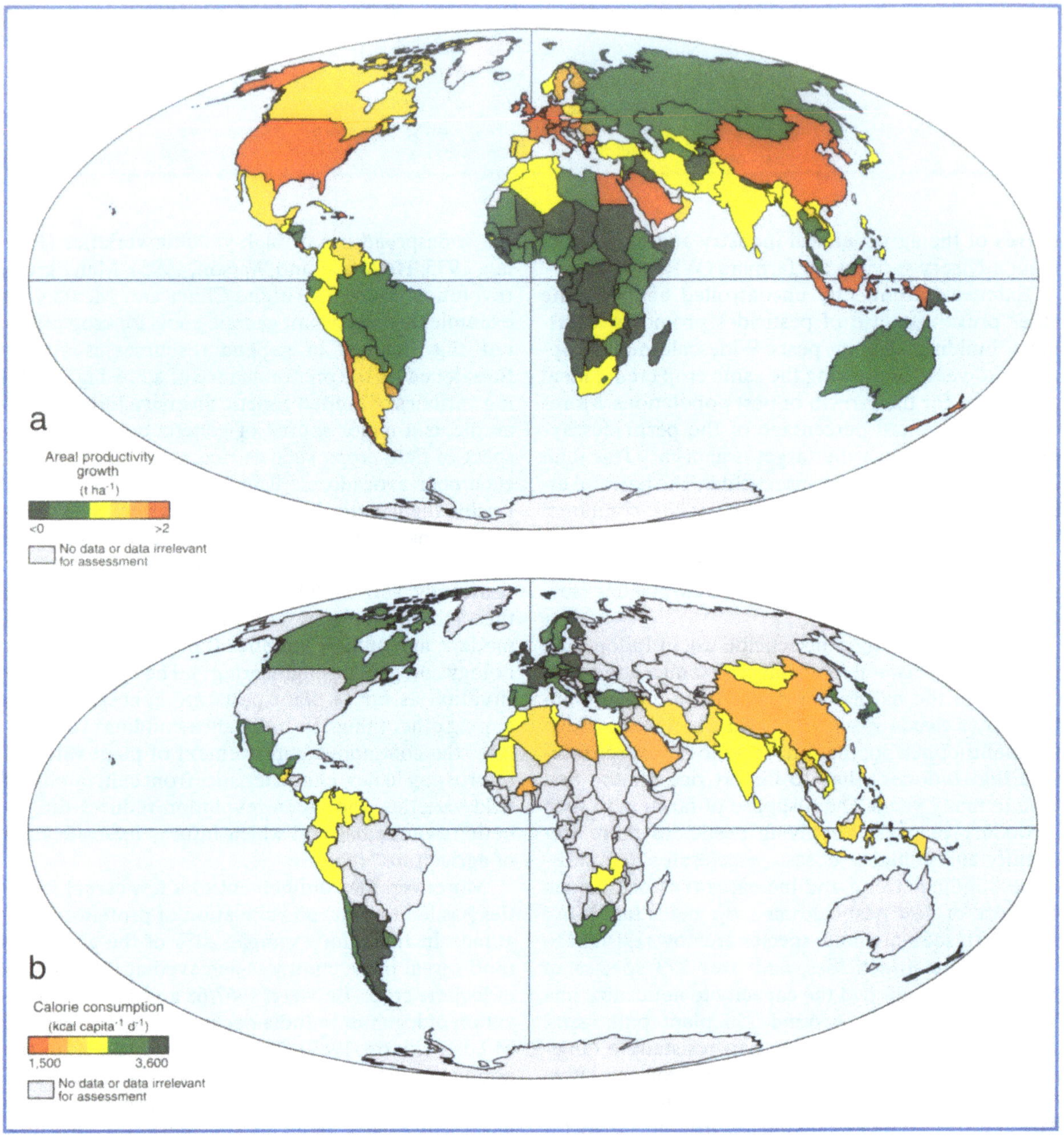

Figure D 3.3-4 a–b
Specific indicators of the green revolution. a) Absolute areal productivity growth of cereal crops, 1960–1990. b) Average food supply deficit in 1961, in calories per person.
Source: BMBF "Syndrome Dynamics" project, PIK Core Project QUESTIONS and WBGU

3.3.2.2
Syndrome intensity indicators

In gaining a global overview of how widespread the Green Revolution Syndrome is, it is essential to have an indicator for the incidence and intensity of the syndrome. The first step in developing such an indicator is to identify those countries in which a process has occurred according to the green revolution pattern and then, in a second step, to examine how pronounced the syndromatic and therefore problematic nature is in these countries.

EMERGENCE OF THE GREEN REVOLUTION

On the basis of the mechanism described in Section D 3.3.2.1, the emergence of the green revolution between 1960 and 1990 is examined with the help of

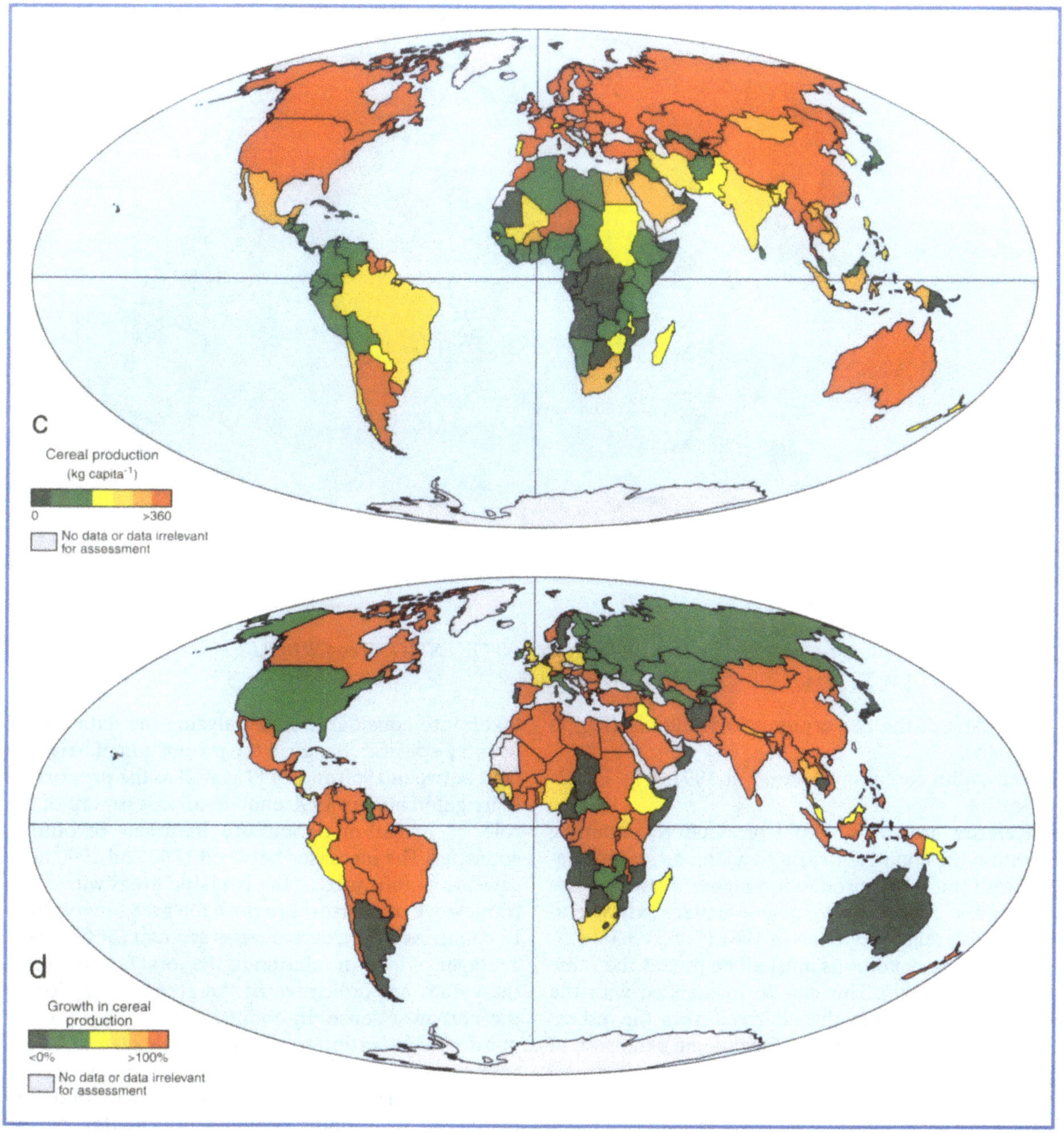

Figure D 3.3-4 c–d
Specific indicators of the green revolution. c) Per capita cereal production in 1991. d) Relative growth in cereal yield remaining in the country, measured as the difference between cereal production increase and the rise in cereal exports relative to cereal production in 1961.
Source: BMBF "Syndrome Dynamics" project, PIK Core Project QUESTIONS and WBGU

economic indicators to determine whether the requirements for the following argumentation are met:

If (1) substantial growth is observed in the areal productivity of cereals *and* (2) a food shortage exists at the beginning of the period under review *and* (3) the cereal-growing sector of the country can make a significant contribution to the food supply *and* (4) there is a growth in total cereal production, which remains in the country, *then* a green revolution is taking place in the country concerned.

The following basic indicators are used to measure the above conditions at national level (FAO, 1993, WRI, 1992):

1. Absolute rise in areal productivity in cereal production from 1960 to 1990 (Fig. D 3.3-4a).
2. Average food supply deficit in 1961, measured on

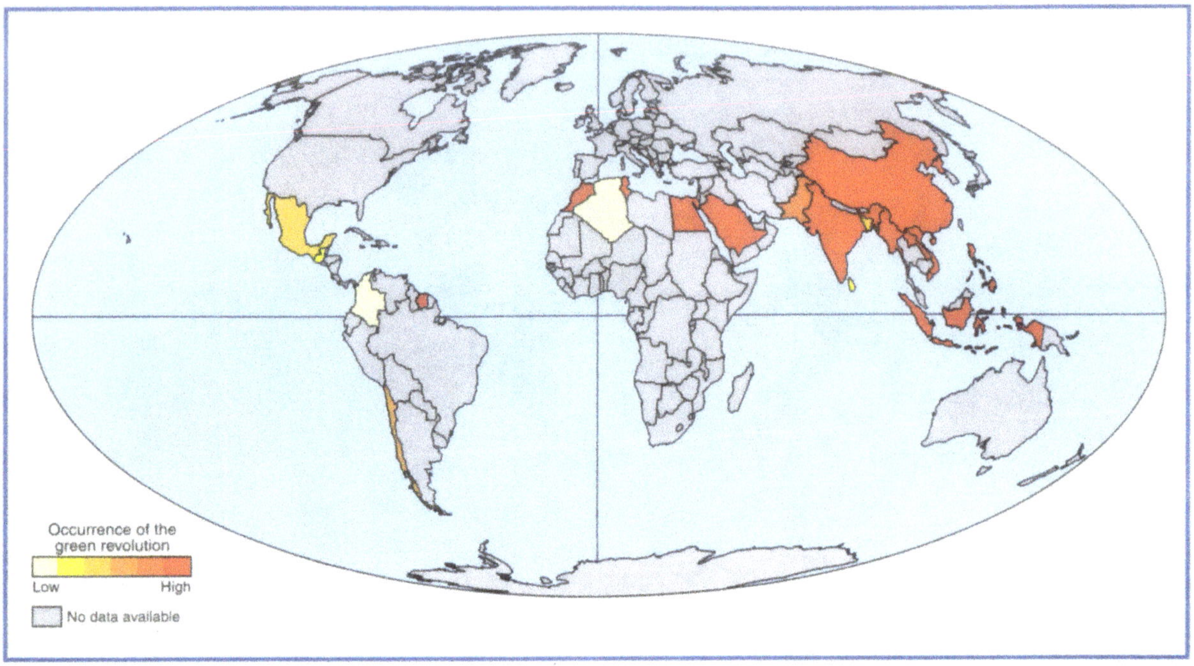

Figure D 3.3-5
Occurrence of the green revolution.
Source: BMBF "Syndrome Dynamics" project, PIK Core Project QUESTIONS and WBGU

the basis of the per capita calorie supply (Fig. D 3.3-4b).

3. Per capita cereal production in 1991 (Fig. D 3.3-4c).
4. Relative growth in cereal production remaining within the country, measured as the difference between the cereal production increase and the rise in cereal exports in the period under review relative to cereal production in 1961 (Fig. D 3.3-4d).

These four conditions must all be met at the same time (AND links). This can be formalized with the help of certain truth values derived from the indicators introduced. To do this, a fuzzy-logic extension of the classical AND link (formation of the minimum) is used. The result is an indicator that states whether a green revolution has taken place in a country (0 = did not take place with certainty, 1 = took place with certainty). These basic data sets are shown separately in Figs. D 3.3-4a to d and the result in Fig. D 3.3-5.

Syndrome effects in green revolution countries

The countries identified in this way are now examined and assessed with regard to syndrome features. The first group of countries (high intensity values) has an intensity indicator for the "occurrence" of the green revolution of between 0.6 and 1. In the group of countries evidencing a "limited green revolution" (low intensity values), the intensity indicator is 0.3–0.6. Countries with values below 0.3 were not taken into consideration. Analyzing the data on the country-specific change in the proportion of irrigated land between 1960 and 1990 as well as the proportion of irrigated land in 1990 enables an assessment of the role of irrigation agriculture in green revolution countries. The increases between 1960 and 1990 indicate the significance of the irrigated areas within the framework of measures to push the green revolution. In countries where cereal crops account for high percentages of land in relation to the total farming area, the values are indicative of the green revolution in the narrower sense. In countries like India, the expansion of irrigation to cereal crops is seen to be excessive.

The country-specific percentage of degraded land under cereal cropping provides information on the state and rate of soil degradation. The data on the state of soil degradation also include degradation processes that took place *prior* to the green revolution (e.g. due to the colonial export crop economy in India), while the continuation of soil damage, measured on the basis of the soil degradation rate on cereal cropland, can clearly be attributed to the green revolution (Oldeman et al., 1990).

To assess the global relevance of the Green Revolution Syndrome to the ecosphere, the total area of degraded land attributable to the syndrome in each country and the land affected by irrigation-related degradation are taken into account. The change in the number of people living under the poverty line in

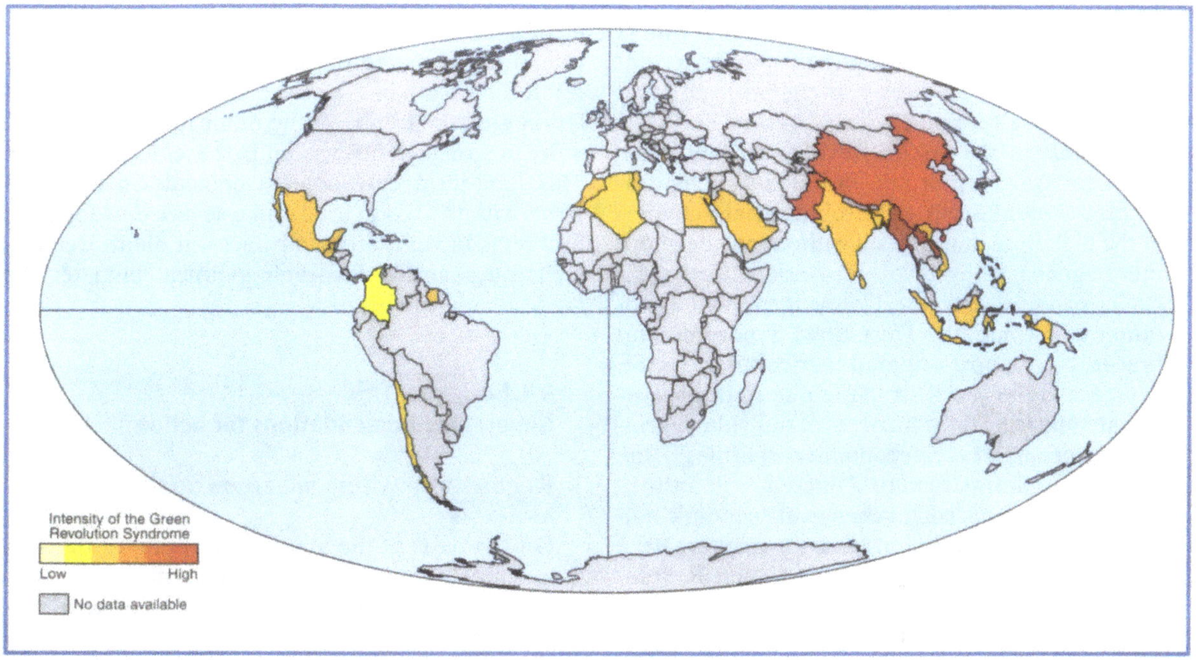

Figure D 3.3-6
Occurrence of the Green Revolution Syndrome.
Source: BMBF "Syndrome Dynamics" project, PIK Core Project QUESTIONS and WBGU

rural areas in relation to the absolute number of rural poor (1985–1992), and the change in the proportion of export earnings used for debt servicing (1970–1990) provide indications of the social dimension.

The analysis shows that various forms of soil degradation, such as water and wind erosion, play a major role in the Green Revolution Syndrome. Furthermore, forms of degradation that are typical for irrigated land, such as salinization and waterlogging, are evidently widespread and advancing, particularly in more arid regions. This causes mounting, long-term, and in some cases irreversible damage to production areas. Salinization and waterlogging of soils in China are of immense global relevance, though the sum total of water-related soil degradation in other countries is also of global significance.

In the following, an indicator will be defined for the occurrence of the Green Revolution Syndrome on the basis of the material considered thus far. In contrast to the green revolution itself, the impacts typical of the syndrome will now be taken into account. The country-specific soil degradation rate will be applied to cereal-growing areas as a measure of ecospherical degradation within the Green Revolution Syndrome. The growth of rural poverty and the ratio between debt servicing and export income for each country is the basis on which the socioeconomic impacts of the green revolution are assessed. Both dimensions are then evaluated with equal weighting and "added". The partial indicator thus obtained is limited in the next step by comparing it with the intensity of the green revolution. If the latter does not play a major role in the respective country, any socioeconomic and ecospherical degradations that emerge can only be partially explained by the green revolution. The resulting indicator for the occurrence of the Green Revolution Syndrome is shown in Fig. D 3.3-6.

3.3.2.3
Syndrome linkages and interactions

Linkages to other syndromes are shown as gray clouds in the network of interrelationships for the third stage (Fig. D 3.3-3). It is not until these interactions are taken into consideration that the full import of the problems becomes clear:

- Due to the intensive irrigation needs of the green revolution, greater importance is attached to large-scale dam projects with all their negative implications for society and the ecosphere. To this extent, there is a direct interconnection between the Aral Sea Syndrome and the Green Revolution Syndrome (see Section D 3.4).
- The Favela Syndrome is given added force by the Green Revolution Syndrome, because migration to urban agglomerations due to "mounting poverty" increases (see Section D 3.5). A "drift" toward

the Favela Syndrome occurs when small farmers are demoted from landowning producers to tenant farmers and landless workers with no local employment alternatives.

- In contrast to the Green Revolution Syndrome, the Dust Bowl Syndrome includes the environmental consequences of agroindustrial activities at the highest technical level without any development agenda being involved. National self-sufficiency plays a minor role. Unlike the Green Revolution Syndrome, the Dust Bowl Syndrome embraces the export-oriented agricultural sector. Changes in the social structure due to disproportionate impacts on the incomes of individual farmers ("widening of socioeconomic disparities"), the creation of infrastructure ("increase in institutions") and the paradigm change in economic policy ("international indebtedness") may create a Dust Bowl Syndrome out of the Green Revolution Syndrome.
- The displacement of farmers to marginal locations ("widening social and economic disparities") may generate a Sahel Syndrome (WBGU, 1997) from the Green Revolution Syndrome.
- The green revolution causes a rise in labor productivity and increased value added in the agricultural sector. In addition, agroindustry can develop in the areas of supplies and services, capital goods and marketing. These factors induce greater capital accumulation in the industrial sector and a concomitant loss of employment. In this way, the Green Revolution Syndrome may favor the emergence of the Asian Tigers Syndrome.

Figure D 3.3-7 illustrates the syndrome linkages over time. The graph shows the relationship between the Green Revolution Syndrome and other syndromes. As can be seen on the time axis, the individual syndromes began at very different points in time. In some cases, trends that played a major role in the emergence of the syndrome had been established during the colonial period or even earlier. The Favela Syndrome, for example, is coupled with the Green Revolution Syndrome through the trends "widening of socioeconomic disparities" and "population growth". The Favela Syndrome is a post-war phenomenon, whereas the latter two trends have been significant since the days of colonialism. The Green Revolution Syndrome began in the mid-1960s, whereas the three syndrome-specific core trends were established in the pre-colonial and colonial periods. The "intensification of agriculture" trend already existed in the "hydraulic societies", e.g. during the Mogul period in India and the British colonial period, when irrigation systems were further developed. Social polarization processes ("widening of socioeconomic disparities") were already established – in India, for example – during the period of feudal rule, then cemented during the colonial period and reinforced again in the post-colonial era. The "population growth" trend, on the other hand, did not acquire major significance until the colonial period. This long-term development coincide in the mid-1960s with the rise in food crises, severe development problems in rural areas, advances in plant-breeding techniques and a favorable political climate: the green revolution begins.

3.3.2.4 General recommendations for action

RECOMMENDATIONS BASED ON THE SYNDROME ANALYSIS

On the basis of the above description of the syndrome mechanism, its primary interactions and linkages with other syndromes, a number of general recommendations can be derived.

- The intensification of agriculture under the green revolution has caused a loss of genetic diversity. The pool of genetic resources shrinks as a direct consequence. This means that an increasingly narrow base is available for research, thus inhibiting progress in biotechnology and genetic engineering (see brown sub-linkage in Fig. D 3.3-3). At the same time, "resistance building" demands constant improvement of varieties by crossing in new characteristics from cultivated and wild varieties. For these reasons, it is essential that supportive measures are taken to stop or at least slow down the rapid process of genetic erosion. The action needed includes not only gene banks and botanical gardens (ex-situ conservation), but also in-situ protective measures (e.g. on-farm conservation, etc.), which have to be implemented as part of an integrated strategy for the preservation of genetic resources.
- As food production rises, dependence on cereal imports declines. However, this is accompanied by a growing dependence on imported agricultural technology, agrochemicals and seeds (know-how and technology transfer). The "international indebtedness" trend is reinforced because these imports have to be financed with scarce foreign exchange. As a consequence, the ability to import agricultural inputs is reduced and an inadequate supply for the national markets may result. Agricultural production becomes more vulnerable to crises due to possible fluctuations in supply. Since the complementarity of inputs is no longer guaranteed, the "intensification of agriculture" trend weakens. Declining yields are the consequence (see red sub-linkage in Fig. D 3.3-3). However, the

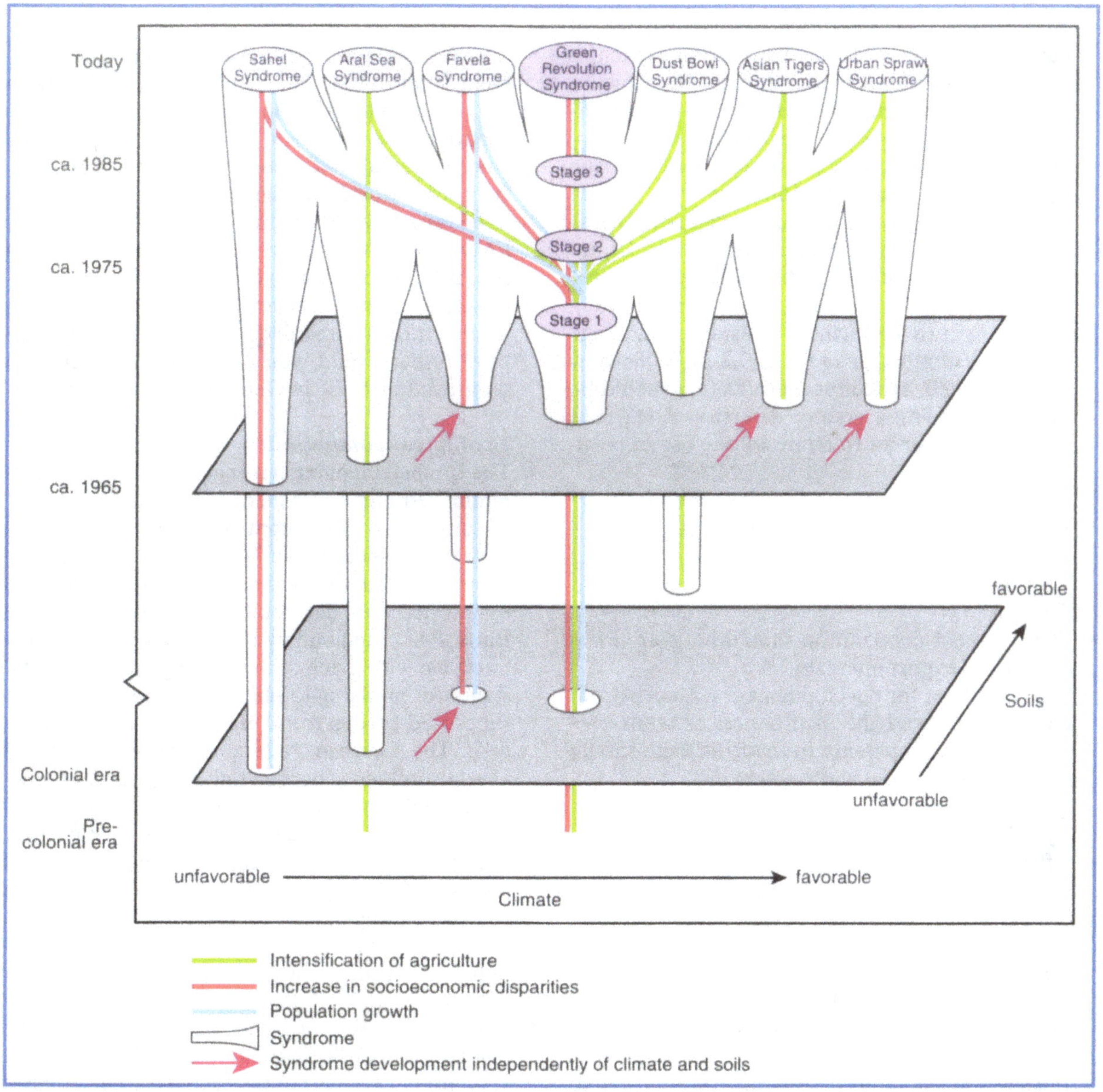

Figure D 3.3-7
Syndrome links over time.
Exposition factors of the Green Revolution Syndrome:
- Increasing food crises
- Rural development problems
- Innovations as prerequisite for intensification of agriculture
- International policymaking

Disposition factors of the Green Revolution Syndrome:
- Soil and climate conditions unsuitable for the green revolution
- Polar distribution of land ownership
- Feudal relations

Source: WBGU

pressures caused by mounting debt can also be reduced by increasing the cultivation of export crops, the result being a progressive intensification of agriculture (WRI, 1994). With cultivation of export crops accounting for a major portion of pesticide consumption, pesticides come to play an increasingly significant role as an instrument for maximizing crop yields and competitiveness on the international markets, and as a threat to the natural environment and human health. This interrelationship underlines the importance of building local research and production capacities in the countries concerned and of adopting low-input strategies. Germany has made valuable contributions in this respect in the context of development cooperation, and in future should pursue such work with renewed vigor.

- Due to the "resistance building" trend, new pesticides or greater amounts of existing pesticides are needed ("know-how and technology transfer") and applied ("intensification of agriculture"), leading in turn to further "resistance building" (see green sub-linkage in Fig. D 3.3-3). In the view of the Council, agricultural production should be focused on using a greater diversity of varieties and species in order to minimize the use of pesticides.

RECOMMENDATIONS FOR DEVELOPMENT COOPERATION

General

- Development cooperation should be geared less to national export interests.
- The guidelines for development assistance should consistently avoid the distribution of seeds protected by foreign patents. Instead, the focus should be on local potential and resources.

Efficiency

- Efficient systems for the provision and use of machinery and supplies are of paramount importance for the success or failure of agricultural development measures. In the view of the Council, it is essential to have institutions which ensure that operating requirements and information are supplied promptly, in sufficient amounts and with nationwide coverage. The withdrawal of the state from this area in the context of structural adjustment measures, opening it up to private sector involvement instead, is a step in the right direction. However, it is essential here that policymakers preserve or create a suitable conditional framework and that the private sector takes over the necessary functions. The transition must be carefully monitored and given institutional support wherever necessary.
- The Council welcomes the food security swaps, referred to in the Global Action Plan of the World Food Summit, as instruments for promoting food security (see also Box D 3.3-3). A key prerequisite here is the transparency of the instrument, including the co-determination rights of the affected country governments and the NGOs involved. In deploying this instrument, water issues should play a key role in the analysis of ecological compatibility.
- The neglect of agricultural research and development support that can be observed worldwide is not in keeping with the role of agriculture in economic development, combating poverty, and environmental and resource protection, particularly since worldwide agricultural production has to be raised by 60% by the year 2010, according to FAO data. In the field of development policy, this key role needs to be given the attention it deserves. The Council regrets the sharp decline in the level of funding provided for international agricultural research and would very much welcome a significant increase in Germany's contribution.

Ecological compatibility

- The Council recommends that use of pesticides be minimized and controlled within the framework of integrated pest control methods (WBGU, 1995). This should also be included as a binding commitment in the guidelines for support applied by international development institutions (World Bank, FAO, etc.) and by individual donor nations. Last year's lowering of the binding environmental standards applying to plant protection in projects supported by the World Bank is a cause for concern. The German Federal Government should exert its influence here so that a high level of environmental standards is maintained.
- Extensive pasture and livestock farming is a better alternative to intensive crop farming in many arid and semi-arid regions because it matches local conditions. Regional development schemes for rural areas should be reviewed with this aspect in mind, especially where marginal lands are involved.
- The Council supports the 1995 decision by the Consultative Group on International Agricultural Research (CGIAR) to base its research activities on an ecoregional approach and recommends active participation on the part of target groups as well as integration of culture-specific knowledge into the research and development process. Greater focus should be placed on some less popular crops that could play a key role in this respect, such as sorghum, millet and sweet potatoes.

Social compatibility

- Structural adjustment measures generally tend to have positive impacts on agriculture. On the one hand, fertilizer and other subsidies enable many people to earn a livelihood. On the other hand, the relatively low cost of production factors reduces incentives for wise and economical use. Moreover, the financial resources needed to provide subsidies must be generated elsewhere in the economy, with no direct indication of who has to pay for this additional burden. A reduction in fertilizer subsidies and the introduction of full-cost charges should be striven for, but should not take place until the subsistence level can be safeguarded by other means. Structural adjustment measures should include social compatibility in the definition of their goals, as agreed at the World Summit on Social Development in Copenhagen in 1995.

RESEARCH

- Our knowledge about the trajectory of the green revolution in this age of globalization is still fragmentary. The impacts of the international debt crisis and subsequent structural adjustment measures have been inadequately studied with regard to their interactions with the green revolution. There is a need for research in this area.
- The public participation frequently called for in connection with regional development programs has not been sufficiently specified. The conditions on which successful participation depend need to be studied in detail (Box D 3.3-2).

3.3.3 Water-specific syndrome description

Artificial irrigation is becoming a factor of increasing importance for the food security of humankind. Only 17% of the world's croplands are artificially irrigated, yet this area yields nearly 40% of the global harvest (FAO, 1996b). The significance of water as a resource for the green revolution is reflected by the fact that 50–60% of the yield increases achieved by developing countries since the 1970s can be attributed to irrigated agriculture (Barrow, 1995). Many countries, e.g. China, India, Indonesia and Pakistan, obtain more than half of their food production from irrigated crop farming. Egypt would be able to produce very little food indeed without water from the Nile (Postel, 1993). In Mexico, 80% of public spending on the agricultural sector has flowed into irrigation projects since 1940. In China, Pakistan and Indonesia, irrigation swallows around half of all public investments in the agricultural sector (FAO et al., 1995). Great importance is attached to irrigation in the context of development cooperation as well. In the 1980s, 30% of the credit volume granted by the World Bank went into irrigation projects.

The yield increases that constitute the principal goal of the green revolution can only be achieved in semi-arid and arid regions with the help of intensive irrigation. The high-yielding varieties (HYV) that are planted display a very high assimilation capacity when adequately fertilized, but always in conjunction with substantial water losses and concomitantly high water demand (Fischer and Turner, 1978; Schulze, 1982). The water requirement of a high-yielding wheat variety, for example, is three times higher than that of wheat varieties traditionally used in India (Shiva, 1991). Furthermore, the shorter growth period of the high-yielding varieties enables double or even triple harvests. In most regions, however, additional harvests are only feasible with at least intermittent irrigation, while another factor is that shorter growth periods result in reduced yield potential under similar conditions (Schulze, 1982; Donald and Hamblin, 1976). The return on the high level of external inputs necessitated by double or treble harvesting and intensive irrigation is a relatively small net additional yield.

3.3.3.1 Water-specific syndrome mechanism

The syndrome-specific network of interrelations highlights the significance of water in the Green Revolution Syndrome (Fig. D 3.3.-3). The negative impacts vary according to the region concerned. On the one hand, "intensification of agriculture" often reinforces the "lowering of water tables" due to pumpfed irrigation, thus triggering "changes in local water balance and particulate load". On the other hand, the "intensification of agriculture" by means of irrigation systems may promote waterlogging and salinization due to inappropriate drainage systems and rising water tables.

3.3.3.2 Water-specific network of interrelations

SALINIZATION

The most frequent cause of degradation of irrigated land is salinization. The soils of semi-arid and arid regions are by nature very saline. Salinization of topsoil may occur here due to the conversion of natural vegetation into cropland, regardless of the intensity of farming or the specific crop varieties that are grown (see Section D 1.3). Improper irrigation of the soil accelerates and reinforces this process. A basic problem is that even freshwater contains dissolved salts that may become more concentrated prior to watering as a result of evaporation. The dissolved salts are flushed out of the topsoil (rooting zone) into deeper horizons only if there is sufficient precipitation or supply of irrigation water, and the soil is well drained. The factors typically responsible for salinization are inadequate irrigation, excessive irrigation, inadequate drainage and infiltration of water from

BOX D 3.3-2

Participative methods of data collection and project planning in development cooperation

Nearly all the world summits and convention negotiations of recent years have called for more public participation without specifying this objective in any greater detail. New participative approaches to development planning have been used by development practitioners for some time now and have long been an issue within the field of development cooperation.

All development cooperation projects rely on baseline surveys for analyzing problems and elaborating development strategies. In recent years there has been mounting criticism of conventional data collection methods, such as standardized interviews, because the problems and conflicts identified in this way did not necessarily coincide with the target population's view of the problems, and because little success was achieved by projects based on such surveys. Criticism was leveled in particular at the amount of labor input and time consumed in generating mountains of useless data. Moreover, the results produced by external "experts" were often incongruent with the realities as experienced by the population in question, because the poor were easily overlooked, objective information was given priority over personal information, and because the results were acquired, analyzed and used as the basis for decision-making by outsiders only.

The mistakes of the past have meanwhile been identified, and active involvement of the people concerned has been recognized as a major condition for success. The pioneering work was carried out by British researchers who developed the first participative appraisal and planning methods for development cooperation (Chambers, 1992; Schönhuth and Kievelitz, 1993). The most important families of participatory approaches include Rapid Rural Appraisal (RRA) and Participatory Rural Appraisal (PRA). These two methods are aimed at identifying needs as perceived by rural population, defining priorities for project work and recognizing the conflicting interests of population groups. They also serve to focus conventional baseline surveys on the most relevant concerns. RRA and PRA are also used in preparing feasibility studies and evaluating projects.

Rapid Rural Appraisal is an approach by which essential information on existing development problems and potential solutions in developing countries can be collected at local level within a relatively short time using non-standardized methods and by integrating the knowledge of the local population. RRA is seen as an alternative to conventional survey methods wherever the focus is less on obtaining accurate figures as on making the most realistic appraisal of all issues related to the development goal.

The Participatory Rural Appraisal approach evolved from the RRA approach and places greater emphasis on bottom-up strategies. The people concerned are actively involved in analyzing problems and development planning, while external persons, usually an interdisciplinary team, are confined to a *catalyst* role. Priority is given to the interests of local communities. Appraisal results are intended to help the population understand its own situation and to serve as the basis for self-reliance and development projects. In the PRA approach, the village or urban community residents are also "experts", in that external people and the local population jointly analyze the local situation on which joint planning and action are then based. This process is referred to as *sharing realities*. PRA focuses more on understanding complex circumstances than on collecting quantitative data, with the main emphasis placed on those aspects of greatest relevance for decision-making.

PRA is based on a number of key concepts. One of them is "triangulation" – the validation of survey material by using different sources. Here it is important to ensure interdisciplinarity and a balanced gender ratio by having several teams of changing composition. The principle of *learning in the community* ensures that the study team, in which members of the local population are also represented, becomes familiar with the problems from the perspective of the people concerned. The function of the team is to support self-determined development. To improve the "inside" view, the PRA team lives within the local community for the entire appraisal period and takes part in everyday life. The principle of "optimal ignorance" or "appropriate inaccuracy" means that the appraisal should cover just enough information and detail as necessary. Data collection and analysis should be carried out with appropriate instruments, using simple diagrams, graphs or figures. Appropriate means, for example, counting seeds to arrive at quantitative data. An important condition for success is "visual sharing". In contrast to a conventional survey, the appraisal results are presented to the participants in a clear and accessible form, e.g. as graphs and diagrams on a display board. Participants can input their own proposals and detect weak points in the problem-solving process right from the start. Such local analyses are supplemented by local presentations in order to discuss the results publicly. This greatly reduces the risk of planning errors. Regular follow-up meetings highlight the progress made in implementing the development project. The principle of self-criticism serves to identify people who have been forgotten (e.g. the poor), things which have been overlooked (e.g. peripheral regions) and one's own mistakes. A PRA takes place in three steps: workshop, study and evaluation. The study and evaluation phases last between ten days and four weeks. PRA approaches have been used by *Save the Children* in the Gaza Strip and by the *Aga Khan Rural Support Program* in Pakistan.

The strength of PRA, which was developed by practitioners in the field, is its informal, experimental and open nature. However, PRA is no substitute for project-supporting research and long-term studies. The future focus of PRA will be on training development experts. If PRA methods were included in university training, it would be possible to carry out case studies into specific problems of the PRA approach, something that has only been tackled in isolated cases so far. Another key factor concerns the future reception of such unconventionally obtained PRA results by the developers of social science theory (Schönhuth and Kievelitz, 1993).

unlined irrigation channels. These factors may cause waterlogging of the soil and a rise in the local or regional water table. If the latter rises to within 1.5 m of the surface or less, water can rise to the surface through capillarity and evaporate. The salts that are dissolved and mobilized in the water are left behind and accumulate in the rooting zone or at the soil surface (Pereira, 1974; Barrow, 1994). Salinization is accompanied by soil clogging, which cannot be eliminated even with high capital input. Salinization and waterlogging of the soil also cause damage to soil flora and fauna. Furthermore, the frequently high salinity of drained irrigation water impairs the quality of the receiving waterbody and may lead to conflicts with downstream riparians of the river system ("changes in loads of particulate and dissolved substances").

The total area of salinized land is not known to any degree of precision. It is estimated that at least 15 million hectares in developing countries produces significantly reduced harvests because of the high salinity of the soil. India, China and Pakistan are particularly affected. Studies by the World Bank have shown that soil waterlogging and soil salinization reduce Egypt's and Pakistan's crop yields by around 30% (Postel, 1993).

Use of fertilizer

In 1994, more than 150 million tons of mineral fertilizer were consumed worldwide (application of nutrients in the form of nitrogen, phosphorus and potassium), about 66% of that in the developing countries. Nitrogen fertilization is particularly relevant in connection with "waterbody eutrophication" and "damage to health" (nitrate in drinking water). In the green revolution countries, annual turnover in nitrogen fertilizers per hectare of cropland rose on average from 4 kg N/ha in 1961 to 91 kg N/ha in 1994. In Europe, by contrast, average nitrogen fertilizer consumption has declined since the early 1990s from over 110 kg N/ha and year to less than 90 kg N/ha (FAOSTAT, 1997).

The consumption of nitrogen fertilizers in most green revolution countries is still well below 100 kg N/ha (Table D 3.3-1). However, North Korea, Egypt and China have already reached fertilization intensities (>200 kg/ha) far higher than those of some European countries. Slurry, manure and biological fixa-

Table D 3.3-1
Mean annual consumption of nitrogen fertilizer in the green revolution countries, 1994.
Source: FAOSTAT, 1997

Country	N-fertilizer consumption (kg N ha^{-1} of cropland)	Country	N-fertilizer consumption (kg N ha^{-1} of cropland)
Albania	17	Mexico	45
Algeria	7	Morocco	15
Bangladesh	83	Myanmar	9
Chile	48	North Korea	310
China	207	Pakistan	78
Colombia	53	Philippines	44
Egypt	208	Saudi Arabia	52
Guatemala	66	Sri Lanka	63
India	56	Surinam	59
Indonesia	54	Tunisia	9
Laos	1	Vietnam	132

tion of nitrogen must also be added to nitrogen input. Eutrophication of surface waters because of excessive and/or improper use of fertilizers and, in particular, due to nutrient loads in drainage water can be expected in future. In most developing countries, nitrate contamination of groundwater has yet to reach problematic levels. In Haryana (India), however, nitrate values of 114–1,800 mg/l have been measured in well water (national reference value: 45 mg/l) (WRI, 1994).

Use of pesticides

Surplus irrigation water transports not only salts and fertilizers to surface waters and groundwater, but also pesticides. For example, even minute quantities of the herbicide atrazine in streams, ponds and rivers are enough to damage ecosystems. Atrazine inhibits the growth of algae and plankton, thus eliminating the food base on which fish and other organisms depend for survival.

Lowering of the water tables

In many semi-arid and arid regions, the only way to provide sufficient irrigation water is to use groundwater. If the depletion rate exceeds the recharge rate, the result can be a "lowering of the water table"; this, in turn, has impacts far beyond the irrigated area and may generate a potential for conflicts between users. Dry rice farming in northern China, for example, requires around 600 mm of water per harvest. Transpiration from the plant amounts to 150–200 mm, i.e. the remaining 400 mm evaporate from the soil or from irrigation channels. Thus, if precipitation amounts to 300–400 mm, an additional 200–300 mm of irrigation water are needed. In India, the number of water pumps used in agriculture rose from 4.33 million (1980–81) to 9.1 million (1991–92) within a period of 10 years. A study of the Ludhiana district (India), a semi-arid region dependent on artificial irrigation from wells, recorded a drop in the water table of about 0.8 m annually from the middle to the end of the 1980s (WRI, 1994). In cereal-growing regions in the northern Chinese plain, the water table is sinking at a rate of around 1 m per year. From Tianjin, China, there are reports of a "lowering of the water table" at an annual rate of 4.4 m (Postel, 1984 and 1989). When non-renewable fossil groundwater reservoirs are mined, the problems described above acquire a special relevance. A disastrous consequence of water-table lowering in coastal areas is the intrusion of seawater into the aquifer, making the groundwater unusable for drinking and irrigation (e.g. the Gujarat aquifer in India). However, it is not just direct use of groundwater or river water that has a negative impact on the water balance of affected regions. The high transpiration rate of crop plants and the frequently high evaporation losses during irrigation reduce runoff and the renewal rate of groundwater resources (Wilber et al., 1996).

Summary

Water demand rises rapidly when farmers switch from traditional to high-yielding wheat varieties and when millet and maize are replaced by rice. "Water pollution" and "eutrophication" are aggravated due to fertilizer loads to surface waters, sometimes with direct impacts on the health of the rural population ("increasing damage to health"). This is because irrigation water and drinking water often share the same water cycle. Loads to groundwater are less serious, given the quantities currently applied (Advisory Research Council at the BMZ, 1995).

All three hydrosphere trends reinforce "freshwater scarcity", which has an attenuating effect on "increasing food production". Competition between cities and the surrounding rural areas over water resources intensifies as a result.

3.3.3.3 Water-specific recommendations

The "intensification of agriculture" according to the green revolution pattern has several negative impacts in the hydrosphere. These include "lowering of the water table", "eutrophication" and "changes in the local water balance". All three trends aggravate "freshwater scarcity" and thus have negative impacts on irrigated agriculture. In this way, "freshwater scarcity" slows down the increase in food production, exacerbating the threat of a production-based food crisis (reinforcement of the "poverty/hunger" trend). "Freshwater scarcity" may also boost the "widening of socioeconomic disparities". In many cases, "freshwater scarcity" is caused by the widespread use of motor-driven pumps to raise groundwater from boreholes, to such an extent that "lowering of the water table" may occur. In the worst cases, neighboring farmers may no longer be able to access groundwater using their traditional wells. This causes the "poverty/hunger" trend to intensify further. The "increasing importance of national economic strategies" comes to the fore with the aim of agricultural intensification according to the green revolution pattern. This closes a positive feedback loop, a major feature of the syndrome's dynamics (see blue sub-linkage in Fig. D 3.3-3).

As indicated in the foregoing, "freshwater scarcity" performs a key function in the dynamics of this syndrome. This means that, if water-related aspects are taken seriously from the very outset when engineering strategies for agricultural development, the gains in terms of sustainability can be enormous by comparison. Within this process, consideration should also be given to social compatibility aspects and to culture-specific factors.

WATER-RELATED RECOMMENDATIONS

Efficiency

- The coordination between donor countries, NGOs and developing countries commenced during the Water Decade by the Collaborative Council for Water Supply and Sanitation must be continued and intensified.
- Just how urgently international action is needed is seen by the fact that as much as 60% of water withdrawals for irrigation fail to reach the crop (FAO, 1996a). Many irrigation systems are in a poor state of repair. Nearly 150 million hectares worldwide, nearly two thirds of the world's total irrigated area, needs some form of upgrading (Postel, 1993). In the view of the Council, the key issue today is to improve existing irrigation systems rather than create new ones.
- Inadequate drainage, over-intensive irrigation, poor maintenance and seepage losses from water conveyance systems are the most frequent human-induced causes of waterlogging and salinization of soils in irrigated agriculture. Here again, improvement and rehabilitation of existing irrigation systems as well as comprehensive training for farmers are the most urgent tasks to be tackled through development cooperation.
- The introduction of water pricing to promote efficient use is desirable in principle, provided that small farmers are assured a secure livelihood when no alternative employment is available. The creation of water markets must always be accompanied by schemes to secure the basic needs of the population.
- If the price for surface water is too high, more use is automatically made of groundwater reserves. This necessitates the coordinated management of surface water and groundwater resources.
- A sustainable food security policy is not possible unless investments are made in water infrastructure. Water development projects and water management systems must form an integral part of every regional development program. With regard to sustainability, the Advisory Council believes that priority should be given to small-scale water development programs, since experience has shown that locally adapted technologies can be maintained more easily by local operators and thus on a long-term basis. The wider context should not be neglected, however, particularly where potential conflicts over use are involved.

Ecological compatibility

- The Council welcomes the CGIAR action program adopted in 1995, particularly the definition of the five major research thrusts: increasing productivity, protecting the environment, saving biodiversity, improving policies and strengthening national agricultural research. The topic of water could be dealt with on a cross-cutting basis, especially with regard to environmental protection and policies for managing natural resources. Of great importance in this context is the provision of continued support to the world's two leading institutes for research on water-saving technologies for agriculture, the International Rice Research Institute (IRRI) in the Philippines and the International Irrigation Management Institute (IIMI) in Sri Lanka (both CGIAR sub-organizations). The International Food Policy Research Institute (IFPRI) in the USA should also be involved in this work, specifically by covering socioeconomic issues.
- In the rural development context, the primary focus must be directed in future at the selection of native varieties, agroforestry and multicropping systems. The relatively small areas of land required by these agricultural systems, combined with the high diversity of their produce (carbohydrates, proteins, fodder, firewood, etc.) are essential for securing subsistence. Here, the Council recommends that support be provided to two CGIAR institutes, namely the International Crop Research Institute for the Semi-Arid Tropics (ICRISAT) in India, and the International Institute of Tropical Agriculture (IITA) in Nigeria.
- The energy required to produce animal protein is much higher than for plant production. Given the finite resources available, the implication is that a secure protein supply for the world population can only be achieved in many cases by encouraging the cultivation of protein-rich legumes (with the exception of soybeans).

Social compatibility

- Many governments subsidize irrigation water as part of their food security policy. The objective of such policies can be to achieve a large measure of independence from food imports, or to secure food for the poor by making food generally cheaper or enabling food to be grown. If the foreign exchange is not readily available, many of these countries may find themselves unable to compensate for food cereal deficits by means of imports. Moreover, complete dismantling of subsidies for irrigation water would threaten the subsistence of many small farmers who produce mainly food crops. Nevertheless, generalized subsidization of irrigation water leads to a reduction in incentives for economical water use. Another aspect is that water subsidies benefit all consumers, regardless of their income and the products in demand. To secure an adequate food supply, therefore, it is necessary to redistribute resources for the benefit of the poor, or to make basic foodstuffs cheaper. If subsistence farmers with no employment alternatives are to maintain their livelihoods, water benefit should be provided to needy target groups. It is extremely important in this connection to identify the most vulnerable groups.

- The World Bank (World Bank, 1992) and the International Food Policy Research Institute (Rosegrant, 1995) have already pointed out the importance of secure land tenure and clearly defined water rights for efficient resource use. They promote a long-term orientation in soil management and water use, and provide an important basis for accessing formal agricultural credits. Enhancing legal security for farmers (especially tenant and small farmers) is a contribution to resource protection and an appropriate means of realizing the right to food and water laid down in the International Covenant on Economic, Social and Cultural Rights. For this reason the Council would very much welcome more detailed specification of these water rights and the establishment of institutions for enforcing them (see Box D 3.3-3).

RESEARCH

- The baseline irrigation water requirement for securing nutritional subsistence must be determined for each specific country and region.

THE NEW GREEN REVOLUTION: ASSESSMENT ON THE BASIS OF THE SYNDROME ANALYSIS

The "new green revolution" approach presented by the FAO at the World Food Summit is an agricultural development strategy aimed at avoiding the errors made in the "old" green revolution (FAO, 1996a). The focus here is on reducing the use of external inputs, participation by target groups and the creation of an enabling framework ensuring broad-based effectiveness (ownership rights for establishing planning security and guaranteeing efficient use of resources, income distribution, etc.). In addition, greater efforts are to be made to prevent post-harvest losses. The following objectives are pursued:

- revival of national extension services and research institutions,
- an environmental focus on the part of agricultural research and extension, especially the promotion of low-input systems and biodiversity in agriculture,
- cooperation with international agricultural research institutions within the framework of the CGIAR system and a clear focus on eradicating poverty,
- opening national markets,
- ensuring a high priority for food security policy.

Biotechnology is regarded by the FAO as an important option for the future, but the requisite increases in production output are not expected from it alone (FAO, 1996a). The Council shares this cautious assessment.

Nevertheless, it is evident that genetic engineering has achieved great advances in the following areas of plant production:

- product quality, including changes in the quality of fats and oils (e.g. rape), changes in the composition of amino acids (e.g. methionine concentration in maize and soybeans), vitamin content as well as changes in starch concentration and composition (e.g. potatoes),
- perishability (e.g. ethylene metabolism in tomatoes, Flavr Savr tomato),
- tolerance to herbicides (e.g. tolerance of soybeans and maize to BASTA total herbicide),
- viral resistance (e.g. tobacco, potatoes and rice),
- resistance to bacteria or fungi (e.g. tobacco, potatoes),
- insect resistance (e.g. to *Bacillus thuringiensis*),
- male sterile plants (e.g. tobacco, rape).

Expectations have not been met with regard to the following:

- increased yields,
- higher water-use efficiency or drought resistance,
- salt tolerance.

Future production increases will mainly be attainable on the basis of improved agricultural systems. Securing subsistence for most small-farmer producers requires cropping systems with a high diversity of produce providing for a balanced diet and covering the demand for firewood and animal fodder.

The potential offered by molecular biotechnology for a new green revolution is very limited, in that genetic engineering has only succeeded so far in intervening in the "secondary metabolism" of plants. The application of genetic engineering methods has concentrated on modifying single genes and thus certain monogenetic characteristics of plants in order to achieve immediate effects – on the resistance of plants, for example. The yield of a certain crop plant is the result of very many environmental factors (temperature, light, nutrient and water supply, competition, food, parasites, diseases, etc.) as well as genetic, physiological and morphological characteristics (photosynthesis, leaf area, growth rate and period, harvest index, etc.). Many of these factors are positively or negatively correlated with and dependent on one another (Donald and Hamblin, 1976; Schulze,

BOX D 3.3-3

International legal aspects of food security

THE RIGHT TO FOOD AND WATER IN INTERNATIONAL LAW

The right to food and hence the right to water form part of the social human rights. As far back as 1948, the UN General Assembly adopted the Universal Declaration of Human Rights, Article 25 of which states that

"Everyone has the right to a standard of living adequate for the health and well-being of himself and of his family, including food, ...".

However, declarations of the United Nations General Assembly do not have binding legal force, so the various sub-committees of the UN have been endeavoring since the 1950s to codify these general principles in binding international conventions. In 1966, two International Covenants on human rights were put forward for adoption, one covering political and civil rights, the other the economic, social and cultural human rights. Both Covenants entered into force in 1976 and are now binding on most of the world's nations.

Just like the Universal Declaration of Human Rights, the Covenant on Economic, Social and Cultural Rights (the "Social Covenant") refers explicitly to the human right to food. Article 11 states that

"The States Parties to the present Covenant recognize the right of everyone to an adequate standard of living for himself and his family, including adequate food [...]".

As with the Universal Declaration of Human Rights, this includes the right to water: accordingly, every human being has a right to sufficient usable water to ensure an adequate standard of living. This includes both drinking water as well as enough irrigation water to ensure an adequate supply of food.

The right to food and water does not belong in the category of civil and political rights and freedoms; rather, it implies an obligation of the state towards its citizens. However, there are problems of delineation here as well, as is often the case in the human rights regime: the human right to food and water also means that the individual citizen has a right to defend him or herself against the denial of food and water. This is particularly relevant in domestic armed conflicts, for example, where food security of the civilian population must be guaranteed by all the conflicting parties, or in the international ban on food embargoes recently proclaimed in the UN Global Plan of Action adopted by the UN World Food Summit.

SPECIFYING THE HUMAN RIGHT TO FOOD AND WATER

The human right to food and water means that assistance must be given in all regions in which this right is violated, either because water is not available in sufficient quantity or quality, or because restrictions are imposed on the population's access rights. This involves a duty on the part of states for their own national territories: according to Article 11, paragraph 1 of the Social Covenant, they shall take appropriate steps to safeguard the human right to water and food.

However, the duty to international cooperation is also a component part of the human right to food and water: the Social Covenant explicitly reaffirms the "essential importance of international cooperation based on free consent".

The basic duties of states are defined in more detail in paragraph 2 of Article 11, according to which the states shall take, individually and through international cooperation, the measures which are needed to improve methods of production, conservation and distribution of food (including water) by making full use of technical and scientific knowledge.

It is precisely here that the Council sees a continued and considerable need for action. The mission of the Social Covenant, which has been in force since 1976, namely to implement the human right to water and food with appropriate measures, has not been adequately fulfilled as yet and requires greater efforts (see Section D 5.5). The International Covenant on Economic, Social and Cultural Rights commits the states parties not only to supply food and water, but specifies one essential measure as "developing or reforming agrarian systems in such a way as to achieve the most efficient development and utilization of natural resources".

In order to improve implementation of the human right to food and water, the Council considers it imperative to promote efficient techniques for water use as well as water-saving crops and agrarian production systems. The participation of small farmers is essential for the success of such programs.

The follow-up process to the UN Food Summit

The food security swaps referred to in the Global Plan of Action adopted by the UN World Food Summit (1996) hold great promise as an instrument for promoting efficient systems of resource utilization, as required by the Social Covenant. Such swaps involve buying up the foreign debts of developing countries and providing debt relief to the respective countries on condition that they spend an appropriate amount in national currency on implementing active food security policies. The Council welcomes this instrument as a means of promoting food security. A key prerequisite here is the transparency of the instrument, including the co-determination rights of the affected country governments and the non-governmental organizations involved. Water-related aspects should play a central role when conducting environmental impact assessments in the framework of such an instrument (see Section D 3.3.3.3).

In the follow-up process to the UN World Food Summit, it is essential that the Global Plan of Action signed by the international community in Rome is being honored by all parties. Special emphasis must be given to the development of an internationally accepted code of conduct. In the view of the WBGU, such a code of conduct should include a ban on food dumping, the targeted promotion of locally appropriate and water-efficient production systems, creating an enabling environment for farmers' organization and improving the access rights of small agrarian producers to productive resources.

1982). Until now, yield increases have predominantly been achieved by increasing the harvest index, by morphological changes permitting improved light interception and denser stocks, as well as by higher inputs (fertilization, irrigation) and new land management methods (Donald and Hamblin, 1976; Schulze, 1982). Since the potential of the harvest index is virtually exhausted (values over 50% can now be attained), higher yields can only be expected by increasing the total biomass (Hay, 1995; Donald and Hamblin, 1976). There are also substantial constraints on improving the water efficiency of crops. On the one hand, higher yields are necessarily linked to higher water consumption (see Section D 1.3.4); on the other, water use efficiency is essentially determined by two factors that cannot be manipulated by genetic engineering, namely the ambient water saturation deficit and the primary carboxylation enzyme for photosynthesis (Fischer and Turner, 1978), which is governed by a co-regulation of all enzymes involved in primary metabolism (Stitt, 1994). The drought resistance achieved in crop varieties to date is primarily the result of premature blossom formation (Fischer and Turner, 1978). It is estimated that only minimal improvements in water efficiency are possible by means of classical cross-fertilization (Farquhar et al., 1988).

Despite its undisputed successes, genetic engineering cannot be hailed as an instrument for mitigating, much less eliminating the potential environmental and socioeconomic threat posed by monocropping, cost- and energy-intensive farming systems or economic dependencies. On the contrary, some of the ecospherical and socioeconomic repercussions already experienced in the "old" green revolution may be amplified. When a particular crop plant is made resistant to total herbicides (e.g. BASTA) by genetic engineering, application of the herbicide kills all other plants, without differentiation, and only the selected crop survives. While this may reduce yield losses, it also encourages the unfettered use of herbicides, with all its implications. Genetically modifying plants so that they are resistant to insects (e.g. genes of the *Bacillus thuringiensis*) or pathogens will help initially to reduce pesticide consumption. But if resistant varieties are cultivated as monocultures, there is a risk of resistance building among pests as well, as has already occurred through the intensive application of pesticides (Section D 3.3.2). In effect, the genetically engineered resistance of crop plants would be eliminated, and farmers would be compelled once again to rely on new varieties boasting new resistance genes. Genetic uniformity and genetic erosion as a consequence of monocropping and the failure to grow and cultivate local varieties pose the same threat as in the "old" green revolution.

The use of genetically modified seeds is likely to increase the dependence of small farmers on external inputs in a manner hitherto unknown, and surpass by far the well-known drawbacks of the old green revolution. Seeds and agrochemicals are offered as a "package deal", with individual varieties specially tailored to the application of very specific plant protection agents. Unless consumers are adequately informed about the possible consequences of this linkage between seeds and agrochemicals, there is a risk that a large number of producers will become involuntarily dependent on monopolistic manufacturers holding the respective patents and the power to dictate prices. In this way, the use of genetically modified seeds would greatly increase the vulnerability of agricultural systems to crises. Production losses in the context of such highly specialized agrochemical farming methods would have far-reaching consequences.

Since the resistance achieved by genetic modifcation of crops is effective only in respect of specific pests or herbicides (Snow and Palma, 1997), multiple infestation of crops must still be combated by applying different plant protection agents, which means that the environmental and financial advantages of "built-in" protection are narrowly limited. In spite of these reservations, the potential applications of genetic engineering for sustainable agriculture must be examined carefully and implemented wherever appropriate.

In line with the syndrome analysis above, it is important to highlight the following three constituents of the new green revolution, which have a major role to play in reducing future syndrome development:

1. The use of culturally and socioeconomically appropriate technologies, especially in irrigation.
2. The creation of farming systems featuring greater diversity of varieties and species (agroforestry, multiple-cropping systems).
3. The use of low-input systems as well as local production potential.

Specifically, the Council recommends:

- Further support should be given to an analysis of the potential benefits and risks of biotechnology in agriculture. However, it is essential to develop parallel control mechanisms and regulations in order to comply with biosafety requirements, e.g. through German support for the Biosafety Protocol to the Biodiversity Convention (WBGU, 1996). Special efforts should also be made to avoid making the same mistakes as the green revolution.
- The problem of nutrition cannot be reduced to the shortage of food. For many people, poverty is the major driving force behind chronic malnutrition and hunger. Increased production and rural development must therefore go hand in hand. In addition to promoting agricultural production, the Council also recommends developing the small business sector, the craft trades and markets in order to revitalize rural areas.
- In the view of the Council, agroforestry and multicropping systems should play a significant role in the new green revolution, because their spatial needs are lower for the same yield than is the case with monocropping. Protein-rich legumes should also be a central concern.
- Land management methods such as agroforestry and multiple cropping cannot be established on a large scale without start-up assistance. Such innovations require suitable institutions and support mechanisms aimed at specific target groups. This cannot be expected from the private sector alone. The individual countries must therefore commit themselves to rural development and provide assistance in effecting the necessary transformation in agriculture.
- The new green revolution needs an institutional framework for regulating the private sector and organizations in a way that restricts the externalization of costs and the dysfunctional effects of subsidies, while ensuring that minimum needs are met.
- The new green revolution should not be planned on a large scale and should only be implemented by the state where necessary. Limited knowledge about the best way to achieve the goals envisaged and about the possible adverse impacts that even the new green revolution may have means there should be scope for a diversity of frameworks at national and international level to ensure that a broad spectrum of experience can be gathered and utilized.
- If adverse impacts are to be avoided, the new green revolution should be continuously monitored by means of an early warning system. Such a monitoring system would have to identify existing vulnerabilities and resilience, taking unexpected developments into account.

3.4 The Aral Sea Syndrome: environmental degradation due to large-scale damage to natural landscapes

Large-scale water development projects – Water for agriculture – Energy production using hydropower – River regulation – Dams – Social and environmental impacts – Syndrome links – "Aral Sea" case study – "Three Gorges" case study – Indicator of syndrome intensity – Assessment criteria for large-scale projects – Integrated watershed management

3.4.1 Definition

People have always used and interfered with water resources for energy production, food production, flood control and water storage. Examples include drainage systems as well as dikes, dams and canals for controlling and supplying water. As far back as 2,500 years ago, water development constructions were erected in China, for example the Tu-Kiang dam and the Emperor's Canal, which had great importance for irrigated agriculture and as transport infrastructure. In central Europe, large-scale drainage of wetlands and construction of dikes – especially in The Netherlands – were carried out in the 16th and 17th century to protect against floods and to reclaim land for crops.

In the course of this century, major irrigation schemes and new, multifunctional megadams have been having severe impacts on water resources. The number of dams over 15 meters in height has risen to around 40,000 since 1950; each day, a new dam is put into operation (ICOLD, oral communication, 1997). The total area of reservoirs worldwide is around 400,000 km^2 – more than the area of Germany. Dams are a major influence on the runoff regime: 77% of all runoff in North America, Europe and the former USSR is diverted by dams or other large-scale water resource development projects (Dynesius and Nilsson, 1994). The scale of interference with the balance of nature is evident from the fact that the total volume stored in reservoirs (10,000 km^3) is five times that found in all the world's rivers. The cumulative influence of reservoirs due to the redistribution of mass has even become a measurable factor causing geodynamic changes in the Earth's rotation and polar drift (Chao, 1995).

Large dams were mainly built in the industrialized nations at first, but today they are primarily constructed in the developing countries. All large-scale water development projects are inevitably bound up with intensive interference with the social and ecological structure of a region, resulting not only in the intended benefits, but also in a number of undesired consequential impacts which – if they are taken into account in cost-benefit analyses – may reduce the benefit significantly (McCully, 1996).

The Aral Sea Syndrome describes the problems of centrally planned, large-scale water development projects. Such projects are essentially ambivalent: on the one hand, they provide the additional resources that are needed (water for food security, renewable energy), or they protect existing resources (flood control); on the other, they have severe impacts on the environment and society.

In keeping with their magnitude, the impacts of such construction projects are usually not confined to the local or regional area, but may also take on international dimensions, for the simple reason that the riverine systems affected and their catchment areas are very large and often transboundary.

During the planning phase, the hope is that several development problems can be solved simultaneously with one large project: the increased demand for renewable, non-climate forcing energy sources, for production increases in agriculture through the expansion of irrigated areas, not to mention the protection of people and property against the devastation of floods.

The negative consequences of such projects include not only the direct interference with the water and sediment balance of the catchment and the resulting impacts on the natural systems to which they are coupled. Consideration must also be given to the social consequences, which may range from resettlement of the local population, to intensification of economic disparities, to domestic and international conflicts (McCully, 1996; Pearce, 1992; Goldsmith and Hildyard, 1984).

The inadequate consideration given in most cases to the social and ecological impacts results from the technical bias of planners and their inability to assess and control the effects of a given project. This is further compounded by the frequently one-sided interests of decision-makers with respect to the prevailing technology-centered development paradigm.

This multiplicity of positive and negative effects of major projects gives rise to several possible links to other syndromes of global change (see Section D 2; WBGU, 1997). Supplying water to expansive, intensive agriculture is the most important interface; in particular, there are strong links to the Green Revolution Syndrome (see Section D 3.3) and to the Dust Bowl Syndrome. Construction of a dam, for example, may involve not only compulsory resettlement of local inhabitants, but also a change in land-use rights, usually at the expense of population groups that are already marginalized (Sahel syndrome). Providing resources promotes industrialization, which may itself have severe consequences for the environment (Asian Tigers Syndrome) or induce pull effects that favor urbanization (Favela Syndrome).

The simultaneous occurrence of positive and negative impacts generated by large-scale water development projects and the specific nature of these effects necessitates a detailed cost-benefit analysis for each single project, taking into account all economic, ecological and social impacts and the respective gaps in knowledge. Such an analysis cannot be conducted here.

Nevertheless, it is possible to give recommendations on which to base the evaluation of such projects (see Section D 3.4.5). The approach taken is the Council's syndrome method (WBGU, 1997), which permits the definition of an action framework and subsequent evaluation on the basis of the guard rail concept. The assumption is that it is possible to define marginal zones between undesirable or dangerous areas of non-sustainability and an acceptable action space.

Because the direct impacts on the hydrosphere are always the central focus of this syndrome and the starting point for more extensive indirect effects, water-specific aspects will not be discussed separately.

3.4.2
Water-specific syndrome mechanism

The Aral Sea Syndrome refers to the environmental degradation that may occur as a result of the large-scale reshaping of the landscape that is brought about, in turn, as a secondary effect of large-scale technical projects (megadams, irrigation schemes, river canalization, etc.). In addition to the damage that is directly inflicted on the natural environment, indirect effects of potential relevance for global change may also occur. The social consequences (compulsory resettlement, damage to health, international conflicts, etc.) may also result indirectly in environmental degradation. In order to describe and analyze the complex interwoven mechanisms that make up this syndrome and are relevant for global change, the syndrome-specific network of interrelations (Fig. D 3.4-1) is firstly developed. The secondary effects already mentioned may also be produced through linkages to other syndromes, indicated in Fig. D 3.4-1 by the "clouds" for the respective other syndromes (Fig. D 3.3-3).

3.4.2.1
Core trends at the people-environment interface

An important driving force behind large-scale water development projects is the expectation that the key trends of "increasing food consumption" and "increasing consumption of energy and raw materials" will intensify. For various reasons (see Section D 3.4.2.2), the preferred solution in most cases is the construction of large-scale technical projects. It often appears feasible to achieve effective flood control ("growing threat of natural disasters") and "improved transport infrastructure" simultaneously by means of such large-scale projects. "Centralization of economic strategies" is a prerequisite for project implementation, in that small, regional administrations have excessive difficulty in managing a large canal project or a dam, due to the costs alone. The immediate impact of such a project after its completion is "changes in the local water balance", especially changes in streamflow. Particularly in the case of dams, this invariably involves changes in sedimentation characteristics ("changing loads of particulates and dissolved matter").

These interrelations comprise the core mechanism of any large-scale water development project. Although they are an essential component of the Aral Sea Syndrome, they do not constitute it as such. The entire mechanism, especially the negative interactions and consequences that give a large project its syndrome properties, are summarized in the network of interrelations (Fig. D 3.4-1) and are described below. This network of interrelations necessarily involves generalization at a high level of aggregation, so it cannot be typical for each individual project. Depending on the type of water development project (dam, canal, large-scale irrigation project, etc.), the climate (humid, semi-arid, etc.) or other regional conditions, additional trends, interactions or individual aspects may differ or be entirely absent. In general, however, the network of interrelations describes a pattern of causes and effects that is typical as a "global syndrome" for many large-scale water development projects.

3.4.2.2
Driving factors

The motivation behind large-scale construction projects is rarely confined to one objective. Many dams, for example, are typical multi-purpose projects aimed at improving the basis for development (Baumann et al., 1984). In the following, a number of driving factors behind the planning and implementation of such large-scale projects are described. In some places a distinction will be made between industrialized and developing countries, because the structure of motivation may vary accordingly.

ENERGY PRODUCTION

The desire for "industrialization" and economic growth is accompanied by "increasing energy consumption" for households, industry and export. Power generation is therefore a primary objective of dam building – nearly a fifth of the world's electricity is generated by hydropower. Examples include the dam projects on the Paraná, which now covers almost all of Paraguay's electricity needs as well a large portion of the demand in southern Brazil and Argentina (Seager, 1995). No other form of power generation operates as cost effectively as large hydroelectric power plants (Schmidt-Kallert, 1989), so there is a powerful incentive to cope with "increasing energy consumption" by means of hydropower, especially since alternative ways of producing energy are either absent in many developing countries or may lead to dependence on imports. In many cases, a substantial amount of the energy thus generated is exported – not so much as electricity, but indirectly through the export of locally produced, energy-intensive goods, above all aluminum (Gitlitz, 1993).

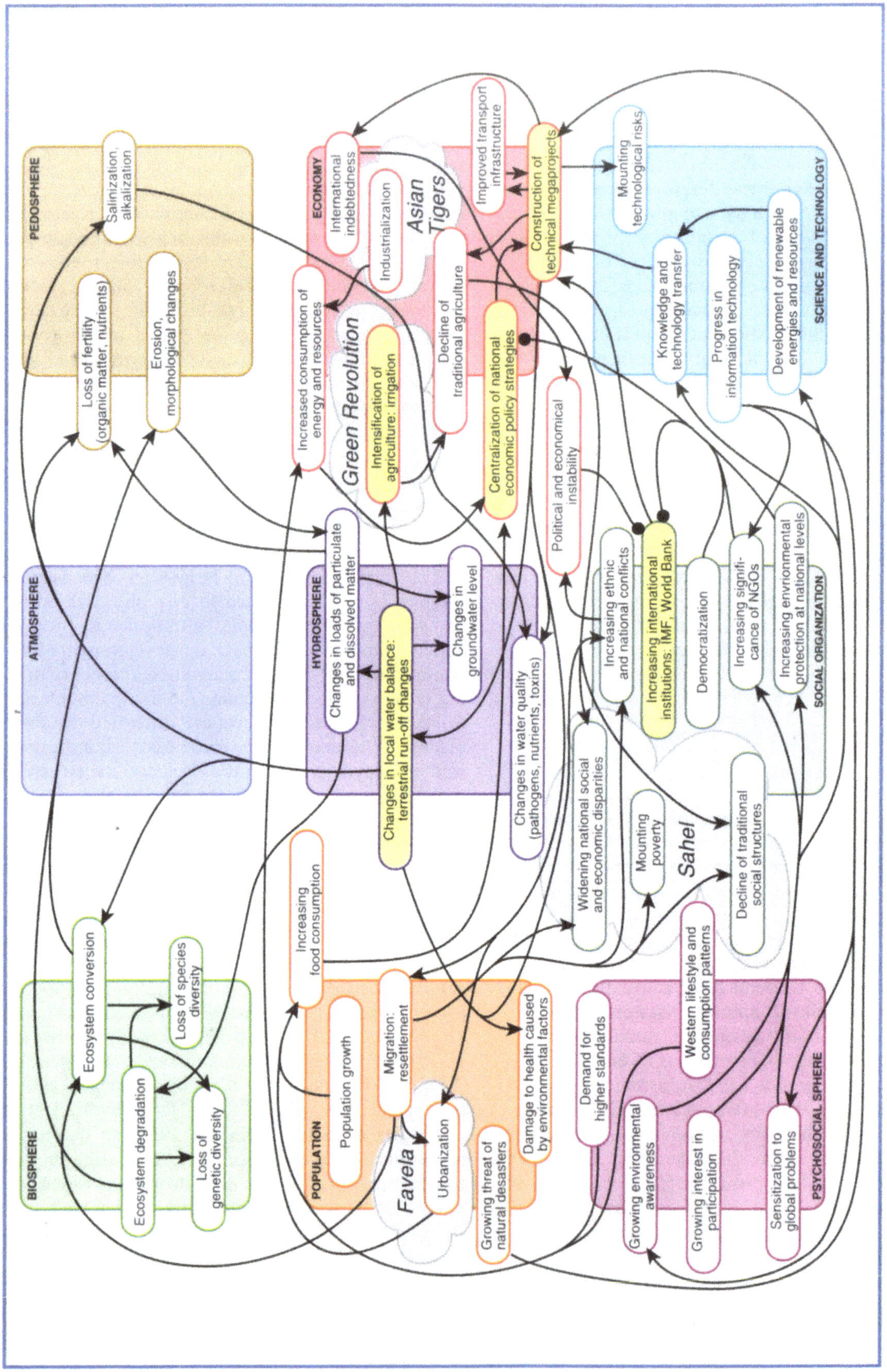

Figure D 3.4-1
Network of interrelations for the Aral Sea Syndrome.
Source: BMBF project "Syndrome dynamics", PIK core project QUESTIONS and WBGU

Nowadays, megadam projects in industrialized countries are likely to be implemented only when unused and thinly populated land is involved and the dam can be used for (cheap) power generation, without the threat of conflicts with owners or users of the inundated areas or with conservationists, or when the groups affected have little leverage (e.g. the James Bay Project in Canada).

Supplying water to large-scale irrigation systems

Population growth compels governments in developing countries to increase agricultural production in order to meet the growing demand for food. As a consequence, many large-scale water development projects are aimed at increasing the area of arable land by providing a reliable source of water, or by improving the yield of existing cropland with the help of irrigation schemes. Examples include the projects in northwest Africa for greening the Sahel Zone. Some projects also make a small contribution to food security in the form of fisheries in the reservoir. Dams in industrialized countries have regional importance where intensive agriculture is carried out and the agricultural yield can be substantially increased with a higher water supply (e.g. California; Pearce, 1992; McCully, 1996). In contrast to developing countries, the main driving force in industrialized nations is less related to food security than to the production of high-quality agricultural produce.

Regulation of riverine systems for flood control and navigability

Another motivational factor is the protection against periodic floods ("increasing threat of natural disasters") afforded by river regulation projects. Special emphasis is often placed on flood control as justification for dam projects, because humanitarian and economic reasons can then be highlighted (e.g. Three Gorges Project, see Section D 3.4.3.2). Using rivers as transport routes ("expansion of transport routes") is another major factor. Improving the navigability of the Yangtze, for example, is an important goal in China's plans for transport infrastructure.

Other factors

The main driving forces mentioned above do not necessarily imply that megadams are the only way to attain such goals. There are usually alternatives, in the form of several smaller projects, for example. The trend to large-scale water development projects was greatly encouraged by the long-prevailing assumption in development policy that development processes can or must be centrally planned and controlled. Large-scale projects create the impression that development goals, such as increasing energy production, can be achieved rapidly and at extremely low cost. In developing countries especially, major projects are often central elements of economic strategies to promote specific sectors or regions.

However, the financial scope of the projects usually precludes any self-financing. For this reason, dams have long numbered among the projects receiving the most assistance from international financing institutions (World Bank, IMF), whose growing significance has further boosted the implementation of large-scale projects. The World Bank alone, the most important public institution for financing dams, has provided $58 billion for this purpose (1944–1994, 1993 dollar value).

This is closely linked to the interests of construction companies and consultants in industrialized countries, who similarly favor large-scale projects. This is certainly another reason why the development policies of industrialized nations provide special support for large-scale projects. The Swedish Agency for Development Cooperation (SIDA), for example, estimates that up to three fourths of the money it lends for hydroelectric projects flows back to Swedish companies (Usher, 1994 quoted from McCully, 1996).

For governments or heads of state in the respective countries, dams have often been used as symbols of economic independence (Schmidt-Kallert, 1989). Furthermore, especially in developing countries, the prestige value of gigantic construction projects for domestic policy (visible evidence of progress and modernity) and its stabilizing effect for the state or government are rated very highly, with the result that priority is frequently given to large-scale projects rather than to small-scale or decentralized alternatives.

3.4.2.3 Impacts on the ecosphere

Impacts on the ecosphere are generated not only by the direct or primary effects of the project, but also by the secondary effects produced indirectly through people's responses. Aside from the relatively small-scale impact of the dam construction work, one can identify several types of environmental degradation caused by terrestrial runoff changes or by the construction of technical megaprojects.

Rivers transport a substantial amount of sediment, particularly in the tropics (Table D 3.4-1). By reducing the flow velocity of rivers, dams act as sediment traps: suspended matter settles behind the dam wall.

It is estimated that about 1% of the storage capacity of reservoirs worldwide is lost every year because

Table D 3.4-1
Sediment loads of selected rivers.
Source: after WWI, 1996b

River	Catchment area ($1,000\ km^2$)	Sediment load (million $t\ year^{-1}$)
Huang He	752	1,866
Ganges/ Brahmaputra	1,480	1,669
Amazon	4,640	928
Indus	305	750
Jangtse	180	506
Orinoco	938	389
Irawady	367	331
Magdalena	240	220
Mississippi	327	210
Mackenzie	1,800	187

of this process, and around 20% of the global storage capacity has "silted up" since 1986 (IRN, 1996; Mahmood, 1987).

Precise forecasts of reservoir sedimentation require long-term data series, which in many cases are not available. The result is that sedimentation rates are almost always underestimated, leading to some projects ending in fiasco. Sedimentation continues to be the biggest technical problem for dams (McCully, 1996).

The effect is manifested below the dam in a reduced load of particulate matter in relation to the original state, thus altering the sensitive sedimentation equilibrium in the lower reaches. The consequences are increased bank erosion and deepening of the streambed, which leads to changes in the groundwater level. Lowering of the water table may be reinforced by the absence of floods. On the other hand, given appropriate soil conditions, a rise in the groundwater level caused directly by changes in streamflow may also occur in the area of the reservoir. The impact on groundwater is particularly important in semi-arid and arid regions because the groundwater there is mainly replenished by rivers. The opposite is the case in humid and semi-humid climates.

The flow velocity is reduced by the dam, turning running waters into standing waters, with all the hydrological consequences this implies (water chemistry, temperature, sedimentation). The waterbody has less oxygen, and the accumulation of contaminants in sediment and their resuspension are reinforced. Depending on tailrace design, some of the cold, deoxygenated deep water is discharged into the river. This leads to higher concentrations of contaminants in the water that remains ("declining water quality"). Another knock-on effect may be caused by the use of the stored water in irrigated agriculture, because the return flow then has a high salt concentration due to evaporation and transpiration ("salinization"). Streamflow also undergoes profound changes; not only is there a decline in total volume, but also a change in the dynamics of streamflow – e.g. surface runoff after rainfall is held back by the dam.

The primary trends in the biosphere are the conversion and degradation of ecosystems. Conversion obviously occurs at the construction site of the dam itself, where installation of the requisite infrastructure causes the destruction of ecosystems during the construction period. The conversion into standing waters has severe repercussions for the biological communities in the river and may even result in the extinction of endemic species ("loss of species diversity"). A particularly serious problem is the fragmentation of the riverine ecosystem by the insurmountable dam walls, which cut off populations from each other. Fish ladders are only a partial remedy (Bernacsek, 1984).

Changes in the water balance also have profound effects further downstream and may cause the loss of wetlands due to the altered sedimentation equilibrium described above. The drastic change in the temporal streamflow characteristics, combined with the relative increase in pollutant load due to reduced streamflow (Fig. D 3.4-2), result in a similar way to loss of wetlands and ecosystem degradation. Due to the reduced sediment deposition in coastal areas, erosion and salinization of aquatic ecosystems in coastal regions are frequent occurrences, as are severe impacts on coastal fisheries (Rozengurt, 1992). The dams on the Indus, for example, have caused the disappearance of nearly the entire 250,000 hectares of mangrove forests in the river delta (Snedacker, 1984), and both the Nile and the Mississippi delta are shrinking at a drastic rate. A direct consequence of such interference with ecosystems is the loss of genetic diversity and even, in some cases, the loss of species diversity.

Another complex involving major impacts on the biosphere concerns the clearing of forests when people forced to resettle by the new reservoir move to areas upslope from the reservoir. This conversion of natural ecosystems causes greater soil erosion, which in turn increases the sediment load into the reservoir ("changing loads of particulates and dissolved matter").

The regulation of streamflow reduces the amount of sediment transported by the river, which often leads to a slow but steady loss of soil fertility in the lower reaches of the river. When excessive water withdrawals are made, as in the case of the Aral Sea, or when the return flow from large irrigation systems is inadequate, salinization may occur downstream (see above) and spread to more distant regions due

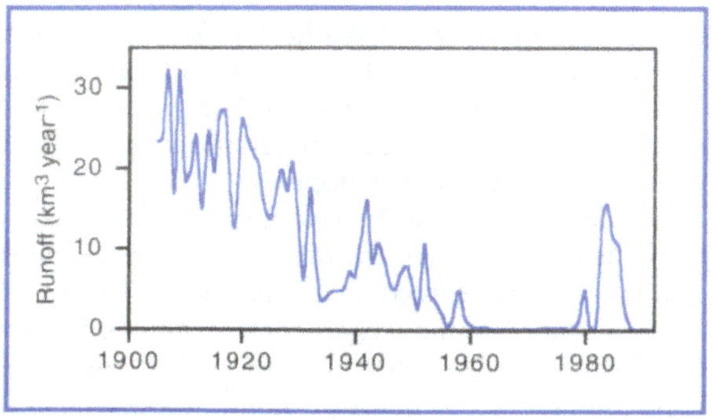

Figure D 3.4-2
Discharge of the River Colorado downstream from the dams (1905–1992).
Source: adapted from WWI, 1996b

to wind erosion and repeated salinization (Dech and Ressl, 1993).

Microclimatic effects, in the form of local cooling, greater fog formation or reduced temperature variability, are of particular importance for the atmosphere. More far-reaching effects are only likely with very large reservoirs, e.g. as a result of changed albedo or evaporation. A substantial increase in evaporation may occur due to the reservoir, especially in arid regions, and this may amount to a serious reduction in regional water availability. Evaporation losses at the reservoirs on the Colorado, for example, amount to a third of total streamflow (Dynesius and Nielsson, 1994).

Dams are not always climate-neutral energy producers, especially when large amounts of biomass are inundated by a reservoir and anaerobic degradation generates large quantities of methane (CH_4), a greenhouse gas. These offset the CO_2 reductions achieved with hydropower. According to model calculations, these emissions may be equivalent to those of power stations fired by fossil fuels during the first 50 years of use, and sometimes much higher (Balbina Dam, Brazil: Fearnside, 1995; Rudd et al., 1993).

3.4.2.4
Impacts on the anthroposphere

In contrast to the impacts on the ecosphere, the consequences for the anthroposphere are linked not only to "changes in surface runoff", but also to the "construction of large-scale technical projects".

The World Bank and other international organizations may have helped countries to realize water resource development projects, but by the same token have contributed to their international debts. This is reinforced by the rocketing costs that are usually incurred when dams are actually built. A rule of thumb is that the incremental costs of a project increase in proportion to its size. The World Bank calculations reveal that cost overruns average 30%, although cases of several hundred percent are not exceptional (McCully, 1996).

Nor can one ignore the effects triggered by the enormous influx of capital, which may reach huge proportions, especially in the case of megadam projects. The final cost of the Chixoy Dam in Guatemala ($944 million), for example, represented nearly 40% of the country's total external debt, while Brazil guaranteed loans of $16.6 billion (1990) for the Itaipú Dam, which made up nearly 14% of the total state debt (McCully, 1996).

The consequence of large inflows of capital for construction projects is often a short economic boom followed by inflation. Corruption is also encouraged in many cases; in 1990, for example, Argentina's President Menem called the Yacyretá Dam a "monument to corruption". Other prominent examples include the Turkwell Dam (Kenya) and the Pergau Dam (Malaysia), where experience was similar. In combination with national debt, the two effects may lead to political and economic instability in the country, which in turn could make further loans difficult to obtain.

Nowadays, megadams are typically built in remote areas of developing countries. The integration of such peripheral regions into the world market through dam projects leads to an extensive transformation in social conditions for the indigenous population, which in most cases exhibits traditional forms of social organization ("decline of traditional social structures"). In the eyes of the local population, dam projects (like mining projects) are the work of alien powers seeking to destroy the ecological and social integrity of their habitat. In their view, land, nature and the Earth are holy and therefore have a completely different value from that perceived by planners, be they government representatives or employees of multinational corporations. Forced resettlement, the lack of participation in planning, decision-making and above all in the benefits of the projects are factors that may turn the local population into marginalized victims of development (as exemplified by drastic cases in India and China). According to estimates, around 30–60 million people have been forced to leave their homes to date because of dam projects, and most of them are in a worse situation after forced resettlement than before. The distribution of the benefits gained from dam construction (power generation, irrigated agriculture) is very uneven and reinforces the trend towards "widening social and economic disparities". Tensions may build up as a result, which can escalate into a national conflict (Bächler et al., 1996).

Large-scale water development projects are frequently aimed at increased irrigation, resulting in the "expansion of land area used for farming" and "intensification of agriculture" trends. On the other hand, the construction of the dam itself, e.g. the associated infrastructural measures and forced resettlement of the local population, as well as changed farming conditions and land-use rights lead to a "decline in traditional agriculture". This in turn may further reinforce two trends described above, namely the "decline of traditional social structures" and "widening social and economic disparities".

The significantly reduced flow velocity behind the dam, in the newly created irrigation channels or in the downstream flow leads to a greater risk of water-related diseases, in particular vector-borne diseases, and to increasing damage to health in general. In recent years, for example, major bilharziosis epidemics have occurred as a consequence of new water development projects (Volta reservoir, Diama Dam on the Senegal, Nile Valley). In addition, given no other changes, the standing waters enhance the living conditions for the *Anopheles* mosquito, the vector for malaria.

The Euphrates, Indus, Ganges and Jordan basins provide examples of the potential for conflict associated with large-scale dam projects, which may escalate into an increasing number of international and national conflicts. Although the conflict potential in semi-arid regions is more pronounced due to the dependence of such regions on only one river in some cases, there is at least one example in which such a conflict can also be observed in a humid climate: the joint Gabcikovo/Nagymaros project between Hungary and Czechoslovakia (at the time of the initial planning; today the Slovakian Republic) for regulation of the Danube, which after Hungary's refusal to continue construction is currently being negotiated before the International Tribunal (see Section D 4.1).

Finally, mention must be made of the growing technological risks associated with dam structures, including dam breaches, which are by no means rare and number among the worst man-made disasters. Dams are particularly hazardous when the weight of the reservoir generates tectonic effects and triggers earthquakes (Gupta, 1992). This phenomenon was probably the cause of the breach in the 261-meter Vaiont Dam in northern Italy in 1963, in which 2,600 people were killed by the floodwave. In the case of the Three Gorges Project, this risk is one of the central issues under discussion, since a breach in the Three Gorges dam would rank as the greatest man-made disaster of all time (Williams, 1993a; see Section D 3.4.3.2).

Awareness of the negative consequences for people and nature is one of the reasons why international organizations (e.g. the World Bank) have begun in recent years to consider the environmental and social aspects when evaluating projects. The growing influence of non-governmental organizations in the field of environment and development (from local citizens' action groups to the International Rivers Network) is manifested in the discussions and hearings that take place in this context. The brunt of their criticism is leveled at the lack of consideration given by the official cost-benefit analyses, which are often kept confidential, to many economic, ecological and social disadvantages of the projects – from losses for coastal fisheries and costs of resettlement to liability risks and costs of decommissioning when the service life of the dam has ended – while the benefits are usually overestimated.

The political pressure thus exerted has produced results: the World Bank, for example, is currently subjecting its own past support practices to critical assessment and now takes a more cautious approach to megadam projects. The World Bank no longer queries the necessity of fully internalizing the ecological and social costs in cost-benefit analyses (World Bank, 1995). The discussions over projects such as the Three Gorges or the Narmada Project are an indication that the two aspects – environmental debate and the political economy of the situation – have induced major international credit institutions to refrain from granting loans (see Section D 3.4.3.2).

Even in developing countries, pressure from environmental organizations is prompting a critical examination of projects; such tendencies act to mitigate the centralization of economic policymaking. In the final analysis, it is trends in the psychosocial sphere, such as growing environmental awareness and the mounting interest in participation displayed by groups and individuals, institutionalized here in the form of non-governmental organizations, that are beginning to affect policymaking. The abandonment of the Okavango Project in Botswana, for example, can be attributed to intensified environmental debate since the 1980s (Pearce, 1992), facilitated by increased networking on the part of environmental organizations through the development of information technology (e.g. access to environmental information via the Internet). In the industrialized nations, there is a noticeable tendency towards enhanced environmental protection at national level and improved opportunities for individual legal recourse, e.g. rights of appeal for individuals and minorities.

The transfer of technology and know-how plays a major role, in that the know-how necessary to construct a large dam is usually not available in developing countries and has to be imported through experts or on-site training. This type of transfer is given added impetus by the fact that hydropower as a renewable energy source is acquiring a growing status in the context of sustainable development. Sometimes a "dilemma of environmental groups" results – particularly in industrialized countries; on the one hand, growing environmental awareness and a sensitization to global problems (such as the enhanced greenhouse effect) lead to a fundamental preference for renewable sources of energy while, on the other hand, the construction of large hydroelectric power stations is rejected because of the environmental impacts at local and regional level.

3.4.2.5 Syndrome coupling

The distinguishing feature of the Aral Sea Syndrome is that it is driven by numerous other syndromes, on the one hand, and that, on the other, it may itself act as a trigger and accelerator of other syndromes of global change (for a definition of the syndrome links see WBGU, 1997). In general, however, the syndromes are so closely linked to each other that it is virtually impossible to differentiate between driving forces and impacts. In Fig. D 3.4-1, the points where linkages exist to other syndromes of global change are depicted by "clouds". There are major interfaces to the Favela, Sahel, Dust Bowl, Green Revolution and Asian Tigers Syndromes. Both the degree of linkage and the scope of historically observed interactions vary.

GREEN REVOLUTION SYNDROME AND DUST BOWL SYNDROME

The Aral Sea Syndrome is very intensively coupled to these two syndromes. The link is generated by the intensification of agriculture, particularly by the interest in expanding irrigation discussed elsewhere in this Report (see Section D 3.3). In the case of major irrigation projects and dams, the ecological consequences arising from the linkage to the Green Revolution Syndrome are of the same order of magnitude as the direct impacts within the Aral Sea Syndrome itself.

THE FAVELA SYNDROME

The main linking trend is urbanization, which summarizes the typical processes of the Favela Syndrome as they relate to the Aral Sea Syndrome (Section D 3.6). Of special relevance here is the increased demand for energy as a result of urban agglomeration, particularly for electricity. The boost in growth and development triggered by power supplies may create a pull effect for in-migration of the rural poor. However, it is also possible that the linkage is first generated by remedial action against the Favela Syndrome – such as setting up a power supply based on renewable energy sources to improve living conditions in the city. The power supply for the residents in Ascunción, for example, is closely connected with the water development projects on the Rio Paraná. Within the framework of a large-scale irrigation project, land-use rights are changed to such an extent that the local population is forced to migrate to the cities.

SAHEL SYNDROME

The coupling mechanism to the Sahel Syndrome consists in the fact that the local population in many cases is prohibited from using or unable to use traditional farming methods once the project has been completed, or traditional land-use rights are altered. Traditional cultivation of some floodplains in northeast Nigeria, the so-called Fadamas, for example, had to be discontinued after the construction of two dams – the employment that remained on the new cash-crop farms was inadequate to compensate for the loss of income (WBGU, 1997; Cassel-Gintz et al., 1997). The case study on the Three Gorges Project also provides clear indications that the Sahel Syndrome was triggered or exacerbated on marginal slopes due to resettlement (see Section D 3.4.3.2).

ASIAN TIGERS SYNDROME

Rapid economic growth, which is frequently followed by corresponding improvements in environmental protection, can only be achieved with greatly increased inputs of energy and water. As examples in Brazil, Malaysia and China show, large hydroelectric plants are installed where geographical conditions are appropriate in order to meet the higher energy demand resulting from industrialization.

3.4.3 Examples

3.4.3.1 Aral Sea

The salinization and desiccation of the Aral Sea is a complex case of severe ecological devastation unleashed by a gigantic irrigation project (Létolle and Mainguet, 1996). Implementation of the planned measures over the last 30 years has led to the decline of a fertile, densely forested and species-rich region in which the population lived primarily by fishing and agriculture, ending in large-scale desertification and devastating consequences for the economy and society (Giese, 1997).

The Aral Sea, once the fourth largest freshwater lake in the world, has a catchment basin measuring

2 million km² and is situated in the arid and semi-arid region of central Asia. The latter comprises the young independent republics of Uzbekistan, Tadzhikistan as well as parts of Kazakhstan, Kirghizia, Turkmenistan, northern Afghanistan and northern Iran. The lake's two main feeders are the Amu Darya and Syr Darya rivers, which have their source in the mountainous regions of central Asia and Kazakhstan.

In the 1950s and 1960s, the USSR hoped that this megaproject, which was geared exclusively to maximum agricultural yields, would increase production and create new sources of foreign exchange, primarily from cotton exports. No studies were carried out regarding the morphological, ecological, economic and sociocultural impacts of such a large project (Kasperson, 1995).

Since the 1960s, expansion of the irrigation system to supply the growing areas of irrigated cropland involved heavy withdrawals from the rivers feeding into the Aral Sea, reducing the inflow of water by 94% (FAO, 1996c). This, in turn, has altered the water balance of the Aral; the salt concentration increased from 12 to 33‰. At the same time, the volume of the waterbody has declined by two thirds (Fig. D 3.4-3). The surface area of the Aral Sea has been halved and 30,000 km² of saline soil on the sea floor has been exposed. The entire flora and fauna of the lake, including 266 known species of invertebrates, 24 fish species and 94 species of higher and lower plants has been decimated, among them four sturgeon species numbering among the world's oldest genuses of bony fish (Kasperson, 1995). The saline water that remains and the exposed lake bottom – a salty desert – offer a meager habitat for plants and animals and cannot be used any longer for either agricultural or fishing purposes. 60,000 jobs were lost in the fisheries industry alone.

The shrinkage of the sea has changed the climate of the Aral region. The reduced moderation of temperature fluctuations has induced a trend towards a more continental climate, with hotter summers and colder winters. Storms transported salt from the area formerly covered by the sea to the surrounding regions, causing soil degradation there.

High water tables produced by intensive irrigation on 50–90% of the cultivated area (Kasperson, 1995) lead to soil salinization in arid regions due to the high evaporation of capillary water and thus to declining yields and crop quality. After the initially high yield increases of 67% in the first 15 years of the project, yields have dropped by 15% since 1975–1985, even though the use of fertilizers and biocides was far above the average for the USSR and the area under cultivation was expanded more and more.

Excessive use of the soil beyond its natural potential and the growing of cotton cash crops in monocultures produced an extensive web of social and economic damage.

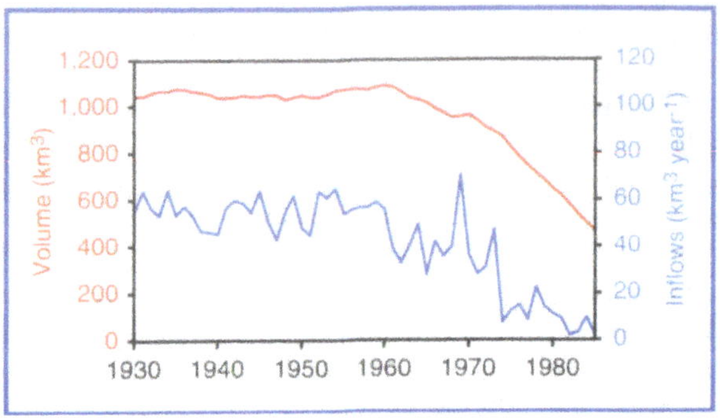

Figure D 3.4-3
Total inflows and volume of the Aral Sea (1930–1985). Source: Gleick, 1993

The health of the population, which had risen to 50 million in the meantime, deteriorated because of the declining quality of the water and the environment. This is mainly attributable to the lack of wastewater treatment and the pollution caused by agriculture. Contact with wastewater contaminated with pesticides (DDT and defoliation agents), particularly in labor-intensive cotton production, led to a 15-fold increase in mortality due to cancer, tuberculosis, typhoid and other diseases among the women and children working there (Glazowsky, 1995).

Estimates of the consequential costs of the Aral Sea disaster show that the former economic benefits of the projects have meanwhile turned into 15–30 million rubles of direct economic losses at the Aral Sea, around 100 million rubles in the catchment area of the rivers feeding the lake and a total of 37 billion rubles in subsequent damage to the environment, health and the economy (Kasperson, 1995).

The example of the Aral Sea shows that the impacts of large-scale water development projects are difficult to control and that they may result in serious damage to the global economy and the global environment. Long-term restoration of the Aral Sea's ecosystem would require that water withdrawals for irrigation be reduced to a fifth of the current level (FAO, 1996c). Achieving this would mean having to abandon marginal areas under cultivation, reduce cotton and rice farming and enhance irrigation efficiency. Because even then the water volume would not be enough to meet all demands, the industrial and settlement structure would have to be changed, something that could only be achieved through political agreements between the countries affected. Complete restoration is now considered impossible; one current World Bank project aims only at rehabilitating certain sections (Whitford, 1997).

3.4.3.2 The Three Gorges project

A dam of colossal proportions is currently being erected on the Yangtze (Chang Jiang) River, at 6,300 km the world's third longest river and the largest in China (1.8 million km² catchment area). The Yangtze begins in northern Tibet and flows eastwards through China. In the Sichuan and Hubei provinces, the river forces its way through a 200-km series of deep and narrow gorges called the Sanxia or the Three Gorges. It is here that China's largest hydroelectric power station will be built, the biggest water development project in the world today.

The plan to contain the river in the Three Gorges with a huge dam dates back to the 19th century. After the Great Flood of 1954, which claimed the lives of 33,000 people and left a million homeless, planning commenced for a multipurpose dam that would not only protect against floods but could also be used to generate electricity and to improve navigation on the river.

During the 1980s, a US study warned against the tectonic and military dangers, and assessed the project as unprofitable in economic terms. Instead, an alternative plan involving a series of smaller dams was developed, but the Chinese authorities continued to favor the megadam version. A Canadian feasibility study concluded that construction was technically, economically and environmentally feasible (CYJV, 1989). In China itself, a 400-strong committee produced a parallel but secret feasibility study, which drew similar conclusions (Barber and Ryder, 1993).

The costs will amount to more than US$ 20 billion, with construction taking about 18 years. The reservoir behind the dam will be 600 km long, inundating several 10,000 hectares of land, including 23,800 hectares of cropland and 5,000 hectares of fruit groves. The region has been settled since prehistoric times. According to the Chinese authorities, the waters will submerge 13 cities (including Wanxian with 140,000 and Fuling with 80,000 inhabitants), 140 villages, 657 factories as well as more than a thousand cultural and archeological sites (Beijing Review, 1992, quoted from Freeberne, 1993; Barber and Ryder, 1993; Aksamit, 1996). Moreover, one of China's most beautiful landscapes, part of China's cultural heritage and one of the country's most important tourist attractions will be submerged forever. More than 1.1 million people will have to be resettled (Ex-Im Bank, 1996).

The project is highly contentious even within China on account of its environmental and social impacts. In response to public protest, hundreds of delegates to the National People's Congress voted in 1989 for the project to be postponed to the next century. A few months later, the "political spring" was over, the protest movement silenced by the events on Tiananmen Square. The leaders of those resisting the dam were put under arrest, and critical publications censured (Dai Qing, 1994).

Within a year, planning had recommenced. In April 1992, without prior debate, the issue was again put to the vote in the National People's Congress. A majority voted in favor of the project, but 30% voted against or abstained – a sign of considerable dissent within the ranks. Building work on the mammoth project started in December 1994.

The benefits of the Three Gorges project

The Three Gorges project on the Yangtze in China is planned as a multipurpose installation, one of the main functions being to protect people living in the flatlands between Yichang and Shanghai against floods (Lin, 1994). The fact that the power station will be the biggest hydroelectric power station in the world, with an output of 17,680 MW, leads to the eyes of the world focusing almost exclusively on energy production as the project's main objective. The Chinese authorities, however, will usually emphasize the absolute necessity of the project for the safety of the downstream population (Lin, 1994), referring in this context to the impacts of previous flooding along the Yangtze (Table D 3.4-2).

The city of Yichang lies directly at the end of the Three Gorges, and the tremendous volumes of water that flow when the river swells caused breaches in the dikes below this city every two to three years between 1499 and 1949. Streamflow measurements have been carried out since 1922 (earlier figures are estimates). The floods of 1931 and 1954 inundated areas the size of The Netherlands and caused very many deaths. Flood control therefore seemed to be the primary reason for building the dam.

Flood statistics based on very old records from Yichang show that a high-water event with a recurrence time, T, of 100 years amounts to 86,300 $m^3 sec^{-1}$, while an event associated with T = 1000 years amounts to 105,000 $m^3 sec^{-1}$. The critical reach of the Yangtze, 300 km long and with dikes 10 to 16 meters high, is the middle section between Yichang and Shanghai. A breach of the dikes in this section would endanger more than 15 million people and 1.5 million hectares of cropland. Since the revolution in 1949, the People's Republic of China has made enormous efforts to improve and raise the existing dikes. However, it is technically impossible to cope with a maximum streamflow of more than 60,000 $m^3 sec^{-1}$ through this section. To ensure safety against flooding with streamflows between 60,000 and 80,000 $m^3 sec^{-1}$, it is necessary to inundate the old flood basins, which today are densely populated. The most impor-

Year	Streamflow at Yichang ($m^3 sec^{-1}$)	Flooded Area (km^2)	People affected (Mio.)	Deaths
1870	105,000			
1227	96,300			
1560	93,600			
1153	92,800			
1860	92,500			
1788	86,000			
1796	82,200			
1613	81,000			
1981	70,800	–	–	–
1954	66,800	32,000	18,9	30,000
1931	64,600	33,000	28,9	145,000
1949	58,100	18,000	8,1	5,700
1935	56,900	15,000	10	142,000
1983	53,500	–	–	–

Table D 3.4-2
Historical streamflow maxima in the Yangtze River at Yichang (below the Three Gorges dam) and extreme flood levels during the 20th century.
Source: Lin, 1994

tant flood basins that would then have to be used have a population today of more than 350,000. These people would have to be evacuated within three days – an almost impossible task to organize. Were the river level to rise even higher, breaches in the dikes of the middle section would be almost inevitable. The number of people who drown accounts for only part of the total death toll, since many people die as a result of other impacts of flooding (epidemics, etc.). Engineers in China point out that the number of people threatened by a major flood is much higher than the number who have to be resettled in order to build the dam.

The Three Gorges dam is therefore planned to absorb the peak streamflow associated with T = 100 without having to inundate flood basins; this necessitates a reservoir capacity of 22.5 billion m^3, while the maximum streamflows for T = 1,000 are supposed to be diverted with the help of the flood basins, without human life being endangered and damage being caused.

The reservoir needed for this purpose is gigantic – measuring approximately 175 meters in height and with a storage capacity of 39.3 billion m^3. The reservoir will fill the Yangtze valley along a 600 km stretch. It will improve navigation, but this will require a series of four locks each 34 meters in height to overcome an elevation of more than 130 meters. Development of transport infrastructure, as the precondition for economic development, is an additional motivational factor behind construction of the dam. Navigation on the Yangtze River would be improved above all in the Three Gorges region, which has been a dangerous stretch hitherto. The transport capacity is expected to increase from 10 to 50 million tons per annum (Bosshard and Unmüßig, 1996).

Investigations have been carried out in Wuhan and Peking using very large hydraulic models to study siltation processes in the reservoir – Chinese engineers are anxious to avoid at all costs a reservoir siltation fiasco such as occurred at the Sanmenxia dam on the Huangho River.

The Three Gorges dam is planned to produce 18 GW of electricity – equivalent to the output of twelve large nuclear power stations. It can be anticipated that the rapid economic development in China, predominantly in the industrial sector, will result in a substantial increase in energy demand (WBGU, 1996). China currently produces about three quarters of its electricity in coal-fired power stations. To reduce impacts on the environment, and especially on climate, generating power from renewables is particularly desirable. However, there is a contradiction between the goals of flood control and power generation. The higher the water level, the more energy can be produced, but this also means that there is less extra storage capacity to act as a buffer against floods. This conflict of objectives is relatively weak in the case of large reservoirs, however, since a small drop in water level results in only slight loss of energy, but in much additional storage capacity.

The negative impacts of the Three Gorges project

The project is controversial because all these undisputed benefits are acquired at enormous expense to society and the environment, and because of doubts about how realistic the planning figures are (Fearnside, 1988; Barber and Ryder, 1993). Environmental organizations are not the only ones to raise objections. The United States Bureau for Reclamation, for example, a special authority with experience in dam-building, has adopted a critical stance towards the project (IRN, 1996). After a two-year evaluation, the Export-Import Bank of the USA concluded in May 1996 that the project in its current form fails to comply with the Bank's environmental guidelines (Kamarck, 1996). The World Bank refused some time ago to lend its support on account of the incalculable financial risks involved (IRN, 1996). The Federal Government has taken a very different approach: in late 1996, Germany approved Hermes export guarantees for construction of the dam in order to ensure that German exporters do not suffer a competitive disadvantage compared with competitors from other countries (Bundestag publication no. 13/5348). Following this decision, Switzerland and Japan announced similar moves to grant export credit guarantees.

Critics of the project make frequent reference to the many environmental, economic and social problems and impacts it involves. They question whether the explicit goals of the project are realistic, including reservations about the design and technical aspects of the project (with regard to safety, flood control and sedimentation), but doubts are also raised about the economic benefits. The method of cost-benefit analysis is similarly a target of criticism, because it gives too little consideration to environmental and social costs, or because the measures planned to mitigate these impacts appear inadequate.

The cost-benefit analyses in the various feasibility studies provide clear indications of the major economic risks implicit in the project. Financial calculations, for example, were based on a discount rate of 15%. A more realistic value would be 12%, as applied by the World Bank when assessing similar projects (such as the Narmada project in India). The Canadian feasibility study concedes that the net benefit of the project would then fall by 59% (CYJV, 1989). Any cost explosion during the project harbors major risks. Estimates ran to US$ 4.5 billion in 1986, but had risen to US$ 26 billion by 1993.

One of the greatest problems when operating dams is control of sedimentation. The Yangtze transports more than 500 million tons of sediment per year, much of which will be deposited in the reservoir due to the reduced flow rate. Other dams in China (Sanmenxia and Gezhouba) show that the rate of sedimentation is mostly underestimated, altering the relation between the costs and benefits of a dam to such an extent that it becomes unprofitable. Nevertheless, the feasibility study comes to the conclusion that the reservoir's life span is virtually unlimited under proper management. However, the assumptions on which the sedimentation analysis in the feasibility study were based have been called into question by American experts (Williams, 1993c). Backwater sedimentation can also hinder navigation in the upper reaches of the reservoir.

When dams disrupt the delicate balance between sedimentation and sediment flow, large-scale irrigation systems downstream may be damaged, and dikes along the river exposed to increased erosion. In the Yangtze estuary, the shoreline will erode and intrusion of salt water will increase, with corresponding impacts on ecosystems and on fisheries yields. These impacts were obviously left out in all feasibility studies.

A breach in the Three Gorges dam would rank as the greatest man-made disaster of all time (Williams, 1993c). Dam-bursts are no rare occurrence in China, and have happened to 3,200 of 80,000 dams since 1950. Earthquakes can also lead to dams collapsing. In the event of war, a military assault on the dam is very likely. During the planned 18-year construction phase, secondary dams must be built that would probably be unable to withstand a 1-in-100-year flood. The feasibility study itself assesses the likelihood of such an extreme event to be 1:20 (CYJV, 1989, quoted in Barber and Ryder, 1993), which seems unacceptably high considering the gigantic scale of damage that might be caused.

In addition to the risks above, the area covered by the reservoir contains a number of waste and toxic sites which should not be submerged unless there is a clean-up operation beforehand. Standing waters must comply with higher standards with regard to the treatment of industrial and municipal wastewater; these follow-on costs must be included in the calculations. The danger of methylmercury being released from flooded land and accumulating in the food chain up to humans (CYJV, 1989) has been given inadequate consideration so far.

The scale of planned resettlement is enormous, with more than 1.1 million people affected (Kwai-cheong, 1995). Resettlement costs are stated as 32% of total costs. Large-scale resettlement programs in China have rarely been carried out in a satisfactory way, and have mostly involved great distress (Freeberne, 1993). On the other hand, the population carrying capacity of the new settlement area, an environmentally very fragile, mountainous terrain uphill from the reservoir, was already exceeded in 1985 by 15% (Gao and Chen, 1987, quoted in Kwai-cheong, 1995). 46% of the cropped land has an inclination of 25° or more. The rates of deforestation and erosion in the region are already very high (Shi et al., 1987, quoted in Kwai-cheong, 1995). Settlement in this area will probably be an exposition factor sufficiently powerful to trigger or further intensify the Sahel Syndrome (for detailed treatment of the Sahel Syndrome, see WBGU, 1997). It will be difficult to find a form of land-use for this region that is environmentally sound in the long term. The high erosion rates already evident will increase still further, producing sediment loads that will curtail the project's life.

It will not be possible to cushion in any adequate form the economic or the social and psychological effects of compulsory resettlement on this scale. A prerequisite for the success of such drastic interference with the social fabric of a region is timely participation of the population in transparent planning and execution of the resettlement program. There is every reason to doubt whether such frameworks exist at all in present-day Chinese society. One indicator of this is that resistance to the dam within China is being suppressed, and critical texts may not be published (HRW, 1995).

In addition to the degradation of ecosystems already mentioned, both downstream and as a result of resettlement, the river would be turned into standing water along the 600 km stretch of reservoir, with inevitable and radical consequences for abiotic conditions (e.g. reduced flow rate, anoxic conditions in deep water, release of nutrients, increased biomass production) and hence for the biological community of the river. Migratory species will be prevented from changing their habitat due to the dam, and could become extinct. There is a high probability that the small populations of endemic river dolphins (*Lipotes vexillifer*, the stock of which is currently less than 200 individuals; Renjun, 1991) and alligators will be wiped out.

Megadams are not the only way to use hydropower or to ensure protection against floods. There are viable alternatives to the Three Gorges project that would enable a more environmentally sound use of hydropower and help reduce the threat of flooding in the lower reaches of the river.

The potential for decentralized small and medium-sized power stations along the Yangtze is considered to be high, and is far from being fully exploited. Moreover, energy efficiency in China is very low, which means that investments in efficiency improvements should be compared to investments in energy production. Alternative methods of flood control (more investment in the existing flood control system, preventing erosion, reforestation) should be examined.

In an overall cost-benefit analysis of the Three Gorges project, it would have made sense to carry out a parallel assessment of alternative solutions in the form of a set of decentralized measures, such as those recommended in a study carried out by US experts in the 1980s.

Recommendations for policy action

Germany's approval of government guarantees ("Hermes" credit guarantees) relating to the Three Gorges project was not based on any cost-benefit analysis accessible to public scrutiny; the statement issued by the Ministry for the Economy referred exclusively to competitive factors as the rationale behind the decision. One reason put forward was that the granting of Hermes credit guarantees in Germany has never required compliance with environmental or social criteria.

The Council recommends that government credit guarantees for large-scale projects having such immense environmental or social impacts be linked in future to compliance with environmental and social standards, and that public access be granted to the relevant studies. In the case of the Three Gorges project, the recommendation to the Federal Government is that it desist from supporting the project unless a prior assessment is carried out into the social and environmental impacts according to the criteria referred to below (see Section D 3.4.5.2).

3.4.4 Indirect measurement of syndrome intensity

As already explained, the Aral Sea Syndrome is of relevance for global water problems for two reasons. Firstly, there are elements in the syndrome mechanism that have a direct effect on water quality and quantity. These include the influences on groundwater, the change in river physics and chemistry (sediment loads, temperature, oxygen concentration, etc.) or possible evaporation effects. The second and more important factor, however, is that the syndrome involves the indirect environmental degradation and social impacts associated with large-scale technical projects for water resource development.

By measuring intensity, the attempt can be made to specify regions in which the syndrome can already be observed (Schellnhuber et al., 1997). The intensity of the Aral Sea Syndrome thus provides an indication of ecologically and socially counter-productive water development projects. Accordingly, recommendations for action will be aimed more at cure than at prevention. Determination of the disposition, by contrast, serves to identify the regions threatened by the syndrome (Schellnhuber et al., 1997). In the context of the intensifying global water crisis and the hydropower potential that is still unexploited, it can be expected that large projects will continue to be planned and built in the future. The degree of disposition is therefore of crucial importance for focusing policy on preventive measures to counteract the Aral Sea Syndrome.

Unfortunately, currently available data does not permit a completely independent determination of the two indicators. For this reason, a middle way is taken here, aimed at determining the degree of vulnerability. This does not refer to the susceptibility of a region to the entire syndrome, but only to its primary negative impacts (see Section D 3.4.2.3 and D 3.4.2.4). This serves to identify those regions in which implementation of a large-scale water development project involves a significant risk of damage. In combination with a map showing the distribution of large dams on the world's major rivers (ICOLD, 1984 and 1988), the determination of vulnerability permits an indirect measurement of intensity: vulnerable regions with numerous dams in the upper reaches of rivers are usually affected by the syndrome because "successful" dam projects still represent the exception to the rule and have not been achieved until recently.

3.4.4.1 Measurement of the core trend "changes in surface runoff"

As described above, the "changes in surface runoff" trend is the key interface between implementation of a large-scale technical project and its subsequent impacts. Therefore, it is critically important in determining the intensity of the Aral Sea Syndrome that this trend be identified appropriately. Unfortunately, there is insufficient data to describe the specific impacts of a dam or a similar water development project completely, since data on the precise location are not available for all projects. For this reason, the first step was to count the number of dams measuring more than 30 meters from foundation to crest in a given country, on the basis of the World Register of Dams (ICOLD, 1984 and 1988); the figures for some large countries (Australia, Brazil, China, India, Canada, Mexico, South Africa and the USA) were broken down according to provinces or states.

The Aral Sea Syndrome differs from other syndromes in that a geographic link exists between the causes and effects due to the natural course of the rivers: the main effects of intervention in the water balance are found in the lower reaches of the river as well as the actual site where intervention occurs. To determine the dependence of environmental impacts on the volume of flow and of particulate or dissolved matter, a simple model based essentially on graphic processing of a global digitalized river network and validated by a sufficient number of measuring stations was developed to describe these flow volumes (Petschel-Held and Plöchl, 1997).

On this basis, the first step was to calculate the number of dams in a country or province as an aggregated indicator of the influence on the riverine systems of a region. This indicator is shown in Fig. D 3.4-4 as a function of the area covered by regions through which a larger river flows. The high values in California and Colorado (USA), southern Europe, China and Japan are particularly striking. Other regions that stand out by virtue of the subnational resolution include Kwazulu-Natal in South Africa, Tamil Nadu in India and Victoria in Australia.

The number of dams per area with flowing waters was combined with the global river network to produce a *dam impact indicator* for the expected number of dams at the local site itself and in the upper reaches of rivers in relation to the respective annual flow volume. The indicator thus describes the changes in surface runoff that have occurred through the impact of all dams in the region, but it cannot be used to analyze the consequences of specific large-scale projects (Fig. D 3.4-5).

3.4.4.2 Measuring vulnerability

On the basis of the syndrome-specific network of interrelations, the following effects can be identified as relevant for the determination of vulnerability and verified with indicators:

- Changes in loads of particulate matter: with the help of a global data set (resolution 0.5° x 0.5°) for sediment deposition in a region (Ludwig and Probst, 1996) and the above-mentioned river model, the natural sediment transport was estimated (i.e. without taking existing river-regulating structures into consideration) and used as a measure of the vulnerability of the region to the effects of the change in sediment load due to dams.
- Impacts on terrestrial and marine ecosystems: the geographically explicit mapping of wetlands according to Matthews and Fung (1987) was used to develop indicators of the sensitivity of terrestrial ecosystems. Based on the high degree of dependence of marine and coastal ecosystems on the sediment loads of nearby rivers, the total substance discharge by the respective river into the sea was used as an indicator of the vulnerability of these systems.
- Health hazards: large-scale water development increases the risk of infections due to schistosomiasis. On the basis of work carried out by Martens (1995), who studied epidemic potential as a function of temperature, and an aggregated indicator for an adequate supply of water and sanitation facilities (WRI, 1996), an indicator of health risks was developed. The absence of digital data on the propagation area of the carrier snails – the hosts for the parasite – makes the indicated risk regions appear larger than they currently are. On the In-

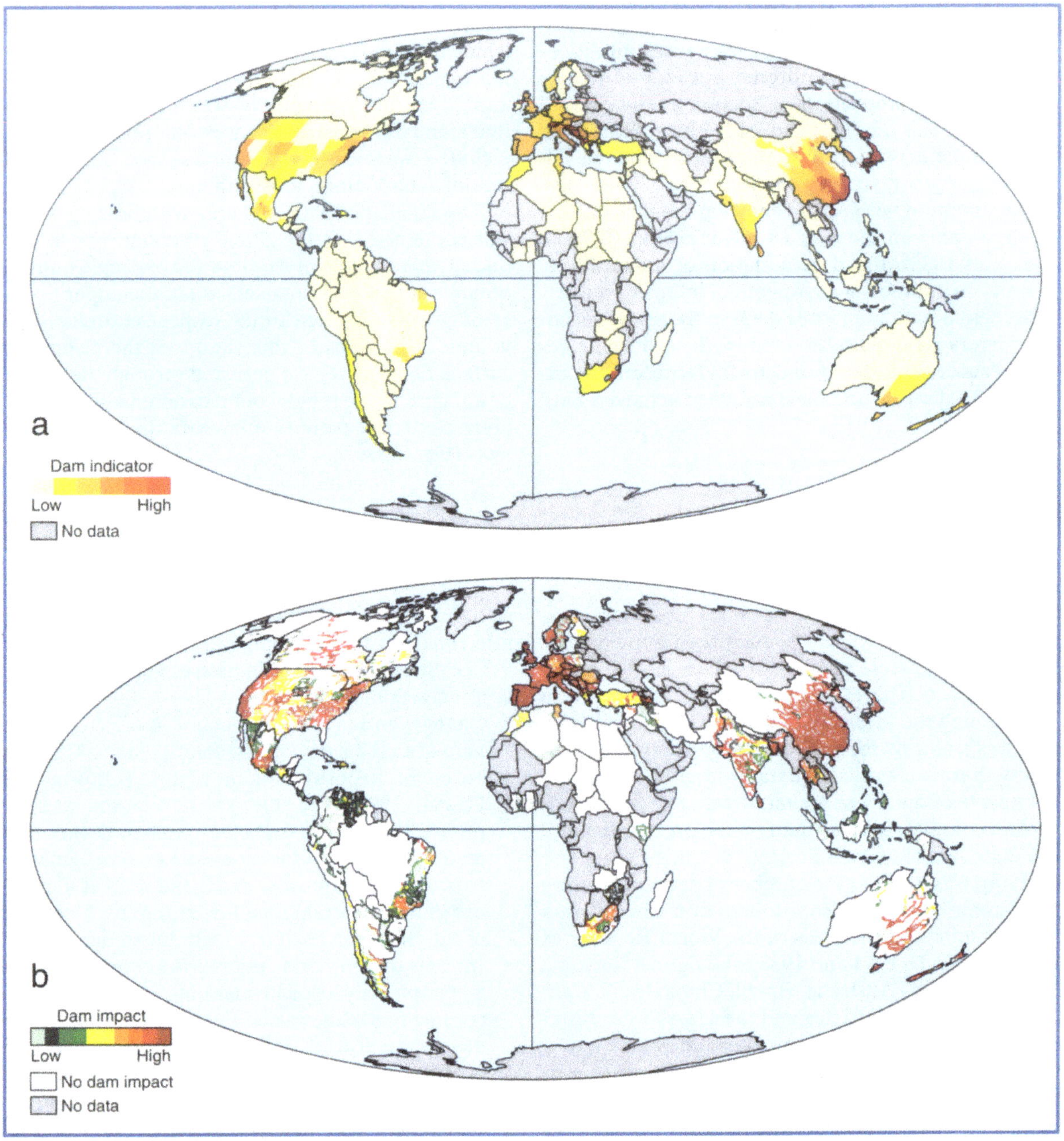

Figure D 3.4-4
a) Number of dams in a province or country in relation to the total river length in the respective region.
b) Dam impact indicator, i.e. expected number of upstream dams per km³ of annual flow volume.
Source: BMBF project "Syndrome dynamics", PIK core project QUESTIONS and WBGU

dian subcontinent, for example, one can find both suitable climatic conditions and extremely poor sanitation facilities. However, since the carrier snail of the pathogen for schistosomiasis is at present hardly encountered there (WHO, personal report), only a few cases of infection are known.

- International conflicts: with the help of the flow model mentioned in Section D 3.4.4.1 and an internal assessment by Council experts, the risk of an international conflict being triggered by large-scale water resource development projects was determined on a geographically explicit basis. This risk is not integrated into the overall vulnerability, but entered as a risk in the intensity map described in Section D 3.4.4.3.

Section D 3.4.2 shows that numerous other categories of damage appear besides those mentioned. On the one hand, however, as in the case of land lost

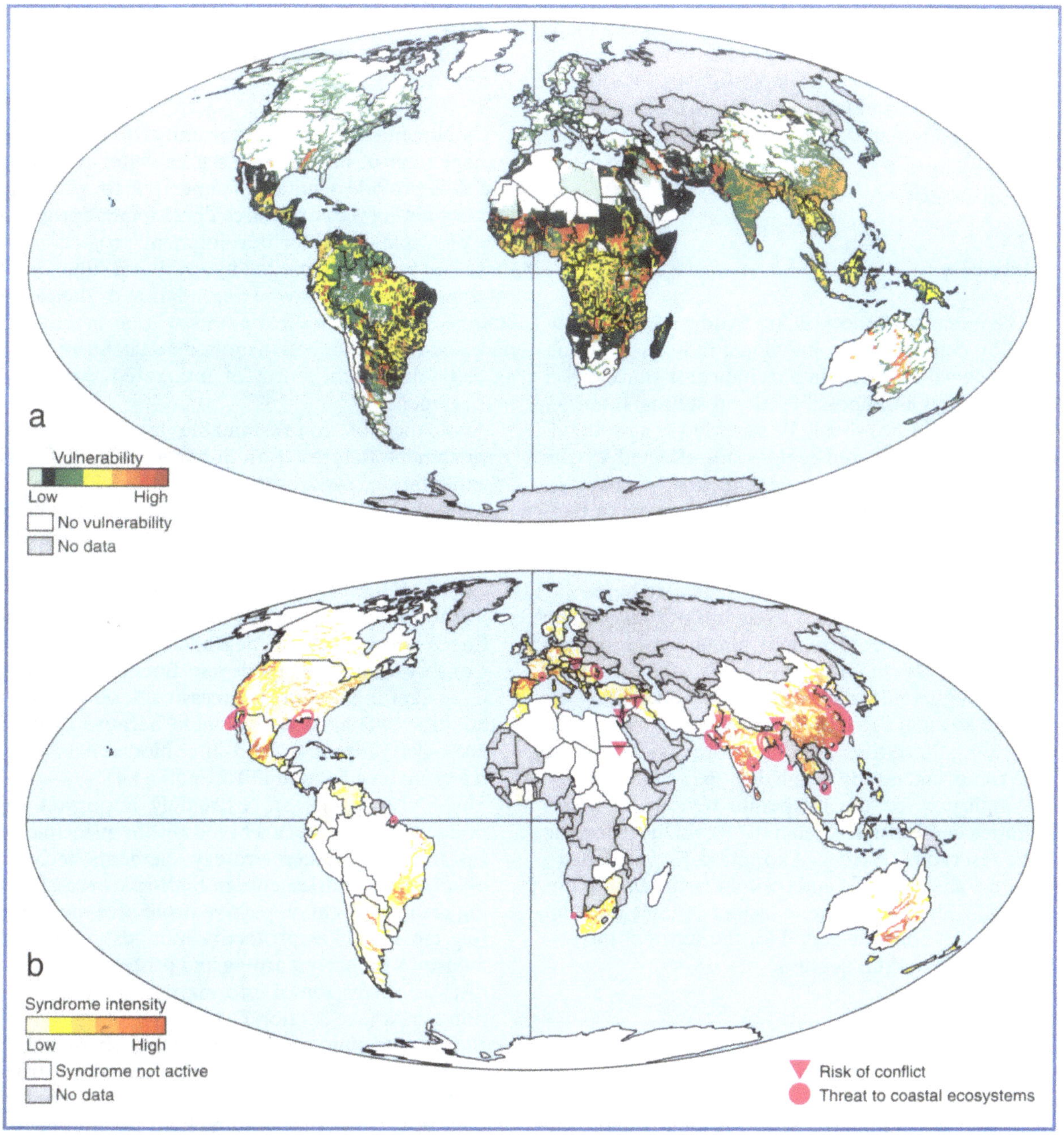

Figure D 3.4-5
a) Vulnerability to severe damage to nature and people due to the construction of large dams. Vulnerable coastal areas are marked by the shaded circles.
b) Intensity of the Aral Sea Syndrome. Because of the uncertainty as to the data regarding number and site of large dams, no assessment was conducted for the countries of the former USSR.
Source: BMBF project "Syndrome dynamics", PIK core project QUESTIONS and WBGU

through inundation by reservoirs, the spatial reference is directly correlated to the dam and thus need not be explicitly included in a first approximation. On the other hand, there is a serious lack of data on the socioeconomic damage at the regional level. Therefore, the overall vulnerability derived from the indicators mentioned and depicted in Fig. D 3.4-5a must be interpreted as an underestimation, i.e. the actual vulnerability is probably higher as a rule.

The degree of vulnerability in the tropics is usually higher than in temperate zones. On the basis of the system of indicators we have developed, this can be explained by the much more unfavorable climatic water balance, the higher risk of contracting water-related vector diseases and the higher sediment loads of the rivers. The river valleys of the Nile, Niger and Mississippi are striking examples of this, in addition to extensive areas in India and China. Due to its sediment load, the Amazon is also indicated as partially vulnerable.

3.4.4.3 Intensity

The intensity of the Aral Sea Syndrome was calculated by combining the individual indicators of vulnerability with the dam impact indicator. Since direct measurement is impossible, the resulting intensity map (Fig. D 3.4-5b) should be regarded as a probability distribution, i.e. red regions are affected by the Aral Sea Syndrome with a higher degree of probability than regions colored green. Overlaps were handled in such a way that the indicator provides an underestimation, i.e. the actual intensity of the syndrome is likely to be higher than indicated here.

Southern China and large sections of India can be identified as the core regions of the syndrome, although they also include provinces or states in Mexico (Guanajuato, Zacatecas), Brazil (São Paulo, Minas Gerais) and Australia (New South Wales, South Australia). Since they are only characterized by the core trend, and not by a high degree of vulnerability, the indicator for the temperate regions of North America and Europe is generally weak. In Africa, the large reservoirs (Assuan, Akosombo, Koussou, Kariba and Cabora Bassâ) and their downstream regions are easily identified. Due to lack of reliable data, indicators were not calculated for the states of the former USSR (marked in gray).

3.4.5 Recommended action

In the case of the Aral Sea Syndrome, the recommendations for avoiding or mitigating the social, ecological and economic consequences of large-scale water development projects are broken down into three categories. This split reflects a ranking of decisions, i.e. the Council favors, with a view to sustainable development, the reduction or avoidance of the disposition to large-scale water development projects involving severe environmental or social impacts. Should such projects prove necessary, however, they should first be subjected to detailed assessment on the basis of full internalization of costs, and only at the end should thought be given to mitigation measures for existing projects.

3.4.5.1 Reducing the disposition to the Aral Sea Syndrome

Catchments are the natural units for sustainable management of increasingly scarce water resources and thus provide a suitable framework for planning and evaluating water resource development projects. The principles of water development projects must therefore be to preserve the integrity and function of catchment areas and avoid degradation of their ecosystems and soils. Several points will be mentioned here and then described in more detail in Section D 5 as examples of this form of integrated watershed management.

- Avoidance of non-sustainable land use in the catchment, deforestation in particular (see Overexploitation Syndrome; Section D 3.2), use of steep slopes for growing crops and other methods that accelerate erosion (see Section D 3.4; WBGU, 1995).
- Conservation or restoration of forests in the watershed may significantly mitigate the threat of floods and prevent soil degradation.
- Conservation of wetlands and floodplains is also important in preventing excessive flood peaks. In addition, wetlands are valuable because of their ecological functions and the biodiversity they maintain (see Section D 1.2 and D 4.4).
- The above points are extremely important for flood control strategies based on the principle of sustainability. Other strategy elements include: avoidance of settlements in high-risk areas, placing greater weight on passive protective measures (e.g. circular dikes, protective mounds), improved measures for early warning and protection against disasters, provision of information to the population at risk (see Section D 1.6).

Before planning specific projects involving major interference with the environmental and social structure of a region, an analysis should be conducted of traditional systems of water and land development.

Firstly, such systems form the basis of sustainable methods of land management that are centuries old in some cases and which may provide suitable alternatives after further development and modification in accordance with the principles of participation and subsidiarity.

Secondly, knowledge about traditional systems is a prerequisite for assessing the impacts of a project. The participation of the local population in planning and implementation is imperative for the success of all these measures (see Section D 5.3).

Finally, it must be kept in mind that many catchments straddle national boundaries. Strategies for preventing international conflicts will be discussed in Section D 4.1.

From the perspective of sustainable development, integrated watershed management must be closely linked to appropriate strategies for improving the efficiency of water and energy use (Section D 3.4.2.2):

- More efficient water use, e.g. by means of modern irrigation methods, by avoiding water losses due to leaks in distribution or recycling systems, or by installing water-saving facilities. Specific references are made to this at various places in this report (e.g. Section D 4.3 and D 4.5). Given the rising agricultural demand for water, the close links to the Green Revolution Syndrome and the action recommended in this connection must be taken into account (see Section D 3.3). Demand in urban areas is influenced by the Urban Sprawl and the Favela Syndromes – recommended action for the latter is provided in Section D 3.6.
- Enhancing the efficiency of energy use offers a number of opportunities for reducing demand, above all in developing and newly industrializing countries in which there is still a considerable potential for increasing energy productivity.

3.4.5.2 Evaluation of large-scale water development projects

If strategies for efficiency enhancement are no longer adequate for meeting the expected increases in demand, the construction of large-scale projects is a possible option requiring due assessment. What is needed in this context is a comprehensive and site-based analysis of the consequences and expectations associated with the project. The Council is not in a position to carry out an evaluation, not even for specific large-scale projects such as the Three Gorges dam. However, recommendations can be given in two directions: first of all, it is possible to establish the guard rails or boundary conditions that must not be exceeded. *Guard rails* are understood as "knock-out criteria", i.e. failure to meet them cannot be compensated for, no matter how well other criteria are met. Secondly, recommendations can be elaborated with respect to the process of analysis and the demands to be leveled at such a process.

GUARD RAILS

An evaluation of water development projects addresses many spheres of impact. On the positive side are the direct beneficial effects, such as power generation, irrigation, flood control, the creation of new habitats for aquatic animals and flora, as well as improved recreational facilities. However, indirect aspects also play a role, e.g. planning security, economic development, job creation, substitution of environmentally unsound energy sources, export earnings, technology transfer and the establishment of efficient institutions.

On the negative side are the direct impacts, such as costs of construction, operation and decommissioning the technical facilities, resettlement problems, job losses (e.g. in fisheries), loss of endangered ecosystems and species, health hazards, safety risks as well as indirect impacts in the lower reaches and in coastal ecosystems due to insufficient sediment loads, increased water pollution, social uprooting and widening economic and social disparities. Inclusion of all relevant aspects in a knowledge-based and balanced process of analyis means considerable expenditure and is not possible without specialized knowledge. Therefore, such efforts cannot be made for all planned projects in general. For this reason, two selection criteria should be applied:

- If one ascertains that only minimal changes are produced in the environment and society, an analysis is not necessary and the project may be carried out, provided it has a positive cost-benefit balance (trivial case).
- If one determines that a project is rated so poorly on the basis of a relevant evaluation criterion that compensation by means of a beneficial criterion does not appear possible, then the project should not be pursued without a detailed analysis (violation of a guard rail).

If cases of triviality or rejection do not exist on the basis of one of the guard rails, then a detailed analysis of the advantages and disadvantages is necessary. Since the manifestations of the Aral Sea Syndrome always take place in connection with projects having far-reaching impacts, the first case of triviality can be ruled out. Thus, it is important to conduct an analysis for all cases connected with this syndrome in accordance with the guard rail principle in order to clarify whether an analysis of the respective advantages and disadvantages is required at all.

Just what are the guard rails, however? In the view of the Council, the guard rails used as criteria should comprise quantitatively or clearly measurable minimal conditions from the perspective of ecology, economics and social acceptability. If one of the minimal conditions is not met, then the project must be rejected, unless all other action options (including that of not taking any action) also violate minimal conditions. In the following, these minimal conditions are formulated for the Aral Sea Syndrome:

Ecology:

- *Biodiversity:* Irreversible conversion of an ecosystem classified as extremely valuable (for example,

with internationally recognized protective status as a World Cultural Heritage region; see Section D 5.6.1), loss of particularly important ecological functions or one or more species regarded as meriting special protection.
- *Environmental degradation:* High probability of irreversible destruction or significant reduction in use potential of water and soil due to salinization, sedimentation or similar processes.

Economics
- *Net benefit*: The net benefit of the project should point to a positive result with a high degree of probability on the basis of a rudimentary cost-benefit analysis.
- *Solidity*: Uncertainty about the expected rate of return may not be greater than a specified percentage of the total investment amount.
- *Management*: A lack of or deficiency in the ability of organizations in the project region to manage the expected secondary impacts and risks appropriately.

Social acceptability
- *Risk*: Graduated risk according to scope (recommendations of the US Corps of Engineers, including a risk aversion factor):
  - $1x10^3$ for up to 100 persons
  - $1x10^4$ for up to 1,000 persons
  - $1x10^5$ for over 1,000 persons.
- *Health*: Expected increase in mortality by a specified factor.
- *Participation*: Minimum of participation and approval on the part of the affected population.
- *Potential for conflict*: Particularly high probability of military conflicts as a consequence of the project.

For every project one must ask whether one of these guard rails has been transgressed. If the answer is yes, one must examine further whether "business as usual" at the same or another place similarly fails to comply with one of the guard rails. Should this be the case, it must be determined whether alternatives exist for steering the project, other projects with equivalent benefits or the status quo out of the "disaster zone" (with an acceptable level of effort and expenditure). If this is possible, the solution with the least expected costs should be taken. If it is not possible, then an evaluation of all options that violate at least one guard rail must be conducted. To ensure that this evaluation can be rationally applied, the guard rails must be defined in such a way that they encompass minimal requirements only.

Procedural recommendations

If none of the guard rails is violated or the proposed project is assessed as complying better with the guard rails than the status quo or another feasible option, then a comprehensive analysis is advisable. This analysis should take into account all the dimensions mentioned above. Given the complexity of these dimensions and the virtually unforeseeable consequences, the Council suggests a multidimensional method of evaluation (Baumann et al., 1984). This involves separate determination of economic criteria (cost-benefit analysis), social criteria, environmental criteria (environmental impact assessment) and risk criteria, and their subsequent organization in matrix form. On the basis of these analyses, the evaluation of the respective project only makes sense in relation to possible alternatives. Assessment must therefore cover the "status quo" option as well as one or more options with equivalent benefits (e.g. decentralized water development projects, or construction of natural gas power stations). Aggregation of the various dimensions is not a task for the analysts, but requires the setting of political priorities. In democratic societies like Germany, discursive forms of weighting should be employed for this purpose.

The outcome of such a process is not predetermined. The prerequisites for such evaluation processes taking effect include the following metacriteria:
- reliable data and results for comparing options,
- qualified statements regarding the uncertainty of the forecasts,
- transparency regarding the decision-making process (who, what, when),
- access to and participation of the major stakeholders (to test the willingness to participate),
- openness of the process,
- necessity of substantiating the specific weighting applied.

These metacriteria are always valid, regardless of whether they are used as guiding principles for the country carrying out the project or as a decision-making aid for the donor country.

Conclusions for the Federal Government
- The first step is to determine whether the first selection criterion (triviality of secondary impacts) is met. If affirmative, a positive decision should be made in favor of support, provided that the cost-benefit balance is positive and the project can clearly be deemed eligible for funding.
- The second step is to examine whether one of the guard rails above is violated by a project with non-trivial secondary impacts. If this is so, the project should be rejected, unless the status quo also violates one or more guard rails. In the latter case, priority should be given to modifying the original

project, or other options with equivalent benefits should be sought for which a violation of the minimal conditions is not expected.

- If the project, in modified or original form, does not violate any of the guard rails, checks must be carried out to determine whether sufficient documents are available to perform a meaningful evaluation according to the above metacriteria.
- If these documents are available, then the Federal Government should either request that an analysis be conducted and disclosed by the country making the application and then evaluate this analysis, or conduct such an analysis itself.
- It will never be possible to obtain all desirable data or meet all participation demands within the framework of the evaluation. Here each individual case must be examined to see whether the requirements for complying with the metacriteria have been met, at least in principle.
- If all these assessments arrive at a positive result, financial support can be recommended. However, the pledge of funding, technical aid or guarantees should be linked to compliance with the respective conditions.
- Since the measures specified in the plans for the execution of large-scale projects (e.g. for the protection of ecosystems or for the involvement of the people affected) are often not implemented at all or only to an inadequate extent, the financing plan should couple the respective payments to compliance with specified target data by certain dates. Here it is important to integrate target cutoff points into the project that make it possible to discontinue the project if serious deficits exist.
- When funds or technical assistance are granted, the current balance of the respective country with such large-scale projects should be included in the analysis of advantages and disadvantages of the project. Even projects that are beneficial in principle may lead to unacceptable risks if institutional deficits (such as a high degree of corruption) exist.

3.4.5.3 Mitigating the impacts of existing large-scale water development projects

The Council holds the view that large dams constitute major interference with nature, the impacts of which are difficult to foresee and almost impossible to mitigate. The hydrological consequences induced by changes in streamflow and sediment balance, as well as most subsequent impacts on the environment, remain unaffected by the measures listed below.

- Measures aimed at integrated watershed management not only can reduce the disposition to the Aral Sea Syndrome, but may significantly enhance projects already implemented. These include, above all, sustainable methods of land use, e.g. reduction of erosion through greater afforestation (see Section D 3.4.5.1).
- Increased sluicing of water from the reservoir has the purpose of raising streamflow and imitating the natural streamflow variability. These measures are expensive, especially for hydroelectric power stations, and can be measured directly in monetary terms; this explains why operators resist them so ardently. Regulations of this nature imposed in the USA have already led to the abandonment of some dam construction projects. Many old dams are not equipped to sluice large quantities of water within a short period of time, as would be necessary to simulate natural conditions in any realistic manner. Initial experience from an experiment in Glen Canyon (Colorado) has been positive in that the river bed has been approximately restored to its natural state (Collier et al., 1997). Further experience should be gained in this respect (see research recommendations in Section D 3.4.6).
- The quality of water flowing out of the reservoir can be positively influenced: examples include oxygen enhancement before or while passing the turbines, or temperature control by exploiting the thermal stratification of the reservoir.
- Attempts to compensate for the negative effects on migrating fish species by means of fish ladders and fish farms must be subjected to critical review. The negative consequences of a dam can be mitigated in this way only for a small section of the biotic environment. One of the prerequisites is a detailed environmental assessment of local conditions, since experience with other dams in other climatic zones cannot be applied without qualification (see research recommendations in Section D 3.4.6).
- "Rescue campaigns" to take wild animals out of flood areas must be regarded more as a public relations maneuver than as a conservation measure, because with the loss of their habitat the animals also lose their life-support system. Relocation of animal populations in neighboring regions is of dubious benefit from an ecological perspective.
- After renewed evaluation of the advantages and disadvantages of an existing project, decommissioning (particularly of dams) may become necessary in extreme cases. Relatively little experience is available in this connection, so further research is needed (see Section D 3.4.6).
- The direct health risks due to large-scale water development projects, in particular for irrigation, are closely linked to the inadequate provision of clean

drinking water and sanitary facilities for the local population (see indicator for schistosomiasis in Section D 3.4.4.2). Wherever possible, such facilities should be upgraded – as part of an overall improvement of the health system – in order to reduce the risk of infection (see recommended action in Section D 4.2).

The evaluation of remedial measures referred to above should be considered in the case of projects receiving development assistance or credit guarantees from Germany; the expenditures for installation and operation should be included in the cost-benefit calculations.

3.4.6 Research recommendations

A substantial and wide-ranging body of knowledge now exists on the impacts of large-scale water development projects. Translating this knowledge to the context of individual case studies is difficult, however; the assessment of environmental costs and their internalization in cost-benefit analyses involves considerable methodological problems. The Council sees an urgent need for research in this context. This applies especially to compliance with the first two metacriteria mentioned in Section D 3.4.5.2, namely "reliable data and results" as well as "qualified statements on the uncertainty of the forecasts".

In support of such research, it is useful to review again the success of previous dam projects – measures funded by the Federal Government in the framework of development cooperation, for example – and to compare this assessment to the analysis of costs and benefits prior to construction. The environmental and social impacts of projects should be observed on a long-term basis in order to improve the basis for assessments and be able to plan remedial measures more successfully.

The Council sees a major research gap with regard to solving the technical and financial problems that arise through the decommissioning of dams. Dams have a limited service life due to sedimentation and structural ageing, for example, or to a subsequently amended assessment of the cost-benefit relation. The decommissioning of dams has been given little attention in studies to date, and the costs of such undertakings have not yet been taken into account in cost-benefit analyses. Fundamental research needs to be carried out in this area, whereby Germany is in a position to gain a lead.

Gaps in knowledge also exist with respect to mitigating the environmental impacts of dams. The technical measures designed to reduce the obstacles to the migration of animals (e.g. fish ladders) are successful in certain cases, particularly for salmonids, but remain inadequate in many other cases. Techniques developed primarily in the USA are often applied to other continents and climates without sufficient investigation of biological and environmental aspects, with unsatisfactory results the inevitable outcome. Experiments with the natural situation of duplicated streamflow regimes may help to reduce adverse environmental effects in lower reaches.

3.5 The Favela Syndrome: Uncontrolled urbanization, impoverishment and threats to water resources and the environment in human settlements

Megacities prone to collapse – Push and pull factors – Failure of governance a driving force – Informal sector ensures survival – Rising water demand in agglomerations puts excessive strains on regional freshwater reserves – Indicators of the syndrome – Wastewater treatment methods – Recommendations

The world's water resources are subjected to pressure by yet another syndrome, one that will have mounting importance for global change in the future – the Favela Syndrome. Evident in virtually all developing and newly industrializing countries, the Favela Syndrome has dire consequences for the people affected, severest impacts on the balance of nature at regional and global level, as well as major indirect implications for the industrialized countries. Rapid urbanization, partially driven by migration from rural areas to the cities, combined with a lack of planning and control capacities at local level, lead to an overloading of urban systems (living and working conditions, transport, supply and disposal services) that in some cities has taken on catastrophic dimensions. The water needs of these agglomerations are staggeringly high, yet the infrastructure for waste and wastewater disposal is totally inadequate. At the same time, large sections of the population do not have sufficient access to safe drinking water and basic sanitation facilities. Such conditions lead to a whole series of typical diseases that can spread to other regions of the world in the course of globalization. After a brief definition of the Favela Syndrome (D 3.5.1), this section analyzes its general mechanism (D 3.5.2), then focuses on the water component and its social and ecological impacts (D 3.5.3) before making recommendations on action to cure the syndrome and, in particular, its water-related problems (D 3.5.4).

3.5.1 Definition

The Favela Syndrome refers to a special development problem resulting from the uncontrolled growth of human settlements – either in specific parts of the core cities or, more frequently, in existing and new centers. Because of the rapid pace of this informal urbanization process and the failure of governance, there is usually little or no state control in the form of land-use and development planning, regulatory control or establishment of the essential infrastructures for supply and disposal. It is this rapidly growing disparity in the urbanization process that forms the actual mechanism of the syndrome, with its characteristic pattern of growth and environmental degradation, and the resultant hazards for people and the natural environment.

In Portuguese, the word favela (English: slums, squatter settlements; Spanish: barrios) refers to an urban settlement typically constructed by rural refugees without proper building materials or official approval. However, it is not just the type of dwelling that defines the Favela Syndrome.

Nor is it confined to those cities in the world that have a high proportion of such settlements. The core element of the syndrome is a mechanism of imbalanced urbanization that can be identified in very different cities, embracing not only settlement types, but also economic and social relations, interconnected flows of materials, substances and energy, as well as natural impacts and feedback effects. This functional unity is what is meant by the term *Favela Syndrome of Global Change* as used in the following. It occurs in cities where one might expect to find the settlement patterns mentioned above – e.g. in São Paulo or Lima. However, inner-city slum areas are also a typical element of global change and part of the Favela Syndrome, even though they are built of solid building materials. Furthermore, the syndrome also includes all those segments in "more developed" cities that display the characteristics shown in the network of interrelations (see below).

"Uncontrolled urbanization, environmental hazards and impoverishment" do not constitute the only syndrome of global change in the urban context, however. Other syndromes besides the Favela Syndrome can usually be found in varying degrees of intensity in urban environments, e.g. the Asian Tigers Syndrome and the Urban Sprawl Syndrome (see Section D 3.2). These syndromes refer to specific patterns of damage, each with their own causes, mechanisms and resultant problems for humans and nature. Their focal aspects, by contrast, are the problems that result from the lifestyles specific to affluent societies or the breathtaking pace at which some "newly industrializing" states are developing. In contrast to the Favela Syndrome, their intensities are not linked to rapid population growth. This does not mean that the Favela Syndrome is purely a developing country phenomenon – manifestations of the syndrome can also be observed in the urban centers of the industrialized countries, and there is a risk that the pressures on megacities will cause them to "collapse", triggering migratory flows that are most likely to point in the direction of the highly developed nations.

In the view of the Council, the process of urbanization as such does not represent a negative trend within the Earth System. The history of humankind is full of examples which show how urbanization has furthered the development of civilizations. From an ecological standpoint, too, concentrating living and production provides a number of benefits (e.g. less land needed per person, more efficient use of infrastructure, higher energy efficiency). Nevertheless, one of the specific features of the urbanization process in the Favela Syndrome is that it places excessive strains on and causes damage to human civilizations and ecological systems on account of the unsustainable trends which characterize it. The focus of our analysis is on this complex pattern of damage. Table 3.5-1 illustrates the quantitative dimensions of uncontrolled urbanization.

Different assessments have been made of the global extent of favela formation. The UNFPA estimates that around 600 million people worldwide live in favelas (UNFPA, 1995). Projections based on the United Nations and UNICEF statistics range from 0.77 to 1.5 billion people who are unable to satisfy their basic needs (food, shelter, health).

There is every indication that the syndrome will grow in significance: by the year 2010 about 3.3 billion people (approximately half the world population) will live in cities; it is estimated that the urban population in 2025 will account for two thirds of the world population (UNPD, 1994). Cities with the highest number of inhabitants in the year 2000 and a pronounced vulnerability to the Favela Syndrome include Mumbai (Bombay), Buenos Aires, Jakarta, Cairo, Calcutta, Manila, Mexico City, São Paulo, Shanghai and Teheran (see Box D 3.5-1). However, not only existing megacities are susceptible to the Favela Syndrome: numerous examples on the African continent, such as Dar-es-Salaam in East Africa and Bangui in Central Africa, show that marginal lands used for agriculture are particularly susceptible to uncontrolled, "explosive" urbanization. The syndrome also includes the vulnerability of settlement structures which – as in the case of Nouakchott, the capital of Mauritania, twenty years ago – have reached a stage of development that can be described

Country	Population (million)	Of which in favelas (percent)	Population of megacities (million)	
Brazil	161	6.9	São Paulo	15.9
			Rio de Janeiro	10.4
Egypt	63	15.1	Cairo	7.7
India	945	7.7 (regionally up to 36)	Calcutta	11.7
			Mumbai	15.1
			Delhi	9.9
Indonesia	200	6.9	Jakarta	11.5
Philippines	69	27.6	Manila	9.3
South Korea	45	14.4	Seoul	11.6

Table D 3.5-1
Inhabitants of favelas.
Source: Oberai, 1993; UNPD, 1994; Kidron and Segal, 1996

as "no longer a village – not yet a city", and which contain the seeds of further uncontrolled and rapid urbanization.

3.5.2
General syndrome diagnosis

There follows an analysis of the symptoms and interactions of the Favela Syndrome, grouped according to the main complexes. The network of interrelations in Fig. 3.5-1 provides an overview.

3.5.2.1
Rural exodus, decline of traditions and uncontrolled urbanization

The Favela Syndrome is a side-effect of the global urbanization process at the end of the 20th century. Worldwide urbanization is characterized by three main features:
- an increase in the absolute number of people living in cities,
- a rise in the proportion of city dwellers among the national population,
- worldwide increase in urban settlements and lifestyles.

The key factors driving these rapid urbanization processes can be roughly divided into two large groups: firstly, the (push) factors prompting people to migrate from rural areas to the cities; secondly, the (pull) factors that attract people to cities (Bähr and Mertins, 1995; Campbell, 1989; Flanagan, 1990; Hardoy and Satterthwaite, 1989).

PUSH FACTORS
- Overpopulation of rural areas.
- Scarcity of cultivable land due, *inter alia*, to large-scale technical projects (dam construction).
- Decline in soil fertility and other natural production factors, degradation of agricultural soils.
- Increasing exploitation of agriculture and soil for market gains (e.g. cash cropping).
- Socioeconomic framework (e.g. dominance of large landowners, lack of rural reforms).
- Demand for higher standards.

PULL FACTORS
- Hopes of securing a livelihood.
- Hopes of finding differentiated labor markets and employment opportunities.
- Expectations of higher incomes.
- Better educational and training opportunities.
- Less social control (higher degree of individualization).
- Better social and urban infrastructure (health care, social security systems, transportation, etc.).
- Better access to information.
- Modern "urbanity" model.

When individual persons, families or entire village clans decide to leave the rural area where they live and work, a specific mixture of cultural and regional factors is usually involved. An extreme case of involuntary rural exodus is when the landless poor are forced by violent action to migrate to cities, whereas the temporary outmigration of young people in order to acquire better education or training represents the other end of the scale. Subjective perceptions and evaluation of objective features and situations (e.g. assessments regarding the future development potential of a region as compared with the city) play just as important a role as these objective factors.

The driving force behind the rapid urbanization process in much of the world is the disparate development of urban and rural areas. Furthermore, most megacities have a functional primacy, meaning that political-administrative, economic, social, cultural and research functions are concentrated in the centers of these urban agglomerations, but are absent in rural regions (Bronger, 1996). In the mid-1980s, Mumbai, Calcutta and Delhi (or 3.9% of the Indian population in 1991) accounted for
- 12.7% of the students,
- 15.5% of the hospital beds,
- 18.3% of industrial output in dollar terms,

BOX 3.5-1

Examples of the Favela Syndrome

The degradation afflicting the urban environment is characterized by the interplay of many different factors. The high concentration of factories and consumption causes high emission levels of air pollutants, made worse by the poor efficiency of most factories. In many cases, local climate conditions prevent the escape of these pollutants from the urban atmosphere, with the result that contamination spreads not only to the soil under the cities, which is already degraded by waste and surface sealing, but also to groundwater resources. Many industrial facilities in these urban agglomerations do not even take minimal precautions to ensure a basic level of safety, proper disposal of waste and wastewater, or emission reductions. Motor vehicles usually lack any kind of exhaust-reducing equipment.

Even at low growth rates, the settlements affected cannot put additional infrastructure in place, because no funding is available and because the local government does not operate schemes for planning and resource management (or the latter are unable to cope with rapidly expanding needs). Existing facilities can only cover a small part of the additional needs, causing the urbanization process to get increasingly out of control and triggering specific people-environment interactions typical of this process – leading to uncontrolled accumulation of domestic and industrial waste, lack of safe water supplies and wastewater disposal, high concentrations of contaminants in soil, air and water, collapsing transportation systems, increasing surface sealing and the acute health risks these factors imply.

MEXICO CITY

6,000 tons of air pollutants are emitted daily to the (urban) atmosphere, constituting a severe threat to the health of Mexico City residents. One refinery and two power plants alone emit 330 tons a day of sulfur dioxide. Soil erosion and the lack of vegetation result in 300,000 tons of dust being generated in the city each year. The wastewater discharged by the Mexico City conurbation is around 50 m^3 a second. Only 70% of it flows into the public sewer system (i.e. for potential wastewater treatment). Domestic sewage enters the environment virtually untreated. As urbanization intensifies, water needs rise. Over the last 20 years alone, there has been subsidence of up to 8 m in the peripheral areas due to excessive water withdrawals (Ezcurra and Mazari-Hiriart, 1996). These factors have severe repercussions on the health of the urban population. In analyses of wastewater, the independent Metropolitan University has identified around 40 pathogenic microorganisms as well as high concentrations of cadmium and lead.

MADRAS

Madras has a population of 3.79 million people, 1.5 million of whom live in slums with no sanitation facilities. The infrastructure was designed for only 1 million residents. Around 4,000 tons of waste are produced each day in the city, most of which is disposed of in the swamp areas near the coast (Appasamy and Lundqvist, 1993). The water resources in the region are already under excessive strain. Annual precipitation cannot compensate for water withdrawals, while storage basins are unable to cover the daily freshwater needs of 350 million liters. Groundwater overdrafting (150 million liters a day) causes the intrusion of seawater into the city's aquifer, greatly reducing the low stock of potable freshwater. With 78 liters of water per capita and day (less than half of that figure in years with extremely low rainfall), Madras has the lowest water availability in the whole of India. Despite having a dual supply system (water of high quality for drinking and cooking, water of lower quality for washing and bathing), water shortages are unavoidable. The costs of providing the city with a safe water supply and sewer system are estimated at almost US$ 200 million. As long as this capital is not forthcoming, 26% of the city area will remain without any form of wastewater infrastructure, i.e. untreated wastewater will continue to be discharged into the rivers that serve large sections of the city's population as a source of freshwater.

SÃO PAULO

In 1985, 15.1% of all households had no access to a piped water supply and 51.6% were not connected to the sewer system. It is estimated that only 10% of all wastewater in São Paulo is treated. Water pollution also affects the coastal waters and particularly the beaches near the city. Major problems also exist with regard to the disposal of solid waste, which in some cases is dumped at unsafe and badly managed sites, secret waste dumps, on roadsides and in river beds. The protection of groundwater resources is totally inadequate (Wöhlcke, 1994).

Bangkok

Most of the canals in and around Bangkok are so severely polluted by organic wastes that they are now anaerobic. Most wastewater is discharged into nearby canals via storm drains. Of the 51,500 registered factories, nearly half are classified as water-polluting. The contaminated wastewater is eventually discharged via the canals into the Chao Phraya River which flows through Bangkok. Levels of dissolved oxygen in the river – at least in its middle section – are far below the official standard, significantly impairing the water quality in the lower section of the river as well (Hardoy et al., 1995).

- 30.6% of imports and exports though ports,
- 34.3% of the telephones,
- 39.9% of the private automobiles and
- 43.5% of the total income tax (Bronger, 1996).

These figures provide an indication of the attraction that urban centers exert on surrounding areas – an attraction that is further intensified by the spread of communication media (especially television) as well as by transportation and tourism. Yet in the period between 1985 and 1989, 72% of all new urban households in the developing countries were only able to find housing in slums or squatter settlements. In Africa, the figure was as high as 92% (SEF, 1993).

The problems of megacities should not detract from the fact that most city dwellers in developing and newly industrializing countries live in cities with fewer than 1 million inhabitants (Hardoy et al., 1995). Many of the latter became urban centers only recently. Public perception of the global crisis affecting the urban environment is predominantly based on the spectacular examples of São Paulo, Mexico City, Lagos, Shanghai or Mumbai, rather than Belèm, Ciudad Jurez, Kumasi, Zhuhai or Patna. And yet the Favela Syndrome can also be found in these latter cities.

Demand for higher standards

As mentioned above, there are a whole range of push factors that prompt rural exodus. Syndrome links to other trends and syndromes centered on the use of agricultural ecosystems (Sahel, Rural Exodus, Green Revolution Syndrome) are therefore manifold. The Asian Tigers Syndrome, although concentrated mainly in newly industrializing countries, is an additional driving force. Within the Favela Syndrome, the demand for higher standards plays a major role. In the course of development and modernization, individuals, families and entire social groups in many countries of the world develop a higher level of demands with regard to the material and immaterial aspects of their lives: more self-determined and easier forms of work, higher incomes, family structures permitting greater autonomy, more freedom of choice in the shaping of biography, less confinement by traditional ties, a higher level of education, etc. This applies not only to industrialized and newly industrializing countries, but also – albeit at a lower level – to developing nations (Inkeles, 1996). As long as rural areas offer people below-average opportunities compared to cities, there is a latent motivation to move to the city. When the demands of the rural population rise as well, this willingness can quickly turn into action. The rise in demands contributing to rural exodus is fostered by family members or other villagers who have already migrated to the cities and who then relay the opportunities they encounter to their families and villages back home. This may result in chain migration.

Decline of traditional social structures

Not only do rising demands facilitate rural exodus, they also speed up the decline of traditional social structures. The demands that grow in the course of syndrome development (in rural areas) are not only quantitatively higher, but also of a different qualitative nature than previous ones. Many of them cannot be met within conventional social structures and the systems of norms and values that these entail. Rural areas can rapidly become claustrophobic for people living there, whereas cities appear to be the right place in which to satisfy demands. The transition from donkey to moped, for example, involves more than a simple quantitative growth in horsepower. A change in reference systems occurs – instead of grass and straw produced by the village, gasoline, oil and spare parts from the urban economy are needed, the social status of the user rises substantially, the newfound mobility enables longer distances to be traveled, etc.

The diversity of cultural traditions in the world is closely linked to village settlement structures and to the rhythm of life and work in crop and livestock farming. The greater the extent to which migration to cities prevents these patterns of settlement, livelihood and lifestyle from being handed down, the more these cultural traditions fade. Rural exodus – seen in its entire complexity – is not simply people moving from villages to larger towns and cities, but refers to profound sociocultural changes. These changes involve not only the actors themselves, but

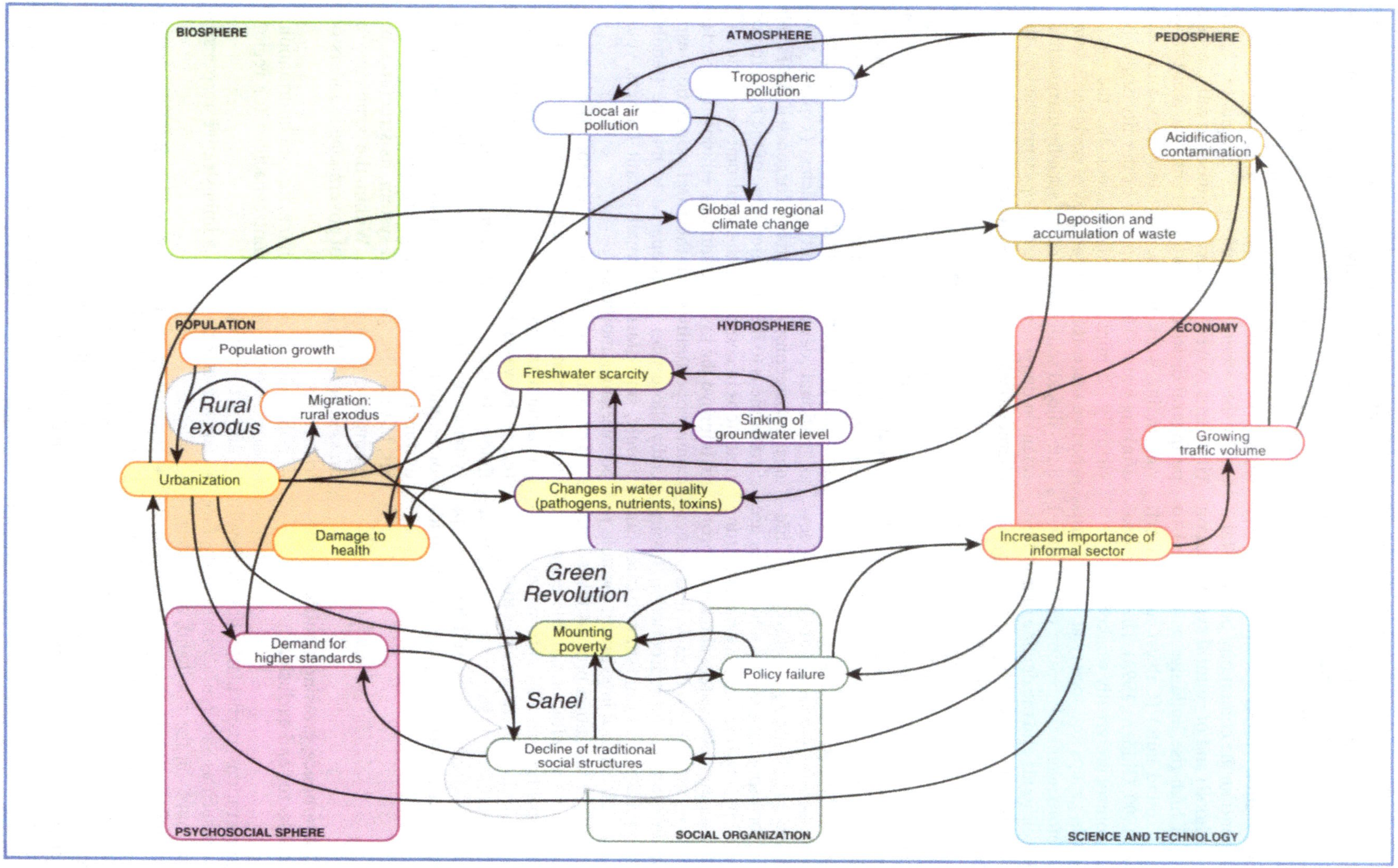

Figure D 3.5-1
Network of interrelations for the Favela Syndrome.
Source: BMBF project "Syndrome dynamics", PIK core project QUESTIONS and WBGU

also the society as a whole, assuming that the migration flows are sufficiently large. The urban system differs from the village in several respects: it follows different rules of economic reproduction (e.g. monetary economy instead of simple barter), of social association (e.g. socialization through employment and friendships rather than the community bonds of family and neighborhood) and of cultural symbolization (e.g. cinema and high-rise buildings as opposed to storytellers and round huts) (Calvert and Calvert, 1996; Flanagan, 1990; Giddens, 1993). The traditional forms of social control are greatly weakened.

For individuals, groups and society as a whole, rural exodus therefore means the erosion of traditional lifestyles and livelihoods, and the decline of traditional systems of norms and values. This process may generate anomic phenomena, such as rising crime, lack of orientation among unemployed youth, or the degeneration of everyday lifestyles across large sections of the population. Such tendencies are already evident in many developing country cities. The growth of religious fundamentalism is attributable in large measure to urban milieus of this kind, although anomie and a propensity to violence are not an inevitable consequence of declining traditions. Old lifestyles and the corresponding norms and values are often superceded by new ones that are a more fitting expression of this new, deliberately chosen environment and the hopes attached to it, as well as by new forms of social integration and reproduction. Invariably, the chances for a peaceful and smooth transition to new forms of sociocultural integration dwindle to the extent that the urbanization process takes place very rapidly, unplanned, without suitable economic opportunities and without an adequate means of political regulation. And yet there are signs that endogenous potential develops in such contexts – in the form of grassroots organizations, self-help groups and fragile networks between the poor and the middle classes.

3.5.2.2
Failure of governance, growing significance of the informal sector and exclusion

Large agglomerations (particularly those in developing countries, such as São Paulo, Mexico City, Lagos or Manila) have, on the one hand, a "modern" sector meeting all or at least most of the economic, social, cultural and urban development criteria that define a global city as exemplified by New York, London or Tokyo. Their inner cities are home to the national headquarters of multinational corporations and possess a functioning local public transport and telephone system, restaurants with national and international specialities, modern stores, etc. This modern sector of the city is also part of the network of international economic and communication links that is in the course of economic globalization not only expanding, but also becoming increasingly integrated (Clark, 1996; Lo and Yeung, 1996).

On the other hand, one finds dilapidated inner-city slums as well as areas of squalor in more or less peripheral districts, where life centers around survival, finding work for the next few days, and clean drinking water. The prosperity of the city center has virtually no impact on these areas and the people who live there. It is this complex of peripheral, sometimes illegal settlements of the socially marginalized, often on dangerous terrain (slopes, marshland), that characterizes the Favela Syndrome.

FAILURE OF GOVERNANCE

The emergence of the Favela Syndrome can be attributed, *inter alia*, to the failure of governance at national level. The urban economy and its infrastructure were promoted one-sidedly and disproportionately, at the expense of rural areas. This "urban bias" of national development policies (Lipton, 1976) needs correcting as a matter of urgency if the Favela Syndrome is to be alleviated. One of the most important steps in combating the syndrome (see below) is to improve the living conditions, employment and income opportunities for the rural population, especially those who are prepared to migrate (landless people, small farmers, small tenant farmers, young people, single mothers).

The syndrome is further exacerbated by the failure of governance at the local level – i.e. by municipal administrations and public-sector management (e.g. in the waste disposal, water supply and transport fields). In this context, "failure" refers primarily to systemic dysfunctions rather than misguided behavior on the part of individual local politicians. These dysfunctions affect not only the political stakeholders, but also the political system, its structures, programs and instruments. Within the syndrome addressed here, the failure of governance is understood to involve

- policymaking based on inadequate information,
- non-existent or deficient capacities for regional and urban planning,
- lack of clarity about administrative responsibilities,
- separation of urban planning and the management of public finance,
- lack or inadequate implementation of environmental legislation,
- underdeveloped or weak degree of local self-government within the national constitution and the implementation thereof,

- splitting of responsibilities for infrastructure provision without consideration of the problems involved (e.g. water supply vs. wastewater disposal, road construction vs. local public transportation),
- corruption and patronage in the public welfare system.

GROWING SIGNIFICANCE OF THE INFORMAL SECTOR

When governance fails, the informal sector grows. The latter term is used as a contrast to the formal sector of the economy, and is characterized by the features summarized in Table 3.5-2. This very broadly defined sector comprises activities (segments) such as small retail enterprises (food, consumer goods), craft trades and small industry (e.g. repair shops, tailors, home workshops), services (e.g. shoe-shine boys, hot food stalls, car wash enterprises), transport, waste disposal, etc. Depending on the definition and legal situation, the distinction between legal and illegal operations (e.g. drug trafficking, prostitution) varies from one country to another. A typical feature of the Favela Syndrome, analogous to the large number of informal settlements, is the enormous importance of the informal sector, since urban planning and resource management is no longer adequate to maintain a basic infrastructure (Manshard, 1992).

The informal sector can be viewed as the urban equivalent of subsistence agriculture. Survival strategies and mechanisms play a major role here (Portes and Schauffler, 1993). Contrary to popular opinion, the opportunities for earning money in the informal sector are relatively good; in many cities, at least, wages there tend to be higher than the minimum defined by the state (Bantle, 1994). The objective opportunities for securing survival, combined with subjective perceptions, largely explain why many people migrate from rural areas to the cities; they are thus a major pull factor in the process of urbanization. Table D 3.5-3 documents the growing importance of the informal sector. Other studies conclude that around 50% of those employed in the city work in the informal sector (ILO, 1993). In some cities (e.g. Lima, Ibadan) the figure is estimated to be over 60% (Stapelfeldt, 1990; Manshard, 1992).

The informal sector is linked to the formal sector in manifold ways (e.g. through suppliers of components or preliminary products for industrial production, or the provision of services) and will be even more closely linked in the future, whereby structural adjustment programs at local government level will play a significant role (movement of jobs to the informal sector, transfer of functions). However, the growth of the informal sector also means that local governments lose resources (e.g. charges and levies) and loyalties, which further restricts their capacity for action. Infrastructural deficits usually grow in proportion to the distance from the city center.

Even though the informal sector has a significant buffering capacity, the Council emphasizes that it cannot provide a long-term solution to the problems of uncontrolled urbanization. In the medium term, it is probable that fewer and fewer people will be able to continue to develop informal settlements on their own. They will come up against barriers that are insurmountable with the means available. This assessment is reflected, *inter alia*, by the intensifying competition for jobs in the informal sector, worsening sewage disposal problems as a result of increasing

Table D 3.5-2
Summary comparison of the formal and informal sectors.
Source: Bähr and Mertins, 1995; Santos, 1979

Dimension	Formal sector	Informal sector
Technology	capital-intensive	labor-intensive
Organization	bureaucratic and market-based	primitive and family-based
Capital	abundant	limited
Labor	limited and qualified	abundant and unskilled
Wages	generally regulated	individual
Prices	fixed	flexible/negotiable
Credit	banks institutional	persons: non-institutional
Relations with customers	indirect, impersonal	direct, personal
Overheads	high	low
Public control	high	minimal
Public assistance	high	minimal
International integration	high	minimal

Table D 3.5-3
Trends in structure of persons employed in the cities of Latin America, 1950–1989.
Source: Bähr and Mertins, 1995

Urban employed	1950	1980	1985	1989
	(percent)			
Persons employed in formal sector	71.1	69.1	67.5	63.9
Of which:				
– in public services	18.8	21.3	24.0	20.8
– in private enterprises	81.2	78.7	76.0	79.2
Persons employed in informal sector	22.2	23.5	27.5	32.1
Unemployed	6.7	7.4	5.0	5.0
Total	100	100	100	100

settlement density, abandonment of agricultural areas and rising traffic volume (WBGU, 1997).

The increasing significance of the informal sector, in combination with poor traffic planning, causes the volume of inner-city traffic to expand. In cities afflicted with this syndrome, the streets are chronically overloaded and local public transport inadequate. Since many areas where the informally employed live are located outside the city centers and away from the factories where they work, the traffic volume in many of these cities rises as the informal sector grows.

Social and economic marginalization

Another major cause of the growing importance of the informal sector within the Favela Syndrome is the trend towards social and economic marginalization in the affected regions or countries. This refers, on the one hand, to all those marginalization processes that act as driving factors in other syndromes (e.g. Sahel Syndrome, Rural Exodus Syndrome), as well as, on the other, to the specific type of marginalization, endogenous to the Favela Syndrome, that drives people into the informal sector. These processes may include (Hardoy and Satterthwaite, 1989; Flanagan, 1990):

- barriers to employment in the formal sector (e.g. educational deficits, lack of credit standing),
- processes of impoverishment without any form of state assistance,
- displacement from districts of origin due to rising rents and property prices, or expulsion by the authorities and
- economic pressure on the formal sector (structural adjustment, globalization of markets).

The failure of governance mentioned above is reinforced by socioeconomic disparities and increases these in turn. The more inequitable the economic and political power structure of a country or city, the greater the probability that the interests of more powerful groups will prevail, and the lesser the extent to which national and local governments possess the capacity and the will to develop and implement a forward-looking policy oriented to the common weal.

Where state bureaucracies operate inefficiently and corruptly, and any focus on the welfare of all is abandoned in favor of egoistic clientelism, urbanization and declining traditions can unleash an unshielded and uncushioned effect on the socially weak and reinforce their marginalization (Bangura, 1996). People in higher income brackets and politically influential groups live primarily in cities, and the politically influential class usually bases its development models and development activities on the urban population (urban bias) – either by gearing production to the needs of city dwellers (e.g. with low food prices) or by using the money siphoned off from rural areas for the creation of urban infrastructures in privileged city districts. On the basis of these economic, political and social privileges and their consequences, such as better infrastructure, food supply and health care, job supply, educational institutions, etc., the cities undergo an internal social split, on the one hand, and, on the other, become centers which attract the underprivileged rural population that is disadvantaged due to the political neglect of the rural regions and low producer prices (Lipton, 1976; Sundrum, 1990).

Marginalization processes are also boosted by the loss of traditions already referred to, when there is no functional equivalent – in the form of homogenous, viable and identity-promoting lifestyle constructs – to replace a world of vanishing, rurally based traditions. For example, children and young people with no educational or employment prospects in the cities are driven onto the streets and sometimes to crime or drug addiction. In addition, the compensatory function performed by the middle classes is not sufficiently developed in stagnating economies or in those whose growth exhibits severe disparities. Middle classes, whether industrial, commercial or administrative in origin, are not only the visible personification of economic transition and social compromise, but also the conveyors of the respective norms and values. Their existence counteracts the pressure of marginalization in material and non-material ways. If middle classes fail to develop, the loss of traditions intensifies the marginalization process. If sections of the middle classes themselves sink into poverty and the informal economy, the syndrome takes on more pronounced contours, as can be observed in sub-Saharan Africa (Bangura, 1996).

3.5.3
Water-specific syndrome description

A characteristic feature of the Favela Syndrome is a settlement structure with inadequately developed water supply and disposal facilities in a situation of rising water needs and water consumption. These problems have many negative impacts on the urban population as well as on ecosystems in and around cities. In this section, analysis focuses on these two aspects.

The depletion of existing freshwater reserves in the course of uncontrolled urbanization is caused by several different factors. Primary among them is the greatly increased and mounting demand for water, which leads to withdrawal levels far in excess of the natural renewal rates of regional water resources. Water withdrawals from rivers, lakes, groundwater and other sources quadrupled worldwide between 1940 and 1990 (WRI, 1996). The scarcity of water resources generates intensified competition between urban, industrial and agricultural users. As a result of rapid population growth, severe shortages of supply are already emerging in parts of China, India and the Middle East.

3.5.3.1
Disparities between withdrawal and supply

Like all agglomerations, favelas are unable to meet the demand for water due to insufficient freshwater reserves in the respective region, and in particular to inadequate infrastructural capacities regarding the supply of water. Since agglomerations were often established on large rivers because of the benefits of transportation, and these rivers are an easily tapped source of water for industry and population, the proportion of water withdrawals taken from surface waters is traditionally high. However, if there are insufficient surface waters, or if they become increasingly polluted with wastewater, then groundwater stocks are mined, or man-made reservoirs are created. Many of the latter are extremely contaminated with untreated sewage and thus highly eutrophic. A well-studied example is the Hartbeespoort Dam near Johannesburg. When groundwater is used as the primary source, the natural rate of recharge by precipitation is usually insufficient to compensate for withdrawals. There are also frequent cases of groundwater contamination by wastewater and landfill leachate.

When the water table in coastal areas is lowered, seawater intrusion and salinization of groundwater stocks may occur – an almost irreversible form of degradation. The extent of possible groundwater salinization can be seen in the tendency for large settlements to be located in coastal regions. Of the world's eight megacities with over 10 million inhabitants, six are situated in a coastal region, while 40% of the roughly 270 smaller agglomerations with a population in excess of 1 million are located in such regions (WRI, 1996). Salinization of groundwater in coastal regions has also been observed in the case of several coastal cities in China and Vietnam. Major aquifers have been severly depleted in northern Africa, India, Southeast Asia and the Middle East due to excessive rates of groundwater extraction.

The damage to the regional hydrosphere caused by lowering and/or salinization of aquifiers may seriously impair the natural vegetation cover and crops in the catchment and in extreme cases may even destroy them (desertification).

3.5.3.2
Water pollution and eutrophication

Uncontrolled discharge of wastewater from settlements and enterprises in the informal sector leads to eutrophication and to an alarming degree of biological pollution in receiving waterbodies and groundwater. The implications for sanitation are alarming, because these waterbodies are used either as sources of drinking or service water or, in the case of surface waters, for recreational purposes or for food production (fishing, aquaculture).

On average, half of all municipal wastewater comes from diffuse domestic sources. Factories in surrounding areas, built without adequate legal requirements and therefore not fitted with equipment for reducing pollutant discharges, are identifiable point sources of pollution. Traffic and uncontrolled waste accumulation also generate large-scale contamination of surface waters and groundwater.

Only a few large cities in developing and newly industrializing countries have adequate sewer systems and treatment plants; usually, only a small portion of the wastewater is purified, even when it is channeled through a sewer system. Existing sewage treatment plants rarely operate satisfactorily. In most cases, wastewater discharges exceed legal and/or hygienically acceptable maxima. The reason for this state of affairs does not necessarily lie in the treatment plants themselves, but in the frequent lack of adequately trained technicians to operate technically sophisticated facilities. Moreover, the high energy consumption and operating costs of modern wastewater treatment plants often prevent them from being used to good effect. As a result, the small number of modern treatment plants in existence are rendered inoperational due to improper use and poor maintenance. In Mexico City, seven of the 14 treatment plants do not operate to capacity and treat only about 7% of the total wastewater. Rather than upgrading and repairing existing facilities, priority was given to the construction of new plants, thus forcing operating costs to rise artificially.

Substantial economic damage is caused by the discharge of minimally treated municipal wastewater into rivers and estuaries in India, China, Venezuela, Senegal and Manila, where they have led to rapid disintegration of the local fishing industry. Regional food problems and migration of the population are direct impacts.

3.5.3.3
Lack of infrastructure and its consequences

Although the number of people in megacities who have an adequate water supply and sanitation facilities has risen in recent years, rapid population growth means that the percentage of people with an adequate supply of freshwater continues to decline. Due to the lack of organized water supply, one favela inhabitant in four must buy water from mobile vendors; this water is 4–100 times more expensive than tap water and varies considerably in quality. The quadrupling of global water withdrawals from rivers, lakes, groundwater and other sources over the last 50 years can no longer be met by regional water resources (WRI, 1996). The water supply system in Cairo, for example, has reached its limits. Settlement policies are being implemented to relieve the situation there, such as forced resettlement of city dwellers in newly created settlements in the desert, whose economic development is to be supported by constructing an 800-km-long canal from the White Nile. Construction of the first, 30-km section has already commenced.

Poor efficiency of existing supply systems

The efficiency of existing supply networks is totally inadequate in most cases:

- Major water losses are characteristic of the water supply networks in Manila, Cairo, Jakarta, Lima and Mexico City, where about 40% of treated drinking water is unaccounted for. The possible causes are leaky pipes, illegal taps (corruption) as well as the unregistered use in informal regions that is tacitly tolerated for socioeconomic and ethical reasons. The annual financial loss due to unaccounted water losses in Central America is \$1–1.5 billion (Black, 1994).
- In Mexico City, the current water price of US\$ 0.1 per m^3 covers only a tenth of the marginal cost of supplying water to the city (WRI, 1994).
- In Rufisque (Senegal), a local project group took on the task of disposing of liquid and solid wastes from the informal sector and improving the economic, social, hygienic and ecological conditions. It undertook further development of appropriate technologies and performed training, monitoring and financing functions. With the help of the population – mostly women and children – as well as municipal and private enterprises, the group developed private sanitation facilities, a sewer system, regulated waste disposal at waste dumps as well as methods for recycling used water and household waste. Ongoing investments can now be made to an increasing extent without external finance, thanks to the amounts paid into the project by the beneficiaries (Gaye and Diallo, 1994).

The scarcity of water resources generates intensified competition between urban, industrial and agricultural users. As a result of rapid population growth, severe shortages of supply are already emerging in parts of China, India and the Middle East. Poor infrastructure gives rise to further deficits in the supply of pure drinking water.

Long-distance water supply networks as one solution

The deteriorating quality of regional water reserves due to pollution and eutrophication by industrial, agricultural and urban wastewater reduces the amount of potable water resources. This is further aggravated by large-scale surface sealing through the expansion of settlements, factories and traffic routes at the expense of green and open areas. Many local governments are looking at the possibility of remote supply as a way out of the crisis, since urban water consumption only accounts for 10% of the total demand worldwide. In Dhaka (Bangladesh), Jakarta (Indonesia) and São Paulo (Brazil) the figure is even less than 1% (WRI, 1996). A remote supply has been installed in the megacities of Jakarta, Bangkok and Mexico City. The latter pipes water from a waterbody 180 km away. Given the current trends in demand, however, this source will no longer be sufficient by the year 2000 (Black, 1994). This type of water supply generates very high costs, not only for the municipal budget, but also for the impoverished population to whom these costs are passed on. Water purchases already account for 20% of the household budgets of the favela poor.

3.5.3.4
Water-specific threats to human health

The combination of high population density, inadequate water supply and poor sanitation in the favelas poses serious health hazards to the inhabitants. Various diseases are particularly prevalent, and the consequential costs for the health care system are considerable. Table D 3.5-4 outlines the water supply and sanitation situation in three cities. Fig. D 3.5-2 compares the mortality rates in Accra and São Paulo according to cause of death and district. The susceptibility to infectious diseases is conspicuously high in Accra and correlates significantly with the sanitation conditions. The chart in Fig. D 3.5-2 also underlines the high degree of exposition and frequency of mortalities in those residential areas with the poorest population.

Diarrheal diseases are primarily caused by inadequate access to water, sanitation facilities and waste disposal, and claimed the lives of 3 million people worldwide in 1995, 80% of them children under five. Around 30% of infections are attributable to the transmission of pathogens via drinking water, and 70% to eating rotten food. Cholera (see Section D 4.2), in particular, has a high affinity to poverty, overpopulation and poor sanitation (WHO, 1996) and can therefore be labeled a typical disease of the Favela Syndrome. Since the 1970s, this disease has been on the rise again, particularly in Africa. Here alone around 79 million people are threatened by cholera, and 120,000 people die of the disease worldwide every year.

Overcrowding promotes the transmission of infectious diseases, such as colds, pneumonia and tuberculosis. Moreover, the psychological stress induced by cramped living space (in the slums of New Delhi, for example, 1.5 m^2 per person) and fear of repression because of illegal housing is substantial.

Rats and fleas are carriers of the plague. *Yersinia pestis* bacteria are transmitted to humans from infected rats by flea vectors. Uncollected waste in the favelas attracts rats, among which the epidemic spreads cyclically. Peru, for example, experienced a severe outbreak of the plague in 1984, as well as minor ones in 1990 and 1992. After "quiet" periods of more than 30 years the plague has appeared again in Botswana, India and Malawi, for example. Outbreaks are reported regularly from countries like Madagascar, the Congo, Bolivia, Brazil, China, Kazakhstan and Vietnam. Around 1,400 people worldwide contracted the plague in 1995, with at least 50 mortalities (WHO, 1996).

Air pollution, due in many cases to the immediate proximity of industrial sites and to exposure to smoke from household fuel combustion, leads to respiratory diseases such as chronic bronchitis and asthma, especially among women and children.

The linkages between the favelas and agglomerations with a high traffic volume mean that these diseases are not confined to the local area. Moreover, since the informal sector has various forms of contact with the formal sector, infectious diseases can spread rapidly and, given the high degree of commercial and tourist mobility, may be transported internationally to regions not otherwise exposed.

Table D 3.5-4
Water supply to urban households (study of about 1,000 households).
Source: WRI, 1996; UNDP, 1996; World Bank, 1996

	Accra	Jakarta	São Paulo
Average per capita purchasing power (US$)	1,000 –2,000	2,000 –4,000	4,000 –8,000
Latrines shared by at least 10 households (percent)	48	14–20	<3
No household water supply (percent)	46	13	5
No waste disposal (percent)	89	37	5

3.5.3.5
Water-centered network of interrelations

The interaction of unfavorable development trends can be seen in the following water-centered network of interrelations for the Favela Syndrome (Fig. D 3.5-3), as may its links to other syndromes of global change.

3.5.3.6
Dynamic measure of intensity of the Favela Syndrome

In the following, we describe those trends and interactions that a) distinguish the Favela Syndrome from related syndromes (e.g. Urban Sprawl Syndrome, Asian Tigers Syndrome) and b) display a high degree of internal linkage (e.g. in the form of self-reinforcing interactions between trends). The following four indicators were selected because they meet both conditions. Linkages between trends are measured, rather than individual trends. The result is a dynamic

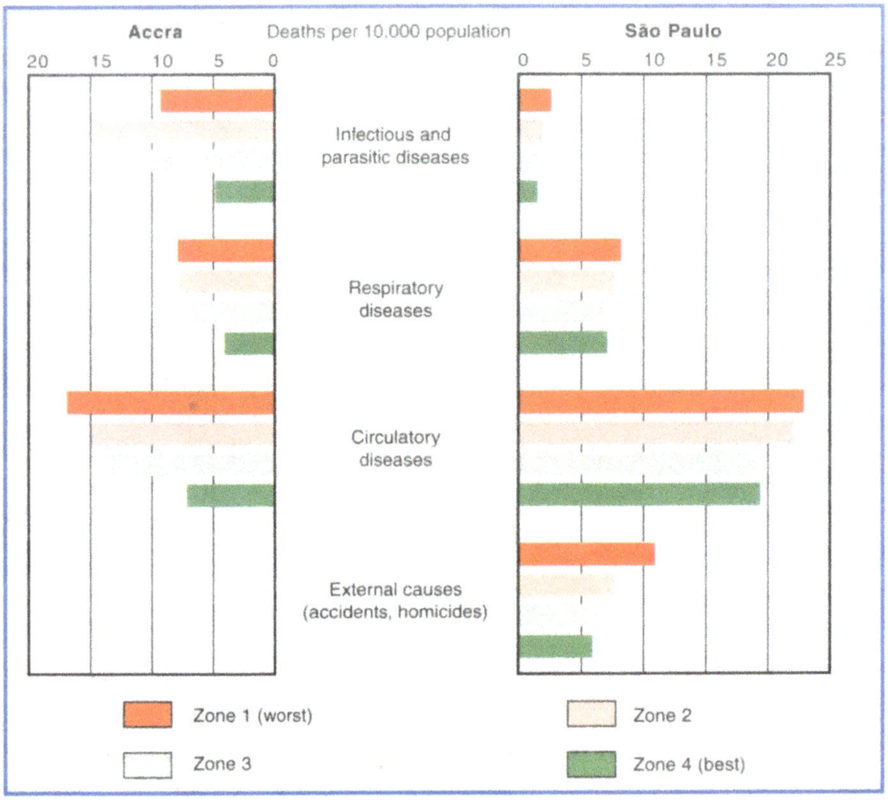

Figure D 3.5-2
Age-adjusted mortality rates by socio-environmental zones in Accra and São Paulo, 1991–1992.
Source: modified from Stephens, 1994

measure of the intensity of the Favela Syndrome that permits an initial assessment of its future development. The following can be identified as key features of the Favela Syndrome:

1. Very rapid growth of the urban population due to migration as well as inter-city migration in a later phase. The interaction of the rural exodus and urbanization trends is a highly syndrome-specific mechanism; this is particularly relevant when assessing the spread of the syndrome to regions that have not yet "contracted" it. The indicator to be used here is therefore the annual rate of urban population growth.
2. Social and economic marginalization represents one of the most important anthropospheric trends within the Favela Syndrome. It fosters the growth of the informal sector and in turn is driven by the latter, through the decline of traditional social structures. The indicators used to measure this key trend are "distribution of incomes" and "number of urban poor".
3. The inadequate infrastructure for providing basic services in urban areas is one of the central features of the Favela Syndrome. This problem is particularly prevalent with regard to water supply and sanitation. Many environmental and social problems result from the fact that the cities affected do not have the financial, organizational and political resources to create an adequate infrastructure to meet the needs of a growing population. The population groups concerned are often forced to take matters into their own hands and make up for the lack of public services. In order to identify, at least roughly, the connection between social marginalization and the mounting importance of the informal sector, use is made of the "growth in public-sector investment" indicator. Stagnation or a minor increase in investment indicates a linkage that intensifies the syndrome.
4. The rapid process of urbanization in many regions overtaxes existing urban structures and capacities in most cases. Moreover, it occurs within a conditional framework of technical, economic and political factors characterized by stark contrasts. Whereas urbanization is accompanied within the Urban Sprawl Syndrome by a relatively uniform access to the urban infrastructure, the Favela Syndrome involves a significant inequity in this respect. The indicator used to identify this syndrome-specific process of marginalized urbanization in the area of water supply is the "proportion of the urban population with continuous access to safe water". When the value of this indicator is low or grows at a rate below average, the syndrome is reinforced.

The spatial distribution of the population is the basis for the mapping of trend linkages. The first question that is raised here pertains to the delinea-

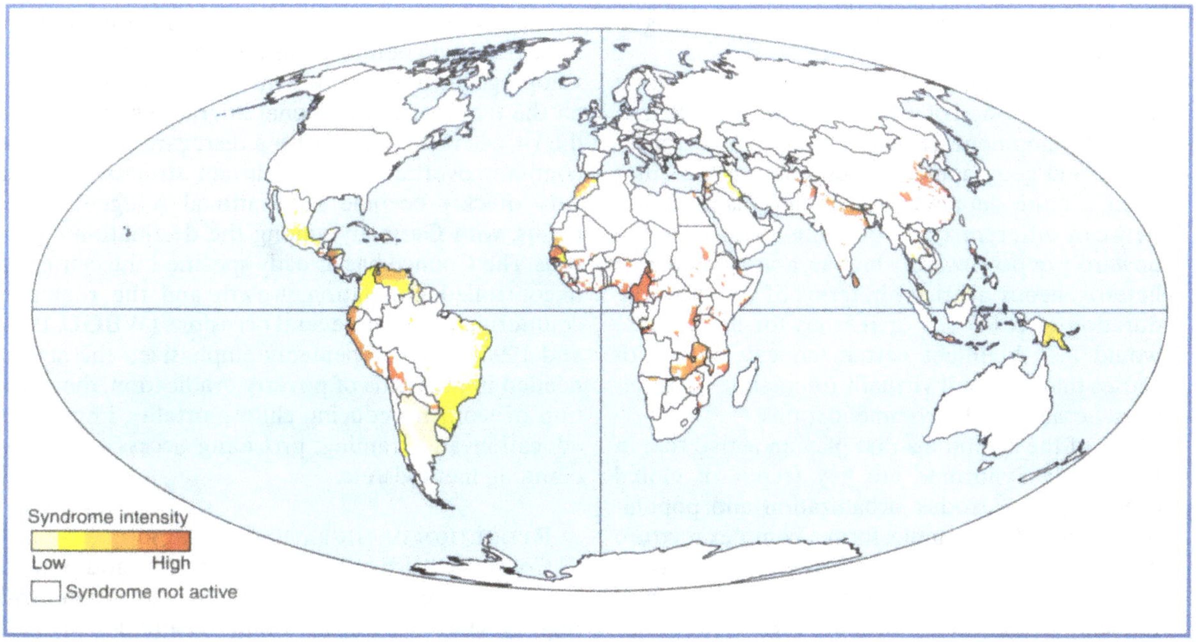

Figure D 3.5-3
Intensity of the Favela Syndrome. Explanations in the text.
Source: BMBF project "Syndrome dynamics", PIK core project QUESTIONS and WBGU

tion of rural and urban areas. Not only megacities, but also smaller urban structures may have a disposition towards the Favela Syndrome, so it makes little sense to measure its intensity on the basis of absolute urban population figures. Another problem is that there is no standardized definition of "city" that could be used for international comparison, so syndrome intensity is measured here by means of population density, irrespective of administrative boundaries.

Using the grid map of population density (grid cells of 5 arc minutes) developed by Tobler et al. (1995), the regions with the highest population density within each country are firstly identified as urban structures, and this on a worldwide basis. The grid cells with the next lowest population densities are also designated as urban structures, until the sum of the people living in the identified regions corresponds to the known percentage of urban population in the respective country.

The extent to which the situation will worsen in the regions thus identified depends primarily on the population growth and on the level of infrastructural investments that cities are able to make in order to maintain or establish a minimum capacity for water supply and wastewater treatment. Hence, the degree of intensity is further determined by a fuzzy AND intersect (minimum function) of the basic map with the regionally specific data for demographic development and public-sector investment (in per capita GNP per annum, in US$). Integration of the relevant data material for 1990 produces a map of syndrome intensity (Fig. D 3.5-3).

3.5.4 Syndrome cure

In the context of the water problems on which our analysis centers, a distinction must be made between three different strategy levels for remedying the Favela syndrome:

- combating causal factors of a more general nature that give rise to the Favela Syndrome and have a bearing on freshwater issues.
- targeted measures to combat water-specific problems affecting the entire network of interrelations or parts of it.
- combating specific water-related problems irrespective of their systemic interrelations within the Favela Syndrome.

3.5.4.1 General recommendations for action

Making recommendations for action in connection with the Favela Syndrome, or deriving basic therapeutic approaches from the syndrome mechanism, is a difficult undertaking for three reasons in particular:

- On the one hand, the definition of the syndrome and the description of its mechanisms indicate that slum development depends on certain socioeconomic and geographical conditions. On the other hand, a more detailed analysis that distinguishes between different types of slums (slums for the upwardly or downwardly mobile, homogeneous or heterogeneous districts) in terms of composition, duration of residence or reasons for living there would also highlight certain concatenations of causes that render it virtually impossible to develop generally valid recommendations.
- Many of the symptoms that play an active role in the Favela Syndrome are key trends of global change. Rural exodus, urbanization and population growth, for example, form a complex mixture of "megatrends" that operate as a crucial driving force behind the Favela Syndrome. Measures of reducing population growth are therefore recommendable in an additional respect as therapy against uncontrolled urbanization. The Council has repeatedly pointed out that effective progress in stabilizing the world population can only be achieved in the long term at best (WBGU, 1994 and 1995).
- As recently as the mid-1960s, efforts were still being made to eliminate the worst conditions in slum settlements by imposing building and health control regulations (Spiegel, 1970). "Slum clearance", "urban redevelopment" and "housing acts" were recurrent terms in the legislation enacted at the time. The social problems were largely ignored in the expectation that they would automatically disappear when the slums were demolished. What actually happened, at best, was that problems were merely transplanted to other areas. Changes arose slowly when planners began to realize that the favelas generally exhibited a relatively stable subculture with a highly developed group mentality, ethical code, intensive family life and neighborly relations. Deriving general recommendations for action should therefore be based on the obvious dichotomy between the suffering and the values of favela life. Clearing a slum does not automatically improve the welfare of its inhabitants.

Reduction of population growth

In light of this assessment, priority should be given to securing an institutional basis for population policy in all states by means of international cooperation and financing programs (WBGU, 1995 and 1996), because the problems generated by uncontrolled urbanization, and hence their elimination, are crucially dependent on population trends. The Council believes that Germany, in common with all other industrialized countries, has a special interest in developing appropriate problem-solving capacities, given the fact that intranational migrations (rural exodus) – when combined with a disregard for the concomitant overburdening of urban structures – can very quickly become international refugee movements, with Germany among the destination countries. The Council has already specified the causes of uncontrolled population growth and the requisite countermeasures on several occasions (WBGU, 1994 and 1995). It has repeatedly emphasized the action needed in the fields of poverty eradication, the position of women, reducing child mortality, improving education and training, providing access to family planning methods, etc.

Reduction of migration pressure

Controlled urbanization requires, in addition to the above, a reduction of migration pressure in rural regions. The basis for this would be a nationally and internationally coordinated regional planning policy aimed at inducing appropriate structural changes in rural areas (WBGU, 1994) and at counteracting the dire effects of urban bias. Polycentric rather than monocentric land-use structures have to be created in accordance with AGENDA 21. Furthermore, it is essential to develop specific guiding principles for regional planning that permit a harmonization between "environment and development" through a balanced mix of land-use structures between city and rural areas as well as within the cities (by maintaining and developing sufficient inner-city green areas).

Sustainable settlements development

The 1992 UN Conference on Environment and Development in Rio de Janeiro played a pioneering role in this context. In particular, the 1996 United Nations Conference on Human Settlements (HABITAT II) in Istanbul laid the foundations for implementing the urban development aspects of AGENDA 21. The HABITAT Agenda provides a global framework for promoting sustainable settlements development. The Global Plan of Action adopted at HABITAT II defines a set of common goals and strategies to be translated into national action plans, the primary objective being to give (greater) consideration to the needs of the poor and homeless as well as refugees and ethnic minorities. Settlements policy and action to combat poverty are indissolubly linked. To improve the data base, it is also necessary to develop a housing information system. Land management and land tenure policy (safeguarding rights to land) must be improved (WBGU, 1997).

In addition to these general recommendations, which pertain to the overall complex of the Favela

Syndrome, the Global Plan of Action contains specific recommendations on water resource management in settlements. This involves, for example: (1) integrated planning of water supply and wastewater treatment systems, (2) measures for avoiding wastage of water and for proper recycling of drinking water, (3) strategies and criteria for precautionary protection and conservation of aquatic ecosystems, (4) improvements regarding the efficiency in allocating infrastructural investments as well as (5) comprehensive, cross-sectoral resource management (Habitat II, Global Plan of Action, 1996). Special research needs exist in connection with the determinants of the carrying capacity of urban structures ("optimal city size") and analytical and forecasting methods for identifying regions with a disposition for intranational migrations (WBGU, 1996).

3.5.4.2 Water-specific recommendations for action

Creation of basic principles for integrated water management

One of the prerequisites for any cure is an adequate information base. This includes inventorizing water stocks, on the one hand, and ascertaining in quantitative and qualitative terms within the framework of a national water strategy the demand for water at local and regional level, on the other. Economic, social and environmental concerns must all be taken into account, in accordance with the goal of sustainability. It is important also to strengthen the capacity of national and local hydrological research institutes and to improve their links to environmental research as well as to institutes in the field of economic and social science research (particularly in the field of urban studies and urban planning). The transfer of knowledge to local government decision-makers must be improved.

Securing a human right to water

Local governments must focus their policies on delivering a basic supply of safe drinking water, especially to the poor, and on treating municipal wastewater. At the international level, the Council has proposed a World Water Charter as an instrument for securing a minimum supply (see Section D 5.5 and Section E 2). These rights must be implemented at local level. Simple methods of water supply and wastewater treatment are not only the sole realistic solution, but operate as well to strengthen the capacity for self-help.

More efficient local government

The failure of governance at local government level should be counteracted by improving the efficiency of public administration at the local level. It may be necessary in some cases to create an appropriate framework at the national level (in the form of a local government constitution). The measures required include the bundling of responsibilities, combining management and financing functions, improving environmental legislation and monitoring, dismantling bureaucratism, decentralizing tasks to the lowest levels possible, increasing the transparency of government activities as well as enhancing the skills of the public administration employees. The privatization of water supply and wastewater disposal enterprises may be a solution, provided that (1) a basic supply and (2) sustainability goals are safeguarded. Systems for delivering water and disposing of wastewater should also be managed as an integral whole, as should surface waters and groundwater resources. Improved management strategies should relate to "real" entities (e.g. water catchments) and appropriate administrative territories.

Appropriate technologies

The development of efficient technologies for water use and wastewater treatment is essential in combating the Favela Syndrome. As far as industrial demand is concerned, the focus should be on achieving the lowest possible water consumption, whereas efforts regarding domestic sewage in the favelas must center on promoting and further developing low-cost, easy-to-handle and, in general, regionally developed technologies with a low degree of complexity. Support measures on the part of donor nations should also foster international know-how and technology transfer in order to optimize research efforts (technical capacity building). Special weight must be placed on minimizing the energy needs of these technologies.

Improved cooperation between local government and the informal sector

The self-help potential of the informal sector should be exploited to a greater extent and integrated with decentralized public administration – not least to relieve the latter of tasks and costs as well as to find more rapid, appropriate and flexible solutions for those worst affected by the Favela Syndrome. A number of NGOs and CBOs (community-based organizations) in many cities of the developing world have taken this route, gaining positive experience in the management of infrastructural institutions and the mobilization of local self-help within the framework of urban supply projects (e.g. the Orangi Pilot Project, Pakistan; Hardoy et al., 1995). Strengthening self-help potential, public participation and individual responsibility may alleviate the syndrome-specific mechanism in which bad governance and the growing importance of the informal sector are mutually reinforcing, and in this way produce productive impacts in the form of synergy effects.

Control of demand (water tariff systems)

The pricing systems operating in most large cities affected by the Favela Syndrome do not reflect the economic and environmental scarcity of water to an adequate extent, so wastage is the invariable consequence. On top of that, many poor consumers are compelled to pay relatively high water prices to mobile vendors because the water supply systems are so inadequate. Existing water charges must be changed so that they reflect scarcities and increase the reliability of supply. To be environmentally and economically viable, water tariff systems should ensure that the costs of collecting and using water are recovered, particularly since the financial resources needed for water management, water-related infrastructure in settlements and resource protection play a role of mounting importance in the cost structure of developing country economies (see recommended action for securing minimum needs in Section D 5.4). A basic supply for low-income user groups must be guaranteed by imposing a basic charge that depends on the ability to pay. Low-income users should be able to reduce the amount they have to pay through active participation in systems of water collection, water supply and wastewater disposal. The demand of major polluters or large consumers should be controlled using the instrument of marginal cost tariffs. Taxing consumption in this way is a financial incentive to use water sustainably (Postel, 1993).

Framework program for inter-city cooperation

The Council urges that a framework program focusing on water be established for German cities wishing to collaborate with cities in other countries. On the basis of standard criteria for financial, organizational, technical and research cooperation and assistance, German and EU cities could apply to make a specific contribution towards mitigating the Favela Syndrome. The geographical focus should be on the syndrome in general, which would extend coverage to surrounding rural areas as well. The GTZ, which has had many years of experience with water supply and sanitation projects, could be commissioned to implement such a project.

Technical measures

The paramount humanitarian aim of the recommendations is to protect the health of the population by improving sanitary conditions in the favelas. If the severe threats to human health are to be reduced, it is imperative to ensure safe and adequate water supply and wastewater disposal. Attaining this goal creates a breathing space for political activities aimed at regaining control over uncontrolled urbanization.

In the medium to long term, action should be taken to adapt the use of water in urban agglomerations (as an economic resource and medium for wastewater disposal) to the principle of sustainability, in order to protect the environment and the natural basis on which all life depends. Neglecting these goals will cause the risk of epidemics to rise, and in the long term will generate increasing numbers of international environmental refugees and other migrants; this migration pressure will be directed primarily towards the industrialized nations.

- *Defining acceptable withdrawal volumes:* Given the finite nature of freshwater reserves and their ubiquitous overexploitation, the primary issue is to determine the maximum acceptable volume of withdrawals, based on renewable water stocks, in order to prevent the total depletion of water resources (see the "guide rail" model, WBGU, 1997).
- *Enhancing the efficiency of supply systems:* The low efficiency of many water supply and wastewater disposal systems should be improved by upgrading and repairing them. In many cases, this generates greater benefits than building new facilities, because time and costs can be saved.
- *Supplying rural areas:* International financial assistance should be channeled first and foremost into supplying water to rural and low-income urban areas in the poorest countries of Africa and southern Asia. The population there has minimal access to drinking water and sanitation. In Ethiopia, Mozambique, the Congo, Pakistan, India and Papua New Guinea, 80–90% of the people do not have an adequate supply of water (World Bank, 1996).
- *Groundwater protection and replenishment:* Overdrafting groundwater reserves can be countered by enhancing groundwater renewal. This is done, for example, by setting up protected water areas with appropriate regulations governing land use on soils with a high permeability to water and by creating water reservoirs for artificial groundwater recharging in accordance with flood and dry periods. This applies in particular to coastal regions, where shrinking groundwater reserves may be severely damaged by saline intrusion.
- *Seawater desalination:* One option for water production in some coastal regions is to desalinate brackish water and seawater. However, the drawback of this method is that it requires a high level of technology, capital investment and energy in-

BOX 3.5-2

Methodology for selecting appropriate methods of wastewater treatment

The suitability of different wastewater treatment systems depends on the size of settlement, geological conditions, topography, water supply, pollution, climate, the sanitary, socioeconomic and technical conditions, the receiving waters to which discharge is planned (river or irrigation system) as well as the size and quality of the sewer system (Arceveila, 1996).

Additional criteria relevant to wastewater disposal techniques include degree of economic efficiency, reliability, easy handling ("low technology"), manufacture using indigenous resources, local responsibility and integration of the community. The life cycle analysis (LCA) method can be applied to assess environmentally relevant impacts. Relevant factors include sanitary efficiency, energy flow, phosphorus flow and nitrogen elimination as well as nutrient loads from terrestrial to aquatic habitats, qualitative and quantitative changes in water supply, use of non-renewable energies and materials (pump energy, energy loss in the form of methane, infrastructure and operating equipment), economic costs and the amount of reused water.

Once these criteria have been met, preference must be given to the method with the lowest areal requirements, a simple design and an economic utility value.

In dense settlement areas in hot climates, the best systems for wastewater treatment are those involving a combination of aerobic and anaerobic systems and dry deposition of solids, "low technology", low energy consumption and which can also profit from the high temperatures. The World Bank has developed a suitable method for selecting low-cost methods of wastewater treatment in developing countries with hot climates (Arthur, 1983). Preference is given here to wastewater treatment in anaerobic ponds, where around 70–80% of the organic substance can be decomposed at 25 °C with minimal odors (given a maximum daily input of 350 g of degradable substances per m³ of wastewater). The pretreated wastewater is then safe to use as irrigation water, and can be channeled to fields via reservoirs. Research projects in this field should be aimed at minimizing the land area and energy required by the various methods in use.

A number of conceivable approaches to solving wastewater treatment problems in humid to semi-arid regions is presented in Table D 3.5-5.

If the sewer system can be separated according to different qualities of wastewater, the recyclability of wastewater is enhanced:

- wastewater containing feces is hygienically unsafe,
- wastewater without feces is unpleasant but not hazardous, and therefore suitable as irrigation water,
- treated wastewater without feces is suitable for many uses as service water, in some cases as drinking water.

A number of water-saving or dry sanitation systems, usually on a decentralized basis, provide an alternative to water-based disposal. They include low-cost types that can be installed with local resources, e.g. the "pour flush latrine" and the "ventilated improved pit (VIP) latrine". The former consumes only 2–3 liters as opposed to the 8–20 liters of water needed for flushing with a conventional WC (Hardoy et al., 1990).

These systems have a number of advantages compared to the undesired and inefficient mixture of different types of waste caused by water-aided transport of human, domestic and munici-

Table D 3.5-5
Wastewater treatment in humid to semi-arid regions.
Source: WBGU

Type of facility	Costs	Areal needs	Energy needs	Complexity	Efficiency
Plant-growth beds	low	moderately high	low	moderately low	temp.-dependent
Settling ponds	low	high	low	low	temp.-dependent
Trickling filter	moderate	moderate	average	moderately low	average
Activated sludge	average	low	high	high	high
Multistage treatment	high	low	very high	high	high

pal wastes: in contrast to water-based methods, they do not transport nutrients and foreign substances from terrestrial to aquatic systems, and due to separate deposition the different wastes can be used as fertilizer after appropriate treatment.

Yet systems of this kind have been used to only a minimal extent so far. Special purification processes are necessary to treat toxic wastewater, which usually comes from factories. Only small quantities of organic toxic substances can be decomposed by microorganisms using conventional methods, and heavy metals can be removed only by adding special chemical substances (chelating agents). In temperate climates, heavy metals in composted sewage sludge can be absorbed by worms and stored in humic substances (Protopopov, 1995).

put; where solar energy is used, the area of land needed is considerable.

- *Water pricing:* The measures required to enhance groundwater renewal, etc. involve financial outlays that must be covered by prices or charges reflecting the true value of good water.
- *Water tax:* Industrial water consumption and wastewater discharge will increase radically in the urban centers of the developing countries, as will the competition for water. Taxing the use of water resources should be based on the "polluter pays" and precautionary principles applied elsewhere by environmental policymakers.
- *Water recycling:* Measures aimed at reducing the strain on water resources include the use of technologies that protect water resources, especially industrial water recycling, promoting the recycling of water-intensive products and the use of appropriate wastewater treatment methods that ensure compliance with the relevant requirements.
- *Water monitoring:* Municipal monitoring of pollution sources, waterbodies and water resources that serve as receiving waters for wastewater discharge should be carried out on a long-term basis. This requires the establishment of regionally appropriate environmental regulations regarding the quality of treated wastewater.

Separating industrial from municipal wastewater enables toxic substances to be treated more efficiently. Pretreated or separately collected wastewater is better suited to reuse, e.g. in irrigation. A method for selecting suitable wastewater treatment processes in developing countries was developed in the context of the HABITAT II conference (see Box D 3.5-2).

HEALTH-RELATED ACTIVITIES

A key function in the elimination of health risks, besides the provision of safe drinking water, is performed by wastewater treatment, which must prevent the survival and circulation of pathogens. Appropriate systems must be selected by analysing local conditions with respect to the transmission of disease. Technologies matching the particular context are essential, as is the participation of the local population in improving sanitation facilities. Excellent results have been achieved with decentralized systems in which users participate in the design stage and where active involvement of the local population in the construction phase is encouraged. Identification with the project is fostered and secured on a long-term basis through communication, training programs and participation (the "wastewater neighborhoods" in Brazil, for example).

Since pathogens occur almost exclusively in excrement, sanitation systems that separate urine and feces reduce wastewater contamination, and enable feces to be disposed of as dry substance instead. Less water is therefore needed for flushing.

The WHO (1995) describes the percentage reduction in the risk of diarrheal infection as follows:

- Improved water quality: 16%.
- Improved water availability: 25%.
- Combination of improved water quality and water availability: 37%.
- Improved excrement disposal: 22%.

In regions with a low water supply, treated wastewater can be used for irrigation. The WHO has issued guidelines for safe use of service water. These address the various ways in which diseases can be transmitted, depending on the type of pathogens, their survival in wastewater, the transmission methods, the crops grown and the exposition of rural workers and consumers, as well as body hygiene and immunity. To ensure health protection, however, it is essential that wastewater be treated. In hot climates, anaerobic ponds (sedimentation basins) are highly effective in destroying pathogens. Where land prices are low, these represent a low-cost wastewater treatment method that is easy to build and operate.

Waterborne infectious diseases are a characteristic feature of the Favela Syndrome. Given the rapidity with which these diseases propagate, combating epidemics with a reactive approach – sending doctors and aid to the hot spots as required – has its obvious limits. Pathogens, particularly in the form of new mutants, can be spread throughout the world very quickly by tourists and business travelers, migrants and exported goods. For this reason, future efforts must be concentrated more on preventive measures, such as improving water quality. A paradigm shift might be achieved in this context, since investments in a regulated drinking water supply and sewage treatment promise one of the highest possible "health gains" in the fight against many diseases. With these impacts, such investments are also important measures for stabilizing population size and combating poverty (see Section D 4.2).

4 Key issues

4.1 International conflicts

Water conflicts – Ataturk Dam – Jordan Basin – Gabcikovo Dam – Great Lakes in North America – Different lines of conflict and utilization claims – Global dimension of regional conflicts – Protecting the world natural heritage – Persistent Organic Pollutants (POPs) – The right to water – Conservation of freshwater resources as a multilateral task

4.1.1 Basic elements of conflict analysis

In studying international conflicts, social science has developed and empirically tested a variety of theoretical approaches that can also be used to analyze conflicts over the allocation and utilization of freshwater resources. Within such approaches, conflicts are classified according to the specific policy fields to which they relate, which permits conclusions to be drawn about the way in which the respective conflicts are fought out, as well as appropriate means of resolving them. Another scale on which to evaluate conflicts is the likelihood of their being managed on the basis of ad hoc cooperation between the states involved, or regulated by well-established "international regimes". The term "regime", which originates from the field of jurisprudence, refers in the discipline of international relations to a set of implicit or explicit principles, norms, rules and decision-making procedures around which actors' expectations converge in a given issue area (Krasner, 1983; Rittberger, 1993; on environmental regimes: Young, 1994). "Rules" are specific delimitations on what states may, must or must not do that must largely be complied with in practice. For an international regime to exist, the activities of states in a specific issue area must be explicitly guided by binding rules.

A conflict typology could distinguish between conflicts of interest, conflicts over means or conflicts over values (e.g. Aubert, 1972).

- A conflict of interest between two actors results from scarcity of some kind: two actors want the same object, but there is not enough to satisfy both.
- A conflict over means arises when there is dissension over the right way to achieve a common goal.
- A conflict over values is based on dissension regarding the status of an object in terms of categories such as security, power, dominance or territoriality (e.g. spheres of influence, borders).

A further distinction may be drawn between absolutely and relatively assessed goods. Absolutely assessed goods, e.g. an adequate supply of clean water, are those which acquire their value independently of how much the other parties to the conflict possess. Relatively assessed goods, in contrast, obtain their value by virtue of the fact that one or more actors have a greater share than other parties.

By applying these distinctions it is possible to assess the conduciveness to cooperative conflict management and to regulation by international regimes (regime-conduciveness). The social sciences provide substantial research evidence confirming that conflicts of interest over absolutely assessed goods show a high propensity for international cooperation, whereas conflicts over means are less conducive, and conflicts of interest over relatively assessed goods are much more difficult to regulate. Conflicts over values, i.e. security, dominance, spheres of influence and similar, show the lowest propensity for regime formation (Fig. D 4.1-1).

According to this typology of conflicts, pure conflicts over freshwater in which the issue relates "only" to a scarce resource are more likely to be resolved by cooperative means, as the cases cited demonstrate (see Section D 4.1.3). Such conflicts are more liable to produce a consensus regarding allocation of the resource than conflicts over arms threats or status, which are assessed in relative terms by the parties concerned. In cases where conflicts over freshwater resources involve political goals such as the wielding of power, national security, the exercise of political influence or similar, the prospects for

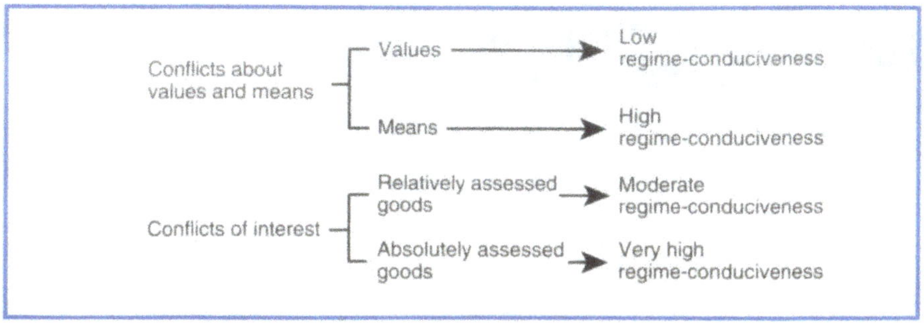

Figure D 4.1-1
A typology of conflicts.
Source: Efinger et al., 1988

cooperative and consensual conflict management are much less.

4.1.2 Pathways to conflict management

The first step towards settling a conflict is for the parties to be willing and prepared to communicate with each other. It is of critical importance, in achieving a consensual agreement or compromise, that the opposing parties abandon at least some of their standpoints and enter a cooperative discourse. Within such discourse, the participants act communicatively. In the ideal-typical form, communicative action is strategic action aiming at mutual understanding (Habermas, 1981). Strategic action is rational, purposive action in interdependent decision-making situations, in which the actors are aware that the result of their actions depends not only on their own actions, but also on those of others (Keck, 1995). Actions are rational when the objectives are achieved in the optimal way. However, these principles are not sufficient in themselves to explain the behavioral changes that are induced in actors by, for example, regimes in international relations (see Section D 5.5.3) or by decentralized decision-making and agreement as a result of mediation (see Sections 5.3.2 and D 5.5.4). These actions are based instead on the goal of understanding and agreement. The actors intend by their actions to arrive at agreement on the situation so that they are able to coordinate their activities on a mutual basis (Habermas, 1981). Action based on or aimed at mutual understanding enables solutions to the conflict to be found through argumentation. Politics, and international politics in particular, consists primarily of speech interactions. Borrowing from Müller (1994), we can identify three points that are decisive for actors to attain their goals through cooperation:

1. mutual trust in the sincerity, reliability and credibility of the other's language,
2. agreement on the definition of the situation and the normative framework; and
3. compromise on allocations.

In social situations where action is geared to agreement, the very preferences of actors are open to change (Risse-Kappen, 1995). In the process of reaching an understanding, argumentation is the discursive element by which claims can be asserted, i.e. everyone engaged in this process must be prepared from the outset to be persuaded through the arguments of the other. The operation of such processes in intergovernmental negotiations can be seen when the issues relate not only to the authenticity of language, but also to arguments derived from convictions that are based for their part on what are accepted as facts or as normatively desired (Zangl and Zürn, 1996). Of critical importance for the outcome of the communicative process is that no one party in this situation has superior power resources, so that a voluntary agreement to cooperate arises solely through the persuasive strength of the better arguments (see Box D 4.1-1).

4.1.3 Regional water conflicts

In the following, we examine various examples of international conflicts over freshwater resources with special reference to the way they are handled politically. This will indicate the range of possible and empirically identifiable patterns of cooperation, from very cooperative forms of conflict management, such as those between Canada and the USA, to totally unregulated conflicts. According to Müller (1993), cooperative arrangements are in place for two-thirds of the world's 200 transboundary watercourses, 30 of these being administered or governed by special organizational bodies. Each of the case studies begins with a brief introduction, followed by an assessment of the escalation risk.

4.1.3.1 The Ataturk Dam on the Tigris-Euphrates

Turkey is currently engaged in one of the biggest water development projects in the world and the

BOX D 4.1-1

Game-theoretical modeling of conflict situations

Game theory can be applied in order to model conflict situations or negotiations in which actors take decisions. Game theory analyzes the structure of strategic decision-making that is generated by the rational calculations of actors. The behavior of actors and the course of the conflict, as well as the decisions of actors are simulated in such a way that possible future outcomes can be predicted. Game theory (see, for example, Axelrod, 1987; Putnam, 1988; Zürn, 1992; Keck, 1993) distinguishes between cooperative and non-cooperative games. In other words, situations are modeled in which there are possibilities of conflict with confrontational interests, on the one hand, and opportunities for cooperation with mutual interests, on the other. The advantage of game theory lies in its simplification of real-life situations, which means that complexity is reduced and conceivable outcomes can be predicted, but this is also the source of the theory's deficiencies. Critics point out that, by reducing the complexities of the real world to a small number of players and simplified situations, game theory provides a poor reflection of the multi-layered structures in international relations and the multitude of factors in operation. Real-world conflicts frequently involve a large number of protagonists with divergent interests, which are lumped together or simply ignored by game-theoretical approaches.

People exert influence on each other through their behavior, and different types of behavior change the world in which actors operate. Game theory is a method that can be used empirically for describing and analyzing situations in which the value of a decision-maker's objective function depends not only on his or her own decisions, but also on those of others. Actors may choose cooperative or non-cooperative means, to solve economic, political or military problems, for example. It is assumed that their behavior is rational, and that they will try to maximize their rewards. The name "game theory" derives from the fact that mathematical formulation of situations with several decision-makers – individuals, groups, enterprises or nations – bears similarities to strategic board games. In game theory, decision-makers are called players, and their objective function is a pay-off or a utility function. This function defines the pay-off or the benefit for a player if he plays a certain strategic combination. A game is described in terms of the number of players, the set of strategies defining the range of possible combinations of the players' strategies, the utility or pay-off function, and the rules of the game. Each player chooses a strategy – bearing in mind that the potential benefit is determined not only by his own decisions, but also by those of third parties. The strategies tell the player what action he must choose in a certain game state in the light of available information in order to maximize his expected benefit.

Game theory is a useful tool for analyzing the impacts of rules governing interpersonal relations. The simplest, "classical" game-theoretical situation is the Prisoner's Dilemma, a lucid demonstration of conflict potential between individuals and the importance of rules. In the situation as originally invented, from which the game takes its name, two people are caught for a minor offense, and possibly a serious crime to which they have not (yet) confessed. Different pairs of responses – one confesses, both confess, neither confesses – lead to different sentences, and it can be shown that individually rational behavior leads to a relatively bad outcome for both if there is no contact between them.

Environmental problems can frequently be modeled as Prisoner's Dilemmas. The following situation can serve as an example: in manufacturing certain products, two states, A and B, discharge contaminated effluent into a shared waterbody. The production of these goods provides the inhabitants with a quantifiable benefit valued at 8. If the effluent is not treated, the water becomes so polluted that neither side is able to derive any further benefit from it. If both states implement measures to treat the effluent, the improved quality of the water generates a benefit of 6 for the inhabitants of both states, whereby the costs for treating the effluent are valued at 4. If only one state implements such measures, this produces a benefit of 3 for both states due to the improved water quality.

		B Strategy 1	B Strategy 2
A	Strategy 1	I 10, 10	II 7, 11
A	Strategy 2	III 11, 7	IV 8, 8

Cells I–IV in the matrix show the size of the net pay-off to the inhabitants of A (left) or B (right). If both states refuse to treat their wastewater, their pay-off from the production of goods will be 8 in each case (cell IV). If both states implement measures to treat their wastewater, each society will have a net pay-off of 10 (cell I): the net pay-off from production is reduced by the costs of wastewater treatment from 8 to 4. This loss is offset by the net gain from the utility value of the purified, shared water, which amounts to 6 for each state.

In the situation as described, a dominant strategy will emerge for both A and B if neither knows the strategy adopted by the other: regardless of the behavior of B, A will always choose strategy 2 (no wastewater treatment), and regardless of the behavior of A, B will always select strategy 2 (no wastewater treatment). If country B decides to treat its effluent, the population of A will obtain a benefit of 11 by refusing to treat its effluent. The benefit would be only 10 if A opted to implement wastewater treatment as well. If B decides against treatment, the people in A are still better off if they adopt strategy 2 (no treatment) (benefit of 8 instead of 7). Regardless of B's behavior, the benefit is greater for the people in A if they decide against purifying their wastewater. The same applies for the people in B. Since both states will act rationally – in the absence of cooperation – and decide against wastewater treatment, the situation in cell IV will ensue. However, the people in both states would be better off if both A and B were to treat their effluent. Each society could then achieve a benefit (utility) value of 10. To bring about this situation, it is necessary to have rules that effectively reduce the uncertainty about the behavior of the respective other. Unless there are binding rules governing their behavior, both states will act in such a way that the outcome produced is that of cell IV – the situation neither side wants.

most ambitious development project in its history – the series of dams referred to as the Southeastern Anatolia Project (Günedogu Anadolu Projesi – GAP). One indication that this gigantic water development project is likely to cause major tensions with neighboring states are the surface-air missiles that Turkey has installed to protect the project from military attack. However, this does not necessarily imply that the dam would automatically lead to open warfare between Turkey as the upper riparian, Syria as the middle riparian and Iraq as the lower riparian. Even though no trilateral negotiations for allocating the waters of the Tigris-Euphrates are being conducted, there is no acute risk of war at present.

Unilateral vociferation of needs and uncertainty about the intentions of other riparians together promote an atmosphere of mistrust and generalized wishful thinking in the region, which in the medium term could severely exacerbate the existing problems concerning resource distribution and the environment in the Tigris-Euphrates basin. If burgeoning growth of water demand is compounded by unexpected dry spells, political relations between the three riparians may deteriorate rapidly. Armed conflict, triggered off by different perceptions, cannot be wholly excluded, particularly since the Near and Middle East will probably be regions with severe water stress in the next century.

The real source of animosity is the Ataturk Dam, the heart of the large-scale water development project in Turkey that is supposed to turn southeast Anatolia into a prosperous, dynamic economy. The storage capacity of all the dams taken together will amount to about 115–120 billion m^3 in the year 2000. Syria and Iraq, the downstream riparians, are therefore anxious that the planned projects in the upper reaches, in addition to those already constructed, will impose serious restrictions on their freshwater requirements, which are growing substantially. The streamflows for the Tigris-Euphrates basin are in fact known, so the gap between demand, on the one hand, and current and forecast requirements, on the other, can be assessed with some precision. Deducting the water volume consumed in the upper reaches results in an average streamflow at present of approximately 700 m^3 per second. In bilateral negotiations with Syria and Iraq in 1984 and 1987, Turkey guaranteed a streamflow at the Turkish-Syrian border of 500 m^3 per second, which Syria and Iraq agreed at their own discretion to share in a ratio of 58 to 42. If the "equitable" shares the parties have accorded each other are compared with the existing water volume of

700 m³, then the volume actually in dispute is around 200 m³ per second. In other words, the issue revolves around the allocation of 21% of the Euphrates waters.

A consensus solution is impeded by development targets and strategic interests that have led in recent years to major divergence between positions. These include the consumption targets of the three riparians, which have been set very high or even exaggerated in each case, given the focus on agricultural modernization and the locating of new industry, in order to achieve a better bargaining position. Both present-day requirements and future consumption targets are far in excess of the river basin's capacity.

Another factor is the complexity of security issues in the Tigris-Euphrates basin, with a specific constellation of conflicting forces emerging due to the geostrategic location of the riparians as well as cultural, socioeconomic and environmental differences. There is a long-standing tradition of regional conflicts – the Ottoman Empire dominated the entire region right up to the First World War, so Turkey's current dam projects awaken bad memories of Ottoman hegemony in Syria and Iraq. The megadam project must also be seen in the context of domestic problems in southeast Anatolia; the project stands symbolically and within development policy for the integration of east Anatolia, and hence the Kurdish people, into the Turkish state.

However, Turkey is able to mobilize strong economic and environmental arguments in support of its claim, as upper riparian, to be responsible for water use in the entire basin. It may indeed make good economic sense to operate irrigation schemes where they are most productive and consume scarce water resources most sparingly. But a regional "division of labor" along these lines would increase the vulnerability of the downstream riparians; despite the higher costs involved for them, Syria and Iraq are endeavoring to engineer a national supply basis that is as independent of Turkey as possible. Moreover, the majority of the Syrian and Iraqi population comprises small farmers for whom these waters are an essential basis of their livelihood.

Nevertheless, Syria and Iraq have divergent interests – for Syria, in particular, the Euphrates is the central source of water for industry and agriculture. Because precipitation falls to as little as 200 mm a year in the inland areas, agriculture is only possible with irrigation. If Turkey implements its plans for the upper reaches of the Euphrates, Syria might face sustained water scarcity. For this reason, Syria proposes that demand and supply be calculated separately for all three countries. Iraq, on the other hand, has a stronger position vis-à-vis Turkey. Firstly, it is not exclusively dependent on the waters of the Euphrates, but now has the Tigris virtually all to itself. Secondly, it is able to couple the flow of water from Turkey to the flow of oil into Turkey. Moreover, Iraq lays claim to an "historical right" to irrigate additional areas, in that some irrigation systems have been in operation since the ancient Sumerian dynasties. But it is precisely this aspect that makes Iraq vulnerable, because irrigating 1.95 million hectares of cropland creates the highest level of water consumption in the entire region.

Within this complex web of conflicts, Syria has a middle position both geographically and politically. It has been exposed to increasing pressure from both sides. Because of its own high level of water consumption and the history of hostilities between the two nations, Syria would be very unwilling to join sides with Turkey against the downstream riparian, Iraq. But a joint approach by Syria and Iraq is inconceivable since the second Gulf War, certainly not with the political systems currently in place in the two countries. For all the differences in positions and interests, Turkey and Iraq might possibly manage to reach bilateral agreements to the detriment of Syria.

The different positions of the three riparian states are also expressed in divergent interpretations of international law on freshwater resources (see Section D 5.5). Whereas Syria and Iraq seem to focus their policy of "fair" share on the legal concept of "shared resource", Turkey insists as upper riparian on a fundamentally different analysis; it rejects both neighbors' claims to two thirds of a resource that is situated in 88.7% of Turkish territory. The same argument is advanced in respect of the Tigris. The position taken by Iraq, which wants to safeguard 83% of the Tigris waters for itself and concede only 13% to Turkey, even though 52.8% of the water is on the latter's territory, is rejected as inequitable. Turkey favors the concept of "equitable and reasonable utilization" of transboundary watercourses, which according to its view, takes general precedence over the "no harm rule", thereby favoring the interests of upstream riparians. Furthermore, it has to be noted that Turkey's reliance on the principle of "equitable and reasonable utilization" depends highly on its own perception of the rule. Consequently, Turkey was one of three States that opposed the adoption of the "Convention on the Law of the Non-Navigational Uses of International Watercourses" on May 21, 1997 by UN General Assembly Resolution 51/229 [36 I.L.M. 700 (1997)]. The Convention, which was negotiated on the basis of the International Law Commission's Draft Articles on this subject matter (UN Document A/49/10), lays down a different basic understanding of the principle of "equitable and reasonable utilization" and strikes a different balance between the two fundamental rules, which Turkey perceived as unacceptable.

Opportunism feeding on geographical factors is only one reason for implementing a comprehensive water regime in the region. Such a regime would have to give adequate consideration to the complex and interdependent roots of potential conflict in the various river basins.

Another dimension of the water conflict is its instrumentalization by the strategic interests of other nations. The different interests of the three riparians relate not only to the water resources themselves or to technical aspects, but are rooted primarily in historical and political developments.

Turkey, for example, sees itself more and more as a global player in the political arena ever since the disintegration of the USSR. As a member of NATO, a candidate for EU accession and for full membership in the WEU, Turkey is striving for a bridging function between East and West that would permit it to operate as a regional superpower in all directions. The country has considered itself since 1989 as one of the main pillars supporting the establishment of a stable architecture for pan-European security. On the other hand, it is located geographically in the midst of a circle of acute conflicts in the Balkans, the Caucasus, in Central Asia, the Near East and the Middle East. The claim to power deriving from such a bridging role provokes unpleasant memories among Arab neighbors of the period prior to the Ottoman Empire's demise.

Governments in the region link solutions to the Kurdish problem directly and indirectly to the water issue. Direct links are generated above all by national issues within Turkey. Observers assume that Turkish politicians are very unlikely to invest billions in a region they believed might become autonomous or achieve independence in the middle to long term. On the contrary, GAP stands symbolically and within development policy for the integration of east Anatolia and its inhabitants into the Turkish state. These efforts to integrate the region economically are an indication, in turn, that any policy change involving recognition of the Kurds as a national minority is inconceivable for policymakers. The armed conflict that has flared since 1984 between the Kurdish Workers' Party (PKK) and the Turkish military has repeatedly delayed work on the project, and discouraged many foreign investors.

There is very little likelihood, in the present at least, that the water problems in the Tigris-Euphrates basin will escalate into armed conflict over the distribution of water resources. However, if the riparians continue to insist on unilaterally defined consumption targets, the security risks will mount. Any analysis of future threats must recognize the roles played not only by typical security policy and competing geostrategic interests in a scarce resource, but also by environmental security issues. Indeed, it is precisely the radical transformation of the basin's environment, mediated through the socioeconomic problems of the rural producers in all three countries, that may destabilize international security. Conflicts over resource distribution within the three countries will promote the marginalization of the rural population. The Ataturk Dam is more likely to boost rural poverty than mitigate it.

Turkey's Three-Staged Plan for Optimum, Equitable and Reasonable Utilization of the Transboundary Watercourses of the Tigris-Euphrates Basin could provide a basis for compromise. The first stage envisages the exchange and evaluation of meteorological and hydrological data, as well as the calculation of water quantities and water losses at agreed measurement sites. The second stage comprises an inventory of arable land, the evaluation of soil quality and assessments of use patterns and irrigation systems. The third stage requires the integrated evaluation of land and water resources, with the objective of minimizing water losses and water consumption by identifying and implementing advanced irrigation systems. These activities are supplemented by the evaluation of total water consumption by the three states, the measurement of evaporation rates, and more.

The plan is a good basis for a regional water regime. However, an essential requirement is that Turkey take into consideration the objections raised by the two downstream riparians regarding the political leverage that Turkey would gain and their anxieties about being dependent on Turkish goodwill. Including the degree of external vulnerability as a further criterion would be one way to do this. The objective would have to be a reduction in the possibility of one party causing harm to another. The number of people per unit of flowing water could be used as an indicator in this context. Negotiations must be based on the accepted principle that the economic and environmental costs generated by a purely unilateral or even confrontational approach would greatly exceed the size of concessions needed in the middle term in order to engineer a cooperative solution.

4.1.3.2 The Jordan basin

The conflict in the Middle East between Israel and its Arab neighbors relates first and foremost to national security and territorial claims. However, there is a further dimension, a "conflict of welfare" involving disputes over a limited resource, which has been gaining in significance (Schmid, 1993). The conflict over finite water resources can be traced back to the very beginnings of the Middle East conflict, but has become increasingly important in recent years. On the one hand, the fact that Israel obtains 50% of its waters from the disputed territories means that the overall peace process in the region involves not only security issues but also Israel's water supply. On the other hand, the growing relevance of the waters in the Jordan basin can be attributed above all to population growth in the region, the overexploitation of water resources – especially through the explosive growth of irrigated agriculture (Renger, 1995) – and water pollution (Libiszewski, 1995). The waters of the Jordan and Yarmuk are crucial to the existence of Israel and its Arab neighbors (Durth, 1996; Dombrowsky, 1995).

Solving the water conflict will not suffice to resolve the overall conflict, but it is an essential component of any solution. The rights to watercourses have played a decisive role since the Israeli state was first established. Johnston's Unified Plan of 1955, which contained proposals for sharing the waters of the Jordan and Yarmuk among Israel, Jordan, Lebanon, Syria and the West Bank, failed to take hold due to the lack of consensus from Syria and Lebanon. Exploiting this unregulated situation, Israel was drawing almost twice as much water from the Jordan in the early 1990s as envisaged by the water-sharing regime in Johnston's Unified Plan (Lehn et al., 1996). The dispute between Israel and its neighbors is closely linked to the question whether the allocations according to the Unified Plan have effectively been shifted in Israel's favor as a result of the latter's territorial gains (Soffer, 1994). However, the security conflict over territory and borders that has been waged in numerous wars has been going through a phase of consolidation since 1990. In the current situation, the conflict over use and allocation of the water resources must be seen as a "conflict of interests" as opposed to a conflict over values, with water being assessed absolutely and not relatively. This means, comparatively speaking, that there is now a "very good" chance of regime-formation if research findings in the social sciences are anything to go by.

Initial steps towards formation of a water-sharing regime are contained in the Israeli-Jordanian Peace Treaty concluded on October 24, 1994. The treaty contains detailed provisions concerning quantitative allocations, protection of water quality and the requirement to manage and develop their water resources in such a way that no harm is done to those of the other party. The substantive content of the peace settlement is based on the doctrines of customary international law.

An unusual feature of the treaty, and one which relates the specific circumstances faced by the two countries, is their recognition that existing resources are not sufficient to meet their needs and that the parties must therefore cooperate in order to develop new water resources (Alster, 1996): expanding Jordan's supply of water could only be achieved otherwise if Israel were to limit its own share – a concession that would be unacceptable to the Israeli population.

Unfortunately, very little progress in implementing the treaty commitments has been achieved to date. Moreover, a bilateral agreement of this kind must become an integral part of a wider, basin-wide agreement among all the riparians if the conflict is to be resolved. The Declaration of Principles on Interim Self-Government Arrangements concluded by Israel and the PLO in October 1994 is a further element of such an accord. A possible forum for settling the conflict is the working group on regional water resource management set up within the framework of multilateral peace negotiations. The background to the working group was the realization that certain issues can only be dealt with in a regional and international context (Renger, 1995). Participants in the working group negotiations include not only the immediate protagonists, but also other states from the region as well as leading nations in world politics and the global economy.

Johnston's Unified Plan of 1955 is often cited in the literature as an example of a non-zero sum game. This is because it involved not only the riparian states but also external third parties who could mediate between the conflicting parties, provide technical and financial assistance, or apply political pressure on the parties (Eaton and Eaton, 1994). There were therefore obvious parallels to the working group on water resources within the multilateral peace talks.

The working group on water has been unable so far to reach any significant decisions due not least of all to the integral role played by the water issue within the process as a whole. Some results have been achieved in connection with water harvesting issues, where the focus has been concentrated on the search for technical solutions to water scarcity. But what is really needed is a political solution. Political successes to date have been rather modest: for the first time, Palestinians have been accepted by Israel as participants in this forum.

4.1.3.3 The Gabcikovo Dam on the River Danube

International conflicts over transboundary waters can be settled not only through direct negotiations, but also by bringing them before international courts or tribunals. In the latter case, states no longer have the last word regarding the outcome, which is determined on the basis of the relevant regional treaties or customary international law. Precisely on account of the uncertain outcome in such cases, the jurisdiction of international tribunals has played a secondary role in international politics.

A contrary example is the conflict between Hungary and the Slovak Republic over dams on the River Danube. In 1993, the two states brought their long-standing conflict before the International Court of Justice (ICJ) in The Hague, after mediation by the EC had failed to engineer a settlement. As far back as the 1950s, Czechoslovakia and Hungary had given thought to a joint hydroelectric dam project on the Danube; in 1977, the two states signed a treaty to this effect, in which they agreed to build a hydroelectric dam at Gabcikovo, on the Czech side, and to create a reservoir upstream from Gabcikovo at the Hungarian town of Dunakiliti. The Gabcikovo dam was designed for peak-load operation, i.e. the turbines were not to be driven by the natural streamflow of the Danube. Instead, the river was to be held back for 18 hours a day in order to achieve a higher output from the turbines (720 megawatts) during the remaining six hours of the day in which the water would be allowed to flow. This would have enabled power generation to be synchronized with peak demand for electricity. On the other hand, this system requires an additional, downstream reservoir for evening out the daily fluctuations in water level – essential for navigation, for example. According to the 1977 treaty, this lower dam was to be built at Nagymaros in Hungary.

In 1989, Hungary decided to stop work on the project and asked Czechoslovakia to follow suit, particularly in view of the environmental hazards induced by the project. Prior to this step, there had been considerable opposition from the Hungarian population, supported by the Hungarian Academy of Science. Czechoslovakia did not share Hungarian misgivings about the environmental impacts and continued with construction of the Gabcikovo dam. To substitute for the Hungarian dam at Dunakiliti, it drew up an alternative plan that envisaged a reservoir at Cunovo, on Czech territory, and then proceeded to implement the plan ("Variant C"). Hungary responded by rescinding the treaty on May 19, 1992. The Cunovo dam enabled Slovakia to operate the hydroelectric power station at Gabcikovo using the natural streamflow of the Danube.

As Klötzli (1993) has shown in a study conducted as part of the Environment and Conflicts Project, the Gabcikovo project contains the germs of a conflict going much further than the purely environmental context. The Danube project, designed under the socialist regime, endangers not only the balance of nature in a unique floodplain landscape, but is being increasingly instrumentalized by both sides in connection with the national minorities issue. Both sides are misusing the environmental conflict in the conflicts over the rights of minorities .

That said, Hungary and the Slovak Republic took the step of bringing the issue before the International Court of Justice in 1993. It is interesting to note that Hungary is basing its position before the ICJ primarily on environmental arguments. It is anxious about losing biodiversity, about disruption of the balance of nature in the region, and about deterioration in the quality of Danube water; the latter would jeopardize the supply of drinking water to Budapest, which covers most of its needs from wells sunk into the sand and gravel beds beneath the river's floodplain. The Czech side responds to these objections either by rejecting them outright, or by referring to possible counter-measures.

The case, which the ICJ decided recently on September 27, 1997, was very complex, both legally and practically. In its judgement the ICJ found that Hungary, for example, has violated the legal principle of pacta sunt servanda by unilaterally suspending its contractually agreed operations at Nagymaros and Gabcikovo. The Court clarified that a "state of necessity" as invoked by Hungary and laid down by the International Law Commission in Article 33 of the Draft Articles on the International Responsibility of States (Yearbook of the International Law Commission, 1980 Vol. II, Part 2, p. 34ff., to which the ICJ referred; meanwhile the draft articles have gone through a second reading: see GAOR 51, Supp.10, p. 125, to be found at http://www.un.org./law/ilc/chap03.htm#doc74) could not permit the conclusion that the treaty obligations had ceased to be binding upon it. A "state of necessity" would prohibit the incurrence of international responsibility only. It is worthwhile to point out that the ICJ acknowledged expressly that ecological concerns – as expressed by Hungary – could indeed relate to an "essential interest" of a State required to invoke a "state of necessity". It stressed the great significance it attaches to respect for the environment, not only for States but also for the whole of mankind. However, the Court found that Hungary had not convincingly shown that a real, "grave" and "imminent" "peril" to the environment as an essential interest to Hungary had existed in 1989.

The Court continued by finding that Czechoslovakia was entitled to proceed, in November 1991, to the "provisional solution" as described in the Special Agreement. Hungary's reaction, the termination of the 1977 Treaty and related instruments, did not therefore have the effect of terminating them. On the other hand Czechoslovakia was not entitled to put the "provisional solution" into operation from October 1992 onwards. Finally, the Court decided that both Parties have to compensate each other for the damages incurred by their unlawful acts and that Hungary and Slovakia must negotiate in good faith in the light of the prevailing situation, and must take all necessary measures to ensure the achievement of the objectives of the Treaty of 1977, in accordance with such modalities as they may agree upon. Part of it will be the establishment of a joint operational regime in accordance with the 1977 Treaty.

The future is likely to see more international conflicts ensuing from divergent assessments of the environmental impacts of a nation's activities, especially since environmental awareness and environmental standards vary from one state to the next. The ICJ advanced and strengthened international environmental law by acknowledging an "environmental state of necessity". Nevertheless, the decision shows the restrictions that environmental protection by law faces: uncertainties in fact cannot wholly be compensated by law. The willingness of Hungary and Slovakia to bring the controversy before the ICJ should be viewed as a notable example of conflict settlement.

4.1.3.4 The Great Lakes in North America

The US-Canadian regime governing transboundary watercourses shows that conflict over freshwater resources can be solved by peaceful means. In 1909, the two states signed the first Boundary Waters Treaty, to which many amendments and additions have since been made. The regulations relating to the Great Lakes are seen as a pioneering achievement in the monitoring of water pollution in a shared river basin and its ecosystem. The main focus is on monitoring the riverine ecosystem, not just the avoidance of conflict. The regulations are aimed at preventing disputes before they arise. Special bilateral institutions exist in which the US states and Canadian provinces play an integral role. When a conflict arises, procedures are set in motion to investigate the facts of the case with the involvement of technical experts and the general public. The procedure ends with a recommended settlement for the disputing parties.

The Boundary Waters Treaty of 1909 includes principles and mechanisms for preventing or settling disputes, especially when they relate to water quality. The treaty also contains one of the first ever provisions for preventing water pollution. The Joint Commission established under the Treaty is a dispute-settlement instrument that enables either party to refer an issue or grievance to the Commission for investigation. The Commission may then issue a report containing recommendations, but this does not have the same status as a formal dispute settlement verdict. This procedure is known as "reference" and is the instrument used for many years to settle disputes. The procedure is based on investigations, hearings and expertises. Since 1909 there have been 52 references to the Joint Commission. Three references have been made in the last twelve years: the 1985 Flat Head River Reference, the 1986 Great Lake Levels Reference and one reference to obtain public statements on the US-Canadian air quality treaty of 1991. The Great Lakes Levels Reference of 1986 is particularly important, because the Commission chose an ecosystem approach, is itself permanently involved in monitoring the water level, and puts forward proposals to the parties. The Joint Commission has also played a major role in settling transnational conflicts through mediation, or in applying the results of technical expertises.

In 1964, in response to a record low level in the Great Lakes and serious concern about water pollution, the USA and Canada submitted a reference to the Joint Commission. The Commission set up two joint investigatory boards comprised of technical experts recruited from the governments of the two countries and the affected federal states and provinces. After the joint investigatory boards had reported on serious problems with phosphate contamination and eutrophication of the waters, the parties negotiated the Great Lakes Water Quality Agreement of 1972. It contains joint water quality targets and a procedure for monitoring pollution. In addition, a formal institution was established, namely a Water Quality Board charged with responsibility for coordinating the efforts of US federal states and Canadian provinces to monitor and control pollution. A Research Advisory Board was also set up to give scientific advice to the Joint Commission and the Water Quality Board. Finally, a joint regional office was set up with its headquarters in Windsor, Ontario.

In 1978, the 1972 agreement was amended and extended in scope, while maintaining the institutional structure. The objective of the new agreement was to restore and preserve the integrity of the ecosystem in the Great Lakes basin. This was a pioneering goal, in that it focused on the entire ecosystem of the basin and envisaged a pollution control system based on ecosystem management. This meant that not only direct emissions were taken into account, but also pollution from other sources. In 1987, a Protocol amending the Great Lakes Water Quality Agreement was negotiated. The Protocol relates to the various causes of pollution in the Great Lakes and attempts for the first time to implement an ecosystem approach aimed at effectively protecting water quality throughout the Great Lakes basin.

There has been little dispute regarding implementation of the Great Lakes Water Quality Agreement. This is partly attributable to the instruments it provides for promoting cooperation between the parties in monitoring pollution.

In addition to the international treaties between the two countries, there are other major agreements at sub-national level between the relevant US states and Canadian provinces. In terms of conflict mediation, their major efforts to cooperate on pollution prevention and other issues of relevance to water have created a level of mutual trust in the wish of both sides to preserve the resource. The precautionary measures in place have certainly helped to minimize and resolve cross-border conflicts. They are important because they involve a high level of public participation. In addition to the presence of many NGOs demanding protection of the Great Lakes, the government bodies have involved citizen groups through hearings, information exchange and through various working groups, at the respective national levels and at the federal state or province level. This, too, has had a positive impact on the resolution of potential conflicts over the Great Lakes.

Given the close relations nurtured by the USA and Canada, the rather informal investigative procedures ending with recommendations to the parties have proved effective. Experience acquired in connection with the Great Lakes and with water controversies between the two nations provides useful insights that could be of value for water management and environmental disputes in other regions of the world.

Both states view disputes over transboundary waters as pure conflicts of interest, in which the issues at stake are the utilization, conservation and protection of the waters and their ecosystems. These conflicts of interest have not been instrumentalized in the foreign policies of the two countries, nor linked to foreign policy values such as security, power, spheres of influence or similar. It therefore comes as no surprise that the USA and Canada have cooperated for almost ninety years when freshwater disputes have arisen, and have succeeded in firmly establishing an efficient bilateral regime governing transboundary waters.

4.1.4 Degradation of freshwater resources as a global problem

The case studies show that utilization of scarce water resources has been the source of many regional conflicts between states. These have been resolved by cooperative means when they were pure conflicts of interest, but inadequately regulated if the conflicts of interest were supplemented by conflicts over the choice of means, or security issues and spheres of influence. In the eyes of the Council, however, the degradation of freshwater resources throughout the world is not merely a regional problem.

Although the problem of freshwater resource degradation does not involve interdependencies as complex as the destruction of the stratospheric ozone layer, climate change or the contamination of the oceans with persistent organic pollutants, the latter problems involve environmental damage in one state impacting directly on environmental assets in another, which has led in many cases to various forms of effective regulation by international regimes (on conventions in these areas, see WBGU, 1996; on the formation of global environmental regimes, see, for example, Simonis, 1996; Biermann, 1994; Breitmeier, 1996). Water is not an identical resource throughout the world – stocks vary considerably in quantity and quality, and are regionally finite – and this aspect generates different lines of conflict and different claims to the resource (BMZ, 1995 and 1996; SEI, 1996; Gleick, 1993).

The Council sees four aspects that lend a *global* dimension to the degradation of freshwater:

1. regional water conflicts may escalate and cause destabilization at global level;
2. certain waters form part of the world natural heritage and therefore need the protective mantle of the whole international community;
3. the pollution of freshwater impairs the marine environment and thus part of the global commons;
4. the degradation of freshwater resources is posing a mounting threat to the realization of the human right to food and water, which was codified (for the contracting parties) in the International Covenant on Economic, Social and Cultural Rights of 1966 and which imposes a commitment on the international community according to the respective capabilities of each state.

4.1.4.1 Regional water conflicts as a threat to world security

In many cases, international water conflicts are confined to the region in question, as is apparent in the specific issues described above. But disputes like those over the waters of the Euphrates and Tigris are also indicative of the major destabilization potential bottled up in such regional conflicts. More than 200 watercourses, innumerable lakes and groundwater reservoirs have transboundary basins. Very few conflicts over utilization rights have ever been brought before the International Court of Justice.

A number of other conflicts have been regulated by international regimes in certain regions, in which the utilization rights of each state are precisely defined. However, regional regimes of this kind have mainly been established between states that cooperate relatively closely, as in North America or Europe. On the other hand, conflicts in Western Asia over the Euphrates and Tigris, or in East Asia over the Mekong (where the upper riparian has stayed aloof from the regime of lower riparians) show that, in the absence of cooperative regulations, water conflicts may persist and even escalate. The likelihood of the latter is magnified by the very fact that pressure on the remaining resources increases and that more and more people are potentially affected by water crises.

The Council sees two main solutions: firstly, conflict regulation and conflict avoidance mechanisms must be improved for states affected by water conflicts; this includes, in particular, the support provided by the UN Convention on the Non-Navigational Uses of International Watercourses recently adopted by the UN General Assembly (UN GA Res. 51/229), and perhaps the creation of a specialized mediation and negotiation body, such as an international mediation center for water conflicts (see Section D 5.5).

Secondly, every opportunity must be exploited to prevent local and regional "water crises" from arising in the first place and later from leading to international conflicts. If the present-day trends described in this Report continue, existing water conflicts will worsen and new ones will be added. The threat to global security inherent in the escalation potential of regional water conflicts places the protection and conservation of freshwater resources firmly on the global agenda. Nations have a similar interest in combating regional water crises on the global level, so there is a strong need for a regime, ideally a global regime, for mitigating and preventing "water crises".

4.1.4.2 Freshwater resources as part of the world natural heritage

Some inland waters qualify as part of the world natural heritage by virtue of their unique biodiversity, or their outstanding scientific or aesthetic value. To protect this heritage, an international Convention concerning the Protection of the World Cultural and Natural Heritage was signed by the Member States of UNESCO as early as 1972. This convention commits the parties, *inter alia*, to place certain sites designated as part of the "world natural heritage" under special protection. Furthermore, states may ask the international community for assistance if they do not have the resources themselves to carry out the necessary measures in an adequate manner. The World Heritage Fund was established when the convention entered into force, and partly consists of compulsory contributions made by the signatory states, making it the pioneering fund for similar instruments in later years, such as the Multilateral Ozone Fund under the Montreal Protocol and the Global Environment Facility (GEF) of 1990/1991 (Ehrmann, 1997). Like the Multilateral Ozone Fund and the GEF, the main function of the World Heritage Fund is to give developing countries the assistance they need to preserve the world heritage on their own territory.

Where certain inland waters may be deemed part of the world natural heritage, protecting them is the responsibility of the international community when the states concerned do not have sufficient resources of their own. This is a further illustration of the global dimension to the use and protection of freshwater resources (see Box D 1.2-1).

4.1.4.3 Inland waters and marine pollution

70–80% of marine pollution is land-based, more than half being discharged into the sea from rivers. More oil, for example, is discharged into the world ocean via rivers (and coastal cities) than from tankers (WBGU, 1996). In this context, too, there is a global dimension to local and regional use of inland waters. As the Council explained in its 1995 Report (WBGU, 1996), protection of the seas requires above all a comprehensive range of land-based measures to protect the environment. To date, developing countries have treated only 5% of their effluent; with rising population densities in coastal zones, combined with industrialization and intensification of agriculture, the pressures on inland waters and coastal waters is developing into an environmental problem of ever-greater significance.

The Council believes that only global solutions can provide the answer. It explicitly welcomes the resolution adopted by the Washington Intergovernmental Conference on Protection of the Marine Environment from Land-based Activities of 1995 to commence negotiations for a global convention for the reduction and ultimately a ban on persistent organic pollutants. The first round of negotiations will probably start in 1998. Given the obvious similarities between the problems arising from POPs and from ozone-depleting substances, the Council recommends that the Montreal Protocol on Substances that Deplete the Ozone Layer, generally assessed as an outstanding success for global environmental policymaking, be taken as a model for the planned POP Convention. However, global reductions in "normal" wastewater must remain a key issue alongside the discrete problem of POPs, because their accumulation in the coastal areas threatens to cause severe damage to biodiversity. Already, 10% of the world's coral reefs has been destroyed; a further 30% is at risk, especially in coastal waters (Biermann and Hardtke, 1997).

4.1.4.4 The "human right to water"

The use and protection of freshwater resources have a global dimension, being indissolubly linked to the international protection of human rights. In accordance with Article 25 of the Universal Declaration of Human Rights (1948) and Article 11 of the International Covenant on Economic, Social and Cultural Rights of 1966, every person has a right to food, which includes drinking water (McCaffrey, 1992). States that signed the Declaration of the World Food Summit in Rome in 1996 reaffirmed that "every man, woman and child has the right to be free from hunger and malnutrition" and that "access to nutritionally adequate and safe food is a right of each human being".

The United Nations Conference on Human Settlements (HABITAT II) includes, in the Istanbul Declaration of 1996, an almost identical reaffirmation of the human right to food and explicitly the right to "adequate water". All these documents were supported by Germany.

However, these social rights should not be viewed in a strictly legal sense, since final declarations by UN summits or the Universal Declaration of Human Rights are not binding on states, even when they are based on a broad consensus. However, there is no disputing that the social right to food and water involves an obligation on the part of governments to do everything in their power to ensure that this right is realized.

According to Article 11, para. 2 of the International Covenant on Economic, Social and Cultural Rights, this obligation applies to the international community as well, which should support as far as it can all those states that are unable with their own resources to guarantee the human right to food and water. Within the human rights regime, major regional water crises are therefore more than a regional problem, but are immensely relevant to the interest of all states in the realization of social rights. To that extent, providing all people with a basic adequate supply of water for drinking and sanitation can be seen as a task for the international community as a whole.

4.1.5 Summary

The utilization and distribution of water resources may lead to conflicts between states that add up, especially with increasing global scarcity of water resources, to a dangerous potential for escalation. International conflicts over water distribution may be regional at first; however, due to their potential to endanger peace, the international community has an interest in preventing local and regional water crises and in peaceful settlement of regional water conflicts. A global interest is also present where part of the world natural heritage is threatened with elimination as a result of regionally confined activities. The entire international community is similarly affected when local and regional activities pollute the world's oceans. Even when water crises are confined to one specific region and have no direct impacts, such as migratory flows, on neighboring states, they can nevertheless impinge upon a global interest of the international community if people are no longer assured a basic supply of water – the "human right to water".

Brought together, these various aspects show that the protection and conservation of freshwater resources are the responsibility of the international community in many cases, which in other contexts has led to the creation of international regimes. The UN Secretary-General, in a study produced in 1997 with the UN special organizations, concluded that the various action programs adopted to date were insufficient, and that a new "Global Consensus" among the international community is needed for worldwide protection of freshwater resources. The Council agrees on the whole with this assessment by the UN Secretary-General, and stresses the need for greater international cooperation (for more detail, see Sections D 5.5 and E 2).

In regions where tensions surround transboundary water resources, the Council recommends that peace-promoting and peacekeeping pilot projects on sustainable management of water be initiated in order to de-escalate the potential for conflict. Involvement in such projects should be open to all parties. Shared use of the resource should be characterized by fair and equitable cooperation and allocation. Pilot projects of this kind could be encouraged as part of development, economic and environmental cooperation activities (for further recommendations on peace-promoting activities, see Sections D 5.5 and E 2).

4.2 Spread of waterborne diseases

Half of the world's population suffers from water-related infections – Developing countries particularly affected – New pathogens in industrialized countries – Hosts include mosquitoes and snails – Malaria spreading – Problem areas include mobility, resistance, climate change, poverty, growth of irrigated farming, rats

In the first half of this century, many infectious diseases were thought to be in decline, at least in the industrialized nations. Unfortunately, many parts of the world are now witnessing a resurgence of these diseases. This situation is a result of numerous and disparate factors that influence the incidence of infectious diseases: rapid population growth, dense human settlement near forests and swamps, high mobility, global trade, inappropriate use of pesticides and antibiotics, adaptation of pathogens to environmental conditions, social and political disintegration and, finally, perturbations of regional climate.

In global terms, waterborne infections are still one of the principal causes of disease and death, particularly in developing countries in the tropics and subtropics. Such infections are gaining significance in the industrial nations as well, mainly due to highly resistant parasitic pathogens. About half the world population currently suffers from water-related diseases. For this reason, regulated water supply and sewage disposal meeting the water hygiene criteria of the World Health Organization (WHO) represent some of the most effective preventive measures for combating these diseases worldwide. Investments in this area are likely to provide the highest possible "health gains". While safe drinking water may prevent many diseases, vaccine programs will continue to be a second major dimension of preventive medical efforts.

The United Nations designated the 1980s as the "International Drinking Water Supply and Sanitation Decade", in the course of which 1.3 billion people received a new drinking water supply and 750 million people sanitation facilities. Nevertheless, 1.2 billion people were still left without access to clean water and 1.7 billion without proper sanitation. According to United Nations estimates, nearly 900 million more people will fall into these categories in the 1990s due to population growth alone, simply because investments in infrastructure cannot keep pace with population growth (Fig. D 4.2-1). Such a rise is also evident in the European region of the WHO (including Russia). On the other hand, the susceptibility to infection has declined markedly in countries that have succeeded in providing hygienically pure drinking water and sewage disposal systems. Other successful measures for combating waterborne infections have included vaccinations and health education, especially for women who were reached by simple therapy concepts.

4.2.1 Diseases related to water use

In addition to poisoning by contaminants in water, there are numerous infections that can be transmitted by the use of water. A distinction must be made here between infections transmitted

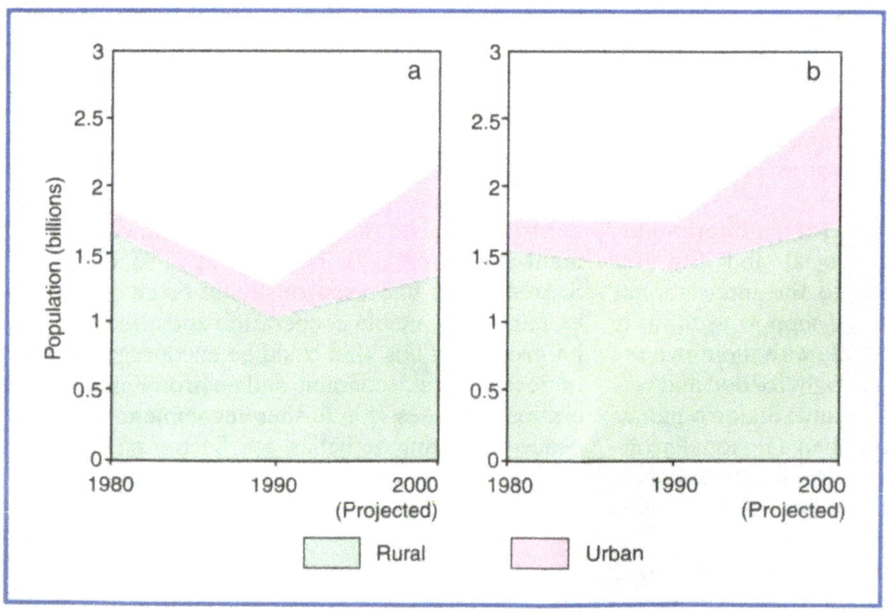

Figure D 4.2-1
a) Population without access to clean drinking water.
b) Population without access to sanitation facilities.
Source: Gleick, 1993

1. by the use of contaminated drinking water or through skin contact with contaminated freshwater or seawater, or
2. by animal hosts or carriers that transmit the pathogens in the area surrounding their freshwater habitats.

Table D 4.2-1 provides an overview of the major waterborne diseases, listed according to a systematic biological classification of pathogens. The numbers of people contracting or dying from these diseases are sometimes contradictory, depending on source, so the figures in the table should therefore be taken as an approximate reference only.

4.2.1.1
Use of contaminated drinking water

Local communities need an infrastructure that is technically and epidemiologically intact in order to ensure an adequate level of water hygiene. Even in developed countries, epidemics are often caused by the consumption of drinking water contaminated

Table D 4.2-1
Water-related diseases.
Source: PAHO, 1994; WHO, 1994, 1995 and 1996; Michael and Bundy, 1996

	Pathogen	Disease	Vector	Endangered persons (million)[a]	Incidence (1,000 per year)	Deaths (1,000 per year)	Growth through climate change
Viruses	Polio viruses	Poliomyelitis		no data	110	5	
	Dengue viruses (DEN virus)	Dengue fever	e.g. *Aedes aegypti* (mosquito)	2,400	560	23	++
	Yellow fever viruses (YF virus)	Yellow fever	e.g. *Aedes aegypti* (mosquito)	450	200	30	++
Bacteria	Pathogenic *Escherichia coli*, *Shigella* et al.	Diarrheas		no data	1,200,000 –1,800,000	3,000 –4,000	
	Salmonella typhi	Typhoid		no data	16,000	600	
	Legionella pneumophila	Legionnaire's disease			no data	no data	no data
	Vibrio cholerae	Cholera		no data	380	120	?
Protozoa	*Entamoeba histolytica*	Amoebiasis (amoebic dysentry)		no data	no data	no data	
	Cryptosporidium parvum	Cryptosporidiosis		no data	no data	no data	
	Giardia lamblia	Lambliasis		no data	500		
	Plasmodium sp.	Malaria	*Anopheles* (mosquito)	2,400	300,000 –500,000	2,100	+++
Trematodes	*Schistosoma* sp. (flatworm)	Schistosomiasis or bilharziosis	freshwater snail	600	200,000	20	++
Nematodes	*Wucheria* sp., *Brugia* sp.	Lymphatic filariosis	mosquito	1,094	117,000	no data	+
	Onchocerca volvulus	Onchocerciasis (river blindness)	blackfly	123	17,500	no data	++
	Dracunculus medinensis (guinea worm)	Dracunculiasis	Crustaceans (water flea)	100	100	no data	?

+ = probable ++ = very probable +++ = highly probable ? = unknown
[a] Projections for population growth, based on 1989 figures.

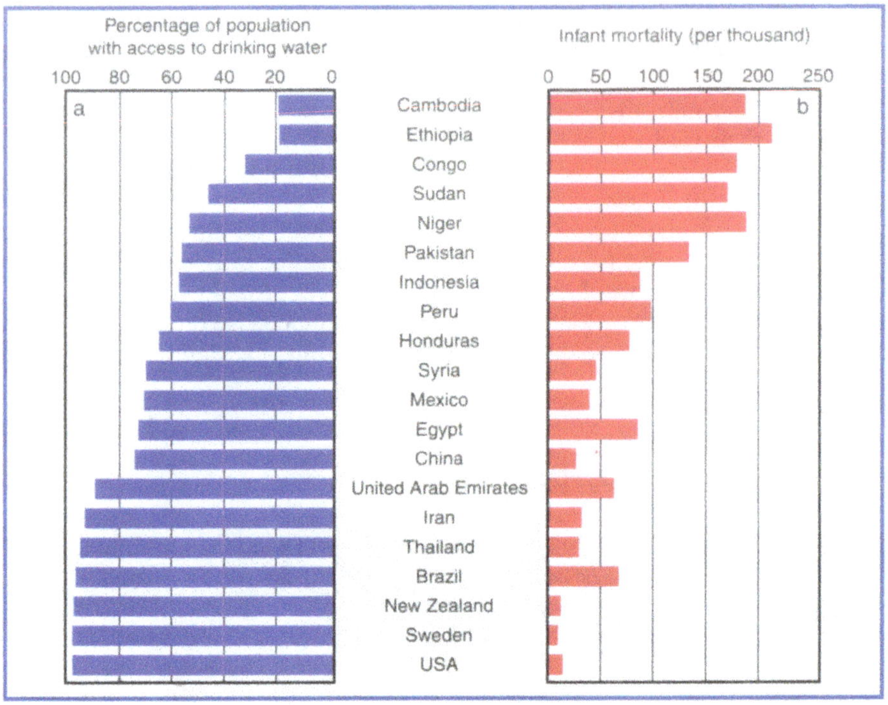

Figure D 4.2-2
a) Population with access to clean drinking water.
b) Infant mortality.
Source: Engelman and LeRoy, 1995

with sewage as a result of damage to the pipe system (Usera et al., 1995). Most at risk are large settlements with a central water supply. The threat is compounded by new pathogens that have appeared in recent decades. In 1976, for example, cases of people with cryptosporidiosis, triggered by the *Cryptosporidium* parasite in drinking water, were reported for the first time. The biggest cryptosporidiosis epidemic occurred in the USA in 1993. This problem arises in both developing and industrialized countries. One example from Russia has shown that hygiene regulations tend to be ignored during cold periods, leading to an increased incidence of waterborne enteric infections (Kartsev, 1995).

Infections transmitted through drinking water are particularly widespread in developing countries where fecal-oral transmission is a consequence not only of poor drinking water quality, but also of insufficient water for sanitation purposes (Bangs et al., 1996). 25 million people in the developing world die each year from drinking infested or contaminated water. Pathogens or carriers in water are responsible for 99% of the diseases transmitted by drinking water worldwide, while only 1% occurs as a result of chemical pollution. The infections are mainly due to protozoa (*Giardia*, Cryptosporidia), bacteria (e.g. *Escherichia coli*, Salmonellae, Shigellae, *Campylobacter* sp., Yersiniae, *Vibrio cholerae*) or viruses (e.g. rotaviruses). These pathogens are able to enter the drinking water system and cause acute enteric diseases or even systemic infections because drinking water and sewage systems are inadequately separated.

Globally, acute diarrheal diseases are the second most frequent cause of infant mortality, after acute respiratory illnesses (WHO, 1996). Of the total 3.1 million total deaths per year, nearly 80% involve children under the age of five. There is an obvious relationship here between access to clean drinking water and infant mortality (Fig. D 4.2-2). Cholera alone claims 120,000 lives annually. This infectious disease has re-emerged in many places recently (Fig. D 4.2-3) where it was previously assumed to be wiped out. Settlements with a very high population density and a lack of basic sanitation facilities are repeatedly the breeding ground for epidemics. Several dramatic outbreaks of cholera have occurred recently in refugee camps and urban slums (see Section D 3.5). Since 1995, Europe has again been witnessing several thousand cholera cases a year. Cholera can be transmitted not only through drinking water, but also through brackish water, when people swim in the sea, or through the ingestion of contaminated fish. As well as the high salt content, the sea provides other conditions that are favorable for cholera pathogens, for example an alkaline environment engendered by industrial wastewater and with concomitantly high populations of microfauna, as found along the coasts of South America and in the Bay of Bengal. New varieties of cholera bacteria make the disease difficult to combat in the initial phase. In contrast to *Vibrio cholerae* O1, which does not leave the intestine in the course of the infection, the cholera pathogen *Vibrio cholerae* O139, known since 1992, can become invasive through capsulation and lead to generalized in-

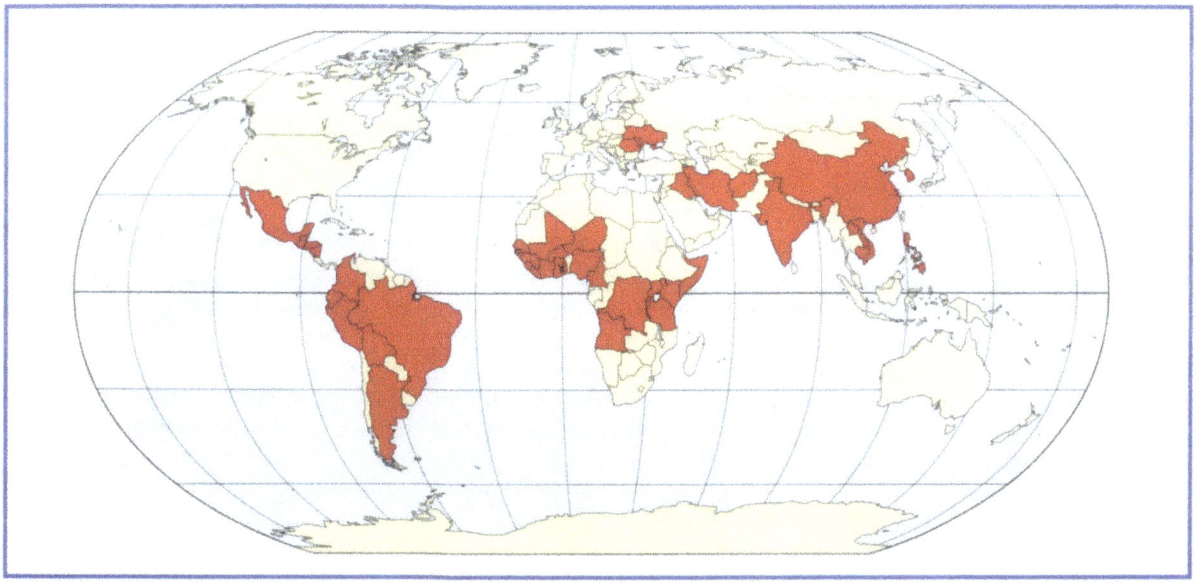

Figure D 4.2-3
Outbreaks of cholera in 1995.
Source: WHO, 1996

fections (spread of germs to the entire body). The epidemiological significance of this new pathogen cannot be assessed at the moment. Until now, however, it has been confined to Asia. More recent findings on the survival of *V. cholerae* outside humans and between pandemic periods indicate the need for a systems approach to studies in this field. There are evidently complex relationships between the spread pattern, the appearance of new strains and anthropogenic environmental changes, such as oceanic eutrophication and rising surface water temperatures, which foster phytoplankton growth (McMichael et al., 1996).

Approximately 4 billion cases of diarrheal diseases occur annually. Diseases associated with a lack of water hygiene, such as bacterial, viral and parasitic enteric infections, are frequently contracted in developing countries undergoing rapid urbanization and slums formation, as well as during large-scale migrations resulting from wars or natural disasters (see Section D 3.5).

Somewhat less frequent are parasitic infections, such as amoebiasis (infection with *Entamoeba histolytica*) and lambliasis (infection with *Giardia lamblia*), which are contracted in the same way and cause acute or chronic health disorders: amoebic dysentery, amoebic liver abscess as well as Lamblia enteritis with complicated malabsorption syndrome (deficient absorption of vital food components via the intestine). Mention must also be made of cryptosporidiosis, which is contracted through infection with oocytes of *Cryptosporidium parvum*. The largest outbreak to date, in which over 400,000 people suffered from diarrhea, took place in Milwaukee, USA, in 1993. This outbreak was caused by drinking water contaminated with *Cryptosporidium*. The direct and indirect costs are estimated at over $100 million (Exner and Gornik, 1997). Dracunculiasis, an infection with guinea worm larvae (*Dracunculus medinensis*) that are ingested together with microscopically small freshwater crustaceans contained in drinking water, has only regional relevance.

Of the infections that attack the entire body, Typhus abdominalis, hepatitis A and hepatitis E are the most frequent. Their frequencies correlate negatively with increasing quality of the water supply (Perez et al., 1996). The most important of the salmonellosis diseases, which are directly related to inadequate hygiene standards for drinking water and food, is typhus with 16 million cases and 600,000 deaths a year, predominantly in Asia (WHO, 1996). Lax monitoring of drinking water may lead to such waterborne infectious epidemics in industrialized countries as well (Yatsuyanagi et al., 1996).

A special case is Legionnaire's disease, which is caused by the bacterium *Legionella pneumophila* and contracted via aerosols produced when warm water is splashed – while showering, for example. This disease poses a threat both in warm countries and in temperate zones wherever there is a central hot water supply or when used water is stored in roof cisterns. Furthermore, unintentionally ingested wastewater may induce infections, e.g. while partaking in leisure-time activities in inland waters or at the sea.

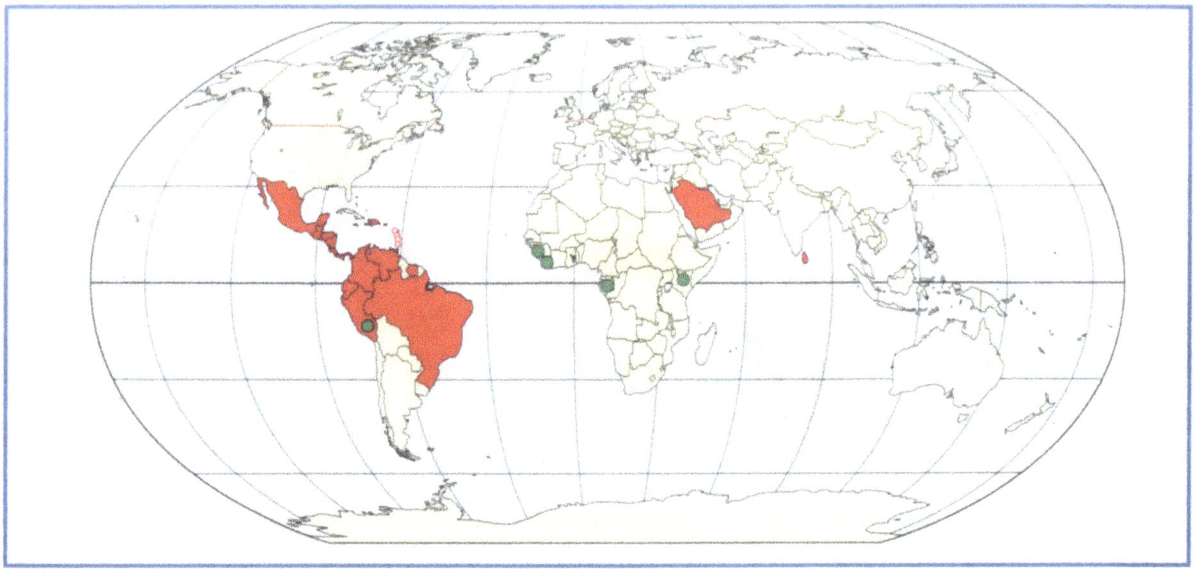

Figure D 4.2-4
Outbreaks of dengue (red) and yellow fever (green circles) in 1995.
Source: WHO, 1996

Poliomyelitis, caused by polio viruses, has declined in significance thanks to successful vaccination campaigns, but continues to occur in African and Asian developing countries.

Leptospirosis is a frequently severe infection resulting in renal failure, jaundice and hemorrhaging. It is contracted through the intake of *Spirilla* (*Leptospira*) via skin wounds and mucous membranes in contact with inland waters, or via rats or mice that excrete the bacteria in their urine. It is of regional significance only.

4.2.1.2
Water-based hosts and carriers of infectious diseases

Mosquitoes

Mosquitoes, whose larvae grow in freshwater, and other arthropods cause numerous viral diseases by transmission of so-called arboviruses (Nielsen et al., 1996). Of the 20 most common, all have caused epidemics. In addition to humans, other species such as apes, birds and rodents provide a natural reservoir for such viruses. Some of these viruses occur almost exclusively in rural regions of the tropics and subtropics; they are responsible for such diseases as O'nyong-nyong fever, Sinbis, eastern and western equine encephalitis, Barmah forest virus and Kyasanur forest disease, Murray Valley encephalitis, Rocio, Rift Valley fever and California encephalitis. Others are also the cause of diseases in larger settlements, e.g. Chikungunya, Ross River, Mayaro virus, Venezuelan equine encephalitis, dengue, yellow fever, Japanese encephalitis, St. Louis encephalitis, West Nile and La Crosse encephalitis.

The majority of infections run their course as feverish systemic infections. If complicated by meningoencephalitis (inflammation of the brain and meninges), death or permanent defects are often the result. Hemorrhagic fevers, characterized by a great tendency to hemorrhages, have the highest mortality rate and can take the form of dengue, yellow fever, Kyasanur forest disease and Rift Valley fever. Dengue (Fig. D 4.2-4), which can easily lead to death when it occurs as dengue shock syndrome, is the most important arbovirus. The infection occurs in over 100 countries which account in total for 2.5 billion of the world population (WHO, 1996).

Due to the breeding habits of the vector insect, the mosquito (*Anopheles* sp.), malaria is similarly confined to freshwater biotopes and is globally the most important water-related vector disease. Of the 500 million people who contract malaria every year, 2.1 million die as a result, including 1 million children. Mortality is almost exclusively caused by Malaria tropica, for which *Plasmodium falciparum* is responsible. Malaria tertiana, caused by *P. vivax* or *P. ovale*, and the rare form, Malaria quartana, caused by *P. malariae*, are fatal in exceptional cases only. Malaria occurs in 91 countries (Fig. D 4.2-5), with 40% of the world population living in these regions. Africa accounts for 90% of all malaria cases (WHO, 1996). Malaria is rarely found in urban areas because the *Anopheles* populations there have declined. In

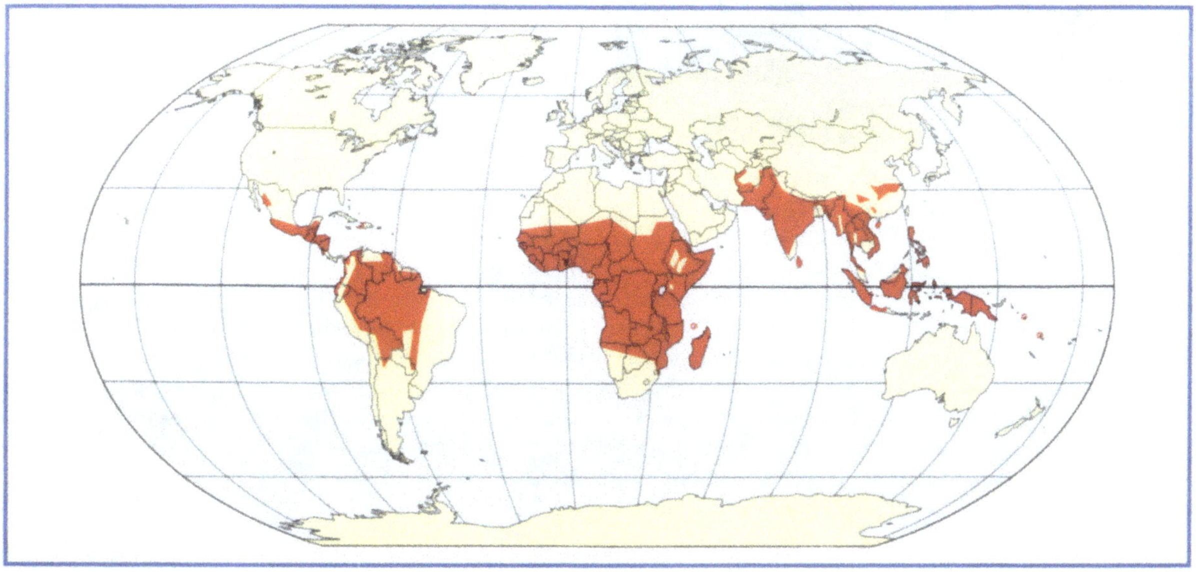

Figure D 4.2-5
Incidence of malaria.
Source: WHO, 1993

Southeast Asia, however, the mosquito is already adapting to city life.

Lymphatic filariosis, a threadworm infection that can also be transmitted by mosquitoes in the form of *Wucheria bancrofti*, *Brugia malayi* and *Brugia timori*, is widespread in the tropics and subtropics on all continents; 120 million people are infected (WHO, 1996). The disease leads to acute and chronic lymph tract inflammations with swelling of the dependent parts of the body (elephantiasis).

Onchocerciasis, transmitted by blackflies that breed in fast-flowing waterbodies or attach their larvae to river crustaceans in slow-flowing waters, is a threadworm infection caused by *Onchocerca volvulus* and feared because of the blindness that may result (river blindness). Around 18 million people are infected, mostly in Africa and to a smaller extent in Central and South America.

AQUATIC AND MUD SNAILS

Aquatic snails with lungs or gills serve as intermediate hosts for schistosomiasis in natural inland waters, as well in as irrigation and dam facilities with suitable aquatic flora. The total number of people infected by this disease, also known as bilharziosis, is currently estimated at around 200 million, with 200,000 deaths occurring each year. Its principal vectors, *Schistosoma mansoni*, *S. haematobium* and *S. japonicum*, use freshwater snails as hosts; these release the larva stage, the cercariae, which are infectious for humans. These larvae can pierce intact skin or mucous membrane and develop into adult flatworms. Acute symptoms, such as cercarial dermatitis and Katayama fever, may appear when larvae penetrate the body and subsequently grow. The eggs produced by the female are excreted with the urine or feces, or cause chronic inflammation, even carcinomas, in various internal organs. *S. japonicum* has another significant reservoir in working animals. *S. mansoni* and *S. haematobium* cycles, in contrast, can only be maintained where excreta containing schistosome eggs are able to enter bathing, drinking or used water sites containing populations of the relevant host snails (Burchard et al., 1996). In the newly built recreational centers in southern Africa, an increasing number of surfers and canoeists contract acute or chronic schistosomiasis.

Water snails of the genus *Bithynia* are primary hosts of the Chinese liver fluke (*Clonorchis sinensis*) and of the cat liver fluke (*Opisthorchis felineus* and *O. viverrini*). Secondary hosts are freshwater fish, which also serve as a source of human infection. Major health disorders include inflammations and blockages of the biliary tract, with biliary tract carcinoma as a further complication. The Chinese liver fluke alone infects about 20 million people.

The large liver flukes, *Fasciola hepatica* and *F. gigantica*, use Lymnaeidae (snails living in mud) as well as primary hosts and shore plants such as grasses and watercresses as secondary hosts. Working animals and people are infected as false hosts via the latter. The infection causes inflammations of the liver and biliary tract, but is regionally confined to South America and northern Africa at present.

BOX D 4.2-1

Rats and disease

Rats have greatly enlarged their natural habitat, a process that has been fostered above all by growing urbanization. They can be regarded as constant companions of human beings. A major factor behind the spread of infectious diseases is the fact that rats live preferentially in sewers. However, diseases occurring outside of the sewer system are frequently transmitted by rats living in the contaminated area. To this extent, rats, especially the Norway rat, represent a link between sewage contaminated with infectious germs, on the one hand, and people and pets, on the other. This major epidemiological risk increases in correlation with the concentration of germs in sewage and high rat populations in the sewer system. Rats are carriers of a great number of parasites (lice, fleas, ticks, worms) as well as bacteria (*Leptospira*, *Borrelia*, *Salmonella*, *Yersinia*, etc.) and viruses. The expansion of sewer systems in many cities in developing countries is an indication of the potential risk. From ports, rats spread into all areas of human settlement and develop skillful strategies for adapting to the prevailing environmental and living conditions. Greater attention and research efforts must be focused on rats as a possible multiple vector for infectious diseases.

4.2.2 Trends in the spread of waterborne infections

INCREASING MOBILITY / TOURISM

Air traffic has grown by 7% annually in the past 20 years, while forecasts for the coming 20 years assume further annual growth of 5%. More and more tourists and business travelers are seeking out remote areas in all corners of the world. Parallel to this trend, migration flows are on the rise, with around 120 million people living outside their country of birth and millions each year going to other countries in search of better living conditions. These are the preconditions that foster the rapid spread of infectious diseases, both harmless and extremely dangerous. Countries in temperate zones are directly affected by this development and are forced to take preventive and protective action. Epidemic control measures at ports and airports are no longer sufficient – what is required is a comprehensive preventive strategy at the local level.

One example of an infection spread by long-distance tourism is the emergence of schistosomiasis in Malawi or South Africa (Taylor et al., 1995). Here it must be kept in mind that the host snails for schistosomiasis are also found in Portugal and a permanent influx of the schistosomiasis cycle via tourists is theoretically possible.

Travelers who go from temperate zones to the tropics increasingly contract arboviruses. It can be assumed that around 40 million such travelers come from the USA and Europe alone, with an upward tendency. As intercontinental travel grows, viruses, vectors (organisms that transmit diseases) and reservoir animals will increasingly spread. Dengue virus types in particular have been passed on a global scale in this manner. Individual cases of the severe dengue hemorrhagic fever have already manifested themselves among tourists. Dengue biotopes have arisen temporarily in Texas, with subsequent transmission to humans. The mosquito *Aedes albopictus*, another dengue carrier, has spread from Asia to Europe and the USA, in some cases via residual water in imported used car tires (Gubler, 1996). In addition to dengue, yellow fever is another disease that has been spreading in recent years. Many tropical and subtropical cities are potential sites for urban yellow fever epidemics transmitted by the *Aedes aegypti* vector (WHO, 1996). Rats as well are playing an increasing role in the spread of infectious diseases (Box D 4.2-1).

FORMATION OF RESISTANCE AND NEW PATHOGENS

Mass chemotherapy and chemoprophylaxis, as used to combat onchocerciasis, lymphatic filariosis, malaria and bacterial infections, are accompanied by the development of resistance against the drugs used. Particularly untargeted and uncontrolled use of malaria prophylactics has led to the formation of substantial resistance against chloroquine, pyrimethamine-sulfadoxine, Mefloquin and quinine among malaria vectors. If, however, the administration of drugs for targeted therapy is monitored, such resistance is reversible in some cases, as has been shown in Hainan (China) and Thailand, for example. Negative developments regarding the effectiveness of chloramphenicol, Co-trimoxazol and ampicillin against *Salmonella typhi*, the vector for Typhus abdominalis, have also been observed. Considerable advances have been made in the eradication of dracunculiasis, on the other hand. However, not only the pathogens themselves, but also host organisms are increasingly developing resistance – e.g. to insecticides.

In spite of the successes achieved so far, infectious diseases are as important as ever, even in industrialized countries, and their dynamics remain very unpredictable. More pathogens in drinking water are likely to be discovered or to come into being. Examples include the epidemics of Legionnaire's disease and cryptosporidiosis in connection with inadequately treated drinking water in the USA, the bacterium *Salmonella enteritidis PT4* in Great Britain in 1988, as well as the cholera vector *Vibrio cholerae* O139, which first appeared in India in 1992 and can also cause systemic infections because of its capsulation.

New findings necessitate a critical examination of the criteria for evaluating the significance of pathogens in drinking water and the effectiveness of prevention strategies. They include:

- recent epidemics caused by protozoa in drinking water (*Giardia*, *Cryptosporidium*);
- findings on the reproduction of microorganisms in biofilms in water pipes and household plumbing (*Legionella*, pseudomonads);
- the importance of drinking water for the transmission of viruses;
- the uncertain correlation of these pathogens to classical bacterial indicators for assessing water purity;
- the increase in populations at risk;
- the necessity of using surface waters as a source of drinking water.

(Exner, personal communication. Draft resolution adopted by the conference on "Water and pathogens", Bonn 1996).

Climate change

The geographic spread of vector diseases is dependent on the living conditions required for the growth of carriers and pathogens. The dominant factors are temperature, availability of surface water, soil moisture, water vapor as well as certain forms of vegetation. A shift or expansion of the relevant biota due to a change in climate will thus affect the future distribution of regions at risk. In addition to climate zones, a major role is also played by changes in annual cycles, for example; this applies in particular to the lifetime of the vectors to which the pathogen is adapted (IPCC, 1996a).

The mean global temperature is currently forecast to rise about 2 °C by the year 2100 (IPCC, 1996b). At local scale, warming may lead to reduced agricultural yields and thus malnutrition, which in combination with a rise in ultraviolet and cosmic radiation on the immune system will increase the susceptibility to infections. On the other hand, local cold spells result in a higher incidence of intestinal infections, especially due to the high chloride resistance of pathogens at low temperatures (Kartsev, 1995).

If global warming continues, one can assume that certain diseases transmitted by mosquitoes, including malaria, dengue fever and viral encephalitis, will spread. The major factors responsible for this are higher mosquito reproduction and bite rates and shorter incubation times of the pathogens in the vector. In Texas, populations of Malaria tertiana have already been able to establish themselves temporarily. In Turkey, the incidence of Malaria tertiana has risen dramatically in recent years to a current rate of more than 100,000 cases a year. The carrier of the disease, the female *Anopheles* mosquito, is also found in temperate European zones. Malaria can establish itself rapidly here, particularly if mean temperatures increase. The spread of the disease varies considerably according to region, however. Malaria is declining in areas where pyrethroid-impregnated mosquito nets are employed on a large scale, as is done with visible success in China. In general, however, malaria can be expected to grow in connection with global warming (see Box D 4.2-2).

The pathogenic agent for yellow fever, by contrast, is less sensitive and therefore likely to spread to new regions, even if there is only a moderate climate change, due to the variability of its host selection (Maurice, 1993). The incubation period of the yellow fever virus diminishes with increasing temperature from several weeks to 8–10 days. Dengue fever is transmitted by the mosquito genus *Aedes*. In the past decade, dengue fever has spread again in Central America, and in Columbia *A. aegypti* has extended its habitat from its previous maximum elevation of 1,000 m to over 2,000 m. However, dengue fever epidemics have seldom been observed at mean temperatures below 20 °C. *A. albopictus*, another vector of the dengue virus and even more resistant to cold than *A. aegypti*, has established itself in the USA and, given a rise in temperature, could spread to Canada (IPCC, 1996a). Schistosomiasis will tend to become more widespread as temperatures rise because, on the one hand, the aquatic snails that act as hosts grow and reproduce more rapidly and, on the other, the flukes themselves have better chances of survival in the host at higher temperatures. In Egypt, for example, a decline in the incidence of flukes in snails was observed in winter (WHO, 1990).

The effects of an altered climate on the distribution and quality of surface waters have an enormous impact on the living conditions of the pathogens that cause infections via drinking water. Floods resulting from heavy rain may also create transmission links between sewage systems and sources of drinking water, thus fostering the spread of diarrheal diseases (such as cholera) even in industrialized countries

BOX D 4.2-2

Malaria on the rise

According to WHO estimates, 36% of the world population today lives in regions where malaria is endemic (WHO, 1996). Since the outbreak of the AIDS pandemic, six times more people have died from malaria than from AIDS. In Germany, 800–1,000 cases a year are reported among people entering the country or returning from malaria regions, though the number of unreported cases is estimated to be two to four times higher. Malaria continues to spread on a global scale. Various factors play a role in this process: population growth and migrations, wars, agricultural development, irrigation activities, dam construction, deforestation, growth of slums, short-term weather changes and presumably medium-term climate changes as well. The WHO (1993) specifies 11 geographical zones with different causes for the spread of malaria:

Central America: agricultural development, irrigation and settlement combined with resistance to insecticides.

Amazon rainforest: clearing activities, breaking up of malaria-resistant biotopes.

African cities: inadequate sanitation systems, high resistance to drugs.

Dry savannah and marginal desert areas in Africa: flooding, migration of the population.

Ethiopia: environmental destruction, drought, large-scale resettlement.

Savannas and forests in Africa: growing resistance to chloroquine.

East African highlands and Madagascar: major changes in agricultural land and practices, possible rise in temperature.

Afghanistan: lack of control due to civil war.

Central South Asia: clearing of forests, even in hilly regions.

Cambodia, Laos, Myanmar, Thailand and Vietnam: rapidly growing risk in marginal areas of civilization due to economic activities (e.g. mining). Highest resistance in the world to drugs.

In spite of possibly new and adapted vectors, temperature remains a limiting factor for further spread. While the minimum temperature for the development of mosquitoes is around 8–10 °C and the optimum is about 25–27 °C, *Plasmodium vivax* ends its sporulation at temperatures below 14–16 °C, *P. falciparum* at temperatures below 18–20 °C (Miller and Warrell, 1990). Therefore, relatively minimal changes in temperature can have a significant impact on the spread of malaria. Martens et al. (1994) developed a model calculation that takes into account the impacts of climate changes on mosquito populations and the incubation time of the parasite. Assuming a temperature rise of 3–5 °C by the year 2100, it was calculated that 60% of the world population will live in potential malaria regions in future. Moreover, the Martens model forecasts an additional 50–80 million cases of malaria per year for 2100. Given the complexity of the ecosystem, however, identical environmental changes may have completely different impacts at the regional level (Lindsay and Birley, 1996). Demographic, socioeconomic and technical changes were not taken into consideration, so the results must be evaluated with great caution.

100 years after the discovery of the transmission cycle and after the initial spectacular successes in research and implementation, malaria research is currently in a state of crisis, as are concepts for combating the disease, for preventive medical efforts and for therapy. Malaria continues to spread as a result of population pressure and environmental destruction, possibly reinforced by the global warming and increased precipitation in the tropics.

Source: Diesfeld, 1997

(IPCC, 1996a). Longer warm spells may enhance the survival capacity of numerous bacteriological organisms.

Marginalization / poverty

The number of water-scarce countries is rising (see Section D 1.4). Although there has been an increase in facilities for hygienic water supply and wastewater treatment in recent years, the absolute number of people lacking access to such facilities has risen due to population growth. Nevertheless, under the (very optimistic) presumption that significant advances will be achieved in drinking water supply and wastewater disposal in coming years, one can assume that the incidence of waterborne and other infections will fall considerably within the next 20 years, particularly in the developing countries, while injuries and other non-infectious illnesses are more liable to increase. Diarrheal diseases, for example, which currently cause the second highest rate of cumulative disability (in disability-adjusted life years, DALYs) after respiratory infections, will only be the ninth most frequent health disorder in the year 2020, after heart diseases, depression, disability due to traffic accidents, cerebrovascular diseases, respiratory tract diseases, pneumonia, tuberculosis and the impacts of war (Murray and Lopez, 1996).

VACCINATION

Poliomyelitis is clearly declining. It is the only waterborne infection that is currently taken into account by the WHO's successful "Expanded Program of Immunization" (EPI). Future programs, however, will involve the use of newly developed vaccines to treat yellow fever as well as bacterial intestinal infections. With the help of the EPI, polio was eradicated in the USA five years ago, and this goal will presumably be achieved on a worldwide scale by the year 2001. After smallpox, poliomyelitis will be the second disease to be eradicated worldwide solely through vaccination.

The development of a vaccine against malaria has failed to materialize, but there are some promising new approaches (Butler, 1997). At present, the potential vaccine appears to provide protection only for 60 days, so its usefulness is restricted to travelers for short periods and therefore cannot help the threatened local population. A number of open questions are due to be clarified by field tests in Gambia.

EXPANSION OF IRRIGATION

The spread of waterborne diseases was greatly favored in the past by the sheer number of water development projects. The creation of ponds, reservoirs, irrigation and drainage channels as well as major shortcomings in the water supply and wastewater treatment systems of many cities in developing countries have all promoted the continued existence or spread of a number of diseases and continue to do so. In recent years, new irrigation systems and water reservoirs in central and northern Africa as well as in the Middle East have created ideal conditions for the spread of those species of snails that transmit bilharziosis. In addition to the latter, which occurs when water is taken from slow-flowing waterbodies, there has been a spread of infectious diseases such as malaria, yellow fever and river blindness. More recently, farming practices in Rajasthan (India) have undergone a fundamental change following the construction of the Indira Ghandi Canal. Wheat and cotton are now grown by means of irrigation farming. Many people in search of work have been attracted to these regions. The 445-km-long canal has proved to be an ideal breeding ground for mosquitoes during the monsoon season. Instead of high yields and prosperity, the monsoon rains brought farmers a rapidly spreading malaria epidemic. The occurrence of malaria as well as dengue fever and Japanese encephalitis is not unusual in India. However, the irrigation channels carried the epidemics deep into the country, bringing farmers and workers into contact with the diseases.

No convincing figures are available on the extent to which the economic benefit of such measures is offset or even reversed by importation of the disease (see also Section D 3.4). The PEEM (Panel of Experts on Environmental Management for Vector Control) of the WHO, FAO and UNEP, plus UNCHS since 1991, examines the impacts of water management on the incidence of waterborne diseases and on efforts to combat them, with special reference to environmental aspects. To date, however, it does not appear to have undertaken much beyond planning, meetings and local analyses (Bos, 1997).

The growing incidence of Rift Valley fever and Japanese encephalitis can also be explained by changes in agricultural and irrigation practices. The most important of the arboviruses mentioned is dengue, which together with its major vector, the *Aedes aegypti* mosquito, is widespread in the tropical and subtropical regions of all continents. *Aedes* has adapted well to urban environments by using small reservoirs of water, such as flower pots, car tires, birdbaths, gutters, barrels and even plastic tarpaulins, to live and breed in. The incidence of the disease has risen dramatically worldwide, particularly in connection with the urbanization process in tropical developing countries. More than 50 million cases, 200,000 of them severe, are expected every year. Between 1989 and 1994, the number of dengue fever cases in Latin America increased sixty-fold.

Within the framework of the Onchiocerciasis Control Program (OCP) coordinated by the WHO, substantial progress is being made towards eradicating onchocerciasis by means of mass treatment with Ivermectin and by combating buffalo gnats. The program is now being extended to the whole of tropical Africa. Combating lymphatic filariosis, on the other hand, is less successful at the moment.

4.2.3 Need for action and recommendations

COMBATING CAUSES

Combating pathogens and hosts

The use of chemical agents to combat pathogens and their hosts directly has long been common practice, though with varying degrees of success and with innumerable side effects. Forty years ago it was believed that malaria could be eradicated with DDT within a relatively short time. That was a false hope, however. Instead, this particular insecticide has become an epitome for the use of chemicals in the environment without due regard for the side effects, however undisputed its success has been (Carson, 1962). Malaria continues to remain the tropical disease with the highest incidence and number of fatalities. Mosquito populations and their larvae are still combated with the help of insecticides or, for example, suspensions of *Bacillus thuringiensis*.

Combating mosquitoes or other animal hosts can only serve as one element of a comprehensive strategy. Such a strategy must also include early diagnosis and treatment of infected people, as well as the establishment of local health services and educational campaigns. Moreover, research capacities must be strengthened in malaria-afflicted regions, firstly for monitoring purposes and, secondly, in order to identify the ecological, social and economic determinants of the disease.

Confining habitats

In many countries, swamps have been drained and rivercourses regulated, thus limiting or destroying the habitats of insects. These measures were undoubtedly successful in combating malaria. After construction work to render the Upper Rhine navigable, for example, malaria disappeared from the areas inundated by the river from around the mid-19th century onwards. From today's perspective, however, wetlands are increasingly threatened and are now seen as biotopes worth protecting. Thus, if building measures are carried out to confine the habitats of animal hosts for pathogens, the opportunities for combating disease must be weighed up against the loss of valuable biotopes.

Conversely, the construction of large reservoirs and canal systems creates habitats for host organisms such as mosquitoes and aquatic snails. There are numerous cases (see above) in which outbreaks of malaria, dengue fever and bilharziosis increased after the opening of large-scale water development projects. Environmental impact assessments of such projects must therefore include an analysis of pathogen or host colonization and specify possible countermeasures.

Limiting transport/spread

Growth in mobility is a significant factor behind the global spread of waterborne diseases. It is therefore becoming increasingly important to educate those potentially at risk, such as long-distance and business travelers, about the health risks involved and about preventive measures that can be taken. Public information campaigns are also needed to encourage vaccination in general. It is also necessary to inform the public about graduated prevention in order to prevent the build-up of resistance on the part of pathogens and host organisms at local level.

Limiting exposure

Drinking water supply and disposal

A centralized or decentralized system for supplying clean drinking water and treating wastewater is crucial, in global terms, to the elimination of water-related diseases. A decisive reduction in waterborne enteric infections, for example, can be achieved in this way (Omar et al., 1995). In addition, the incidence of skin and mucous membrane diseases declines when people have clean water for washing themselves. In rural areas of developing countries, predominantly women spend a great deal of time procuring water for drinking and other purposes. Having adequate sources of water near the home, e.g. from pump wells, generates extra working time, which benefits economic development. In urban areas, the infrastructure for drinking water supply and sewage disposal, particularly in slum districts, must be improved. Studies by the World Bank (1993) show that projects for improving water supply and wastewater disposal in slums can cut morbidity rates by 25%, of which two thirds is achieved through better hygiene and one third through clean drinking water.

One of the fundamental questions to be asked is whether a reactive approach to epidemic control, as practised to date, has reached its limits, given the faster pace at which water-related diseases are spreading, and whether more weight should therefore be given to precautionary activities in future, such as improving water quality. A paradigm shift might be achieved in this context, since investments in a regulated drinking water supply and sewage treatment promise one of the highest possible "health gains" in the fight against many diseases. Given their impacts in this connection, such investments are important also in stabilizing population growth and eradicating poverty.

For this reason, the Council recommends that greater support be given in the framework of development cooperation to projects that generate positive impacts on regional health care by improving drinking water supply and the treatment of wastewater. Each project should be intensively examined and assessed for these potential impacts. The participation of the people concerned is a decisive element here, since behavioral changes with respect to the management of water resources can only be effected by means of educational activities and an understanding of the risks involved.

Centralized water treatment

Centralized hot-water supply systems for public buildings, even in industrialized countries like Germany, should be monitored for Legionnaire's disease and appropriate steps should be taken, e.g. heating or chlorination. Monitoring of such facilities is particularly necessary in hospitals, hotels and other public facilities (Walker et al., 1995).

PREVENTIVE MEASURES

Nutrition

As with other health disorders, the course of a waterborne infection depends to a substantial degree on how well the person concerned is nourished. Malnutrition complicates such infections. Therefore, measures for eliminating protein, iodine and vitamin A deficiencies as well as nutrition-related anemia have high priority also with regard to waterborne infections, especially in developing countries.

Education

As part of health education, people can be instructed in self-treatment, in particular about giving rehydration solutions (containing essential salts) to children suffering from diarrheal diseases. In Egypt, for example, a national program succeeded in dramatically lowering the mortality rate in this way. Networks have been formed with various relief organizations under the auspices of WHO in order to combat diarrheal epidemics (WHO, 1996).

Since 1994, WHO and UNICEF have been endeavoring to combat diseases affecting children in a program entitled "Integrated management of childhood illnesses". The diagnosis for a sick child is often inadequate because the symptoms of different diseases overlap. In training courses held in Ethiopia and Tanzania, for example, participants are trained to respond quickly to symptoms of disease (respiratory infections, diarrhea, malnutrition, etc.) and to contact the public health services. Such prevention courses are seen as a highly effective solution for combating the high rate of infant mortality in developing countries (WHO, 1996). The strategies for combating infant mortality are well known:

- vitamin-A supplement,
- targeted efforts to combat malnutrition,
- encouragement of breast-feeding,
- rotavirus vaccination,
- development of vaccines against Shigellae and *Escherichia coli*.

Unfortunately, health service provision of vaccinations and treatment for sick children is very little used in many countries; in Bolivia, Pakistan, Cameroon and Burkina Faso, only 20% of sick children were able to receive this kind of health care. Raising the level of education of the population, especially among women, is one of the most effective instruments for improving family health. Personal and food hygiene, the use of health services, the readiness and ability to pay fees for water supply and waste disposal as well as for preventive health measures and treatment are crucially linked to the level of education of families.

Combating schistosomiasis requires health education first and foremost, and secondly the use of drugs (WHO, 1996). In the case of *Schistosoma japonicum* infection, the additional reservoir represented by working animals is a special challenge. Effective vaccines, which have already been tested in China in the form of radiated cercariae (liver fluke larvae), for example, merit support.

Vaccination

The development of vaccines against waterborne infections must be promoted to a greater degree. Vaccination campaigns have proved to be one of the most successful tools of traditional medicine, also in terms of the cost-benefit relation. Whenever efforts to bring together manufacturers, research institutes and public health facilities, including the WHO, as part of a concerted action actually succeed, then such vaccines are indeed developed, e.g. against malaria, dengue fever, cholera, rotaviruses and enterotoxic *Escherichia coli*.

Greater use of available vaccines, such as those against yellow fever and Japanese encephalitis, should also be supported under WHO coordination. The WHO's Expanded Program of Immunization provides for these and other vaccinations.

Public Health

The changes in the epidemiology of waterborne infections that can be expected in connection with global warming and constantly changing biotopes require interdisciplinary cooperation between doctors, climatologists, biologists and social scientists. The tools to be used in this context include monitoring of the epidemiological situation, integrated geographical and social-scientific modeling, and geographic information systems (Patz et al., 1996).

For Germany, where the epidemiology of tropical infections is still in its infancy, this would mean the forging of links between the Robert Koch Institute as the designated successor of the Federal Health Department, on the one hand, and university and non-university-affiliated epidemiological research and service institutions. Moreover, a network (European, as far as possible) should be linked to international organizations, such as the Centers for Disease Control and Prevention (CDC) in the USA and, in particular, the WHO, and should include the respective

Health Departments. The surveillance system set up in England and Wales is currently an exemplary model. Such steps require the development and implementation of advanced and rapid health information systems (with GIS). Essential tools include national and international infection epidemiology working groups that can detect increased incidences of diseases at an early date with the help of modern techniques such as molecular epidemiology (Usera et al., 1995). They must also be able to discover new types of infections that will continue to be a significant factor in the future. The necessary epidemic control measures must additionally include immediate and long-term relief programs for developing countries. Development of resistance to drugs can be checked only if the latter are used in a targeted manner and use is monitored. This follows from a comparison between Germany and the USA (better-monitored curative therapy in Germany) as well as from the successful measures taken in Thailand and China (Hainan) against development of resistance by malaria pathogens. Monitoring and combating resistances should also be a task of the networks.

Knowledge about waterborne infections is still wholly inadequate in Germany. This knowledge should be imparted to a greater extent in the course of medical training and passed on to the population through suitable media. Another competent contact in this context is the German Association of Tropical Medicine, which currently holds the chair of the European Federation of Tropical Medicine and International Health and actively focuses on waterborne infections in its programs and with the help of numerous individual members.

The individual and public health services in the developing countries must be further developed. This need is shown by the resurgence of sleeping sickness in post-independence Central Africa due to the collapse of the vertical controls implemented by the colonial administration. The services offered are frequently of inferior quality. For this reason epidemics related to drinking water cannot be detected at all or only to an inadequate degree. Early detection, however, is a prerequisite for determining the causes and combating them as quickly as possible. Structural activities (decentralization, innovative financing models, quality assurance) for improving health services have priority over initial and further training of human resources and should be an integral element of development cooperation.

In summary, the Council recommends the following action to combat waterborne infections. The concept of comprehensive quality assurance can be certain of acceptance by the public and thus contribute to internal political stability in the long run.

- Greater support for drinking water and wastewater projects in the framework of development cooperation, giving special consideration to health impacts.
- Installation and support of national and global epidemiological networks to observe and analyze the epidemic situation with respect to waterborne infections and their resistance to drugs.
- Linking food security programs to improvements in drinking water infrastructure, in the framework of development cooperation.
- Targeted training of decision-makers and responsible persons in health and environment agencies working in the fields of drinking water and wastewater hygiene, or water utilities, so as to initiate the necessary structural measures.
- Stipulation of clear responsibilities in the monitoring and control of drinking water quality for water supply companies, authorities and reference hygiene institutes, and installation of the appropriate infrastructure.
- Deployment of an interdisciplinary committee under the direction of the WHO to draw up internationally accepted guidelines for quality assurance and preventive measures.
- Construction of reservoirs and open irrigation systems should not be supported as long as the health impacts of such measures have not been assessed and parallel combating measures cannot be offered.
- Dissemination of health information for the population concerned, particularly women as the main coordinators of the family, giving consideration to self-treatment methods in areas with inadequate medical care.
- Improved implementation and support of vaccinations. Specific programs for combating malaria, lymphatic filariosis, schistosomiasis, onchocerciasis and Legionnaire's disease.
- Use of chemicals to combat mosquitoes and their larvae only when environmental impacts are taken into account.

Research needs

- More intensive development of vaccines against waterborne infections.
- Further investigation of possible pathogens of waterborne infections with regard to verification procedures, ecological and epidemiological characteristics, the impact of water treatment techniques, disinfection methods and reproductive conditions.

4.3
Water and food

Historical corn belts of the world – Grain production increased – Calorie consumption varies regionally – Large-scale decline in production of basic foodstuffs – Rapid rise in forecast water demand by agriculture

4.3.1
Historical background

Providing food for humanity is closely linked to the availability of water. The physical accessibility of drinking water forms a direct linkage, while the impossibility of producing many foods without inputting water is an indirect linkage. In addition to this existential importance, water use is also of central importance in the growth of human cultures. In the large river basins (e.g. the Tigris / Euphrates, the Nile, the Ganges or the Niger), the use of water in agricultural irrigation systems enabled the rise of the oldest civilizations over 8,000 years ago.

The organized production of food began more than 10,000 years ago with the growth of agriculture in various areas of the Earth enjoying favorable environmental conditions (Sick, 1983; Kinder and Hilgemann, 1986). The combining of crop and livestock farming in a system involving the cultivation of wild grain as well as the domestication of sheep, goats and pigs created the basis for sedentarization of the population. Differing environmental circumstances enable a distinction to be made between the tropical savannah farmers who concentrated on vegetative production of tuberous plants, and the steppe farmers with their production and sowing of cereal crops. Various centers of agriculture developed independently of one another in different parts of the globe, such as the Chinese center, the West Asian center and the Andine center (Nentwig, 1995). Early agriculturalists were already familiar with a broad range of foodstuffs, which thus enabled further population growth and societal development. The climate change that began in the middle Stone Age (around 8,500 B.C.) brought about dramatic changes in environmental conditions, resulting in the aridification of large areas in the previously favorable regions in the subtropical-tropical belt of savannahs and steppes in the northern hemisphere as temperatures rose (Sick, 1983). Population growth and intensifying aridification led to migrations to the floodplains of large rivers and to the growth of advanced civilizations on the Nile, Euphrates and Tigris, Ganges and Indus Rivers (Kinder and Hilgemann, 1986).

Within these riverside cultures, the agricultural communities of the steppes and savannahs soon developed into differentiated societal systems with different hierarchies, cultural elements and division of labor. The supply, distribution and range of food underwent changes within and between these different societies. Here we have early evidence that diet is intimately linked to societal development, and thus depends on cultural differences, social status, technological development and mobility as well as on wars and environmental changes.

4.3.2
Population growth and food

Thomas Malthus's famous "Essay on the Principle of Population" was published almost exactly 200 years ago, in 1798. Based on the assumption that population and food production would rise in geometric and linear progression respectively, i.e. at different growth rates, Malthus predicted that the supply of food would become extremely inadequate, and that population growth would be adjusted by famine to the level of food production. Since then, however, the world population has risen more than sixfold to 5.7 billion people today. Exponential growth meant that the time it took for the world population to double became increasingly shorter. Given the current mean growth rate of 1.7%, only about 38 years will be needed for the latest doubling from three to six billion people. Since the beginning of the 20th century, population growth has been primarily concentrated in the developing countries (Fig. D 4.3-1), where the food situation is the most precarious. Although growth rates are declining, most population growth will continue to take place in the developing nations.

Whereas the developing countries accounted for 65% of the world population in 1955, this figure had risen to 78% by 1990 and will grow to 83% by the year 2025, according to UN estimates (WRI, 1994).

This contrasts with negative income trends among developing countries. The world market prices for agricultural produce, which have been falling consistently over the last 15 years, impose an extremely heavy burden on the developing countries due to the immense importance of agricultural production in these economies. In 1991, the poorest 20% of the world population shared a mere 1.4% of global income, while the richest 20% had 84.7% of this income at their disposal. More than 1.1 billion people earn less than $1 a day.

Population growth rates are high in the poorest countries of the world and in water-scarce countries (India, Bangladesh, Kenya, Nigeria, Egypt, Burundi, Somalia, Sudan) (Engelman and Leroy, 1995). The

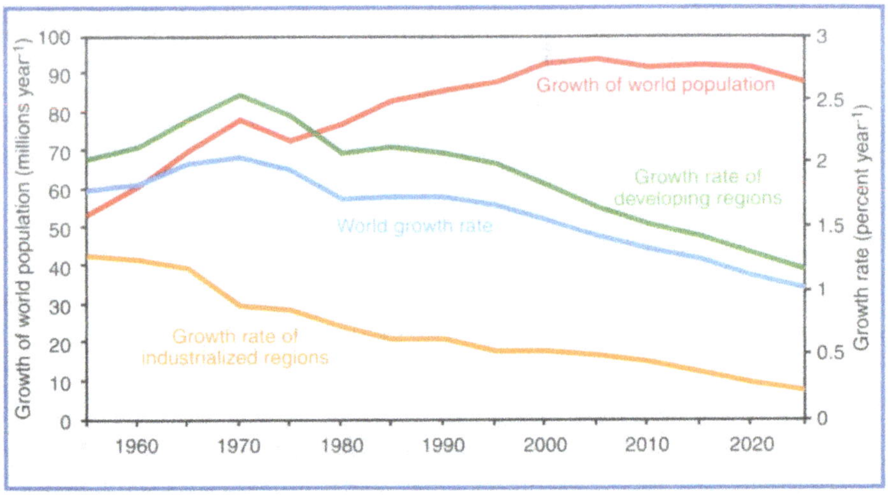

Figure D 4.3-1
Growth of world population (in percent and in million people per year).
Source: WRI, 1994

situation becomes critical in regions suffering from water shortage *and* high precipitation variability. Inadequate economic development, climatic and pedological extremes, water scarcity and pollution as well as population growth exacerbate the food situation in the developing countries dramatically. The situation in the sub-Saharan countries, but also in South Asia, is a source of particular concern, while the problem of malnutrition in the rapidly growing states of Southeast Asia has been largely eliminated.

In the competition between "the stork and the plow", two strategies have always been available for the "plow", namely expansion of farmland and the intensification of crop farming. Until the middle of this century, the focus was on expanding the area of cropland. This method is only practiced where potential arable land is available or where there are no regulations on the use of forests and grasslands. In many regions with potentially usable cropland, agriculture is prevented by water scarcity from producing good yields. Water is a key agricultural resource. Accordingly, intensive efforts to build up irrigation farming were made at an early date. As a result of this process, irrigated areas have increased fivefold in this century (1900–1994) from 50 million hectares to 250 million hectares, while water withdrawals have risen simultaneously by a factor of six (Clarke, 1993; Postel et al., 1996) (see also Section D 1.3). Today, around 75% of global water withdrawals flows into agriculture. Withdrawal and consumption vary considerably at the regional level, however. More than 70% of irrigated land is located in developing countries. Asia has the highest proportion (62%) of the world's total area of irrigated land. 86% of water withdrawals there is used for irrigation purposes, which is equivalent to 40% of global water consumption. In arid and semi-arid developing countries, 80% of the water withdrawals is used by agriculture, while only 40% is used for agricultural production in the humid regions of the industrialized nations (FAO, 1996c).

Improvements in the supply of food, particularly since the mid-1960s, have been brought about through the intensification of agriculture. This was achieved by means of new production techniques – not only the more widespread use of irrigation, but also high-yielding varieties, fertilizers, plant protection agents and greater mechanization. This process, described as the "Green Revolution", generated enormous increases in yield (see Section D 3.3).

Despite the gloomy forecasts mentioned above, the supply of grain to the world population has improved both quantitatively and qualitatively over the last 30 years, with the main boosts in worldwide production occurring between 1969 and 1989 (Figs. D 4.3-2 and D 4.3-3). During these 20 years, the production of root crops grew by 0.8% annually, cereals by 2.6% and milk, meat and fish by 2% (Schug et al., 1996). In absolute figures, grain production roughly trebled from 680 million tons in 1950 to 1,970 millions tons in 1990, which corresponds to a per capita increase from 247 kg (1950) to 336 kg (1990).

In addition, rice production rose from 151 million tons in 1960 to 350 million tons in 1990, corresponding to a rise of 230% or an annual growth in production of 7%. According to estimates by Yudelman (1994), the share of production on irrigated areas is 60% for rice and 46.5% for cereals. Only 17% of the world's farmland is irrigated, though the FAO (1996c) has calculated a 40% share of food production on such land; this reflects the great importance attached to irrigated crops for world food.

In the developing countries, total food production rose by 39% in the decade between 1980 and 1990. Growth rates were extremely high in Asia (50%) as well as in Africa (33%). In absolute terms, food production improved in 101 developing countries (Oltersdorf and Weingärtner, 1996). However, it is

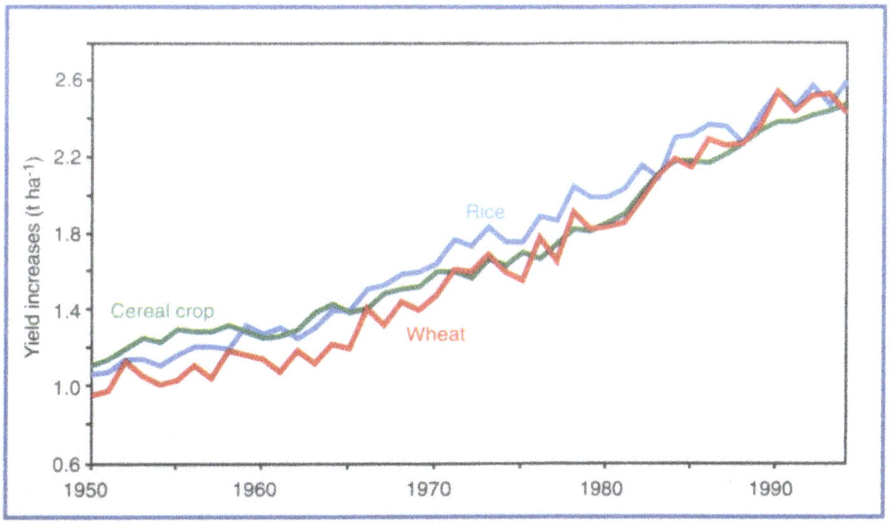

Figure D 4.3-2
Increases in cereal crop yields since 1950.
Source: WWI, 1996a

equally evident that growth rates have fallen steadily over the last 30 years. While the annual growth rate between 1960 and 1970 was 3%, it dropped to 2.3% between 1970 and 1980 and was down to only 2% between 1980 and 1990 (WRI, 1996). From 1990 to 1995, grain yields fell by 12%, corresponding to a per capita decline in supply from 336 kg to 300 kg. Moreover, such growth rates need to be examined at the regional scale. Europe, for example, has recorded constant production growth, while per capita yields in Africa have declined significantly (Fig. D 4.3-4).

The undeniable successes of the "Green Revolution" cannot and should not conceal the severe disparities that continue to exist between North and South in respect of food production and distribution, or the fact that more than 800 million people suffer from chronic malnutrition or hunger (FAO, 1996c). The percentage of chronically undernourished persons in the sub-Saharan African states is 37%, or 175 million people. In South Asia as well, where the proportion of chronically undernourished persons was reduced from 34% (1970) to 24% (1990), there are 277 million people suffering from malnutrition. Even in China, the country with the highest production increase in the last 20 years, 189 million people are thought to be chronically undernourished (Uvin, 1993). At least 82 developing countries have a per capita calorie supply of less than 2,500 kcal, which puts them in the category of countries with chronic malnutrition (Schug et al., 1996) (Fig. D 4.3-5). The figure would presumably be much higher if a statistical differentiation could be made according to social criteria, region and gender. Even if the nutritional subsistence level is assured at the household level, the food security of individual family members may be threatened. For example, meals are often distributed to the benefit of the male heads of the household and boys. Girls and women are therefore more frequently affected by malnutrition (FAO, 1987; UNDP, 1992).

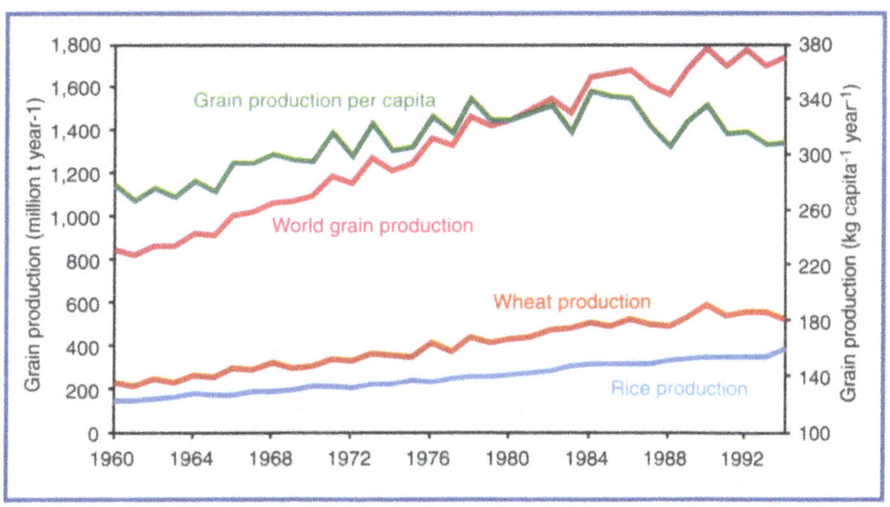

Figure D 4.3-3
World grain production and growth of per-capita production, 1960–1994.
Source: WWI, 1996a

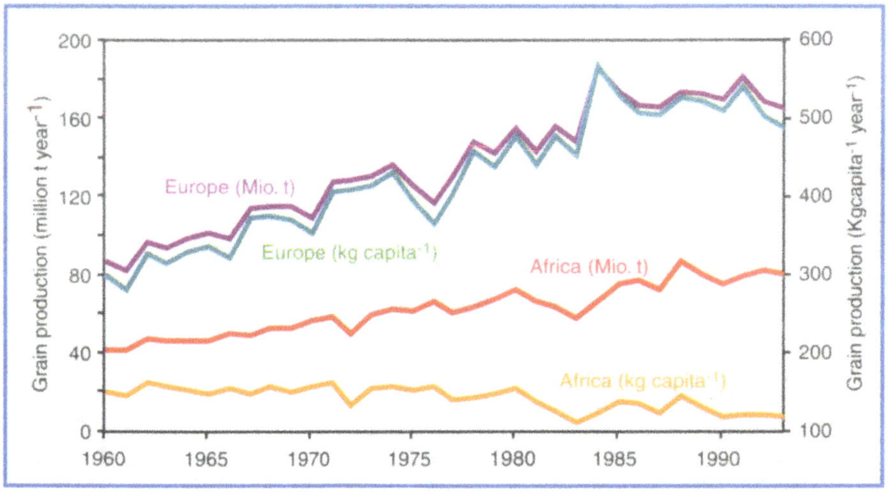

Figure D 4.3-4
Comparison of grain production levels in Europe and Africa (1960–1994).
Source: WWI, 1996a

In Ethiopia, Cambodia and the Maldive Islands, the calorie supply dropped below 1,800 kcal per capita in the mid-1980s. The contrasts become even more obvious if one keeps in mind that in the sub-Saharan region, for example, barely 5% of the food comes from animal sources in contrast to 30% in the industrialized nations. The energy requirement for animal calories is seven times higher than for plant calories. This means that the primary calorie consumption per capita is four times higher in the industrialized countries (9,500 kcal) than in Africa.

These imbalances prompt the question whether a further doubling of the global population is actually possible in view of the food supply situation, and given the major damage caused to the environment worldwide by food production. Such damage includes the increasing degradation of soils on a global scale (WBGU, 1995), the spread of desertification in marginal areas around deserts and the loss of valuable ecosystems through the expansion of pasture and farm land at the expense of forests and grasslands. The situation appears even more threatening if one considers that the production increases of the last decade barely kept pace with population growth, and that current trends indicate a stagnation both in farmland expansion and yield enhancement (Figs. D 4.3-3 and D 4.3-6).

The area of cultivated land used to produce basic foodstuffs is declining, decreasing in the case of grain crops by more than 16 million hectares relative to 1981 (8.5%). In view of the 90 million people added to the world population each year, the annual increase in irrigated area of less than 1% represents a real per capita decline of 12% by the year 2010. In terms of total cropland, the situation is even more daunting; the available per capita area will fall by a total of 50 million hectares, or 21%, despite the increase in total agricultural area (WWI, 1996a). According to estimates produced by the International Food Policy Research Institute (IFPRI), this contrasts with a 56% rise in grain demand and a 74% rise in demand for animal products in the 1990–2010 period (Holtz, 1997).

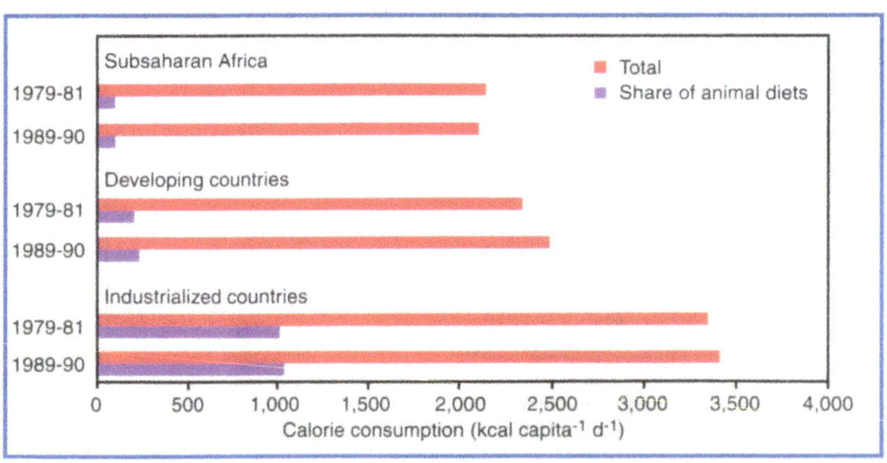

Figure D 4.3-5
Calorie consumption and diets in different world regions.
Sources: Alexandratos, 1995; Oltersdorf, 1992

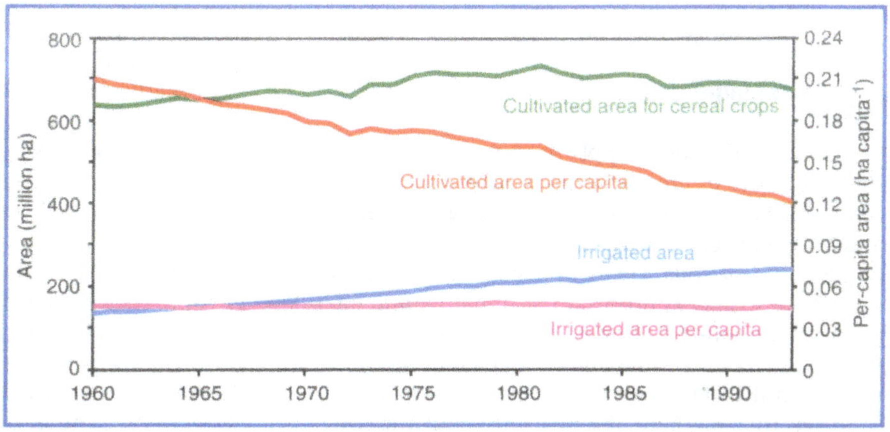

Figure D 4.3-6
Growth in areas used for cereal crops and irrigated farming.
Source: WWI, 1996a

4.3.3 Food and water consumption: current situation and a look into the future

Grain crops account for around 50% of the world's cultivated land. One can assume that about 1,000 tons of water is needed to produce a single ton of grain (WWI, 1996a). This figure includes transpiration from plants and evaporation from soils, but not the losses due to wasteful irrigation methods or the water needed to desalinate the soil. Thus, the quantity of water referred to is more a reflection of minimum requirements for grain production. Based on world grain production of 1.72 billion tons in 1995 (FAO, 1996c), a water volume of 1,720 km^3, or 300 m^3 per capita, is required. The quantity of grain produced worldwide in 1995, i.e. 300 kg per capita, corresponds to exactly the amount necessary to cover the calorie needs of the population. Given 100% efficiency and a diet consisting of plant foods only, 300 m^3 of water per capita and annum would theoretically suffice for to supply the world with food. Current consumption of grain is around 150 kg per capita and annum on a worldwide average and therefore meets only half of humanity's calorie needs.

If the goal for the future is a nutritional basis for all humankind that is equivalent in composition to the current global diet, then 84% of the food would have to be from plant sources and 16% from animal produce. This would correspond to a reduction in the industrialized countries of 13% and an increase in the supply of animal calories in the developing countries of 6%. If the plant portion consisted solely of grain, water consumption would be 570 m^3 per capita and year. The significant rise results from the inefficient calorie turnover in the production of animal produce. An average of seven grain calories is required to produce one animal calorie. Based on the above water consumption, an additional 51.3 billion m^3 (51.3 km^3) of water annually is needed to cover the additional food needs for the growth in population of 90 million people. Given a constant average per capita consumption of food, an additional 1,556 km^3 of water will be required for food production in the year 2025. This is 18 times the quantity of water that flows down the Nile every year, or 58% of the volume that was used in agriculture in 1990. These figures do not take into account the water losses due to inefficient irrigation. If present production conditions in irrigation farming are applied to the year 2025, water withdrawals for food production would rise by a further 20% to 1,867 km^3.

The situation looks even more critical if FAO data are taken into account. In Asia, where 60% of the world population lives, the FAO estimates per capita water withdrawals at 2,000 m^3 per annum for a balanced diet providing sufficient meat (FAO, 1996c). This high figure results from the substantial volume that goes into irrigation for food production and includes the large losses of water due to inefficient irrigation systems.

If the latter withdrawal rates are compared with renewable water stocks (WRI 1996; Engelman and Leroy, 1995) and expected population growth to the year 2025 (mean UN forecast) is taken into account, then in the Asian region China (1,780 m^3), India (1,498 m^3), Pakistan (1,643 m^3), Iran (812 m^3), Afghanistan (1,105 m^3), South Korea (1,158 m^3) and Singapore (179 m^3), with a total of around 3.43 billion people, will fall below this minimum value. Owing to the expected economic development in these countries, water scarcity will be further exacerbated and food import needs will rise. Another problem concerns increasing urbanization, which will lead to distribution problems between cities and the surrounding areas as water resources become scarce. Apart from this quantitative aspect, impacts on water quality can be expected since the pressure on using water for irrigation will result in deterioration of the quality of available water.

4.3.4 Recommended action

To secure the food supply for a world population that continues to grow at a rapid pace, additional efforts will have to be made to improve agricultural and forestry production and to check the progressive degradation of soils. This will necessitate the development of locally appropriate, sustainable and environmentally sound strategies for water use in agriculture and forestry that place substantially more weight on the abiotic and biotic potential of specific locations than has been the case until now, and which preserve or protect this potential by means of suitable measures.

Efficient use of water for crop production poses a major challenge for agricultural engineers and farmers. More efficient irrigation systems can be achieved by improving the technical systems for supplying water via canal and pipe systems, as well as for distributing and dosing water on agricultural areas. In future, greater care must be taken to focus support less on high-tech engineering and to make use of more adapted, low-tech methods, which are often less costly. Moreover, greater consideration should be given to the experience and sociocultural background of the population concerned.

- In regions with scarce water resources, land-based wastewater treatment or the watering of crops with treated wastewater will gain increasingly in importance as the needs and wastewater volume of private households and industry rise. These long-term trends should be taken into account today in the planning of development projects, and steps should be taken at an early stage to effect a change in water use patterns and thus avoid unwise investments. Germany has a long tradition in wastewater treatment. The experience gained in this field should be utilized and adapted to other regions.
- Irrigation farming is frequently coupled with large-scale water development projects. For ecological, economic and social reasons, it is recommended that support be given to the development of small decentralized projects that are often easier to adapt and which build on local structures and experience. In combination with present-day systems for transmitting data and information, even small projects can be optimally managed and monitored.
- To enhance irrigation efficiency, plants must be selected or cultivated that are insensitive to fluctuations in water supply and have a high salt tolerance. The use of such locally adapted plants would reduce the risk of yield losses, impose lower demands on irrigation systems and thus provide for greater acceptance.
- As already stated in the WBGU 1994 Annual Report, rainfed cropping also holds substantial potential for increasing yields in large parts of the world. Supporting decentralized facilities for "water harvesting" will not only increase the volume of water available to plants, but at the same time reduce the degradation of farmland by preventing erosion. As with irrigation, such measures must be accompanied by selection of adapted plants.
- By promoting rainfed farming, one can mitigate the pressure on traditional, intensively irrigated "corn belts", and at the same time counteract further widening of regional disparities.
- In the view of the Council, multiple cropping systems and agroforestry should be given priority over monoculture farming. Within the framework of an integrated agricultural development strategy, the targets of enhanced food grain production and rural development should be pursued in tandem, in order to avoid growth that has no impact on development.

Aquaculture is a water-related branch of the food industry that will continue to develop in the future (see Box D 4.3-1) and which involves the production of animal and plant food. It deserves further support in both its intensive and extensive form. Past mistakes arose from inadequacies in production methods and from a lack of acceptance of the products themselves. A substantial part of the support given to aquaculture projects should therefore be devoted to marketing aspects and to enhancing acceptance.

- Land-use systems and aquacultures based on the exploitation of fossil water reserves should only be supported after meticulous assessment of ecological, economic and social criteria. In many regions, high-quality groundwater reserves are the only "sources" for supplying the population with drinking water. Because these reserves are finite, their use should be analyzed with regard to long-term regional development.
- The growing need for food cannot be met by an increase in agricultural production alone. Instead, effective social security systems are called for that obviate the need to have large numbers of children in order to ensure survival and to create purchasing power for acquiring food. There can be no food security on a long-term basis without social development.
- A key area for increasing productivity is education. It is therefore essential to support school-based education, and to create the necessary infrastructure alongside the establishment of social security systems.

BOX D 4.3-1

Aquaculture - the growing importance of a traditional production method

Integrated aquaculture in the form of fish-rice crops or pig-fish crops has a long tradition of use in Asia, where they have contributed to food security and rising production. Through specialized cultivation of aquatic plants and animals, aquaculture has transformed in the last two decades into an expansive sector of the fishing industry, with growing economic importance and annual growth rates of 10–15% (in 1994, turnover had already reached $40 billion; FAO, 1996c) (Fig. D 4.3-7). Approximately 45% of production and 33% of sales income are generated by freshwater aquaculture (1994). Aquaculture is most widespread in the developing countries, particularly in the Asian region, where around 75% of the worldwide production takes place and 57% of sales is earned, thus improving the food situation and economic development. 70% of sales is from freshwater aquaculture.

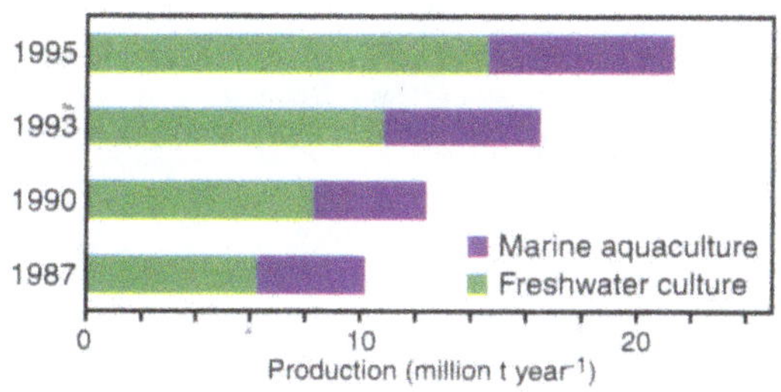

Figure D 4.3-7
Growth of aquaculture production.
Source: FAO, 1996c

While the supply of freshwater fish comprised only 19% of total fish consumption in 1994, 68% of this share was produced by aquaculture (FAO, 1996c). As the population grows and catch quotas on the world's oceans stagnate or decline, it can be assumed that aquaculture will continue to expand its share of total fish production. Because of its high-quality protein content, fish can make a significant contribution to nutritional and food security. 17% of the worldwide supply of high-quality protein is provided through the consumption of fish. In 1994 12 million tons of fish was produced as food by means of aquaculture (FAO, 1996c).

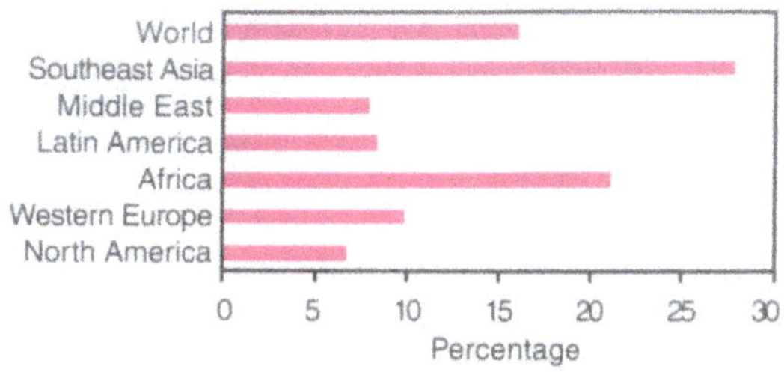

Figure D 4.3-8
Percentage of protein supply obtained from fish.
Source: WWI, 1996a

The percentage of protein supplied as fish is far above the world average in the countries of Southeast Asia and Africa, where the figure is 28% and 21% respectively (Fig. D 4.3-8). In the field of aquaculture, a distinction must be drawn between industrial aquaculture and extensive, often integrated aquaculture. The former focuses on luxury products, such as salmon, lobster or prawns, is export-oriented and cost-intensive. High stock densities frequently lead to eutrophication and severe organic pollution of the waterbody. Moreover, water quality is severely reduced due to microbial contamination, antibiotics and insecticides. On the other hand, low-input aquaculture implemented on an integrated basis in small farming systems in developing countries can make a major contribution to food security by providing a broader food base, reduced susceptibility to droughts or other negative impacts on yield, and to securing or improving operating income through the supply of local markets. Prerequisites include cultural adaptation, availability of land and free access to water.

The FAO supports regional integration of aquaculture into existing farming systems in nine countries of southern Africa with around 20,000 small ponds and a yield potential of 50–200 kg ha^{-1} $year^{-1}$. Aquaculture in the form of fish-vegetable crops also plays a growing role in the peripheral areas around Asian cities. It offers the rural population living near the city an additional source of employment and the constantly growing population in the cities an important source of food.

BOX D 4.3-2

Irrigation systems of the Nabateans

People in the arid regions of the Earth learned at an early stage how to handle water as a scarce resource to their best advantage. A particularly interesting and still exemplary irrigation system was developed by the Nabateans in the Near East. These extremely well-adapted systems existed from 1000 B.C. to 700 A.D., when they were destroyed during the Arabian wars. In regions with 80–200 mm of precipitation a year, this system enabled a sustainable food supply for a family to be produced on a 35-hectare plot for centuries without needing external water supplies. The system was based on the fact that after rainfalls (2–3 mm a day) in desert areas with silty soils, the water flows off from the soil surface, collects in valleys and runs off in fast-flowing streams. This is caused by low infiltration rates due to poor wettability and clogging of the surface soil. If this water is collected in sinks, sufficient water can be provided to plants from the soil through the small-scale storage there, so that longer drought periods can be survived without any connection to groundwater.

The Nabateans reinforced this natural effect of flood water retention through structural measures, resulting in sustainable agricultural management of valley meadows in the Negev Desert for over 1500 years. The structural measures included

- building canals that diverted the flood water along slopes to the farmland,
- collecting stones on the slope to increase the water yield (this was important in years of low precipitation),
- constructing stone walls along the edges of valley meadows to protect the cultivated land against sheet floods from the slope and divert surplus water to a receiving waterbody (important in years of high precipitation), and
- sluices for admitting water to the actual farm.

The farm itself was laid out in the form of terraces, where each plot was surrounded by an earth or stone wall so that the farm area was lower than the wall. The terraces had the function of retaining water and collecting sediment. The system can be described as follows.

At precipitation rates higher than local infiltration, a sheet flood occurs as described above. This water was collected, diverted to the farm via canals and stored there in a system of ponds. Surplus water was conveyed to lower-lying plots in the valley via weirs. The collected floodwater infiltrated the soil in the course of about one week. The stone walls were so designed such that the soil profile was not only thoroughly moistened, but also eluviated. This prevented an accumulation of salt in the rooting zone. After the flood, the soil was hoed to prevent transport of water to the soil surface by blocking the capillaries, because salinization of the topsoil would otherwise have resulted.

The plots were mainly planted with woody plants with a large rooting depth (olive, almond, pomegranate, carob, apricot, peach, plum, pistachio, grape vines). In addition, deep-root fodder (*Atriplex*, grasses) was grown. Cultivation of vegetables and cereals was possible but was not carried out regularly, and then only on certain plots.

The ratio between catchment area and arable land was around 7–5:1. On experimental farms in Avdat and Shivta, harvests were achieved even in dry years with only 80 mm of precipitation. Water use was strictly regulated among the Nabateans:

- there were rights of use in neighboring valleys (indicated by canals extending beyond a single catchment),
- the height of the bordering walls corresponded to the usable soil depth (at the head of the valley , where soils are shallow, the walls are lower than at the foot of the valley),
- surplus water was passed on in a regulated system (of sluices) to neighbors in lower-lying valley areas,
- a receiving waterbody channeled floodwater away and simultaneously filled large cisterns alongside streams.

Setting up a farm of this kind requires relatively little technical input (10 persons, 200 days of manual work). Maintenance of the walls was the most important part of running the farm since they can be damaged in flood years. The yield of a 5-hectare farm with a catchment area of 35 hectares, largely consisting of slopes without vegetation in the desert, was sufficient for the subsistence of one family. Today the fossil remains of Byzantine farms are still used by Bedouins for growing grain – without any signs of salinization. The farms work without a drop of water from man-made irrigation. The yield per tree is higher than with irrigation farming, though the per area yield is lower due to the smaller number of trees.

The idea of a farm watered by collected rainwater could be implemented in all arid regions with non-sandy soil. Since it permits only subsistence farming and not cash crops, this knowledge has not been disseminated to the extent merited by its potential.

4.3.5 Research recommendations

Projects conducted jointly by researchers from Germany and developing countries should be given general support. "*In situ*" research performs an important multiplier function. The cutbacks in funding for international agricultural research should be counteracted with greater commitment on the part of Germany. In particular, it is recommended that the BMZ provide more funds than in the past for promoting research on the wide variety of aspects related to the use of water for food production in the developing countries. These activities should be coordinated with those carried out by the BMBF and BML. Integrating the existing research potential of universities should be strengthened by involving the DFG in the design of appropriate programs .

Specific research needs exist in the following areas:

- Development of methods for distributing and dosing irrigation water that enhance the efficiency of water use.
- Development of irrigation techniques that ensure sustainable use of soils.
- Intensified research into the breeding of salt- and drought-resistant crop varieties. Traditional varieties whose cultivation has been neglected up to now should be included to an increasing extent in such programs.
- Research on the potential for using process water and wastewater in agricultural production.
- Research on new water-harvesting strategies for soil protection and crop production in arid and semi-arid regions (see Box D 4.3-2).
- Further development of the potential offered by aquaculture, and research on the societal conditions for its success.
- Exploitation of the opportunities offered by international research programs, such as IGBP and IHDP, that deal with global water problems and the interaction between crop production and the water budget.

4.4 Degradation of freshwater ecosystems and neighboring habitats

Salinity limits usability – Acidification: more than 4,000 lakes affected – Warmth and pollutants impair water – Non-native species alter biotic communities – Overfishing – Interference with waterbodies and wetlands – Ramsar Convention – Wetlands regulate climate

The degradation of freshwater habitats by pollutant loading, physical damage and direct anthropogenic interference leads to impairment of many of their functions. The type and severity of degradation influences the specific natural, cultural and utilization functions that are weakened or totally lost (Section D 1.1). Anthropogenic disturbances are particularly drastic when they cause freshwater habitats to shrink, for example through interference with the water balance (Aral Sea), lowering of water tables, canalization of rivers, building of dams, building development and peat cutting. Pressures and impairments that alter physical, chemical or biotic factors have impacts on water quality especially (Section D 1.5). Such damage can be induced, for example, by thermal energy inputs in the form of cooling water discharge from power stations, by the increased turbulence and mechanical stresses caused by shipping (physical factors), the deposition of inorganic and organic substances (chemical factors), the introduction of non-native species of flora and fauna, and the excessive removal of biomass (overfishing) (biotic factors). Different types of interference and stress are often coupled with each other and are also linked to impacts from neighboring and distant regions. Mechanical stress produced by ships can resuspend pollutants in sediments, for example, or acid precipitation can lead to metals being leached from soils into groundwater and surface waters. Non-sustainable land use increases erosion, thus endangering freshwater habitats.

The natural features of aquatic habitats can vary considerably, depending on geological, climatic and hydrological conditions. Quite often one can find semi-natural habitats with characteristics that are not dissimilar to those of degraded habitats. The trophic levels of lakes, for example, can show substantial variation, such that lakes that are eutrophic due to anthropogenic nutrient loading are barely distinguishable from lakes with a naturally high concentration of nutrients. The food chains in salt lakes are simple, and the use functions for humans often negligible – characteristics that may also appear in cases of human-induced salinization. Other forms of degradation, such as contamination with persistent organic pollutants, have never occurred under natural conditions. This makes it all the more difficult to assess their impacts or to determine the critical loads (for the critical load concept see WBGU, 1995). The assessment of factors that lead to ecosystem degradation depends first of all on the changes in and losses of functions that they imply. The weighting attached to the various ecosystem functions will determine the significance of their loss (global dimensions? size of area affected? acceptable alternatives for people affected?). There is also the question as to whether these functions may be discharged in some other way, or whether other ecosystems performing the same ecological functions are present. Degradations must be assessed as particularly severe when they are essentially or practically irreversible, i.e. when they cannot be reversed or mitigated either through the self-organization capacity of the ecosystems, or through measures to restore and rehabilitate them. Processes involving the extinction of species, the degradation of unique habitats, long-term loss of use or the destruction of ecosystems of particular importance for climate must also be included in this category. Conserving the natural range of existing ecosystems is therefore an eminently important task. The prevention of degradation in the future and the elimination of present-day causes of degradation are major steps towards accomplishing this objective. In many cases, it makes sense to implement targeted rehabilitation measures (mediation, restoration), or to create new habitats. Preserving ecosystems not only safeguards their present functions, but also ensures that functions required in the future are preserved (see section D 1.2.5). This also helps to safeguard those functions whose existence or significance have not been recognized as yet.

4.4.1 Salinization and desiccation

SURFACE WATERS

Due to the low tolerance towards elevated salt concentrations on the part of most freshwater organisms, increased salinity causes devastating changes in biotic communities and generally involves a decline in biodiversity. At salt concentrations of 5% and more, almost all freshwater fishes lose their capacity to reproduce (Arrow, 1994). Elevated salt concentrations can render water unusable as drinking water or for agriculture (see section D 1.5). Lakes in arid regions with high evaporation rates have high natural salinities. At 270 grams per liter, the Dead Sea has the highest salinity of all the world's major inland waters. Besides bacteria, the only organism to survive these conditions is a single exceedingly salt-tolerant species of green algae (Wetzel, 1983). The diversion of enormous quantities of water from tributaries in order to irrigate cotton plantations has caused the Aral Sea to shrink to half its previous size. The communities of what used to be the world's fourth largest freshwater ecosystem have been irreversibly disrupted as a consequence of the increasing salinity caused by reduced inflow (Plotnikov et al., 1991). This trend has led to economic collapse in many parts of the region (see Section D 3.4).

The discharge of highly saline water from municipal sewage systems leads to elevated salinity levels in many rivers. However, the largest quantities of salt are discharged to surface waters from mining operations, oil wells and agricultural land use. The impacts on lower reaches of running waters are often particularly severe. In central Europe, the salinization of the River Rhine and the River Werra due to potassium mining are well-documented examples. In both cases, the stock of species was drastically reduced (1989). In recent years, the decline in salt concentrations in the river Werra has led to a visible recovery of biotic communities.

SOILS AND GROUNDWATER

In arid areas, salinization and desiccation of groundwater stocks are the consequences of excessive groundwater pumping. In water-poor regions, the increased demand for water generated by long-distance tourism, mostly from the rich industrialized countries, often leads to the total exhaustion of groundwater reserves. In coastal areas where groundwater is pumped on a large scale to provide water for drinking (India, Mexico, the Mediterranean) or for mining purposes (North Carolina, Great Britain), the artificial lowering of water tables has often resulted in the intrusion of sea water into the groundwater.

Regions with salinized soils (Table D 4.4.1) are found in the salt marshes of temperate zones, in the mangrove swamps of the tropics and sub-tropics, and in the salt marshes of salt lakes in arid regions. Salinized soils are common in arid and sub-arid regions, where precipitation does not suffice to transport dissolved salts to deeper horizons. The transport of water is here reversed – from deeper horizons to the topsoil where salinization occurs. Salinization occurs in the vicinity of coasts due to the high salt loads in precipitation. Increased transpiration of crops leads to elevated salt concentrations, even on non-irrigated grazing and farmland. But it is primarily in irrigated crops – not only in the sub-tropics – that salinity reaches a critical level. The salinization process often involves sodium ions, which lead to the formation of soda and to alkalization of soils because ion exchange is blocked by the sodium ions (Section D 1.5.2.2).

The problem of salinization has existed for a long time. Ever since humans discovered agriculture, the demise of cultures was associated in many cases with the salinization of soils (Marschner, 1919). Worldwide, about one-third of all irrigated land is affected by salt and it is estimated that more irrigated land is lost to agriculture through salinization than is gained through the irrigation of new areas (Marschner, 1990; Pereira, 1974). The annual increase in salinized soils

Table D 4.4-1
Salinization phenomena caused directly or indirectly by natural and anthropogenic factors, their regional distribution and forecast trends.
Source: WBGU

Type	Climate and location	Example	Secondary cause	Primary cause	Prediction
NATURAL SALINIZATION					
Salinization of soil	Arid and semi-arid climates	West and northwest India, Egypt, north-west China, Australia, western USA, Namibia	Evaporation greater than precipitation	Climate	Increase with climate change
Impacts on groundwater	Salt marshes on coasts, marshes near salt lakes	North Sea and Black Sea coasts, , Israel, Australia, China	Saline groundwater, inundation	Proximity to coast	Increase with sea level rise
	Sinks, valleys meadows, etc.	Great Plains (USA, Canada), Indus and Ganges plains	Saline groundwater, inundation	Geology, climate	Constant
ANTHROPOGENIC SALINIZATION					
Felling of forests	Mediterranean, temperate	Western Australia	Change in rhizosphere	Land acquisition	Constant
	Continental, boreal	Siberia	Change in transpiration	Land acquisition	Increasing
Felling of low-lying fôrests	Tropical humid	South America	Salt loads from coast	Cultivation of cash crops and farming for export	Increasing
Transformation of steppes and irrigation	Continental, temperate	Kazakhstan	Change in transpiration and furrow irrigation	Cultivation of cash crops and cotton for export	Increasing
Transformation of steppes and irrigation	Continental, temperate	Nebraska (USA), Alberta, Saskat-chewan (Canada)	Change in transpiration, sprinkler irrigation	Cultivation of grain crops and lucerne	Increasing
Irrigation from rivers	Arid and semi-arid climates, tropical	Murray Valley (Australia)	Sprinkler irrigation and furrow irrigation	Cultivation of cash crops for export	Declining
		Indus (Pakistan)	Furrow irrigation	Cultivation of cotton for export	Increasing
Modification of old irrigation methods	Monsoon climate	India	Furrow irrigation	Cultivation of cotton for export	Increasing
		China	Pond irrigation	Rice crops for food security	Increasing

(natural and anthropogenic) is estimated at 0.16–1.5 million hectares per year (Barrow, 1994).

4.4.2 Acidification

Sulfurous and nitrogenous gases emitted primarily from fossil fuel combustion and distributed by atmospheric transport is leading worldwide in areas of high precipitation to acid rain and to acidification of waterbodies and soils. Worst affected are soils and lakes in low-calcium areas with low buffering capacity (Figure D 1.5.2). Emissions of acidification agents containing sulfur have declined in Germany as a result of emission reduction measures; however, there are no signs as yet of the pH values in rain being restored to their natural level (Brown et al., 1995). The decline in pH values causes changes in the cellular ion budgets of aquatic organisms (Klee, 1985) and leads to death when values are extreme (Lenhart and Steinberg,1984).

One consequence of acidification is the enhanced release of toxic metal ions from soils and sediments to surface waters and groundwater (especially aluminum, copper, cadmium, zinc and lead). In aquatic organisms, aluminum and other metals lead to the inhibition of enzyme functions and may cause reduced blood formation (Lampert and Sommer, 1993). High concentrations can be acutely toxic. The release of metals is further reinforced by eutrophication and abnormally high temperatures in waterbodies. Both forms of stress lead to oxygen decline, which increases the solubility of many metals.

More than 4,000 lakes in Norway and Sweden have been impaired by acidification. Salmon and trout species have effectively disappeared from many rivers (Rossland et al., 1986). Similar impacts of acid rain have been observed in the High Tatra and in northeastern parts of North America, in the Adirondack Mountains (New York, USA) and in the province of Ontario (Canada) (Burns et al., 1981; Dickson, 1981).

4.4.3 Eutrophication and pollution

Eutrophication of surface waters

Excess nutrient inputs, of phosphate and nitrate in particular, raise primary production. Phytoplankton biomass increases, and the species composition of biological communities is altered. In extreme cases of eutrophication, dense blue-green algal scums are formed that produce noxious and toxic substances in many cases. The flow of particulates to the lake bottom increases (Lampert and Sommer,1993), and the decomposition of organic material in deepwater layers leads to oxygen depletion and even to complete anoxia. Such anoxic conditions enhance the remobilization of phosphate from lake sediments (internal loading), thus amplifying the impact of eutrophication by positive feedback. The increase in primary production improves fish yields (Hartmann and Quoss, 1993), but species composition also changes. Salmonids, found particularly in waters with low productivity, are very sensitive to high temperatures and low oxygen concentrations. Deoxygenation of deep water in eutrophic lakes means that fish with high oxygen requirements can survive only in warm surface layers, but these do not provide optimal conditions. Salmonids are therefore replaced through eutrophication by edible fish of lesser value belonging to the Cyprinid family (carp) (Carpenter et al., 1996). In highly eutrophic tropical lakes especially, changes in the oxygen budget can lead to dramatic summer fish kills. In lakes of the temperate and cold zones, complete anoxia can occur when oxygen inputs from the atmosphere are blocked by ice cover and the photosynthetic supply of oxygen is minimized by the lack of light (winter fish kills).

In most inland waterbodies, eutrophication is attributable to increased phosphate loads from the catchment. In West Germany, as recently as 1989, this phosphate input came predominantly from municipal sewage (UBA, 1994). In Europe and North America, phosphate loads to many waterbodies have been dramatically reduced through reductions in wastewater discharges achieved by constructing ring sewers, adding a third purification stage to sewage works, and by substituting phosphate in detergents with other water softeners, such as organic complex-forming agents (Burns et al., 1981; Wagner, 1976; Tilzer et al., 1991; Wehrli et al., 1996). In former West Germany, phosphate emissions from municipal sewage works were reduced by 74% between 1985 and 1995 (UBA, 1994). As a consequence, the trophic status of many lakes was successfully reduced. Although mass algal growth close to lakeshores has declined rapidly on the whole (e.g. Lake Constance), the restoration of lakes to their former oligotrophic status is only possible after a considerable time lag in many cases. In shallow lakes especially, phosphates continue to be released at high rates from benthic sediments, where they accumulated during eutrophication, even after such a reduction in external phosphate loading. In Lake Trummen (Sweden), the only way to improve water quality was to remove (lake bottom) sediment that was extremely contaminated with phosphate (Björk, 1985). For remediation to succeed, it is essential to know which nutrients are limiting production, and where these nutrients come from.

Concerted action to rehabilitate the waters of Lake Constance (eliminating roughly 80% of phosphate in domestic sewage throughout the catchment by means of a third purification stage and by constructing ring sewers) led to a drastic reduction in winter concentrations of total phosphate (to less than 25% of the maximum value as at 1995). Despite these measures, annual primary production and phytoplankton biomass have declined only by about 25%. The delayed response of the system to reduced external nutrient loading is still not sufficiently understood (Kümmerlin,1991; Tilzer et al., 1991).

Significant phosphate emissions originate from fertilized soils. Phosphates are removed from soils mainly in particulate form through soil erosion, and are flushed from aerated soils to a lesser extent than the more mobile nitrogen compounds. Nevertheless, emissions were reduced by only about 20% between 1985 and 1995, in contrast to the situation with municipal sewage. The very large loads of nitrogen compounds (nitrate, nitrite, and ammonium) from agriculture (effluents, leaching) have hardly changed. Whereas the atmosphere is of no global significance as a pathway for phosphate loads, the atmospheric deposition of nitrogen, e.g. from agriculture, is substantial and many times greater than what enters the environment via effluent and leachate (UBA, 1994).

In general, only one nutrient limits biomass production, and in the case of inland waterbodies this is usually phosphate. However, in some inland waterbodies featuring special biogeochemical conditions, the most important limiting nutrient is nitrogen, not phosphate. In these cases, nitrogen addition may be eutrophying. In Lake Tahoe (USA), which is extremely low in nutrients and biomass production, more than 65% of nitrogen loads originates from the atmosphere (Jassby et al., 1994 and 1995; see Section D 1.5.1). The increase in these nitrogen loads has caused an increase in phytoplankton biomass, and has increased the turbidity of what used to be exceptionally clear water. The changing ratio of phosphate to nitrogen in the water is now causing the limiting nutrient to shift from nitrogen to phosphate (Goldman et al., 1993).

Running waters are often used as receiving waters for wastewater, cooling water and other inputs. Their rapid renewal rates and good aeration means that running waters can withstand greater pressures than standing waters (Niemeyer-Lüllwitz and Zucchi, 1985). However, their stress-bearing capacity declines when weirs or dams are constructed, because the physical and chemical behavior of such waters becomes similar to that of lakes (longer retention times in water, development of temperature stratification in summer). The same occurs, although to a lesser degree, in the lower reaches of large rivers, where gas exchange is reduced due to lower velocities of flow, and depletion of oxygen induced as a result of heavy organic pollution (Sections D 1.2.1 and D 1.2.2).

Nitrate in groundwater

Enrichment of groundwater with nitrate occurs on a large scale, above all through the diffuse deposition of nitrogen through atmospheric loads (Section D 1.5.1.1), but especially through the leaching of fertilizers from cropland. The usability of groundwater for drinking water supplies can be seriously jeopardized by high concentrations of nitrate. In the EU and in Germany, the maximum value for nitrate in drinking water is 50 milligrams per liter. In order to safeguard drinking water supplies from groundwater, certain areas in Germany have been designated as water protection areas where land use is subject to stringent controls. Fractured rock aquifers, as are found in the karst regions of former Yugoslavia, on the Yucatan peninsula and many Caribbean islands, for example, are particularly vulnerable, and assessing the size of such designated areas requires large-scale hydrological surveys.

Pollution of waterbodies

The range of anthropogenic substances entering the environment is too great to be surveyed in full. There are many hundreds of thousands of chemical compounds, of which only a minute proportion can be detected by waterbody monitoring programs. The various types of pollutants and pressures on water resources include: pesticides, heavy metals, hormonally active substances, radioactive isotopes, pathogens and parasites. Radioactive substances represent a special risk, while the significance of environmental chemicals with hormonal effects, such as pesticides, dioxin and PCB, is still a source of considerable controversy. Even at very small concentrations, the latter substances can have crucial effects on the endocrine system as well as negative impacts on the reproduction and development of aquatic organisms (Froese, 1997).

In standing waters, contaminants are able to accumulate to a high degree due to the lack of mixing. Many pollutants are adsorbed by suspended matter and are actively ingested by organisms. When particles or dead organisms sink to the riverbed or lake bottom, they carry pollutants with them that then accumulate in the sediments. When sediment deposits are stirred by dredging and construction work, by floods, as a result of water movements induced by wind or by shipping, as well as by benthic fauna (bioturbation), or by chemical mobilization (oxygen depletion, changes in redox potential, or fluctuations in salt concentration), pollutants in the sediment are released to the substance cycles of the waterbodies.

The Great Lakes in North America, the largest lacustrine region in the world, have such high levels of pollution due to large-scale external loads that drinking water may be extracted at only 3% of the 8,000-km shoreline. This accumulation of pollutants is further reinforced by the low renewal rate of the lakes (Abramowitz, 1995). In southern Finland and on Lake Baikal, paper and cellulose production are the main source of heavy pollution (Lindstrom-Seppa and Oikari, 1990).

Toxic pollutants impair the normally high self-purification capacity of running waters when they kill or harm the bacteria responsible for the self-purification process. When rivers are polluted as a consequence of large-scale discharge of effluent from industrial centers, or through accidents, a high rate of deoxygenation and the direct impact of toxic substances can lead to the elimination of many groups of fauna, as occurred after the Sandoz disaster on the River Rhine (Lelek and Koehler, 1990). There have been many cases in which the biological communities of rivers have recovered rapidly from such singular events, but this cannot be generalized to the same extent for other waterbodies (see Section D 1.2.5).

Discharged cooling water raises the temperature of adjacent or downstream portions of the waterbody. As a result of these changed conditions, biological communities and the turnover rates of substances undergo changes. Elevated deoxygenation rates as a consequence of higher temperatures is a particularly serious effect that is attributable both to the temperature dependence of the physical solubility of oxygen in water, as well as the higher metabolic rates of aquatic organisms. Thermal pollution control plans have been drawn up for rivers in highly industrialized areas exposed to severe thermal stress; these plans are supposed to ensure that the waterbodies are not heated up by more than 3 °C. Streamflow variations are of critical significance for the capacity of rivers to tolerate thermal pollution. Discharges of cooling water at low water levels have particularly severe consequences, especially in the warm season.

GROUNDWATER

Groundwater pollution does not normally occur until after a certain delay, because the leaching of pollutants from polluted soils at industrial locations and under landfills into interstitial groundwater is normally a slow process. Fractured rock aquifers (granite, lava, karst) with low rates of water renewal are particularly vulnerable to anthropogenic pollutants, which can be very quick in appearing. Even in highly technified industrialized countries, certain groups of pollutants in groundwater (for example radioactive isotopes, phenols, arsenic) prevent any further use of this resource (WBGU, 1995). Groundwater pollution, once it has occurred, is almost impossible to eliminate and must therefore be classified as severe damage to water resources.

4.4.4 Introduction of non-native species

INVASION AND INTRODUCTION OF NON-NATIVE SPECIES

International transoceanic shipping leads to unwanted transfers of non-native species between previously isolated aquatic habitats. In addition, fauna and flora are deliberately introduced into freshwater habitats for commercial purposes. Linking continental catchments via canals provides other new dissemination pathways for local and exotic fauna and flora. The explosive propagation of invasive species, such as the zebra mussel *Dreissena polymorpha* in the Laurentian Great Lakes of North America, is a consequence of high mobility and a low degree of control by natural regulation (Griffiths and Schloesser, 1991). Due to the lack of natural enemies otherwise present in the native habitats (predators and parasites), introduced non-native species can have severe impacts on native species of flora and fauna, and may even eliminate them entirely. The introduction of non-native species can result in the following long-term outcomes for the ecosystems concerned:

1. Permanent establishment is not possible.
2. Indigenous species are displaced.
3. Coexistence with indigenous species occurs, with the establishment of a new equilibrium.

In the Great Lakes of North America, about 130 non-native species have been introduced so far. As a result, the fisheries yield of native species has collapsed to a mere 0.2%. The economic damage is in the order of several billion US dollars (Abramowitz, 1995). In many cases, the explosive spread of non-native species during the initial phase leads to population decline and to the establishment of a new equilibrium within the biological community. When indigenous and non-native species coexist, this is not an identical state to the one existing prior to the invasion of the foreign organism. The accidental introduction of the American waterweed (*Elodea canadensis*) from North America into European inland waterways about 100 years ago led at first to drastic changes in the structure of riparian biocoenoses. In the meantime, the species has become firmly established in the affected ecosystems (see also Box D 4.4-1).

BOX D 4.4-1

The introduction of non-native fish species and the impacts: Two case studies

THE INTRODUCTION OF THE NILE PERCH TO LAKE VICTORIA

The deliberate introduction of the Nile perch (*Lates niloticus*) to Lake Victoria (Uganda, Kenya, Tanzania) in the early 1950s caused mass extinction among the lake's endemic fish species. Of approximately 350 endemic species, 200 have fallen victim to non-native predatory fish and growing pollution, while remaining stocks were heavily decimated. Although an export-oriented fisheries industry for catching the 200 kg and 2 m long Nile perch has been successfully established, the price paid by approximately 30 million local inhabitants was the destruction of their main source of food and livelihood (Abramowitz, 1995).

THE INTRODUCTION OF COMMERCIAL FISH SPECIES TO SOUTH CHILEAN LAKES

Biogeographical isolation has resulted in the biotic communities of South Chilean Lakes being relatively species-poor. Water fleas (*Daphnia*), the most important source of food for plankton-grazing fish in European and North American lakes, are absent. Because the productivity of these waters is usually low, the only species to be successfully introduced have been rainbow trout and brown trout, which feed on benthic fauna. Eutrophication of these lakes due to aquaculture and to intensive land use in the catchments has led to the propagation of larger zooplankton species that are suitable as food for fish, and to conditions similar to those in European and North American lakes. Native fish fauna feeding on small zooplankton (copepods) are likely to be eliminated (Carpenter et al., 1996).

The two examples show that the introduction of non-native fish species can result in major changes in the structure of the food web, changes that can have unforeseeable impacts. In the case of the Chilean lakes, future developments will probably be very much affected by eutrophication.

Besides the accidental introduction of non-native species and the impacts this can have, dramatic changes in the structure of entire food webs sometimes result from the deliberate introduction of non-native useful plants and animals from which high yields are expected. Impacts on native communities vary and are unpredictable in most cases. Effects on the food web are often severe. Successfully introduced fishes are often rapidly growing species that can effectively use and eliminate populations of prey species in order to maintain their high rates of reproduction.

4.4.5 Overfishing of inland waters

The fisheries yields of inland lakes are dependent on the primary production of the waterbody as well as the efficiency with which the potential catch is able to use the organic substance thus formed. Only in the rarest cases do suitable species make direct use of the plant biomass. They are mostly at the end of the food chain: plankton algae are grazed upon by plankton-eating fauna (zooplankton), which themselves serve as food for plankton-eating fish. Energy losses occur at each transfer step that are cumulatively all the greater, the longer the respective food chain (Odum, 1971). The population density of the next grazer population reaches saturation level when the highest possible population density (carrying capacity) has been reached for the given food situation.

Sustainable use of a fish population is only possible when catch size is balanced by the growth of young fish. The most rapid growth is then achieved when the population density has reached about half the maximum carrying capacity. In many cases, however, fish stocks are depleted much more rapidly than this, resulting in a reduction not only in population density but also in growth rates. As a consequence, the fisheries yields are smaller than they could be if catch rates were optimized. An important fisheries management measure is to stock with young fish in order to ensure the minimum level necessary to maintain population size. The maximum population density and hence the total fish yield cannot be increased, however.

4.4.6
Declining area and quality of inland waterbodies due to direct intervention

Many direct forms of anthropogenic intervention and exploitation damage the ecological integrity and hence the quality of inland waterbody ecosystems. Construction projects can even cause the destruction of entire ecosystems. Examples include the drainage of wetlands, the canalization of rivers, diking, and the construction of retention basins and dams. In numerous industrialized countries, the ecosystems of the major rivers have been largely destroyed by a chain of dams (e.g. the Danube and Volga rivers). Some rivers have only a few short stretches where they can still flow freely. Impoundment of rivers and lakes has resulted not only in the elimination of species-rich biota in lakeshore and riverside areas, but also in the destruction of wetlands in adjacent valley meadows.

DAMS

Dams alter the hydrological regime of entire river basins and cause large-scale changes to environmental, economic and sociocultural factors in the areas affected. The ecological impacts can be summarized as follows:

1. Increased evaporation from the water surface due to longer residence times of water.
2. Impoundments produce thermally stratified waterbodies that behave similarly to natural standing waters. The result is the formation of vertical gradients (dissolved oxygen, nutrients), plankton growth, the breeding of insect vectors that spread disease (e.g. mosquitoes responsible for transmitting malaria) and to elimination of running water biocoenoses.
3. Breaks in the watercourse prevent migrations that are essential for the life cycle of many fishes and other organisms. It is feared, for example, that damming the Yangtze River will expose an endemic mammal species, the Yangtze dolphin, to the threat of extinction. Dolphins in the River Ganges face a similar fate (Reeves and Leatherwood, 1994).
4. The natural nutrient balance of entire catchments is altered when valley floors are no longer fertilized by annual floods. Agriculture must then rely on man-made irrigation systems and fertilizers (an example is the Nile Valley).

RIVERS USED AS SHIPPING CANALS

As human settlements have become increasingly concentrated, the need for land resources and for regulated watercourses has had growing impacts on major rivers (Niemeyer-Lüllwitz and Zucchi, 1985). Many running waters have been reshaped into mere canals for transport and irrigation. The semi-natural state of many riverine ecosystems has been utterly destroyed by dredging, canalization, impoundment and maintenance operations (Claus and Neumann, 1995).

New land is wrested from riparian areas and floodplains mainly by diking and by deepening the river bed. Diking leads to higher flow rates, as a result of which the river scours deeper into the stream bed. This causes the water table to sink over extensive areas, leading to wetland destruction and declining yields from cropland.

The reduced retention capacity of regulated river systems during annual flood peaks triggers a race against time to raise the height of dikes. Enormous damage is inflicted on economies by the increasing incidence of devastating river floods, such as those occurring in the catchment of the intensively regulated Mississippi River (USA). Areas of land in river catchments that are used for intensive agriculture (as in Huang He, China) are unable to retain as much water as undisturbed catchments (WBGU, 1995).

NAVIGATION, WATER SPORTS AND LEISURE USES

Waves generated by ships can agitate the bottom sediment and cause it to be resuspended. On heavily navigated waterways, deposited contaminants can thus be reactivated. The mechanical stresses imposed on emergent, floating-leaved and submersed macrophytes by the currents and waves generated by shipping can lead to breaking and uprooting of entire stands and to inhibited blooms. Along intensively navigated waterways there is often a complete absence of littoral flora, which have difficulty establishing themselves due to modified hydrological conditions and partial reinforcement of the river banks.

Fauna are also disturbed by heavy shipping traffic and recreational activities. Birds and large aquatic mammals (e.g. sea-cows) are worst affected (Zhou, 1986).

4.4.7
Impacts of the loss and degradation of wetlands

Wetlands, as the element linking terrestrial and aquatic ecosystems, are particularly exposed to natural and anthropogenic degradation (Table D 4.4-2).

WETLAND LOSS AND DESICCATION

Drainage of wetlands is usually for the following ends:

- River canalization and lowering of water levels.

Table D 4.4-2
Global loss of wetlands. Sources: modified from WRI, 1996 and UNDP, 1994

	Wetlands ca. 1990 (1,000 km²)	Wetland area lost (in percent)	
Europe	154	Italy (1994):	94
		Great Britain (1991):	60
		Netherlands (1991):	60
		France (1980):	10
Middle East	8	Pakistan (1994):	74
Far East	11	India (1994):	79
Southeast Asia	241	Thailand (1994):	96
		Vietnam (1994):	100
		Indonesia (1994):	39
Australia and New Zealand	15	Australia (1990):	95
		New Zealand (1994):	90
Africa	355	Botswana (1994):	10
		Dem. Rep. Congo (1990):	50
		Malawi (1994):	60
		Cameroon (1994):	80
		Nigeria (1994):	80
		Chad (1994):	90
Alaska	325	-	
Canada	1,268	-	
USA	422	(1984):	57
Central America	18	-	
South America	1,524	-	
Global	5,900	-	

- Combating water-related diseases (especially malaria).
- Acquiring land for agricultural, industrial and commercial development.
- Road construction and residential development (surface sealing).
- Direct use (cutting peat for fuel, conversion to parks and gardens).

The conversion of wetlands to farmland, especially in tropical countries dominated by agriculture, has not always led to higher yields. The diversified use of natural wetlands in the Niger Delta, for example, probably generates more income for the national economy than a rice monoculture is capable of doing (Drijver and Marchand, 1986; see also Section D 1.2.4). There is a growing need for cost-benefit analyses that take into account not only the direct income generated, but also the costs of medium- and long-term impacts (e.g. floods, lowering of water tables, environmental damage such as biodiversity loss) (see Box D 4.4-2 and D 4.4-3).

Nutrient budget

High nutrient loads in inflowing river or drainage water as well as precipitation water rich in nutrients lead to enhanced plant growth and in the long term to wetlands becoming overgrown. Many wetland areas are eutrophic and highly productive under natural conditions, and their capacity for converting nutrients to biomass is used worldwide as a simple means of improving water quality (Section D 1.2.4). Nitrate is converted to atmospheric nitrogen (denitrification) in water-filled soil layers with low concentrations of dissolved oxygen. This self-purification process reduces nutrient uptake and the amount of nutrient that is subsequently released (Vought et al., 1994). In many cases, however, neighboring habitats are exposed to emissions from wetlands. Integrated aquacultures, as found in sub-tropical and tropical countries in South Asia (Thailand, India, Vietnam, China and Taiwan), can put pressure on surrounding aquatic ecosystems, to which they discharge a nutrient-rich, deoxygenating load comprising fecal waste, feed and various chemicals (e.g. pharmaceutical preparations or algal growth inhibitors) (Dierberg and Kiattisimkul, 1996). Maximum loads occur when fish ponds are emptied for seasonal harvesting.

Wetlands with low concentrations of nutrients, like the swamps and moorlands of Scandinavia and northern Asia, receive most of their water in the form of rain. Rainwater's low buffering capacity and the ability of peat-forming *Sphagnum* mosses to exchange ions (sodium and potassium ions are exchanged for hydrogen ions) create acidic conditions regardless of the geological base. Bog flora are well adapted to such conditions (Ruttner, 1962). These characteristic features may be lost following changes in the chemical composition of precipitation (see

BOX D 4.4-2

The Pantanal – one of the largest wetland areas in the world – is endangered

The Pantanal in central South America belongs for the greater part to Brazil and extends westwards as far as Bolivia. Until now, extensive use has enabled the preservation of this unique ecosystem, which features an enormous diversity of lifestyles and cultures adapted to the flood cycle (Junk and da Silva, 1995; Heckman, 1994). In addition to the increasing pressures induced by changing use patterns on the floodplain (marked green on the map in Fig. D 4.4-1) and in the catchment (marked white on the map), the "Hidrovia" mega-sized water development project is exposing the Pantanal to the threat of massive environmental degradation with unforeseeable social and economic impacts. "Mercosud", the South American common market of which Brazil, Argentina, Uruguay and Paraguay are members, is planning to develop the river system by making it navigable over a 3,400-km stretch from Nueva Palmira in Uruguay to Cáceres on the upper reaches of the Rio Paraguai.

Covering an area of 180,000 km², the Pantanal features savannahs, forest islands, lakes and rivers surrounded by gallery forest (Tucci et al., 1995). Large parts of the Pantanal are inundated by floods during the annual rainy season. Much of the water evaporates before it can flow back to the Rio Paraguai (Ponce, 1995). The Pantanal therefore has a regulatory effect on the runoff regime of the Paraguai-Parana-La Plata river system and on climate processes in South America.

The Pantanal is of great ecological value because of its size, its almost virgin condition and due to the large number of threatened species, including the jaguar, the marsh deer, the Giant Armadillo, the Giant Otter and Lear's Macaw (Alho et al., 1988). The diversity of habitats existing in this mosaic of tropical floodplain savannahs and terrestrial vegetation is unique among the subtropical to tropical climate zones (UNEP, 1995). The indigenous population, which traditionally lives by fishing and livestock farming, depends on the Pantanal for its resources (da Silva and Silva, 1995).

Following the period of economic growth in the Pantanal up to the end of the 1950s, which was based on cattle ranching, sugar cane crops and timber exports, the economic importance of the region went into progressive decline as a result of increasingly cheaper competition from northeast Brazil (Kohlhepp, 1995). The main cause of this decline, besides inefficient production methods, were the high transport costs conditioned by the poor communication links to the major Brazilian cities and the export ports on the Atlantic coast.

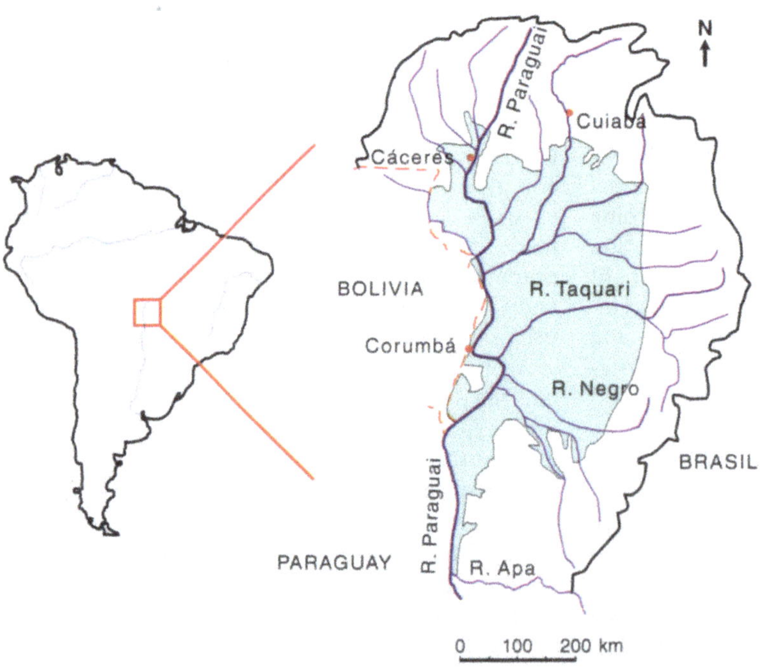

Figure D 4.4-1
Map of the Pantanal.
Source: adapted from Junk, 1993

Today, only a small volume of regional freight is transported on the Rio Paraguai between Cáceres and Asunción. When the river level is low, navigation is brought to a standstill at times. Year-round navigation would require large-scale measures to regulate and maintain the river. The "Hidrovia Project", initiated and supported at international level, is expected to provide a new boost to the regional economy by improving transport capacities, especially for soja beans from croplands to the north of the Pantanal via a waterway that is navigable throughout the year. Modifications to the landscape are planned at more than 200 sites, and include dredging a deeper stream bed, removing rocks and straightening meandering sections of the river. There can be no foreseeing the consequences that the destruction of natural rock barriers at the exit of the Pantanal is likely to have. On the whole, these hydro-engineering measures threaten a severe reduction in the river basin's water retention and evaporation capacity. Accelerated streamflow as a result of dredging is likely to cause some areas of the Pantanal to dry out, as well as increasing the risk and scale of flooding downstream. The adverse economic impacts of building the waterway through the wetlands, such as damage to fisheries and the need for cost-intensive flood control measures, were inadequately analyzed by the planners. This made the waterway appear a more cost-beneficial alternative to expanding the rail network.

Socioeconomic changes in the surrounding areas, such as gold prospecting, the conversion from traditional farming methods to cash crops for export, rural exodus and the concomitant growth of urban agglomerations are reinforcing the negative impacts on the Pantanal (Coy, 1991). The structural transformation into a modern cattle ranching economy induces changes to the landscape, such as diking. Gold mining operations discharge mercury residues to waterbodies, soils and air (von Tümpling et al., 1995). Soil erosion caused by open-cast mining, road construction and the cultivation of soya and sugar cane in the catchment flushes enormous amounts of sediment into lakes and rivers, causing habitats to silt up and the conditions for animal and plant life to deteriorate over wide expanses (Wantzen, 1997).

What needs to be done?

The Hidrovia Project is expected to have severe and large-scale impacts on the society, economy and environment of the entire region (see also Section D 3.4). In particular, the threat to biodiversity in the Pantanal and of severe flooding downstream from the Pantanal are problems of international significance. The decision to support the "Hidrovia" megaproject should be based on an assessment of the social, economic and environmental impacts. In this connection, it is essential to analyze the limits to economic growth within the region (e.g. supply, market conditions) and to consider alternatives like rail transport so that sustainable use and locally appropriate development remain a feasible option for the landscape in the long term (Section D 3.4). The level of acceptance shown towards the Hidrovia Project on the part of international aid organizations such as the Inter-American Development Bank (IDB) is still low.

Analyzing the natural geographical conditions and assessing the impacts of existing and planned human interference on this unique ecosystem require specially targeted tropical research efforts. As part of the SHIFT program (Studies of Human Impact on Forests and Floodplains in the Tropics), the Federal Government is supporting research projects focused on the environmental and socioeconomic problems of the Pantanal region.

Section D 1.5.1.1), resulting in shifts in the range of typical flora and fauna species.

Significance of wetlands for climate

Wetlands regulate the Earth's water and temperature balance. Many display high levels of productivity (Mitsch et al., 1994). Most of these areas contain large amounts of organic material because their primary production exceeds decomposition. Between 20% and 30% of the total fixed carbon and nitrogen in the world's soils are found in moors, tundra and marshes alone (Martikainen, 1996). This is equivalent to about 40–60% of total atmospheric carbon, and about three times the total carbon in the tropical rainforests (Maltby and Procter, 1996). Substantial amounts of these carbon and nitrogen pools have been released over the past 200 years through drainage for agriculture and forestry, and by cutting peat as a source of energy. At present, annual CO_2 emissions from these areas by agriculture alone are equivalent to 3.5% of the total emissions to the atmosphere from fossil fuel burning. Wetlands are effective sinks for carbon dioxide, on the one hand, but they are also sources of other gases, such as methane

BOX D 4.4-3

The Ramsar Convention

On February 3, 1971 in Ramsar, Iran, the "Convention on Wetlands of International Importance Especially as Waterfowl Habitat" (the Ramsar Convention) was signed by 18 states. Wetlands as defined by the Convention are shallow waterbodies (such as lakes, ponds, rivers and coastal zones) and land that is inundated or saturated with water at least periodically, such as marshes, moorland, swamps and floodplains. Protection of these areas must be regulated internationally, not only because they are so important for biodiversity the world over, but also because many animals that live in wetlands or are dependent on them at certain stages in their growth frequently transcend frontiers in the course of migration (Matthews, 1993). This applies to fish, birds and amphibians.

The Ramsar Convention entered into force in 1975 and has been ratified by 101 nations so far. Almost all states in Europe, North America and Oceania have ratified the Convention, whereas the area in Asia and Africa covered by the Convention is only 61% and 42% respectively. In 1975, the Convention was the first multilateral agreement aimed at intergovernmental collaboration in the protection of natural resources and at "wise use" of such resources. To this day, the Convention is the only agreement for the protection of habitats. In the "Ramsar List", the Parties to the Convention have since designated 867 wetlands as protected areas: these cover a total 62.5 million hectares, or about one tenth of all wetland areas in the world. Germany alone has designated 31 wetlands (660,569 hectares) under the Convention framework. Protected wetlands worldwide include marine and coastal areas (rocky coasts, salt marshes or tidal flats like the Wadden Sea), in addition to inland wetlands from rivers to desert oases, as well as man-made wetlands such as reservoirs, fish ponds or saltwater pools. Most wetlands on the Ramsar List are either lakes, lagoons or swamps.

Under the Convention, the Parties commit themselves to "wise use" of all other wetlands not included in the Ramsar List. In this respect, the Ramsar Convention was well ahead of its time in 1975. The concept of "wise use" has been elaborated in the course of time and has pronounced similarities with the more recent notion of "sustainable development". Guidelines for the "sustainable use" of wetlands have been developed through the collaboration of Parties under the Ramsar Convention.

In 1982, the Parties amended the Convention to give more powers to its various institutions: The Paris Protocol adopted in 1982 provided an amendment procedure, under which a permanent Ramsar Bureau charged with supporting implementation of the Convention was set up in Switzerland in 1988. Legal developments up to 1994 included added powers for the Conference of the Parties, creation of a Standing Committee and a separate budget for the Conference of the Parties. The Conference of the Parties meets every three years (most recently the 1996 conference in Brisbane) to discuss the implementation of the convention and to improve its operations by means of resolutions and recommendations to the Parties. One example is the so-called Montreux Record created by the 1993 conference. This is a list of Ramsar sites where change in ecological character has occurred, is occurring or is likely to occur. The aim of the list is to facilitate prioritization under national conservation programs. As soon as a wetland is put on the Montreux Record, the Parties are expected to establish a monitoring procedure that is then eligible for support from the Ramsar Bureau. 1990 saw the creation of a new Wetland Conservation Fund, now known as the Small Grants Fund. The aim of the fund is to provide developing countries with technical support and to finance programs aimed at the "wise use" of wetlands. The Fund is maintained only partially from the Convention's core budget, relying primarily on voluntary contributions. Projects receiving support tend to be small in scale, rarely involving sums in excess of SFR 40,000.

The Convention has one main weakness, however. It has no instruments at its disposal for mitigating or ameliorating external impacts that cause damage to wetlands. It is unlikely that the regulatory powers of the Convention will be extended. Although some regional agreements regulating the use of transboundary waterbodies now contain provisions for protecting riparian ecosystems, it will be necessary in future to intensify the practical integration and coordination of the Convention with other regimes for environmental protection.

and N_2O, which have more powerful climate-forcing effects. The precise emissions of these gases depends on aeration, water level, soil temperature and the "maturity" of the wetland area. For example, drainage of wetlands with sinking water levels causes a reduction in methane production, but increases N_2O emissions from nutrient-rich wetlands. The probable effects of a global temperature rise would include not only modified biota, but also accelerated decomposition processes. It has not been possible so far to produce a definitive assessment of the combined impact on biogeochemical processes of temperature rise, changes in hydrology and possible modifications to biota.

4.4.8 Recommended research and action

RESEARCH RECOMMENDATIONS

- Definition of critical limits for physical, chemical and biotic loads, and investigations into the importance of physical characteristics (morphology, hydraulic properties, temperature regimes, etc.) for the stress-bearing capacity of standing and running waters. Here, the term physical load includes the withdrawal and heating of water, as well as the generation of turbulence. Chemical pollution includes nutrient and contaminant loads.
- Research into climate-forcing processes in freshwater ecosystems (especially in wetlands), and their significance for biogeochemical cycles.
- Further development of concepts for rehabilitating anthropogenically damaged ecosystems, e.g. by supporting the self-purification process and other intrasystemic processes that contribute toward improved water quality. These include measures for controlling food chains, the restoration of deepened river beds and for the re-establishment of nutrient-poor conditions (reoligotrophication). Such activities need supporting research.
- Assessments of sustainable fisheries yields for waterbodies that are important for the local food supply and economy. Before non-native fishes are introduced, it is essential to evaluate the potential impacts on the structure and function of freshwater ecosystems.

RECOMMENDATIONS FOR POLICY ACTION

The Council recommends:

- That pressure be exerted at international level to ensure that municipal wastewater and other potentially harmful wastewater are not discharged untreated into surface waters and groundwater if this implies a threat to habitats or other resources worthy of protection, or would mean that critical stress limits are exceeded.
- The promotion of agricultural production methods that impose minimal pressures on the environment and human health, e.g. through contributions to the optimization of crop cultures against the background of edaphic and climatic suitability, or through support to organic farming.
- That direct interference with freshwater ecosystems, especially the impoundment of riverbanks, the canalization and shortening of river courses and the construction of dams and weirs be seen as ecological damage, the benefits of which must be weighed up against the loss of functions they imply.

4.5 Water technologies: Basic principles and trends

Production, distribution and treatment of water – Water disposal – Adapted technologies

The term "water technologies" refers in the stricter sense to all technical facilities for supplying households, public facilities, agriculture as well as trade and industry with water of adequate quality and quantity, and for disposing of and purifying the wastewater discharged after use. In a broader sense, it also includes technical facilities for optimizing water use in the specific cultural and locational context. The primary objective thereby is to reduce the consumption and pollution of water.

Securing daily water requirements provided the initial driving force for the development of technologies for collecting, transporting and storing water. At a very early stage, ways were also found of accessing the high-quality water resources that are protected against evaporation, namely groundwater. The introduction of crop cultivation and irrigation necessitated raising large volumes of water and moving them through canal systems, as well as flood control systems, and this, too, provided an early impetus for technological advances. The fundamental importance of water for humankind has not only had impacts on technological progress, however, but has also shaped human cultures in the same way. Water has been a key factor in development throughout the past, and will continue to be so in the future, especially since it is becoming increasingly obvious that we are still a long way from the sustainable management of water resources.

Water collection		Water distribution	
GROUNDWATER:	Vertical wells, horizontal wells, infiltration collectors, foggaras and qanats	EXTRACTION:	Free-flowing methods and pump stations
SPRING WATER:	Spring water collectors, infiltration collectors	TRANSPORT:	Distribution networks, channels
SURFACE WATERS:	Pumping stations for lake or river water	STORAGE:	Storage basins in hills and valleys, water towers, cisterns, reservoirs, aquifers
BANK-FILTERED WATER:	Wells		
GROUNDWATER RECHARGE:	Infiltration plants, wells		
SEA WATER:	Desalination		

Table D 4.5-1
Equipment and facilities for harvesting and distributing water.
Source: Rott, 1997

4.5.1
Water supply

Water is supplied by means of technical facilities for collection and distribution, and for treating collected water in accordance with the quality requirements of the respective consumer groups. Table D 4.5-1 gives an overview of the various types of facility used for water collection and distribution (Förster, 1990).

Groundwater, spring water and well water are most effectively protected against contamination by the covering layers and the filtering effects of soil. They thus provide untreated water of high quality that to date has required only minimal treatment, if any. However, the increasing pollution of these resources with contaminants and other unwanted substances is making treatment of groundwater more and more necessary. The purifying effect of infiltration through soil is utilized in bank-filtering and artificial groundwater recharging as a form of semi-natural treatment of surface waters. Seawater is available in large quantities, but the desalination required to produce freshwater is energy-intensive and hence costly.

4.5.1.1
Water collection

Wells are used to collect or withdraw groundwater, bank-filtered water and artificially recharged groundwater. There is an increasing tendency to construct ever-deeper wells with ever-greater capacities. In this case, water collection is generally only possible with the aid of pumps. Depending on the local conditions, various types of wells are in use (Förster, 1990) (Table D 4.5-1): radial collector wells are used in the case of shallow aquifers. These consist of a large well shaft from which horizontal filter lines are driven radially to all sides at groundwater level. The large filter area makes them especially suitable for soils with low permeability. Their advantage over vertical wells is that they produce high yields from a confined space, although this can only be exploited to the full in the case of more prolific aquifers. Besides these two well types, shallow groundwater and water trapped behind impermeable vertical boundaries is collected in shafts using perforated pipes driven into the ground or mountainside. These well types are useful when substantial lowering of the water table is to be avoided, or when the terrain does not permit the construction of wells. Another technology, applied especially in the Near East and northern Africa, is that of so-called qanats, or foggaras. These consist of a series of wells and tunnels that collect groundwater underground and channel it down a slope until it exits further downhill (BMZ, 1995).

When river water is withdrawn, care must be taken to ensure that the fluctuations in water level do not cause the wells or intake structure to dry up. The flow and distribution of bed load or contaminants must also be taken into account (Förster, 1990).

Extraction of lake and reservoir water requires precise knowledge of the waterbody. Depending on size and geographic location, seasonal or daily changes in stratification may influence the distribution of organisms and the chemical processes occurring in the lake. If one is familiar with these natural processes, the withdrawal depth can be selected such that unwanted substances – algae in particular – do not enter the untreated water at all, or only to a minimal extent (Förster, 1990).

In the case of artificial groundwater recharging, the infiltration of surface water into the filter or vadose zone is extremely critical. This is where substances contained in the water but not actually dissolved are fixed, and where biologically degradable substances are broken down through the surface effect of the filter grain or soil. The larger the areas available for infiltration of a specific quantity of water, the lesser the extent to which clogging is a problem. This applies both to above-ground and underground infiltration facilities. The decision as to which infiltration method should be chosen depends not only on the hydrological conditions, but also on other existing or planned uses of the terrain and on the quality of the surface waters (Förster, 1990).

Methods used to desalinate seawater include reverse osmosis, electrodialysis and distillation. However, these technologies are still expensive and require large quantities of energy. Consequently, desalination of seawater will not be a significant source of freshwater supplies, at least in the near future.

4.5.1.2 Water distribution

In rare cases, open canals are used to transport large quantities of water from the withdrawal site to the treatment facility. Closed systems are always to be recommended for the distribution of water after treatment. In addition to pipelines and pumping stations, water storage basins are required in order to maintain reserves, to cope with peak consumption rates, to have water available for extinguishing fires and to ensure constant water pressure throughout the entire region. If there are substantial differences in height, the distribution network is divided into pressure zones in order to maintain the pressure within a permissible range. Computer technology is used to calculate operating states as well as the pressures and flows within the system. Although the fundamental principles of hydraulic engineering are now well established, the accurate determination of water demand and its variation over time, especially the daily and hourly peaks, continues to pose major problems. Furthermore, it is essential to take future trends into account because the service life of water distribution equipment ranges from 15 to 30 years only (Förster, 1990).

The materials used for water distribution pipes vary in their characteristics. Steel, for example, has high strength but is prone to corrosion. Plastics, on the other hand, have a much lower strength but a high resistance to corrosion. Pipes laid up until 1965 were exclusively made of gray cast iron. They have a high resistance to corrosion, but are liable to fracture. The trend away from gray cast iron to ductile cast iron in the course of the 1960s resulted in a pipe with high strength and sufficient ductility; however, it proved to be much more prone to corrosion than had been assumed originally. For this reason, pipes made of ductile cast iron are now fitted with a suitable casing wherever necessary. Asbestos cement pipes are predominantly used outside densely developed regions. The suspicion that harmful amounts of asbestos fibers are released into the water by these pipes was not substantiated in studies conducted by the Federal Health Department in Germany. Prestressed concrete pipes have proved effective for large-diameter transport lines. Plastic pipes are mostly used in large rural supply networks and for household connections. Careful quality monitoring is necessary during the production of plastic pipes. Nowadays, the inner side of metal pipes is protected against corrosion by lining all pipes with cement mortar. Galvanized steel pipes are also used for smaller-scale purposes. Whenever the inner walls of vessels are coated or lined, this coating or lining must be non-porous. Corrosion problems that may arise when waters of varying origin and composition come together can be managed by complying with fixed mixing ratios and, where appropriate, correcting the pH value (Förster, 1990).

4.5.1.3 Water treatment

Facilities for treating collected water use a variety of chemical and biological techniques. Given the intensive human interference with the natural hydrological cycle, and the fact that wastewater discharge and withdrawals of untreated water are converging to an increasing extent, modern water and sewage treatment facilities have become very similar. It therefore makes sense at this point to provide an overview of the methods used in both water and wastewater treatment (Table D 4.5-2). The special procedures for drinking water purification and wastewater treatment will be examined in Section D 4.5.3.2.

An important feature of these processes, besides the way in which they operate, is whether they involve separation – as in the case of physical and some chemical processes – or destruction of substances – as in the case of biological and some other chemical processes. Separation processes remove substances from water. The substance remains unchanged and must therefore be subjected to further treatment or disposed of in some way (physical treatment). When decomposition processes are used, the unwanted substance is acquired in a modified, usually mineralized form, thus obviating the need for further treat-

Methods for water and wastewater treatment		
Physical	Chemical	Biological
SEPARATION METHODS:	SEPARATION METHODS:	DESTRUCTIVE METHODS:
Gas exchange, racks, sieves, sedimentation, flotation, filtration, membrane processes, adsorption	Precipitation/flocculation, ion exchange, distillation	*Suspended biomass:* Aerated and non-aerated tanks, activated sludge process, biotowers and pressure biology
	DESTRUCTIVE METHODS: Oxidation, reduction, photooxidation	*Immobilized biomass:* Soil filters, biofilters, trickling filter process, fluidized bed reactors packed bed reactors

Table D 4.5-2
Processes for water purification and wastewater treatment. Source: Rott, 1997

ment or disposal. In the case of biological processes, biomass is produced that also has to be disposed of.

PHYSICAL TREATMENT

Gas exchange processes are used to remove dissolved gases (e.g. hydrogen sulfide or carbon dioxide) and substances that taste or smell, or when gases such as oxygen are to be dissolved as an oxidation agent in the water. The degree of technical sophistication ranges from cascade aeration, where water flows under gravity from a height over a weir into the tank or basin lying under it, to oxidators into which both the water and the gas for aeration are pumped under pressure.

Techniques used to remove solids by separation (racks, sieves, sedimentation, flotation, filtration) and some of the membrane processes are based on screening and on the use of gravitational, flow and bonding forces.

As a rule, a rack consists of bars arranged parallel to each other and perpendicularly to the direction of flow. This simple design allows solids to be separated reliably and at low cost. Because of the relatively large spacing between the bars, ranging from 10 to 100 mm, a rack is suitable for removing relatively large particles only, and is used in water and sewage treatment only as a pretreatment stage and to protect the following process stages.

To remove particles that are too small to be held back by a rack, sieves with mesh sizes down to less than 0.1 mm can be used. Microsieves with mesh sizes from 5 to 40 μm, such as those deployed in drinking water treatment or more extensive wastewater treatment, permit the removal of very fine suspended matter.

All processes that work on the basis of the different densities of water and the particles to be removed can be designated as gravity processes. When suspended matter with a higher density than the surrounding water is to be separated out, sedimentation processes can be applied. To remove oil drops and grease particles with a lower density than water, the most suitable process to use is flotation.

Sedimentation and flotation reach their limits when the suspended particles are too small, or the difference between their density and that of water is insignificant. To enable extensive separation of solids in such cases, filtration processes are applied.

In addition to the purely mechanical screening effect, filtration can also involve mechanisms for transporting and settling of substances in such a way that the filters can also retain small suspended, colloidally dissolved and genuinely dissolved substances.

A distinction is made between slow and rapid filtration. Slow filtration with a filtering rate of around 0.05 meters per hour is a proven method for drinking water purification if suitably large areas are available. Slow filters are regenerated by stripping the contaminated surface every several months. If large throughputs on a small area are required, rapid filtration is carried out, i.e. filtration at filtering rates of over 5 meters per hour. Fast filters are regenerated by backwashing after filtering periods of one to several days, depending on the amounts of solids retained.

Membrane processes are based on the principle that the water to be treated is forced under pressure through a membrane with a defined pore size. They are used for a wide variety of applications, ranging from separation of suspended particles having a magnitude of around 10 μm by means of microfiltration, through separation of molecules with ultrafiltration, to retention of small monovalent ions by reverse osmosis. Particles of organic and inorganic origin, for example, can be extracted from wastewater by using microfiltration methods. With the help of ultrafiltration, oils or dyes can be recovered from water and reused. Reverse osmosis is used to purify highly saline untreated water for use as drinking water. The technological potential of membrane processes has not yet been exploited to the full. In recent years, its range of application has expanded significantly in all areas of water technology. However, this enormous technological potential is limited by high investment and operating costs and requires highly qualified operating personnel. As a result, membrane processes are currently most widespread in the field of industrial water technology, where they are used for recovering substances and for treating effluent contaminated with substances that are toxic or very difficult to degrade biologically.

Adsorption is based on the physical and chemical bonding forces that cause organic substances in water to accumulate on solid surfaces. To make optimal use of this effect in separating unwanted dissolved substances, the water to be treated is brought into contact with solids whose adsorptively effective inner surface is as large as possible. The most common adsorbents are activated charcoal with an internal surface area of up to 1,500 $m^2 g^{-1}$, lignite coke, bentonite and adsorber resins. Two technical methods are possible: 1) addition of powderized adsorbent to the water, followed by reseparation by sedimentation and/or filtration after an adequate period of contact, and 2) fixing the adsorbent in grain form in a filter bed and allowing the water to flow through this adsorbent bed in a manner similar to filtration methods. Once the adsorption capacity of the adsorbent has been exhausted, it must be regenerated or disposed of.

Chemical treatment processes

Precipitation involves the conversion of a dissolved substance into an insoluble form through chemical reactions triggered by the addition of precipitants. Since it is impossible to separate these resultant insoluble compounds directly from water on account of their very small particle size, the next step usually involves flocculation. The flakes thus formed must then be separated from the water by means of sedimentation, filtration or flotation. Precipitation and flocculation methods are technically refined and are used in the purification of water and the treatment of wastewater to remove many dissolved substances. Recently, however, they have been increasingly replaced by substance-destroying processes in cases where either of the two approaches could be used.

Ion exchange is a chemical-physical process that exploits the ability of surface-charged substances to absorb certain ions from water and in return release an equivalent amount of other, similarly charged ions. This means it is possible to remove specific cations or anions from water with the help of ion exchange processes. Once the capacity of an exchanger has been exhausted, it can be restored by adding a high concentration of the original counterion. Regeneration results in a highly concentrated solution containing the substance to be removed from the water. Ion exchange methods have been technically perfected and are deployed not only in the water purification sector, e.g. for removing nitrates or for softening water, but above all in the area of industrial wastewater treatment, i.e. for treating process effluents and recovering metals from wastewater.

Oxidation and reduction methods are based on the principle that substances in water will react with oxidizing or reducing agents that are added. In photo-oxidation processes, the reaction is triggered by inputting radiative energy in the visible and ultraviolet range, rather than by adding an oxidant. As a consequence, a previously dissolved substance may revert to its undissolved form or to its gaseous form. This means that oxidation, photo-oxidation or reduction methods can rarely be used as stand-alone systems and have to form part of an overall process involving subsequent physical separation stages. Although oxidation, photo-oxidation and reduction processes are deployed in the water purification sector, e.g. for removing iron or for disinfection, they are predominantly used in the treatment of industrial effluents, as a means for removing cyanide and dyes, and for the oxidation of substances that are very difficult to degrade biologically. Depending on the oxidation agent and reactor design used, their deployment requires major investments and operating costs as well as highly qualified personnel.

In the case of distillation or vaporization, the water to be treated is transformed from a liquid to a vapor. Substances contained in the water are left behind if their boiling point is higher than that of water, and this residue must then be disposed of. During subsequent cooling, the condensing water is free of the unwanted substances. Since water has a very high latent heat of vaporization, this process is very energy-intensive. For this reason, further development of technical distillation methods is aimed at recovering, during condensation of the evaporated water, the greatest possible proportion of the energy expended in heating the water and at reusing it in order to heat up additional water. In practice, multistage plants achieve a heat recovery of around 90%. Distillation processes are used for the desalination of seawater, and for the vaporization of industrial wastewater with a high level of organic contamination and which results in a combustible residue.

Biological treatment processes

Biological processes play a key role in the treatment of wastewater and, to an increasing extent, in the purification of drinking water. Technological exploitation of biological processes involves imitating the natural self-purification of waterbodies and ensuring that the degradation of organic compounds through the metabolic activity of microorganisms is controlled and intensified. Microorganisms require an adequate and balanced supply of nutrients – nitrogen, phosphorus, magnesium, calcium and potassium – to maintain their metabolism and growth, so it may be necessary to add certain amounts. A variety of microorganism groups can be used for biological degradation. They may differ in respect of metabolism, the way they produce energy, and the source of carbon required for their cellular development.

In water purification or sewage treatment facilities, different types of organisms form a biocoenosis variously referred to in water technology as flakes, sludge, biological grass or biofilm. In contrast to chemical and physical processes, biological processes do not attain a high degree of effectiveness until after a period lasting weeks or months. Biological processes are employed in the water purification sector, e.g. for biological removal of iron and manganese as well as for denitrification.

In the field of wastewater treatment, biological processes are used wherever degradable organic substances have to be mineralized. They have a very broad range of application and can be flexibly adapted to the specific problem at different levels of technological sophistication. When the decisive criterion is minimized contaminant concentration, and the wastewater to be treated already contains relatively small concentrations (as is the case with household sewage, for instance), aerobic processes are usually used, although these require much energy for aeration and produce large amounts of residual matter in the form of surplus biomass. When the key criterion is the highest possible economic efficiency and the wastewater to be treated contains a high level of organic pollution (such as effluents from the food and paper industries), anaerobic processes are used. Their advantages over aerobic processes are that no energy is needed for aeration, less surplus biomass is created, and methane gas is produced as a useful source of energy.

Drinking water purification

Treatment of groundwater and surface water is necessary when it contains substances that are harmful or undesirable on account of their odor or taste, or may lead to technical malfunctions.

Groundwater contains dissolved inorganic substances and organic metabolic products produced by microorganisms. Groundwater is usually treated with the following methods (Förster, 1990):

- Gas exchange methods for adding oxygen and removing carbon dioxide, hydrogen sulfide methane and odorants.
- Oxidation using microbiological or chemical methods for the removal of iron and manganese. This prevents deposits in pipe systems, an acrid, ink-like taste of the water, and brown spots on household objects and laundry.
- Deacidification by gas exchange or by adding calcium or sodium hydroxide, as well as filtration through alkaline filter materials in order to prevent subsequent contamination of the purified drinking water by the corrosion products of metallic materials containing cement.
- Removal of calcium ions (water softening) by precipitating calcium carbonate or by ion exchange in order to prevent the formation of calcite deposits in pipes (particularly in hot-water systems and household appliances), but also to reduce the need for washing and cleaning agents.
- Removal of dissolved organic substances by means of adsorption on activated charcoal (possibly after treatment with ozone). In groundwater treatment, these processes are only necessary if the water is contaminated with organic halogens or mineral oil products. Naturally occurring humic matter is also removed to a certain extent.
- Disinfection by adding chlorine or chlorine dioxide, or by treating with ozone or ultraviolet radiation, to avoid risks due to pathogens of infectious diseases.

Unlike groundwater, surface water always contains particulate substances as well, such as plankton and suspended mineral substances. Parasites, such as crytosporidia and *Giardia*, are particularly problematic. Additional stages are therefore necessary to treat such water (Förster, 1990).

Surface water can be treated for use as drinking water in a natural manner by means of bank filtration and artificial groundwater recharging (Förster, 1990).

Artificial groundwater recharging has come to play an increasingly important role in recent decades, and in some areas, e.g. the Rhine-Ruhr region, has become an indispensable factor in municipal and industrial water supply. The advantage of this process derives from a combination of two aspects:

- utilization of natural treatment processes with their large buffer capacity against unexpected deterioration of the water quality and
- the possibility of significantly increasing the quantity of water that can be produced by technical means.

The retention mechanisms used in artificial groundwater recharging in order to eliminate inorganic and organic contaminants are basically the same as in bank filtration. However, there are differ-

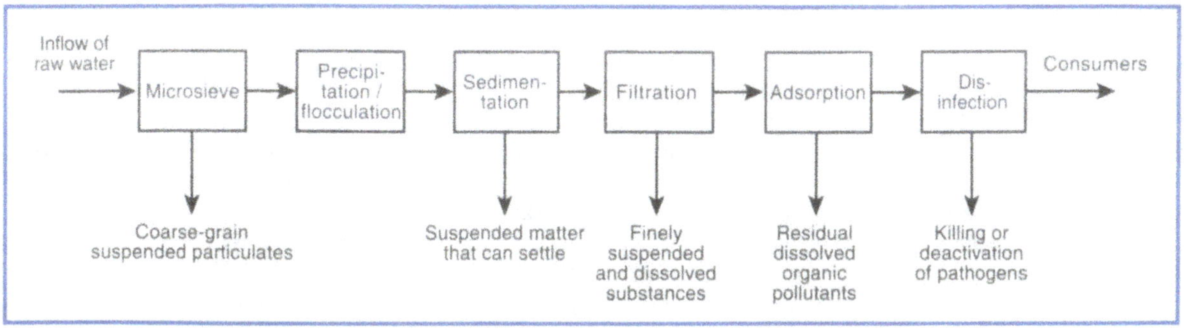

Figure D 4.5-1
Process diagram for a surface water treatment plant.
Source: Förster, 1990

ences in quantities due to the fact that extremely rapid conversion of substances takes place in the biologically very active surface layer, resulting in some cases in greater fixation, but in other cases in greater mobility of harmful substances. Despite the obvious improvements in water quality that can be achieved with these processes, they can only be used in conjunction with physical-chemical treatment methods. Figure D 4.5-1 shows a possible scheme for treating directly withdrawn surface water (Förster, 1990).

Since the mid-1970s, new technologies for producing drinking water from heavily contaminated surface water have been derived from older methods known to the chemical industry. They include the combination of flocculation, sedimentation and high-performance filtration, the use of ozone as an oxidant, and two membrane processes – nanofiltration and reverse osmosis. Water technologies are faced with new challenges due to the increase in halogenated organic compounds and nitrogen pesticides in surface and groundwater (Förster, 1990).

In the field of drinking water purification, efforts were primarily directed for a long time at developing and optimizing chemical processes, but attention nowadays is increasingly focused on the performance and stability of biological purification methods.

In contrast to other biotechnologies and in particular to wastewater treatment, one of the problems that arises in drinking water treatment is the small quantity of biomass. For this reason, biological treatment of such water is only possible with fixed-bed reactors in which the microorganisms are immobilized through adsorption on the carrier material.

What is now needed are biological methods of water purification with high yields per unit of space and time. However, it should also be pointed out that, in contrast to other biotechnological processes for drinking water treatment, relatively tight limits are set due to the costs of secondary purification (removing non-recyclable substances).

4.5.2
Water use

Technical equipment related to water use should also be regarded as water technology, to the extent that it influences water consumption and the quantity and state of wastewater.

Given the widely varying quality requirements that are placed on water in the domestic sector and by other consumer groups, it makes sense to supply water of different qualities. One of the benefits would be to reduce substantially the consumption of high-quality drinking water, which requires costly treatment (Förster, 1990).

In the domestic sector, supplying different qualities of water has been tried out in only a few urban agglomerations to date, e.g. Hong Kong, Singapore and Tokyo (Förster, 1990), since the high investment costs (for the construction of a dual network) as well as reservations concerning hygiene render this a prohibitive option. Facilities for using collected rainwater are more suitable on a decentralized scale (e.g. for toilet flushing, watering plants, or laundry). In isolated cases, systems have been installed for using slightly contaminated wastewater, so-called gray water, for toilet flushing (Lehn et al., 1996). Theoretically, drinking water consumption in households can be reduced by up to 50% using such methods (Förster, 1990; Lehn et al., 1996). Another potential source of savings lies in the further refinement of household appliances that use water, such as washing machines and dishwashers. It is also possible to optimize installation equipment in buildings. Examples include flow limiters, fast-regulating mixer taps and, in particular, economical toilet flushing. Fitting water meters in apartments enables costs to be allocated directly to individual households. This can lead to significant reductions in consumption (Lehn et al., 1996).

In the commercial and industrial sector, supply systems providing different water qualities are already encountered quite frequently. Techniques that reduce consumption are widespread in sectors involving intensive water consumption and wastewater production. For example, water is used several times with the help of counterflow and cascade flushing, or it is kept in the cycle as cooling or process water. An identical production process in the textile processing industry may result in a water consumption of between 40 and 300 m³ per ton of processed fabric, depending on the technology employed. Pollution of the water used can be prevented with methods for minimizing wastage of products and consumables, with dry-cleaning processes and with techniques for recovery materials (Rudolph et al., 1995; Rott and Minke, 1995).

One field where major reductions in water consumption could be achieved is irrigation. Agriculture is the biggest consumer of water worldwide, and the bulk of this consumption is for artificial irrigation. According to the FAO, however, less than 40% of irrigation water actually reached the crop plants (BMZ, 1995; see Section D 4.3). For example, water consumption can be reduced by up to 50%, especially in the arid climate zone, if drip irrigation is applied rather than sprinkler and gravitational irrigation methods (BMZ, 1995). In addition, this has the important side effect that soils irrigated in this way are not salinated as rapidly and can thus be cultivated longer. Since a lower water quality suffices for irrigation purposes, either untreated or minimally treated groundwater and surface water are generally used for this purpose (see Section D 1.5.). More and more consideration is being given to the idea of using treated domestic wastewater. In the early 1990s, for example, most countries in the Middle East and northern Africa had started programs for treating and reusing wastewater not only for irrigation, but also for industrial purposes and for groundwater recharging (BMZ, 1995).

4.5.3 Water disposal

Water contaminated by domestic, agricultural, commercial or industrial use is termed wastewater. In a broader sense, the rain or snowmelt water running off roofs, courtyards, roads and squares can also be included in this category. "Infiltration water" from organized drainage channels, drainage pipes, artificial lowering of water tables as well as from groundwater that flows into the sewer system through leakage points in drain pipes and sewers is included in the calculations of water disposal facilities, i.e. wastewater discharge and treatment plants.

4.5.3.1 Water collection and transport

In the industrialized countries and in urban agglomerations generally, excreta are usually diluted by flushing before being transported in the sewer network. Wastewater and precipitation water are collected in a variety of receptacles from which the water then flows through the sewer network to the treatment plant and finally into the receiving waterbody. The sewer system may consist of one or more networks. A combined sewer is one that transports all types of wastewater flows – domestic and industrial, as well as precipitation water – in a single system. Separate sewer networks involve different flows of wastewater in two or more sewer systems.

Combined and separate systems have various advantages and disadvantages that may be crucial to the planning of the sewer network (Förster, 1990):

- As a rule, the combined system involves lower construction costs than the separate system since only one line has to be laid in the roads. It also requires less flushing of the pipes because they are cleansed by rainwater. On the other hand, it requires larger treatment plants and, in the event of heavy rainfall, must be discharged directly into the receiving body of water due to the resultant overloading of the treatment plants.
- The separate system eliminates the disadvantages of the combined system. It is also easier to expand. Domestic sewer pipes can be laid deeper than rainwater pipes. In the event of heavy rainfall, only precipitation water that is relatively uncontaminated is discharged in an untreated state into the receiving waterbody.

The combined system is more economical for large cities on flat terrain as well as for smaller towns that have limited monitoring facilities. The separate system may be advantageous in municipalities with closed industrial zones. Essentially, however, the costs and thus the selection of the method depend on the terrain and the receiving waterbody. A problem with both systems is that very heavy precipitation runoffs have to be discharged via the sewer system when it rains due to the high degree of soil sealing in densely settled regions. This can be avoided if areas are desealed and minimally contaminated rainwater can infiltrate the ground (Förster, 1990; Lehn et al., 1996).

In many Asian countries, such as China, India, Japan and Vietnam, feces are usually kept separate from wastewater. In this system the excrements are fed back to the nutrient cycle in agriculture either directly, after decentralized anaerobic digestion, or after decentralized aerobic composting. Thus there is no sewer system and no centralized wastewater treatment facilities, at least for domestic sewage. Another advantage of this method is the significantly lower water consumption since no water is used for toilet flushing. In Japan, dry toilets are sometimes provided with collection pits, which are emptied every two to four weeks by suction extraction vehicles. The latter bring the excrement sludge to central treatment facilities equipped with mechanical-biological stages and sludge incineration (Lehn et al., 1996; BMZ, 1995).

4.5.3.2 Water purification

In Germany, wastewater is purified almost exclusively in central treatment plants. These comprise a mechanical, a biological and, where very high purification standards operate, a third chemical-physical treatment stage (Förster, 1990). Similar systems will also have to be applied in the urban agglomerations of the developing countries. Figure D 4.5-2 shows a schematic diagram of such a treatment plant featuring best available techniques.

The wastewater flowing from the sewer system is first fed to the mechanical treatment stage, which is usually composed of racks, sand and grease traps as well as a preliminary sedimentation tank. Coarse screenings are removed in racks with bar spacing from 60 to 100 mm, or fine racks with bar spacing of 10–25 mm. After that, the wastewater flows through the combined sand and grease trap, in which quickly settling inorganic particles are separated by sedimentation, and floating oils and greases by flotation. The wastewater then enters large sedimentation tanks, where it remains for several hours. In this way, even slowly settling particles of organic origin can be removed by sedimentation. The sludge that settles to the bottom of the tank, the so-called primary sludge, is subsequently separated from the wastewater by suction extraction.

The purpose of the biological purification stage that follows is to convert the organic substances dissolved in the wastewater into biomass through the metabolic activity of microorganisms, and to mineralize nitrogen and phosphorus compounds. The most commonly used methods in Germany are the trickling filter and activated sludge processes. The latter is shown as part of the treatment plant diagram in Fig. D 4.5-2. In a final sedimentation tank, the biomass must be separated from treated wastewater. Part of the settled biomass is fed back to the aeration tank as return sludge, while the additional biomass that has grown as a result of nutrients is discharged from the system here as surplus sludge. As a rule, the surplus sludge is returned to the preliminary sedimentation tank, removed from there along with the primary sludge and then subjected to sludge treatment. In many sewage treatment plants, sludge is digested anaerobically in tower-like digestion tanks where the biomass is stabilized and hygienized by methane-producing bacteria.

When discharged effluent must comply with high quality standards with regard to major nutrients, compliance with these requirements necessitates modifications of the biological stage, such as enlarged aeration tanks for nitrification, the creation of anoxic zones for denitrification or anaerobic zones for biological elimination of phosphorus. In the event that these modifications of the biological stage are inadequate, the treatment plant must be extended to

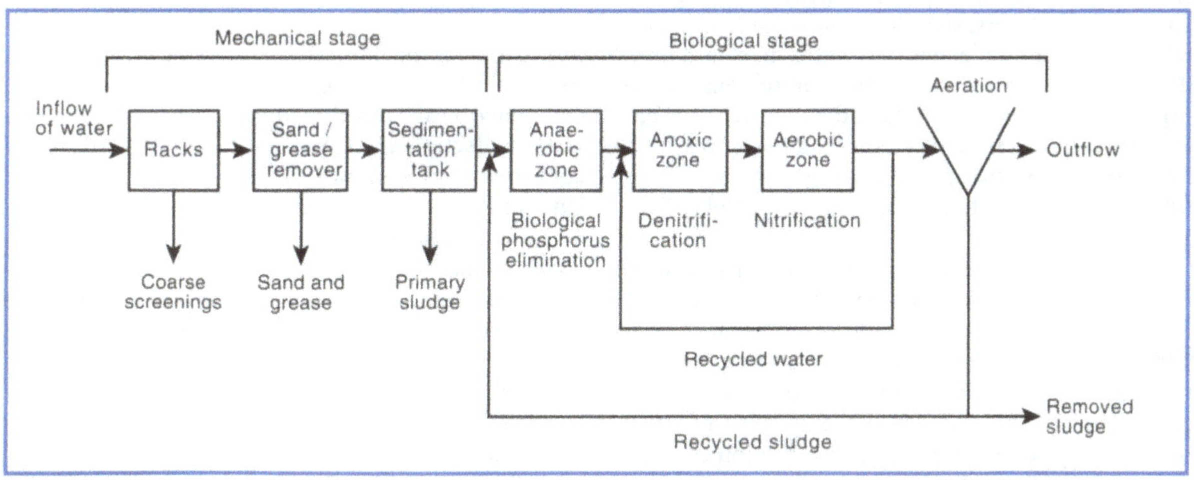

Figure D 4.5-2
Process diagram for a municipal sewage treatment plant with extensive removal of carbon, nitrogen and phosphorus compounds.
Source: adapted from Rott, 1997

include an additional physical-chemical process stage, such as precipitation/flocculation or filtration.

If the wastewater contains relatively high concentrations of refractory substances of industrial origin, addition of a specially adapted physical-chemical stage, such as adsorption or oxidation, may also be considered in individual cases. However, it is preferable from a technical and economic point of view to treat the industrial wastewater within the enterprise itself in a targeted and thus highly effective way, in order to remove or destroy these substances prior to discharge into the sewer system.

Central wastewater treatment plants are virtually inconceivable in the rural areas of developing countries. What is needed instead are adapted solutions, such as separation of excreta from wastewater or sewage fields, non-aerated sewage ponds, wastewater fish ponds or aquacultures, vegetation-covered soil filters and seasonal wastewater treatment reservoirs. In systems such as these, suitable wastewater can be purified with little technical effort. Two more ecologically sound aerobic-biological purification methods are land disposal and treatment in standing waterbodies. In contrast to most other methods, surface treatment or soil use are relatively easy to manage. Here, the wastewater is sprinkled thinly over a slightly sloping surface. The aim is for the plant nutrients that have been converted through oxidation to be absorbed into the metabolism of the plants and transformed into organic material, which is then harvested at periodic intervals. Penetration of the water into groundwater is prevented by selecting dense or artificially compacted soils. This method is likely to produce benefits for tropical and arid regions in particular. Another method that falls somewhere between land disposal and treatment in pools is the use of aquatic plant filters, such as rushes, reeds and cat's tails, in temperate zones, or water hyacinths and duckweed in tropical and subtropical regions. Microorganisms growing on roots and stalks act as a biological filter; this setup functions similarly to a trickling filter, except that plants are used instead of inert filler material (Förster, 1990; Lehn et al., 1996; BMZ, 1995).

Such rooting zone or plant wastewater treatment facilities have recently met with great interest as a low-cost alternative in rural municipalities with up to 1,000 inhabitants, because they are able to meet the minimum legal requirements regarding wastewater purification (Lehn et al., 1996). Compared to conventional sewage treatment plant, construction costs are around 20–30% and the operating costs less than 20% of the usual costs (Hecht, 1992).

Global relationships – developing countries

According to World Bank surveys and UN data, around 1.7 billion people worldwide did not have access to adequate sanitation in 1990 (World Bank, 1992; UN, 1990). This situation is compounded by a large urban-rural divide. Whereas 72% of city-dwellers has such access (corresponding to 377 million people without access to sanitation), the figure for the rural population is only 49% (1.36 billion people without access). The number of people with access provides an initial indication of the inadequacy of wastewater treatment infrastructure, but says nothing about how much of the wastewater flowing through the sewer system is discharged into receiving waters in a purified state.

The UN forecast (WRI, 1990) for the year 2000 shows that the degree to which sanitation services are provided in cities is stagnating or even declining, while it is increasing slightly in rural regions. Due to the high population growth in developing countries, the absolute number of people lacking access to sanitation facilities will rise to approximately 1.9 billion worldwide, particularly in the cities of developing countries. With the exception of West Asia, growth rates are around 70–100%, which means that 633 million people will be living without sanitation in cities in the year 2000.

The urbanization process in developing countries is gaining pace without expansion of water supply systems or increases in wastewater treatment capacity. The result is that wastewater flows untreated or inadequately treated into receiving waters. Domestic sewage is a serious problem in South America, for example. An above-average number of rivers display an extremely high germ count (coliform bacteria). According to estimates of the World Resources Institute, only 2% of domestic sewage in that region is treated (BMZ, 1995). Of 3,119 cities in India, only eight have a fully developed infrastructure for wastewater collection and treatment (WRI, 1996). Even in countries with medium to high income, cities like Buenos Aires or Santiago de Chile (where only 2% and 4% of domestic sewage, respectively, is treated) illustrate the lack of wastewater treatment infrastructure. Another factor is that the wastewater includes not only domestic sewage, but also industrial effluent from a wide variety of production processes. The problem of deficient or non-existent wastewater treatment and direct discharge into local receiving waterbodies is not confined to the developing countries. Even in the OECD states, a third of the population has no access to wastewater treatment (WRI, 1996). An additional problem is posed by the dilapidation of existing infrastructures in the industrialized countries, leading to overloading of the treatment plants and increased discharges of untreated water into receiving waters because of the higher volume of wastewater.

Public investments in water supply and the sewer system account for 10% of the total public-sector investment in the developing countries, equivalent to around 0.6% of GDP (World Bank, 1992). Expenditures for wastewater infrastructures amount to considerably less than 20% of the loans granted by the World Bank, whereby the bulk of this money has been invested in wastewater collection rather than wastewater treatment (World Bank, 1992). Given average costs of US$ 1,500 per household for the collection and mechanical-biological purification of wastewater (WRI, 1996), virtually no developing country in the world is able to afford collection and treatment of wastewater from all households without additional financial support. The sheer expense of wastewater treatment has led to a situation where even in industrialized countries, such as Canada and France, only 66% and 52% of the population, respectively, is connected to wastewater treatment facilities (World Bank, 1992). In view of advancing urbanization, it is advisable to concentrate the investments of developing countries on large agglomerations and to deploy conventional technologies for wastewater treatment in those areas.

Furthermore, there is a great need for low-cost and low-tech methods of wastewater treatment (see Box D 4.5-1). Major technological innovations in recent years include stabilization ponds, which are relatively cheap and easily operated, and the more technologically sophisticated sludge contact methods applied upstream, with which positive experience was gained in Brazil and Colombia (World Bank, 1992). A further option is to increase the reuse of municipal sewage. Wastewater that has been pretreated with mechanical and biological methods can be used in irrigation farming or aquaculture. The organic solids removed in the mechanical purification process can then be used as fertilizer after a composting stage. Different types of biological wastewater treatment traditionally exist in East Asia in the form of aquaculture, which is combined with irrigation farming. In India, for example, 30% of the wastewater in sewage farms is used for raising carp and Tilapia (Lehn et al., 1996). Calcutta disposes of 680,000 m^3 daily in the surrounding wetlands by means of aquaculture. The coliform bacteria count is reduced from 10 million per 100 ml to 10–100 per 100 ml during the flow. The water purified in this way can then be used for irrigation, thus demonstrating the efficiency of a simple user cascade.

4.5.4 Development trends and research needs

Future development should be geared to the avoidance-reduction-recycling hierarchy of objectives and to the realization that only an holistic view will secure sustainable development and use. Future research and development work must therefore focus on the following areas (Kobus and de Haar, 1995; Scherer and Castell-Exner, 1996; Bernhardt, 1993; Wichmann, 1996; Rudolph et al., 1995).

In the field of water supply:

1. Avoidance of anthropogenic contamination of untreated water by further development of technical measures aimed at eliminating problematic substances, such as plant protection agents, nitrogen compounds, heavy metals, drugs and leachate from hazardous waste sites.
2. Prevention of harmful bacteria in untreated water, such as *Cryptosporidia*, *Giardia* and *Legionellae*, by clarifying their source, incidence and behavior during treatment, and by developing reliable methods for eliminating such health hazards.
3. Solving the problems of reinfestation, biofilm formation and corrosion in distribution networks by identifying the interlinkages between pipe materials, on the one hand, and water quality and reinfestation potential, on the other. Developing quality control systems in distribution networks and economical remediation and renewal methods for the network components concerned.
4. Avoidance of water consumption and network losses through the exploitation of all economically acceptable savings potentials in the various consumer groups, as well as optimization of water distribution and water treatment equipment by minimizing network losses and internal consumption. Development of feasible techniques for multiple use and life cycle management, particularly in the industrial sector, as well as improvements in agricultural irrigation techniques.
5. Further development of semi-natural water treatment processes that involve no side effects, with the aim of minimizing the use of chemicals and energy, as well as the volume of residual substances requiring disposal.
6. Further development of biological in-situ treatment methods in order to replace precipitation and flocculation with biological methods and/or with chemical oxidation processes involving the use of catalysts. Further development of adsorption techniques using renewable adsorbents, as well as processes that replace chlorine disinfection with ultraviolet radiation or membrane processes. The greatest potential in this context is seen in bi-

BOX D 4.5-1

Adapted technologies for water supply and disposal in developing countries

A broad range of instruments is available to all industrialized countries and in particular to the urban agglomerations in the developing countries. Further advances will require technical optimization of known processes and more intensified development of high-tech solutions. However, the consequence is that technologies are becoming more complex, are cost-intensive in many cases and therefore require appropriately qualified personnel as well as a highly developed infrastructure.

The situation in the rural regions of the developing countries is completely different. In Africa, where at present only 46% of the population has access to a public drinking water supply, and a mere 34% are connected to a controlled sewage disposal system, water supply and disposal technologies should be culturally and locally adapted, financially feasible, technically simple, and quickly installed through self-help on the part of the population (BMZ, 1995). Examples of such adapted methods are the slow sand filtration system for water treatment, the further development of solar distillation for seawater desalination, the use of anaerobic biological processes for purifying domestic and industrial wastewater, as well as low-yield aerobic methods, such as plant-covered soil filters or aquacultures, for purifying domestic sewage. There is also a substantial need for research into optimizing the purity levels achieved by methods for separating feces from wastewater in order to spread it as fertilizer on cropland, and by methods for utilizing partially purified domestic sewage for agricultural irrigation (Lehn et al., 1996; BMZ, 1995).

An additional effect that can be observed in the developing countries is that losses within water supply networks rise considerably shortly after the establishment of the supply networks. To eliminate such losses it is necessary to improve organizational structures, e.g. by introducing a stringent acceptance test for construction work as well as further development of pipeline technology towards simpler and more reliable sealing techniques. Specific per capita consumption tends to rise dramatically after the introduction of central water supply facilities. Greater efforts must therefore be made, especially in arid regions, to develop and implement water-saving techniques (particularly in agricultural irrigation) and multiple water use. Examples of the latter include the use of partially purified domestic sewage as agricultural irrigation water or for groundwater recharging (BMZ, 1995).

ological processes, membrane processes as well as chemical oxidation with catalysts.

In the field of water and sewage disposal:

1. Further development of production-integrated environmental protection measures for the avoidance of unnecessary water pollution. Pushing forward the construction of mixing and equalizing tanks to provide for uniform wastewater quality. Further development of in-house preliminary treatment with the aim of improving suitability for biological treatment. Examples include the application of membrane processes for recovering reusable substances from wastewater, or the use of anaerobic biological processes and chemical oxidation processes. Special emphasis should be placed here on the development of sector-specific concepts for treating wastewater sub-flows.
2. Reducing the contamination of receiving waters caused by precipitation runoff from sealed areas. Development economical and reliable systems for centralized or decentralized use and infiltration of minimally contaminated runoff from roofs.
3. Upgrading sewers to controlled reaction chambers in which runoff is homogenized and pretreated. Use of membrane processes for improved retention of solids, combined with substantial reduction of spatial requirements.
4. Reduction of bacterial contamination in treatment plant effluents by refining and applying methods such as ultraviolet radiation, ultrasound, membrane processes or subsequent biological treatment stages.

4.5.5 Recommended action

The Council's recommendations for solving global water problems are based on the avoidance-reduction-recycling hierarchy of objectives and on the realization that only an holistic view will secure sustainable development and use. It is crucially important that water resources be protected and used in efficient ways. Specifically, the Council recommends the following action:

- Avoidance of anthropogenic contamination of untreated water by applying legal and organizational instruments against the deposition of problematic substances such as pesticides, nitrogen compounds, heavy metals, drugs and infiltration water from hazardous waste sites.
- Avoidance of water pollution, alongside recycling and multiple use of water by trade and industry.
- Fostering the further development and dissemination of simple water technologies for developing countries.
- Promoting optimized agricultural irrigation systems with the aim of water conservation.
- Support for low-cost and efficient methods for purifying and disinfecting domestic and industrial wastewater.
- Semi-natural methods for purifying domestic sewage, such that residues are minimized.
- Use of artificial groundwater recharge methods involving minimally contaminated roof runoff and purified domestic wastewater.
- Support for projects aimed at using partially purified domestic sewage for agricultural irrigation.
- Reduction of losses within water supply networks and of consumption within the water works themselves.
- Fostering decentralized, adapted, low-tech and low-cost technologies for the treatment of drinking water and the purification of wastewater in rural regions of developing countries.

5 Solutions to the global water crisis

5.1 Guidelines for the "sound management of water resources"

Greatest possible efficiency while observing the imperatives of equity and sustainability – The "guard rails" philosophy – Recent trends in international resource management and international law – The "hydrological imperative"

5.1.1 The guiding principle developed by the Council

The global water problem, like the threat to global soil resources (WBGU, 1995), is a mosaic of regional and local crises that can be classified extensively in terms of the syndromes of global change. Measures for resolving critical situations are therefore most effective when aimed at the elimination ("curing") of the respective syndrome, or at least the mitigation of its water-specific symptoms. This Annual Report puts forward recommendations in this respect.

These recommendations are based, regardless of their context-specific character, on common *normative guiding principles* derived from an umbrella *model for the "sound management of water resources"* in a world in transition.

Such a global model takes account of the fact that current water-related problems exhibit national and international features – in respect of causes (world trade, export of lifestyles, anthropogenic climate change, etc.) as well as consequences (migration, loss of biodiversity etc.). This makes it all the more necessary to generate a global consensus for the management of this, the most precious of all resources.

The model advocated by the Council for the sound management of water resources can be expressed in the following common denominator, namely to

achieve the greatest possible efficiency while observing the imperatives of equity and sustainability.

This dual objective acknowledges the fact that water, like no other environmental resource, is both scarce and essential, in other words an economic asset and a life-giving food. Its properties as essential and life-giving define sociocultural and ecological constraints to economic activity – referred to elsewhere in this Report as guide rails or "guard rails". The sociocultural guard rails comprise the principle of equity, both inter- and intragenerational, the principle of self-determination of human beings and their participation in decision-making, the right of all human beings to their minimum water requirements, and the principle of minimizing as far as possible the risk of disasters, such as floods. The ecological guard rails encompass the principle of conserving the aquatic ecosystems placed under international protection (the world's natural heritage, designated Ramsar areas, etc.) in their entirety, and of securing the key functions of other aquatic ecosystems. These sociocultural and environmental guard rails, so to speak the minimum demands on the part of humankind and nature, define society's action space, the permissible range within which water resource management must operate if it is to comply with the model.

The fact that freshwater is a scarce good demands that water resource management be pursued with maximum efficiency within the guard rails and society's action space. This condition must be met before the transition can be made from "permissible" use to "sound" or "wise" management of water resources. The search for beneficial and maximally efficient management of water resources on the part of society at large and through the operation of economic forces must occur with a minimum of hindrance within the action space afforded to society. However, efficiency can only be achieved if appropriate institutional, technical and educational conditions are met. This is shown in diagrammatic form in Fig. D 5.1-1, where the main constraints on the "sound management of water resources" are portrayed.

At various times in the past, the Council has advocated and explained (e.g. WBGU, 1996) the "guard rails" philosophy as a suitable approach in the shap-

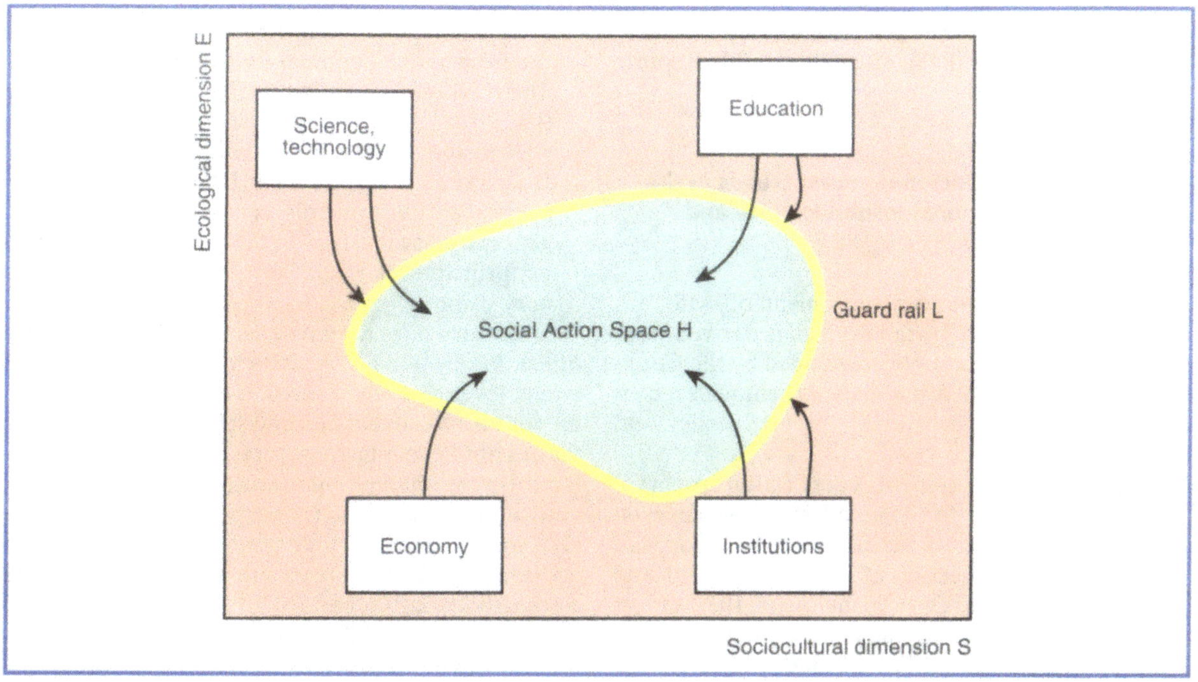

Figure D 5.1-1
The "guard rails" philosophy advocated by the WBGU. The action space for the "sound management of water resources" is shaped and constrained above all by the sociocultural (S) and ecological (E) dimensions. Normative requirements of society define socioenvironmental guard rails or guide rails that demarcate, in the form of contour L (S, E), the permissible action space H as a sub-domain of the possible range of action. Institutional, educational and scientific aspects (the arrows leading to contour L) must be observed in defining and complying with the guard rails. The use and monitoring of freshwater resources must be within these guard rails and the social action space H if they are to be optimized by efficient management, which is supported, in turn, by technical, institutional and educational measures (i.e. the arrows leading into the social action space H).
Source: WBGU

ing of environmental and development policy. In the terminology of integrated assessment, this general approach corresponds to a cost-optimized strategy as opposed to the more desirable but basically unfeasible cost-benefit strategy.

5.1.2 Normative guidelines for sound management of water resources

In what way is it possible to operationalize the model and the sociocultural and ecological guard rails propounded by the Council? On the basis of the analysis in Sections D 1 to D 4, the Council's model can be differentiated in terms of normative principles that constitute a "hydrological imperative":

1. The basic supply of drinking water and water-related sanitation facilities to present generations must be safeguarded.
2. The global stock of freshwater must be conserved for future generations; where non-essential fossil reservoirs are mined, it is imperative to secure the long-term substitution of such sources.
3. Fair access and utilization rights must be guaranteed, also with regard to transboundary freshwater resources.
4. Damage inflicted on other people due to influences on water quality or runoff characteristics (flooding) must be prevented.
5. Cultural identity and the right of political self-determination in the management of freshwater resources must be respected.
6. All aquatic ecosystems under international protection must be preserved in their entirety.
7. The function of other aquatic ecosystems must be secured – e.g. with measures to safeguard the quality of water – as a further condition for sustainable water management of these systems.

Proceeding from these normative principles for the sound management of water resources and from the solutions to the water crisis that the Council elaborates in the sections that follow, the key recommendations for research and action are summarized according to policy field in Sections E 1 and E 2.

Firstly, however, we describe in brief how the Council's model and guidelines are already reflected in recent trends within international resource policy and international law. These recent developments are dealt with in greater detail elsewhere in this Report.

5.1.3 The model as reflected in recent trends in the fields of international resource policy and international law

The model for sound management of water resources outlined here by the Council is derived from the "guard rails" philosophy developed by the Council since its inception; it is also congruent with recent developments in international resource policy and international law.

1. Germany's development cooperation program has been focused for some considerable time on the issues of poverty eradication, education and training, and protection of the environment and natural resources. Over many years, the field of environmental and resource protection alone has accounted for 30% of the assistance granted under bilateral financial and technical cooperation. At the beginning of the International Drinking Water Supply and Sanitation Decade, the Federal Ministry of Economic Cooperation and Development (BMZ) adopted a "Sectoral concept for water supply and sanitation measures in developing countries", in the framework of which numerous water-related projects have since been launched.
2. In 1996, as part of the its Framework Programme in the Field of Research and Technological Development, the EU set up the "Environment-Water Task Force". The aim of the task force is to develop a European strategy for the sustainable management of water resources, to improve the international competitiveness of companies operating in the water resources field, and to concentrate EU policymaking on issues of key importance.
3. In 1997, the United Nations Commission on Sustainable Development presented a "Comprehensive Assessment of the Freshwater Resources of the World" to the International Water Conference in Marrakesh. The report drew attention to the increasing number of countries facing severe water stress.
4. The Council's guard rails approach is also compliant with the principle of national sovereignty within international law. Referring to "guiding principles" and "guard rails" for policymaking respects the sovereign rights of nations, something that is not necessarily the case with definitive recommendations. The model for sound management of water resources takes account of recent trends in international law and could help achieve the urgently needed specification of the legal concepts of "sustainability", "intergenerational equity" and the principle of "optimal utilization" of resources. The principle of maximum efficiency of water use that forms an integral part of the Council's model was postulated as early as 1972 in the Stockholm Declaration on the Human Environment and expressed in the principle of "optimal utilization". This is a clear illustration that operationalizing legal principles is reliant on previous interdisciplinary work.

Water provides a prime example of how resource utilization by human societies has now reached such a high level that any transition to sustainable development is effectively blocked. The main problem so far has not been human impacts on the global hydrological cycle, but on local resources, thus imposing severe limits on the options open to future generations. One way of specifying intergenerational equity in legal terms has already been discussed with reference to transboundary groundwater deposits and "maximum sustainable yield". The idea is a simple one – the rate of withdrawal may not exceed the rate of renewal. "Maximum sustainable yield" would be recommendable as a basic "guard rail" for all types of waterbodies.

The utilization of transboundary freshwater resources also harbors a potential for international conflicts, for example between the upper and lower riparians of a river. The criteria of "fairness" and equitable access to water is enshrined in international law in the principle of "equitable and reasonable utilization". The scope for negotiation when reconciling opposing national interests comes up against limits when one state's interest in covering its essential water requirements conflicts with another state's interest in economic development.

The Council's "guard rail" approach accords with human rights as well. The guide rail here must be the universal human right to coverage of essential water needs, regardless of whether it is the state's duty to provide a safe and adequate supply of water to each individual. When regulating the distribution of water, no government may deny fair access to water to specific regions or population groups. A legal assessment as to whether a state has violated the principle of fair access to water for all must focus on whether or not the state in question has created the administrative and logistical conditions that could be reasonably expected from any state in similar circumstances, under similar natural conditions and with similar capacities. The greater the degree of effective control exercised by a government over its territory, the more stringent the criteria to be applied.

5.2 Sociocultural and individual conditions for water resource management

Water cultures – Significance of water – Culture-forming element – Western understanding of science – "Repair ecology" – Monetary benefit assessment – Water rules – Centralized water management – Water myths – Design element – Individual patterns of behavior – Public perception neglects water – Precaution instead of avoidance – Standards of cleanliness – Lack of incentives for action – Consumption-oriented lifestyles – Invisibility of water pollution – Lack of knowledge – Habits – De-emotionalized water resource management – Consequences of inappropriate behavior barely perceptible

5.2.1 Water cultures: Sociocultural contexts of water resource management

The way in which people manage water resources depends not only on ecological conditions, but also to a considerable degree on the respective sociocultural environment within which they grow up and behave with respect to water – i.e. on their water culture. The diverse significance that has always been attached to water indicates how comprehensively this sociocultural context must be understood (see Box D 5.2-1).

The sociocultural context for water resource management may vary considerably, depending on the location of a society in time and space. In accordance with the specific characteristics of the respective water culture, certain water-related values as well as patterns of perception, assessment and behavior prevail at all levels of social aggregation, from the individual to commercial enterprises and even to politics, with divergent environmental impacts in each case.

The water culture of a society, as the value context that determines to a decisive degree how water resources are managed, is by no means a fixed variable. Rather, it is the product of the manifold interrelations between ecological conditions, on the one hand, and the various sociocultural spheres (politics, economics, religion, etc.), on the other, and is subject to constant change. Water itself, whether as a river or sea, as flood or drought, has always exerted a major influence on the particular manifestations of water culture (see Box D 5.2-2).

The sociocultural value context of water resource management can be differentiated according to various interrelated dimensions that can be observed in different "mixes" in societies that are spatially and/or temporally distinct from each other. The most important of these dimensions, which together constitute the water culture of a society, are

- the scientific and technological dimension,
- the economic dimension,
- the legal and administrative dimension,
- the religious dimension and
- the symbolic and esthetic dimension.

The water culture of a particular society can be described analytically by identifying the specific features in each dimension and by weighting these according to their significance for that society. Viewed in this way, a society's water culture is a value context that codetermines to a greater or lesser extent any water-related activities. However, since the people within a society give little thought to their own water culture, they are usually unaware of it. This makes it difficult to find appropriate solutions to water problems. Only by taking a closer look at other regions and cultures of the world does one realize that social organization can be based on entirely different value systems. In a global analysis such as the one adopted by the Council, this divergence in sociocultural contexts necessitates a differentiated, culturally contextualized instrument mix as part of a sustainable water policy.

5.2.1.1 The scientific and technological dimension

The influence of science on Western society's management of water resources goes far beyond the discussion of findings or establishing the fundamental basis for water technology and water analysis. It also creates generally accepted criteria for decision-making and assessment, and scientists are members of many policymaking committees in the field of water resource management. Science continues to enjoy a high level of credibility within society. The scientific approach is often viewed as the ultimate authority on which all decisions must be based. This creates an incentive for the "scientification" of controversial issues on the part of decision-makers, as a convenient way of ignoring the play of forces within society.

However, scientific experts are not infallible, and their actions are not always devoid of personal interest. In addition, scientific treatment of the complex and interlaced problem areas of global change, including water problems, has substantial structural as well as inherent shortcomings and has proven to be inadequate for its task (WBGU, 1997). The scientific water discourse in most countries of the Western world displays a marked bias toward the natural sciences, while the human dimensions of water resource management, by contrast, are rarely considered.

BOX D 5.2-1

Manifestations and meanings of water

Since primeval times, the diverse manifestations of water have been the source of manifold differences in its meanings that are reflected in the "subjective culture" (values, attitudes, myths, norms, etc.) as well as in the "objective culture" (visible and comprehensible pictures and symbols of water, painting, literature, garden landscaping, urban landscaping, architecture) of a society. Hartmut Böhme brings alive the abundance of manifestations and ways in which water is experienced:

"Water comes out of the earth as a spring, moves as a river, remains stationary as a lake, is the sea in its eternal serenity and endless movement. It transforms into ice and steam; it moves upward through evaporation and downward as rain, snow or hail; it flies as a cloud... It splashes, roars, sprays, gurgles, glugs, whirls, plunges, surges, rolls, trickles, hisses, undulates, seeps, ripples, babbles, reflects, streams, drips... It is colorless and can take on all colors. Through thirst it arouses the most elementary desire, runs refreshingly down the throat; it is tasted, drunk in swallows, gulped down... During the act of swimming it conveys an idea of what floating, gliding, weightlessness are like. The embryo lives in water. Water cleans the body and things, and indeed the soul and spirit... In the transition between liquid and solid it forms unusual zones: slimy, greasy, jellylike, slippery, muddy, boggy, mushy – aggregates, without which we would scarcely know what disgust is, for example... It is shapeless, adapts to any form; it is soft, but stronger than stone. It creates shapes: valley, coasts, grottoes. It shapes landscapes and life forms through extreme scarcity (deserts) or periodic excess (rainy season). It frightens, threatens, injures and destroys people and their facilities by means of floods, storm tides, hail... Thus, water holds death and gives birth to all life... Water is disease (...) and water heals... Water (challenges) human inventiveness: river regulation, dam construction, irrigation facilities, sewer systems, shipbuilding... Water as a traffic route, as a natural trade route... Water as a bulwark against enemies or as a strategic basis of power: Venice, England... Underwater: realm of the depths, of secrets, of the abyss... Water and its treasures: the frozen freshwater reservoirs of the Antarctic; the food resources; the ... underwater resource reserves.... Water and law: water use regulations in cities of antiquity; fishing law in the Middle Ages..., the domestic and international legal regulation of all water on the Earth today... Water and the divine: primeval substance of the Creation; chaos, vanquished by God. The gods and goddesses of the seas and rivers: the water-nymphs, mermaids, sirens... Rhine romanticism: Loreley. Water and femininity... Water, the subconscious and the dreams. "Man's soul resembles water" (Goethe). Water and time: "You cannot step twice in the same river" (Heraclitus). The arts: water and horticulture; water in landscape painting; water poetry" (Böhme, 1988).

The "Western" understanding of science outlined above is also exported to many other cultures through the scientific support of development cooperation and the training of local experts from developing and newly industrializing countries at Western universities.

The culture of technology development and its application in solving water-related problems is closely linked to the prevailing culture of science and research. It would appear that water resource management is virtually impossible nowadays without complex technology, from extraction and treatment to transport and purification. When water problems are viewed from a predominantly technical perspective, this leads to a situation in which misfunctioning water management systems are analyzed and "optimized" primarily with respect to their technical components, while interactions with natural or social factors are effectively ignored. In turn, unwanted or unforeseen consequences of such optimization are then tackled with technological means ("repair ecology"). An example of this is the constant rise in technical effort required in order to comply with quality standards for drinking water. Contamination is mainly countered by means of increasingly sophisticated filtration methods, by connecting to remote water supply networks, or by tapping "fossil" water reservoirs, rather than by focusing on the actual causes of the contamination or even taking precautionary measures to protect drinking water resources. Problems that arise are thus shifted in space and time, but not solved. Therefore, although technical optimization of systems is not automatically good water policy, many water management decisions are still based exclusively on technical criteria.

BOX D 5.2-2

Water as a "culture-forming element"

Throughout human history, water in its numerous forms has had a significant influence on cultural development and has shaped individual as well as collective perceptions, values, and patterns of behavior. This can be illustrated with some examples (Assmann, 1996; Fischer, 1988; Neubauer, 1995; Schua and Schua, 1981; Smith, 1985):

- The transition in humankind from a nomadic to a sedentary existence, which was accompanied by the growing of foodstuffs and the raising of livestock, was only possible along watercourses because these were the only places where the water needed for drinking and irrigation could be found in adequate and reliable quantities.
- In many parts of the world, the development of division of labor in antiquity was characterized by the migration of entire peoples to the valleys of large rivers, as in Egypt (Nile), Mesopotamia (Euphrates and Tigris), India (Indus) and China (Huang He). The growth of locally adapted irrigation technologies, which subsequently influenced the creation of many other technical facilities and shaped societies and social orders, began on the great rivers.
- In ancient Egypt, the water of the Nile was thought to emanate from a godhead and was therefore thought to embody the origins of life. The cyclical floods of the Nile, which guaranteed fertility for the otherwise barren countryside, not only influenced agriculture, but were also the object of religious convictions and the basis for people's perception of time.
- In many cases, such as Bern or Freiburg, cities were established on rivers because, on the one hand, they offered favorable sites for conducting trade, and on the other, the river could be used as a natural protective wall against enemies.
- Since the 9th century, Frisians have been engaged in a struggle with the North Sea to protect their estates against floods and to reclaim land from the sea. The battle with the sea was repeatedly lost, with thousands of people losing their lives in the process. Nonetheless, the technologies required to solve the problems were continuously refined.
- Until recently, there was no freshwater on small islands off the coast of North Frisia, apart from precipitation. This necessitated a complex system for collecting and storing water, which then formed the basis for the artificial hills on which people lived. The permanent state of scarcity shaped the needs and knowledge of the island inhabitants and led to an extremely frugal management of water as a valuable asset, a characteristic that has remained among the older residents despite the improvements achieved in the meantime.

Development policy is also characterized by this technocentric approach. Yet the systems installed in the water supply sector, whether drinking water supply or irrigation facilities, are faced with completely different ecological, social, legal and economic conditions in the developing countries. The transfer of technical systems is accompanied by a transfer of specific values that frequently conflict with those prevailing locally (see Box D 5.2-3). Moreover, the structural deficits of one-sided technically optimized water resource management are more apparent under the less favorable conditions in poorer countries. The worldwide propagation of the water closet, for example, is now considered to be extremely problematic, because enormous quantities of fecal microorganisms flow into rivers via sewer systems and treatment plants, resulting in a substantial spread of pathogens. When rivers in Germany arc excessively contaminated, the result is usually a ban on bathing, but in India, for example, such contamination has a major impact on the supply of drinking water, which is mainly taken from rivers. However, a technical solution to the problems caused by WCs, namely treating river water to produce drinking water, is exorbitantly expensive for poorer countries unless external support is given (Amsel and Lanz, 1992).

5.2.1.2 The economic dimension

Modern economies treat water as a commodity. Here, water is viewed as a resource, a means of satisfying human needs. Since the use and consumption of water benefits people, whether directly as a consumer good or indirectly as an investment good, there is a demand for it. Under certain conditions (if the principle of exclusion does not operate or property rights are inadequately allocated), water resources may be allocated inappropriately. As a consequence,

BOX D 5.2-3

Water without users: irrigation facilities in Peru

The plateau at Lake Titicaca on the border between Peru and Bolivia, the Altiplano, is considered to be marginal, barren land. However, the people living here have always adapted themselves to these conditions by applying their own cultivation methods. In the 1980s, Western agricultural advisors attempted to transform the region north of Lake Titicaca into fertile land by means of artificial irrigation systems. They erected dams, constructed irrigation channels and created terraces. In this process they introduced capital-intensive systems based on agricultural know-how from Europe and America, while planners took no note of the traditional techniques of the native population.

The great majority of these projects are considered failures today. At least nine large-scale irrigation projects north of Lake Titicaca covering a total of over 100,000 hectares are, for all practical purposes, not used at all (Wicke, 1993). The Western planners, who were not familiar with local agricultural-ecological conditions and whose bulldozers covered many terraces with unfertile soil, were responsible for this situation. However, the main reason given was the lack of interest in cultivation of the newly created fields on the part of the rural population, which was entirely incomprehensible to the Western agricultural engineers. In contrast to the scientific, technical perspective of Western nations, for the native Quechua and Aymara the Altiplano is "a cosmic landscape, a structured cultural area full of holy places (huacas), the home of the Mother of the Earth, pachamama. As part of a complex system of values based on reciprocity, the Earth gives the farmers their crops; as an expression of their thanks, the farmers return these rich gifts (pagos) to the Earth" (Erickson, 1992). It is not surprising that Western irrigation technology, which focuses on damming rivers and diverting them to other valleys through tunnels in order to grow unknown cereals imported from the West, is not compatible with such a view of the world.

the water that used to be a free environmental asset becomes a scarce asset in many places, a circumstance that should really be reflected in the price. If external effects are involved, however, water prices may lose their function as indicators of scarcity. Therefore, the basic solution for water problems lies in finding adequate scarcity prices, on the basis of which the mechanism of the market regulates the allocation problem "automatically" and maximizes the overall social benefits.

It is of fundamental importance from the economic perspective that the (monetary) valuation of the benefit of water be based on its various functions for humans. Water and its functions thus take on the nature of commercially available commodities, and alternative uses are assessed according to cost and benefit. This way of thinking must inevitably formulate an abstract definition based on the complex significance of "water", from which economics merely postulates that it is expressed in differing degrees of willingness to pay. An overemphasis of this approach may lead to a suppression of water "uses" that involve more than the mere use of resources (e.g. the installation of artistically designed fountains, or the use of water as an instrument for esthetically pleasing garden and landscape design).

5.2.1.3 The legal and administrative dimension

Rules for the management of water resources number among the oldest social conventions adopted by humankind. The tremendous variety of hydrological, climatic, geological as well as religious and social conditions was reflected in an equally wide variety of water rules and the resultant laws pertaining to water. Not all of these rules and regulations have been codified as written law; many are accepted as unwritten law. Each culture, for example, has its own unique washing and cleaning rituals which strongly influence people's behavior.

In many former colonies, the legal system imposed by the colonial powers displaced the original laws. It is not surprising, therefore, that the application of European systems of water law to countries in Africa, Asia or Latin America led to conflicts with existing values and customs, as well as with the ecological conditions there. In many cases, the new laws undermined the traditional uses of water (see Box D 5.2-4).

Laws are established within a complex network of interests and govern how water is managed by forbidding certain actions and tolerating others. For example, it is of considerable importance whether the contamination of groundwater is strictly prohibited, or whether it is subject to approval and hence sanc-

BOX D 5.2.4

Kenya: From the commons to private property

In Kenya's dry northern region, traditional nomadic lifestyles have developed that use soils, watercourses and water sources on a seasonal basis only. Farming is organized at the village level and is based on shared use of the existing water and soil resources. In accordance with this principle, both land and water are equally available to all members of the village.

Kenya's current water law was adopted in 1951 during British colonial rule. According to this law, a claim to water is based on ownership of land and thus differs fundamentally from the view prior to that. Kenya's population did not know land ownership. The use of soils and water was regulated through access rights. Water was only one of several resources that people had access to on communally managed land. The relationship to water was based on need and use. Traditionally, water was much more than an economic commodity. Emphasis was not placed on rights, but on obligations.

Under British law, control over water in northern Kenya was taken out of the hands of the population and organized by the central state. Use of water was no longer possible without formal official permission. Traditional access to streams and water sites had suddenly become subject to approval, and only water whose use was officially registered enjoyed state protection.

tioned – albeit within limits. By means of a legal act in this case, groundwater turns from an extremely valuable asset that must be stringently protected into a pollutable resource. Although, for example, the German Water Management Act requires the exercise of care in the use of water and waterbodies, water law has increasingly developed into a law that exclusively governs user rights. Requirements to protect and care for resources are regarded more and more as mere obstacles to use and as locational disadvantages, and no longer as an essential basis for human existence, even with a view to future generations. Particularly in developing countries, the requisite regulations requiring the protection and care of resources may even act as barriers to development.

Government administration of water resource management has changed fundamentally as the latter's complexity has grown. Subsequent to the traditional supervision of brooks and rivers by government authorities, the use of water in households has also been increasingly placed under government control. The resultant positive impacts (improvement of the supply situation, simplification of quality monitoring, etc.) contrast with the fact that only a few older people today possess indigenous knowledge about the origin of the water and the special features of the piping and storage system. The centralized water management system that has been installed in many places over the years, along with extensive centrally planned water supply networks, is virtually invisible for the citizens: they cannot see anybody playing an active role. The source of the water flowing out of the tap at home remains unknown. Thus, the administrative structure fosters a perception of water as a technical product that can be bought and apparently flows straight from the tap, devoid of any origin.

In such a management system, water policy decisions can be made "at the expert level", to the exclusion of the public for all practical purposes. The system also facilitates the customary practice of keeping relevant information about use and contamination of waterbodies largely secret. The resulting lack of transparency and cooperation on the part of the authorities has extensively undermined both the knowledge and the interest of the population regarding water issues.

Similar to the structure of water law, the highly centralized approach of the water authorities is found not only in the Western industrialized countries, but frequently in their former colonies as well.

5.2.1.4 The religious dimension

The multi-faceted meanings attached to water in myths and religions across all historical periods illustrate the ambivalent nature of this substance for humanity. In religious contexts there is both reverence and fear of water, since it stands for fertility and for life (as in rain for the fields or water for drinking), or for the threat posed by uncontrollable powers, e.g. in the form of floods and heavy rains, or through its complete absence in times of drought (inter alia, Schröer and Staubli, 1995; Wilke, 1995; in summary form Gaidetzka, 1996b).

In the Creation myths of most religions, water is viewed as the primal element of life, and associated in this way with birth and fertility. In the religious tra-

BOX D 5.2-5

New Zealand: The water culture of the Maori

For the Maori, the original inhabitants of New Zealand, water possesses a deep spiritual meaning. The gods as well as the ancestors are present in it. Rain means that the god of the sky, Ranginui, is crying. He is crying about the Mother of the Earth, Papatuanuku, whose sighs rise as fog. The dead ancestors as well as the mythological gods live in the rivers. For the Maori, the waters of their homeland are part of their identity: whatever happens to the water, also happens to them. This applies both to outside influences and to personal actions.

It is typical that water bears different names according to the circumstances. Waiora, the holy water, falls as rain and comes up from the depths of the Earth as a spring. Only waiora is suitable for ritual ceremonies, e.g. for birth or death. River water is usually waimaori: good and healthy to drink, for catching fish, and as a cure for illness. Polluted or diverted water becomes waikino or even waimate: hazardous water that cannot be used for ceremonies, as a cure or for drinking.

Thus, every river, every lake, every stream is imbued with a more or less curative or dangerous spirit (wairua). Just like all other things – Earth, sky, animals, plants – waters are personified and regarded at the same time as living elements of a larger, universal organism (Taylor, 1987).

ditions of desert peoples, especially, rain is perceived and understood as a blessing sent from God in answer to the prayers and rituals the people offer up. Excessive rain, on the other hand, such as the Flood, is interpreted as a sign of God's wrath at the acts of humans. In keeping with its physical properties, water is associated in a religious context with the removal of impurity, for instance by ritual cleansing or by ascribing healing powers to certain sources. This is closely connected to the conception of water as the water of life, counteracting the passage of time, and in Christian baptism, for example, becomes a sign of conversion, renewal and integration into the congregation of believers. In addition, positive as well as negative images of water can be found in numerous religious conceptions of God, paradise and the hereafter.

In contrast to the almost exclusively "secular" management of water resources in most industrialized countries, the religious dimension of water culture is still maintained in many other cultures of the world. However, the attribution of religious meanings to water does not necessarily protect it against contamination, even in these cultures (see Section D 5.2.3).

5.2.1.5 The symbolic and esthetic dimension

Apart from the religious context, we encounter water in a variety of symbolic and figurative meanings, for instance as the epitome of purity, naturalness and freshness, as a symbol of constant change and transitoriness (*panta rhei*) and as a gentle yet unrelenting force that overcomes all resistance in the course of time (Selbmann, 1995). The various forms of water have inspired innumerable artists – composers, painters, sculptors and poets – to create great works of art, which in turn, as visual or acoustic "water images", have an influence on the way we see and perceive water.

In some cultures, the different meanings attached to water are reflected in the use of different words to refer to the same physical substance, with direct impacts on the respective ways that water is used (see Box D 5.2-5).

The esthetic meaning and significance attached to water has certain affinities with this symbolic dimension. These esthetic elements are expressed, for example, in the use of water as a design element, be it in landscapes, gardens, parks or in urban areas as an artificial lake, waterfall, pond or artistically designed fountain. What is common to them all is that "living" water in all its variations is obviously perceived as having a certain quality that many people are able to appreciate. The attractiveness of water in the environment far exceeds that of other materials or elements (Pitt, 1989).

It would require major interdisciplinary and/or transdisciplinary research efforts to define a given society's water culture in all its dimensions – science, technology, economy, law, administration, religion, symbolism, esthetics – and the respective consequences for water resources. All that is currently available in this context are a number of case studies, many of which are rather anecdotal in character. A systematic analysis of water culture, which has yet to be done for most societies, would highlight the fact that many decisions of relevance to water resources are made within a very tight framework of values, and that little or no thought is given to other important factors, such as future environmental impacts or public participation.

Thus, any sustainable water resources policy must first be aware of the sociocultural value context, i.e. the water culture within which it aims to have an impact. Practitioners in the field of development cooperation should realize that water projects interfere in deeply rooted cultural systems that cannot be understood in technical terms alone. The existing sociocultural framework must be taken into account when elaborating differentiated strategies for the sustainable management of water resources in the various sectors.

Modifying the individual patterns of water-related behavior characteristic of private households requires knowledge of the general factors that condition water-related behavior. This pertains primarily to the role of human behavior as it relates to the problems of *scarcity* (Section D 5.2.2) and *contamination* of water (Section D 5.2.3), as well as in coping with the *threat* that water may pose for people (see Section D 1.6). Most social and behavioral research on these topics has been conducted within the context of a specific water culture as found in Western industrialized countries. This inevitably results in a certain degree of Eurocentrism in the following description, and a confinement in some cases to conditions in Germany. However, this does not detract from the fundamental importance of the factors affecting the way people treat and manage water resources.

5.2.2 Water scarcity and behavior

One of the root causes of water problems is the increasing overexploitation of local water resources, resulting in water scarcity (see Section D 1.4). Such overexploitation may be due to a poor natural supply, excessive consumption, or a combination of the two.

When focusing on human behavior as a cause of water scarcity, it is difficult to isolate quantity-related problems from quality-related problems (see Section D 1.5). On the one hand, a high level of demand on the part of industry and private households for high-quality types of water (e.g. drinking water conforming to strictest quality standards) rapidly leads to scarcity when the quality of untreated water is limited, even in regions with an abundant water supply, like Germany. On the other hand, high water consumption frequently aggravates the quality problem, e.g. because of the large amounts of wastewater that ensue. This interrelatedness should be borne in mind when reading the following sections, where water scarcity and water pollution are treated separately for analytical reasons.

In the perception of the population, water problems play a very subordinate role, both in quantitative and qualitative terms. In the most comprehensive sociological study on the issue of water conducted in Germany in recent years – a large-scale survey in the cities of Frankfurt/Main and Dresden, involving more than 1,000 interviewees (Ipsen, 1994; Glasauer, 1996) – the problems of drinking water supply (quantity) and water pollution (quality) ranked, at 2.1% and 1.2% respectively of all responses to the (open) question as to the most important environmental problem, far behind problems like air pollution (30%), transportation (30%), chemical industry (15.2%) and waste (12.7%). This irrelevance of water problems in the public mind is somewhat surprising, in that the city of Frankfurt had launched a major water saving campaign shortly before, and a broad public discussion of the problem was ongoing at the time of the survey. However, in contrast to the lack of perception of the problem, the same study ascertained that saving water had established itself as a "generally valid norm": 92% (!) of all interviewees felt it made good sense to save water. Nearly two thirds of this group stated water scarcity to be the reason. If the discrepancy that emerges here between a lack of awareness for the problem, on the one hand, and basic social acceptance of saving water, on the other hand, should be substantiated, this would also have implications for the instruments used to achieve the policy goal of reduced water consumption (see Section D 5.3). It would mean that water-saving measures addressing behavior (e.g. providing means and incentives for saving water) are more appropriate for reducing household water consumption than generating an awareness of the problem, e.g. with information campaigns.

In order to investigate the causes of anthropogenic water scarcity, it is first necessary to identify and quantify water consumption behavior (see Box D 5.2-6). If, in a country or region suffering from current or imminent scarcity, private households as "ultimate consumers" are to be induced to make more frugal use of water, then it is necessary to identify the possible driving forces behind varying levels of water consumption (for possible intervention measures see Section D 5.3). Therefore, the systematic approach applied in previous Council Reports (e.g. WBGU, 1994) will be used in the following to examine the various sociological and psychological aspects of water consumption.

Perception

In the perception of people in most industrialized countries, (drinking) water appears to be available in any quantity at any time, provided a price (which is usually relatively low) is paid. Taking the daily water supply for granted is closely related in all likelihood to the invisibility of the underlying infrastructure – water comes into each home "straight from the tap", which is usually the only visible feature of the public water supply. People only become conscious of it when there is a malfunction for some reason.

This was not always the case (Katalyse, 1993; Schramm, 1995). In Germany, for example, the central water supply and wastewater disposal system as we know it today (including WCs and the sewer network) were established only as recently as the 19th century – as a means of combating epidemics. Until then, people could usually experience the scarcity of water resources in a direct way, as members of well cooperatives that emphasized personal responsibility and provided a direct relationship to the resource. Once the water supply system had been centralized, by contrast, the public developed a greater awareness of water quality but delegated responsibility and control over the system to "specialists". Perception of the water supply system became increasingly filtered.

The process of dissociation initiated by the centralization policies of the last century continues along several dimensions, even to this day. Policies requiring a permanent search for new sources of water, as opposed to sound management and precautionary action at local level, enabled a disjunction to develop between the production and use of drinking water – in spatial terms (development of large-scale systems for remote water supply), in temporal terms (consumption of ancient "fossil" groundwater stocks) and in a qualitative sense (physico-chemical production of drinking water from untreated water resources). As a consequence, water problems were shifted to other regions, to other generations and to other environmental media. It became more difficult to perceive interrelations, for example between water consumption in Frankfurt and the desiccation of the Vogelsberg region. No wonder, then, that media reports about "impending water crisis" should cause confusion among the public, when water still pours from the tap, the river bursts its banks, or the fountains in the town continue to splash cheerfully (perceptual ambiguity) (Glasauer, 1996).

Water policies, too, have become increasing dissociated from the people in society. Decisions are made behind closed doors by policymakers for whom citizens are nothing but "trouble-makers" and a nuisance, for example when designating drinking water protection areas. Here, again, at least in a country like Germany with abundant water resources, a situation is created in which public awareness of the water supply does not develop until a conflict or crisis arises.

The situation is different in water-poor countries, where a state of crisis is normality in many cases. In the arid regions of many developing countries, for example, water has always been regarded as a valuable asset, and this in turn led to the development of water-saving techniques in production, irrigation, etc. (Gaidetzka, 1996a). Here, local water availability or water stress are a structural factor influencing, more than all other behavioral determinants, the way that people perceive and manage their water supply.

Knowledge

Due to the lack of suitable data, little can be said thus far about the knowledge of the population regarding quantitative aspects of water consumption. One can assume, however, that ignorance prevails in large sections of the population, not least of all because of inadequate feedback on water consumption and water-saving behavior. One of the results of a survey commissioned by the Federal Environment Agency (BMU, 1996b) on environmental awareness in eastern and western Germany indicates the plausibility of this assumption. Despite the very large range of replies given, only 16% (West) and 25% (East) answered the question about the average daily water consumption in Germany correctly ("101–199 liters").

Attitudes, norms and values

Showers and baths account for around a third of the water consumption in German households (see Box D 5.2-6). This reflects not only the increasing availability of sanitary facilities, but also a change in hygienic needs. Whereas a proper bath was one of the rare luxury goods enjoyed in the Middle Ages, for example, a daily shower or even bath has become a matter of course, and sometimes even a "must" for many people in the industrialized countries today. This shows how social norms, such as cleanliness, a well-groomed appearance and tidiness, can conflict with economical water consumption and/or the corresponding attitude (Glasauer, 1996). The social norms experienced and acquired in one's socialization history and everyday life are major prerequisites for social acceptance and integration. It can therefore be assumed that the individual motivation for water-saving behavior has to be very strong to assert itself vis-à-vis social norms. From both an historical and a comparative cultural perspective, however, standards of cleanliness and/or the associated consumption of water are shown to be very prone to fluctuation (Vigarello, 1988).

Another effect that potentially promotes consumption is the fact that wasteful use of water often has positive connotations (Glasauer, 1996). A good bath relaxes body and soul, and the growing trend to-

BOX D 5.2-6

Water-consuming modes of behavior of private households

In general, water consumption by private households is relatively low in comparison to agriculture and industry (see Section D 1.1). However, in a regional breakdown and in absolute figures significant differences emerge (WRI, 1996). According to International Water Supply Association statistics for 1993, the average consumption of 15 industrialized countries (excluding the USA) ranges from 120 (Belgium) to 316 liters (Australia) per capita and day. In Germany this per capita consumption value declined slightly between 1990 and 1996 at a continuous rate and is currently around 130 liters [Federal Association of the German Gas and Water Industry (BGW), 1997, personal report]. However, indirect water consumption, e.g. through the purchase of consumer articles, is not included in these figures, and usually consumers are even less aware of it than of "visible" consumption (see below). To produce a single car, for example, over 200,000 liters of water is required, and impressive "ecological (water) backpacks" can also be given for other consumer goods (Katalyse, 1993). These hidden water flows are of a global nature: through economic goods (e.g. flowers, citrus and tropical fruits) hidden water flows are imported and exported regularly.

A breakdown of the consumption of households according to types of use indicates the significant savings potential that many experts still see here. According to estimates by Möhle (1983), for example, a third of the drinking water consumption at the beginning of the 1980s was required for toilet flushing, another third was used for taking baths and showers and the last third flowed into use for washing laundry, dishwashing, body care, watering the garden, cleaning rooms, drinking and cooking as well as automobile care. These estimated values must be interpreted cautiously because of, for example, the changes that have been effected in technical standards, the development of water-saving technologies as well as changes in the make-up of households. Nevertheless, it can be assumed that the orders of magnitude and, in particular, the minimal extent to which drinking water of the highest quality is used for appropriate purposes (e.g. drinking and cooking) have changed very little since then.

wards designing domestic bathrooms to reflect a particular lifestyle conforming to the "fun culture" clearly indicates that excessive individual consumption of water is associated with wellness and zest for life (Schramm, 1995).

Incentives for action

Possible incentives for wasteful as well as economical management of water are the price paid for it and/or the rate structure. However, water as an essential commodity usually has a very low price elasticity (Winkler, 1982). Another major incentive for action is to base charges on consumption, both for a water-saving lifestyle and for the installation of water-saving equipment (BMU, 1996b). In apartments anyway, water costs are rarely charged in this way, but are frequently billed at a flat rate like heating costs.

Another experience acting as a "negative incentive" for thrifty use of water resources is that successful water-saving campaigns do not pay off for the population. Instead, they often result in a rise in the water price (Glasauer, 1996). In general, the prevailing incentives favor high water consumption.

Options and opportunities for taking action

The causes of rising water consumption year after year in Germany since the Second World War certainly include the constant improvement of sanitary conditions as well as the growth of water-consuming appliances in the household (washing machines and dishwashers) – in short, in the general availability of infrastructure that promotes water consumption. Consumption levels in the new eastern German states, which are still at a low level because of the lack of baths or shower facilities, among other things, seem to confirm this (Schramm, 1995). These experiences are of importance, particularly with respect to the developing countries and their future water consumption. Materialist lifestyles that emphasize the virtues of consumption seem to lead "automatically" to high levels of (direct and indirect) water consumption.

Perceptible feedback

Even though a central water supply and wastewater disposal system offers a number of benefits, water consumers are frequently not aware of the consequences of consuming or saving water under these specific conditions. Whereas water shortage and quality problems in the well cooperatives of the 18th century were felt directly by the households concerned, today this is often prevented by a lack of feedback on consumption (partially due to the absence of means for determining household consumption), by water prices that do not take consumption into account, as well as by the constant availability of drinking water that is ensured at considerable, but invisible, expense. Lanz and Davis (1995) cite the Greek island of Alonnisos, where a changeover in the supply system from decentralized cisterns to a centralized water supply led to a rise in water consumption, as exemplifying the consequences of an absence of such feedback mechanisms. The direct perception of the impacts of water consumption is important for people's behavior, especially in areas where water is already scarce, e.g. in arid regions.

5.2.3 Water pollution and behavior

In some cases, pollution of standing and running waters as well as groundwater resources is closely linked to the problem of water scarcity. The reasons for this include direct interference in the water resources as well as intentional or accident-related discharge of pollutants into waterbodies. However, the causes of water pollution are increasingly found "far away" from the waterbodies themselves, e.g. soil contamination caused by intensive agriculture (groundwater pollution due to protracted use of fertilizers and pesticides) and leaking landfills, or from air pollution due to emissions of harmful substances from the transport sector and industrial production (problem of spatial and temporal distance) (see Section D 1.5).

There is no other environmental problem for which greater improvements are perceived by the population than for the purity of waterbodies. According to the study by the Federal Environment Agency (BMU, 1996b) already cited, 39% of those surveyed in western Germany and as many as 57% in the new eastern German states see environmental policy as having made the most progress in this field, followed by air, soil, climate, energy and waste. The latter might be due to the fact that interviewees associate the term "waterbodies" more with surface waters (where quality improvements have indeed been recorded) and less with groundwater or drinking water reservoirs. The use of this term may also explain why, in a study conducted by Billig (1994), the issue of "water contamination" ranked third, in response to the question about major future threats, after "hazardous waste sites" and "violence" and followed by events such as "climate changes", "war" and "chemical accidents" (for methodological problems of surveys see WBGU, 1997). However, in the Health of the Planet Survey as well (Dunlap et al., 1993), an extensive survey study comparing different countries, the persons surveyed in 19 of 24 countries designated water quality as one of the three most important environmental problems in response to an open interview question. After air pollution (23 countries) this is the second most frequent reply, far ahead of all other problems referred to. In general, therefore, no clear trend for water pollution can be discerned on the basis of the available studies on the perception of environmental problems.

With the greater focus now placed on the analysis of indirect sources of pollution, including the enormous consumption of water for the production of consumer goods, the relevant patterns of human behavior and their respective determinants are to be found in other areas than behavior directly related to water. To this extent, the following considerations, which refer to behavioral aspects in connection with direct impacts on water, can only cover a small portion of the relevant behavior and must be supplemented by analyses of indirectly water-polluting actions (for determinants of behavior with respect to soil contamination, see e.g. WBGU, 1995).

Perception

Water pollution is not directly perceptible in most cases. Even the clearest water from a "natural" mountain stream – often seen as the epitome of purity – may be ecologically inferior. This makes proper perception of the problem difficult. Whereas the contamination of the Rhine after the Sandoz disaster was actually visible in 1986, when it was colored red over a great distance (though the red color itself was the most harmless thing of all about the contamination at the time), it is rarely possible to determine whether the clear water that comes out of the tap in households is polluted and which substances it contains. Our attention is not drawn to the lead in old water pipes, nitrates or pesticide metabolites until the water utilities inform us about it (or more frequently the media) or, when irritated by contradictory expert opinions, we arrange for private analyses. Apparent contradictions between increasingly alarming media information and personal perception of clear, good-tasting water make people uncertain and arouse distrust, especially towards official, often appeasing sources of information (Glasauer, 1996). Nearly always, however, the "layman" must rely on scientists and politicians to evaluate and differentiate water quality: one fact that is often cited as evidence of successful water policy, i.e. that salmon are swimming in the Rhine again, may be an indication of the improved state of the rivers, but the placatory statement provides no information about the quality of lakes where swimming is permitted or the degree of pollution of drinking water.

Contamination is barely perceptible not only in the drinking water supplied to homes via pipelines. Once wastewater leaving the home loaded with dirt, detergents, cleaning agents and feces goes down the drain, there is no trace of it any more. Open sewers in which a disgusting sewage mixture could be seen and smelled are a thing of the past, and not until it arrives at isolated treatment plants does the wastewater appear again. In general, one can plausibly assume that the individual perception of water pollution has virtually no effect on private, water-polluting behavior.

Knowledge

Even though no direct data are available in this regard, one can assume that the absence of direct perception is linked to a lack of factual knowledge about the degree of purity of the water available to us. Given the reassuring knowledge that providers of water and wastewater disposal services have to meet and monitor the high quality standards for drinking water as well as to make "clean water" out of wastewater in treatment plants, such environmental knowledge was not relevant until recently, particularly since the technical expenditure for the "production" of drinking water and wastewater treatment remained in the dark to a large extent.

Recently, however, the mounting reports of "contaminated drinking water" have induced individual citizens to become knowledgeable, request information from their water utilities and bring their own water samples to be analyzed, also to establish clarity regarding the controversial information that reaches them via the media. In most cases, knowledge about water pollution in itself triggered uncertainty and distrust in the safeness of tap water and has long been a factor behind the decline in drinking water acceptance among the population (Schramm, 1995). In the cited survey on water-related behavior in German cities, over 85% of those surveyed state that they rarely or never drink "water from the tap" (Ipsen, 1994). Supposed or actual knowledge becomes directly relevant to behavior here – the use of tap water as drinking water, which was relatively insignificant anyway, is declining, while more and more people drink mineral waters, which are thought to be uncontaminated.

In view of the large number of potentially harmful substances in water with which we are confronted daily, it not only seems dubious to assume a certain level of knowledge about the water-polluting impacts of individual behavior, but also it appears doubtful that such knowledge can become relevant to behavior. Many water-polluting activities, from washing the laundry to body care, seem to be too intensively governed by habit and too closely linked to values and norms.

Attitudes, norms and values

While religious reverence of water gods may represent a kind of "natural precept of purity" and "holy water" from rivers, springs and wells is kept clean by believers as a matter of course (Katalyse, 1993), such a protective mechanism is lacking in the completely de-emotionalized handling of water as a "lifeless raw material" that can be found in secular societies. However, the attribution of religious significance to water does not necessarily protect it against contamination, and consequently people should be warned against accepting a romanticized view of protective religious norms. From India, for instance, it is reported that Hindus use the holy Ganges River as a waste disposal site with a total lack of concern precisely because of its holiness: a river that can clean everything can clean itself as well (Wilke, 1995).

The unquestioned manner in which water is used as a disposal pathway is characteristic of the attitude towards water that prevails in industrialized countries today (Lanz and Davis, 1995). This is not surprising, given the ubiquity of WCs, sewer networks and the delegation of wastewater disposal to institutions specially designed. The perception of wastewater ends "at the drain", and the act of discharging liquid wastes into whatever type of sewage system has become extensively habitualized. This contributes to a substantial lack of concern for pollution problems. Another factor that operates here was already discussed in connection with water scarcity: the power of competing norms (cleanliness of body, laundry, car, etc.) frequently makes the use and consumption of water both in quantitative and qualitative terms appear to have higher priority than environmental protection.

Incentives for action

It is not without reason that in many economics textbooks water serves as a paramount example of the externalization of private costs in connection with public goods: the incentive structure in the use of waterbodies as disposal pathways almost inevitably encourages their contamination. Faced with the choice of disposing of liquid waste as hazardous waste at great expense or simply "dumping it away", one is tempted to opt for the latter – both in industry and at home. In contrast to soil, where contamination in the form of hazardous waste sites can still be attributed to the polluters years later, water as a flowing element practically offers an open invitation to unpunished pollution.

Perceptible feedback

As with water pollution itself, the interrelations between individual patterns of behavior and their consequences for water quality in the context of a centralized water supply and wastewater disposal system are barely perceptible as a rule. Although almost every use of water is simultaneously connected with contamination, there are no distinctions with regard to the consequences for the polluter, e.g. between a proper bath and the disposal of photographic chemicals down the drain. The mixed treatment of wastewater in the sewer network and treatment plant usually renders it essentially impossible to ascribe individual cases of contamination to polluters. Damage to the treatment plant or receiving waterbody as well as – on the supply side – the cost of purifying drinking water affect individual citizens much later, if at all, through an increase in charges. Pollution, or the absence of same (e.g. due to intensified environmental protection), are not reported to the polluter and thus cannot have an effect on behavior in this way.

In addition to the absence of feedback mechanisms relating to the direct use of water, there is even less awareness of the indirect interrelations between behavior and the pollution of water resources. Causal chains that are separated both temporally and spatially and which are very long and complex in some cases (e.g. consumption of meat from intensive livestock farming, soil pollution due to liquid manure, groundwater contamination) are rarely detected and are difficult to describe quantitatively in relation to individual patterns of behavior. Besides, a strategy aimed at modifying water-related behavior by providing information on "hidden water pollution" would certainly have to tackle the problem posed by the ubiquity of water-polluting behavior.

5.3 Principles and instruments of sustainable water management: Environmental education and public discourse

Environmental education – Water culture as a barrier and potential – Demand-side management – Change in behavior – Effectiveness of intervention measures – Making water perceptible – Promoting a change in values – Avoiding punishment – Offering courses of action – Feedback on behavior – Women and water – Forms of conflict settlement – Consensus building

Each society has its own particular water culture; this can be described in various terms, but is something that many societies fail to perceive in any conscious manner. It emerges in distinct form only in the context of an intersocietal comparison, historical analyses or acute water crises. Nevertheless, as a value context it influences any management of water resources to a greater or lesser extent (see Section D 5.2). In its various manifestations, water culture can act as a barrier, but also as a positive force whenever changes in the management of water resources are made in accordance with the sustainability paradigm. Any strategy planning and selection of instruments for a sustainable water policy must address this sociocultural background and respond to it with a variety of specifically targeted activities. The implication is that there are no instruments for coping with water problems that can be applied worldwide and at any time, and that there are consequences for global policymaking (conventions, etc.) and for water-related development cooperation initiatives, for example. By focusing on a society's water culture, it is possible to derive a wide range of approaches on which to base targeted interventions aimed at improving water resource management.

In the following, psychosocial approaches to behavioral change (referred to in brief as "environmental education") are first examined before looking at the communication processes between various groups of actors ("public discourse"). In practice, it is necessary to coordinate these approaches with legal-administrative, economic and technical measures. However, water-related behavioral patterns cannot be changed at will, because the management of water resources, including associated norms (purity, "desire" or frugality), is learned very early on in the course of a person's socialization and thus usually becomes an established habit that is no longer reflected upon.

5.3.1 Environmental education activities aimed at sound management of water resources

Sustainable water management aimed at changing the way society deals with its water resources – ranging from changes in individual behavioral patterns to changes in lifestyle – can focus primarily either on reducing water consumption or on assuring water quality (see Section D 5.2). Each of these two objectives requires a different set of priorities in the selection of measures. In particular, strategies centered on behavior (e.g. creating incentives and opportunities for action, or giving feedback on the consequences of actions) will differ, depending on whether the behavior to be changed relates to water consumption or to water quality.

Sociological and psychological research on the modification of water-related behavior has focused primarily – as in the field of energy saving – on the quantitative aspect of reducing consumption. This is also the main theme in the following sections. It is important to realize that most research on modifying water-related behavioral patterns is still conducted in the context of "Western water culture". Thus, if research findings are to be transposed to other cultures, it is essential to bear in mind their specific water culture.

The problems involved in securing water supplies on a long-term basis can be tackled with various different strategies. One approach is to try to increase the supply of water, which is achieved in most cases with major technological investments (purification, desalinization, water pipelines from remote sources, boreholes for mining "fossil" groundwater deposits, etc.). This strategy of supply-side management is particularly appropriate for countries with no guaranteed basic water supply. It has its limits when high consumption levels are maintained at enormous expense. Another approach is to try to influence the demand for water with a combination of technical, economic and psychosocial interventions and adjust consumption to the limited supply (demand-side management). In the power supply sector, it was shown that, under certain boundary conditions, demand-side management may be economically more efficient than supply-side investments (least cost planning) – such a strategy certainly appears to be ecologically and socially more sustainable. As far as water supply is concerned, however, such demand-oriented approaches are only gradually starting to establish themselves (Cichorowski, 1996; Schramm, 1995; Winkler, 1982).

On the technical side (supplemented by appropriate legal, administrative and economic measures), demand-side management of water resources requires that people be given the means to take action themselves, e.g. by providing a greater supply of water-saving infrastructure (washing machines and dishwashers, fittings, etc.) or of qualitatively different types of water (drinking vs. industrial water). On the psychosocial side, measures must be aimed above all at people's motivation to adopt particular water-related behavioral patterns, and at changing their perception of problems, their knowledge, attitudes and values as well as the incentives that operate. Marketing and PR strategies enable several of these factors to be addressed simultaneously among broad sections of the population. The goal of demand-side management is to reduce water consumption, supplementing improved technical efficiency by modified water consumption (behavior with respect to taking showers and baths, re-use, etc.).

Compared to the energy-related behavior, which has been studied intensively by sociological and psychological research since the "oil crises" in the late 1970s/early 1980s, relatively few studies have been conducted on the effectiveness of psychosocial interventions with respect to water-related behavior (see Box D 5.3-1). These studies differ substantially in terms of the situative boundary conditions, the behavioral models that are used and the determining factors that are examined, so it is not possible to make general inferences about one field from the other. However, it has been shown that a whole series of factors can contribute to changing water consumption. It is noteworthy here that no one factor – neither a change in prices and price structures, nor the availability of water-saving techniques, nor information campaigns – will achieve the goal by itself (Schramm, 1995; Winkler, 1982). Instead, a combination of different forms of intervention seems to be hold most promise.

On the basis of the taxonomy outlined in Section D 5.2, the next sub-section describes a number of options that, from the point of perspective of the social and behavioral sciences, may be effective – separately or, preferably, in various combinations – in modifying the way people manage their water resources (see also WBGU, 1994).

Perception and knowledge

In order to create an awareness of water and the real and/or potential problems associated with it, it is necessary first of all to make water (once again) a perceptible element within everyday life. This perception pertains to the quality and quantity of available water resources, to the water-supply infrastructure, to the sources of water and its subsequent fate (in spatial and temporal terms), as well as the changes that this resource undergoes through human use.

Potential options include:

- extensive exposure of rivers and canals (this can also benefit the local microclimate and be esthetically valuable),
- ongoing publication of data on quality and quantity by local water utilities, "open days" at borehole and treatment facilities or wastewater treatment plants, and public relations work in these areas,
- environmental education activities in schools and outside, focusing on the local water situation and aimed at making water something that people can perceive and experience,
- establishing a broad-based public discourse on water resources, with specific reference to the type, condition, origin, importance and management of local water resources, as well as the problems involved,
- requiring producers to indicate the amount of water consumed in the production of consumer

BOX D 5.3-1

Effectiveness of psychosocial interventions

Studies on the reduction of water consumption with the help of psychosocial interventions have been conducted since the early 1980s, above all in the USA and Australia (for an overview see Cone and Hayes, 1980; Winkler, 1982; Stern and Oskamp, 1987; Seligman and Finegan, 1990; Gardner and Stern, 1996). On the whole, most studies found the behavioral effects actually achieved to be minimal. This may have to do with the relative inflexibility of water consumption, but may also be due to the selective application of specific measures in each case.

The following factors led to a reduction in consumption in the various studies:

- a perceived social norm of saving water (Kantola et al., 1982),
- low-cost installation of water-saving fittings (Geller et al., 1983),
- a rise in water prices (Berk et al., 1980; Moore et al., 1994),
- communicating the situation as a commons dilemma (long-term benefit of saving water for all people; effectiveness of personal water-saving behavior) (Thompson and Stoutemyer, 1991),
- a combination of information campaigns and a personal duty to help save water (Dickerson et al., 1992),
- a combination of consumption feedback and generation of cognitive dissonance (Aitken et al., 1994),
- water-saving programs involving a combination of educational (information, appeals, consumption feedback, etc.) and regulative elements (temporary consumption restrictions, etc.) (Berk et al., 1980).

The following, by contrast, have proven to be ineffective:

- a general price increase as well as progressive price structure when reduction targets are exceeded ("penalty payments") (Agras et al., 1980),
- discounts for low consumption (Winkler, 1982),
- information on water problems and ways of saving water (Geller et al., 1983),
- "economic education" (short-term economic benefits of water-saving behavior) (Thompson and Stoutemyer, 1991),
- isolated consumption feedback (Geller et al., 1983; Winkler, 1982).

Various authors attribute the relative ineffectiveness of informational interventions (particularly feedback about actual water consumption) to the economic framework in the respective studies, and point to similar effects in the energy sector – the low price for water as well as the usually degressive price structure give a counterproductive signal that precludes any motivation to save water (Geller et al., 1983; Winkler, 1982; Seligman and Finegan, 1990). Nevertheless, such feedback strategies may still be useful, as Hamilton's (1985) findings demonstrate: there is a positive correlation between water-saving behavior and knowledge about personal water consumption.

In addition to the factors mentioned, the various studies found statistical correlations between water consumption and many sociodemographic variables, e.g. age (Kantola et al., 1982), position in society (Thompson and Stoutemyer, 1991) as well as household size and income (Aitken et al., 1994). This underlines the need for interventions aimed at reducing water consumption to address specific target groups.

goods once this exceeds a certain level.

To enhance the effectiveness and efficiency of information activities aimed at effecting a change in the patterns of water-related behavior, appropriate information should be combined with other environmental or social issues (soil, air, consumer behavior, etc.) as far as possible. Such an approach also takes into account the indirect "hidden" interrelation that often exists between water and individual patterns of behavior.

Attitudes, norms and values

A change in values towards sustainable management of water resources must be encouraged in order to create "competition" against existing individual and social norms (cult of cleanliness, fun culture, etc.). This would also include generating acceptance of the need to modify water-related behavior, should this be necessitated by qualitative or quantitative conditions.

The role of prominent personalities or the public sector (e.g. building and investment measures) as models to emulate can be used here to establish water-saving behavior as an "authorized" standard in a credible manner. Publicizing particularly economical behavior and consumer tips as well as offering premiums of a material and symbolic nature for water-saving activities can also be beneficial in this context. To develop "water awareness", activities can be linked to local sociocultural norms or to religiously motivated rules, as well as to the positive associations of a cheerfully wasteful use of water if, for example, water-saving infrastructure is already present in public swimming pools and adventure parks, or "sustainable" behavior there is supported by incentives or behavior modeling (Glasauer, 1996).

In addition, concepts of regional sustainability or efforts to set up a LOCAL AGENDA 21 by encouraging a feeling of local affiliation and responsibility can have an influence on attitudes and values with respect to water as a (local) resource. Fostering a perception of water (problems) and the relevant knowledge in this context is also helpful.

To ensure that water of varying quality is managed on a differentiated basis wherever possible (substitution of sanitary and industrial water for drinking water), it is necessary not only to provide appropriate infrastructural facilities (taps, reservoirs) and to make it easier under building law to use rainwater, but also to conduct appropriate image campaigns. In this connection one can fall back on relevant experience in the advertising industry with activities that have become established as "environmental marketing". The campaigns conducted by individual water utilities to enhance the image of tap water as a foodstuff should also be welcomed, especially in connection with activities aimed at increasing the perception of water in everyday life. However, they must be specifically linked to drinking water quality assurance and transparency in the event of problems, so that no credibility gaps are created.

INCENTIVES FOR ACTION

Incentives for sustainable management of water resources can be based on price structures and on the methods used to bill people for drinking water and wastewater. Preference should be given to consumption-based billing as opposed to the common method of cost allocation. Within this context, rates should be progressive, or at least linear so as not to favor high consumption. Because of the low price elasticity, approaches based on the water rate structure should be implemented, though always in combination with other instruments.

In general, incentives such as premiums, discounts and social recognition (e.g. by awarding "prizes" in water-saving competitions) are to be created or modified so that careful management of water can be worthwhile for the individual citizen as well as for municipalities or water utilities. It is essential here to avoid "punishing" water-saving behavior by imposing subsequent price increases.

OPTIONS AND OPPORTUNITIES FOR TAKING ACTION

Citizens evidently view their own options for saving water as being relatively limited (Glasauer, 1996). This makes it all the more important to create new opportunities and to publicize less known options in the public sector, e.g. by putting up clear and attention-grabbing signs in toilets and showers. In addition, extensive labeling of the appropriate fittings and appliances, as is already the case with some washing machines and dishwashers, can play a useful role here. Creating opportunities also includes promoting innovations within industry in the environmental engineering field.

If the aim is differentiated management of water resources according to different qualities, it is essential to support the relevant infrastructural measures, such as watering gardens with lower-quality industrial water.

Another major instrument for improving the opportunities that people have to take action themselves is to enable public participation in the decision-making processes relevant to water resources, within the context of a public water discourse.

PERCEPTIBLE FEEDBACK

If people are to perceive both the positive and negative impacts of their personal water-related behavior and relate this to their own action, feedback on quantitative as well as qualitative water use and consumption must be improved. This would mean, for example, making it technically feasible to measure household consumption, examining the idea of making such measurement mandatory under building and housing law, and providing for proper presentation of the relevant information, e.g. on noticeboards, displays, etc. The feedback procedures themselves should be designed in a user-friendly manner and be tailored as precisely as possible to the actors in question and their specific behavior. They should also be carried out relatively frequently and enable comparisons to be made.

The consequences of collective behavior can be made perceptible in aggregated form as well, e.g. by regularly publicizing local statistics on pollution in sewage treatment plants, or on water levels in drinking water reservoirs, as well as their respective causes. By making the impacts of water-related behavior visible through display boards, posters or similar media, one can counter the common assumption that personal behavior is of marginal relevance only.

BOX D 5.3-2

The concept of "efficient water use" in Frankfurt am Main

The concept of "efficient water use" was developed for the city of Frankfurt am Main in the early 1990s, with the aim of securing a long-term supply of drinking water. In addition to protecting the city's own groundwater stocks (to secure local resources) and differentiated water use (extensive substitution of industrial water for drinking water), the objective of the concept is to reduce water consumption by 20% between 1991 and 2000, by avoiding waste and losses and by deploying water-saving technology (while maintaining the same standards of comfort and hygiene), as well as by changing consumption behavior. The greatest potential for savings is presumed to be in private households, which consume two thirds of Frankfurt's drinking water.

To change consumption behavior on the basis of this concept, a broad-based water-saving campaign encompassing several elements has been conducted in Frankfurt since 1992, although there is no distinction between different actor groups (Cichorowski, 1996; Koenigs, 1996):

- Creating an awareness of water problems through media advertising (series of posters, radio and cinema spots) aimed at educational and emotional linkage of the issues "water scarcity" and "nature/environment".
- Discussion of problems related to Frankfurt's water supply (primarily based on remote water sources) through events, brochures and newspaper advertisements.
- Providing households with information on the specific options open to them, through brochures, newspaper advertisements, advice centers and a hotline service.
- Implementation of pilot projects in specific city districts (free installation of water-saving fittings, optimization of sanitary facilities, installation of household water meters).
- Honoring "water savers".

A levy on groundwater extractions charged in Hessen since 1992, for example, was used to finance the communication and marketing campaign.

In the period from 1991 to 1995, total consumption of drinking water in Frankfurt dropped by 16%, with per capita savings in daily consumption amounting to 12.3%. Due to the reduction in water consumption that also took place in the area surrounding the city during the period in question, and a water emergency that was proclaimed for the South Hessen region in 1992 and 1993, the reduction in consumption cannot be wholly attributed to the campaign (Cichorowski, 1996; Schramm, 1995). Nevertheless, findings by Ipsen (1994 and 1996) indirectly verify the effectiveness of the campaign. However, these data also indicate that differentiation according to actor or lifestyle groups could even improve its effectiveness.

All these interventions must be designed for the specific use, context and target group, and permit evaluation with respect to their impacts on actual behavior. Due to the potential ineffectiveness of psychosocial interventions addressing only some of the factors mentioned above, it is advisable to tackle several factors simultaneously on the basis of general concepts. One example of such an approach is the concept of "efficient water use" applied in Frankfurt am Main for some years now (see Box D 5.3-2).

Psychosocial interventions in other "water cultures", in the context of development cooperation, for example, must take into account the respective sociocultural contexts. This initially requires a precise analysis of the local water-related conditions of an ecological, scientific-technical, economic, legal-administrative, religious and symbolic nature (see Section D 5.2), as well as of the needs, customs and sensitivities of the local population. The selection and implementation of concrete measures must be based on these basic conditions in order to be able to rule out counterproductive results right from the beginning.

It is of importance, for example, to devise technical measures, the initial priority of many projects, so that they are socially acceptable and to "embed" them in the knowledge and value system of the local population in order to promote acceptance of and identification with these measures. In addition, the installation of water supply technology should be directly linked to support activities in the educational field (culturally appropriate hygiene education, etc.). In developing countries, it is particularly important to provide for suitable involvement of women in water-related measures (see Box D 5.3-3).

BOX D 5.3-3

Women and water in developing countries

In many developing countries, women are the principal actors involved in water resource management, whether as water carriers, managers, users or health instructors. Because of their work in the household, however, they are often the main persons afflicted with diseases related to the use of contaminated water. For this reason, it is important for them to be appropriately involved in the planning and implementation of water supply and wastewater disposal measures.

For many questions arising in the context of water projects, for example, women are often the best informers, promoters and multipliers – from the surveying of local needs to questions of location and equipment and even with regard to acceptance and use of new water supply systems. Their knowledge about natural and cultural conditions relevant to water predestines them as advisors to agencies responsible for the implementation of programs, and in this way ensures the identification of the population with the project. In many cases it is small details that are decisive for the success or failure of water projects, such as the question of the visibility of body hygiene, which plays a role in connection with the choice of location and structural design of public sanitary facilities. Voluntary or paid care, maintenance and management of water systems are also in good hands when performed by women because they have a special interest in an intact system for water supply and wastewater disposal. Thus they are often highly motivated towards water-saving measures. In addition, women have proven to be the most effective instructors as far as sanitary and health issues are concerned (UN INSTRAW, 1991).

5.3.2 Communication and discourse

5.3.2.1 Bases of discursive communication

The importance of communicative action and cooperative discourse from the perspective of the social sciences has already been discussed in Section D 4.1.2. The focus now turns to the various forms of communication, and of discourse aimed at increasing understanding, that operate in social processes. The relevant instruments of communication and discourse are discussion groups, mediation procedures, round-table talks, citizens' participation, etc. at local, regional, national and international level, in which the intention is to bring together all those involved in particular conflicts. The main purpose of discussion groups is to produce an atmosphere of mutuality and social learning; in discourses, the primary focus is placed on collectively binding regulations or self-imposed commitments. Within a given discourse, all participants have the same rights and obligations and voluntarily refrain from exerting strategic influence. With the help of communication and by reviewing statements made, conflicts can be settled either on the basis of a consensus or a compromise that represents a fair and practicable solution for all parties (Renn, 1996a). The capacity for consensus depends on three conditions (Renn, 1996b):

1. A sufficient number of solutions must exist or be created between the extremes (yes or no).
2. The parties involved must acknowledge common rules so that the validity of statements and claims can be reviewed.
3. All parties to the conflict must recognize the precept of fairness. The requirements that must be met in order to achieve consensus are of a higher order than for compromise.

A consensus defines a solution to the problem that all parties to the conflict voluntarily accept on the basis of their inner reason and self-imposed commitment to fairness, and which they themselves may prefer to their original demand.

5.3.2.2 Communicative forms of orientation

Communicative forms of orientation are geared to instigating among the participants of the communication process certain learning processes and insights that are then reflected in modified or adjusted behavior. When an attempt is made to induce behavioral changes through communication, it must always be kept in mind that communication is a reciprocal process. Those who wish to persuade should first become familiar with the perception of the situation and the arguments of those who are to be persuaded. In many cases, traditional cultural conditions can prove to be an ideal support in generating the desired patterns of behavior. For example, if clean water is a culturally established value, it is much easier to communicate the need for a supply of clean drinking water than in cases where this value is not known to the respective culture. In small local discussion groups, the special characteristics of the local or regional culture, the potential social and cultural reinforcers as well as the barriers to change can be identified, and joint efforts undertaken to identify the strategies for promoting the desired behaviors.

One of the difficult problems encountered in communication on environmental issues is the marginal significance they frequently possess for the lifestyle of those concerned. People who must struggle to survive often have little understanding of other people's concern for preserving biodiversity. At the same time, many such people are not fully aware of the direct impacts that environmental problems have on their life, e.g. in the form of damage to health. In order to reach these persons, it often makes sense to associate water issues as "free riders" with other topics within the communication process in which there is strong interest. Elements of water-related communication can be incorporated into situations ranging from rock concerts to the doctor's surgery. The form of event to be chosen here is, of course, the one that has high prestige value in the context of the respective culture and is also suitable for putting forward water issues.

As with discussion groups, communicative processes in general are also a suitable vehicle for consultation and education (see Section D 5.3.1). In water-scarce regions, for example, much can be achieved by setting up advisory centers on water consumption. Such advice could be combined, for example, with advice on field irrigation, on managing a household or with health counseling, or organized in cooperation with water utilities or similar facilities. Tours of water treatment plants are further activities already offered and implemented with success in many industrialized countries. There is no reason why such tours should not be established as a worldwide form of locally based communication. Many people do not have a clear picture at all of the problems involved in wastewater disposal. Having an opportunity to see and comprehend the situation at local level also creates new motivation to change one's own behavior if necessary. Sections of the population that have already been informed can act as multipliers in this connection

As will be discussed in more detail in Section D 5.4, cooperative solutions are superior to state-ordained or private-sector solutions under certain circumstances. However, these must also be organized and supported, for at least as long as those concerned require such assistance. People providing development aid (e.g. Peace Corps workers) and other experts working abroad should be trained to identify the special prerequisites for cooperative solutions and implement the rules and requirements associated with the practical organization and operation of cooperatives. A number of positive examples from South America have demonstrated that the establishment of cooperatives has made a substantial contribution toward helping people to help themselves.

5.3.2.3
Implementation and application of discursive procedures

In the context of increasing globalization of environmental problems, the growing importance of activities and policymaking at the local and regional level highlights the need for more intensive mediation and citizen participation in developing countries as well. The United Nations has issued the general demand that states involve their citizens in the political decision-making process (UNDP, 1993). Nearly all development policy reports and declarations deem it essential that the public be involved in the planning and execution of development programs (see Box D 5.3-4). Effective implementation of even the best programs is impossible if the population resists. Remnants of development policies that have ignored the wishes and concerns of the population affected can be found all over the world. However much consensus may now prevail on the notion that international assistance must provide for public participation in matters involving water supply and wastewater disposal, it is still unclear how this can be achieved in each specific case. Should referenda be organized, or should property rights be granted by public approval? How are participative forms of decision-making to be incorporated into the existing political and social order? How can participative forms be implemented in a country ruled by a dictator? There is no single, universally applicable answer to these questions. The paramount goal should be to avoid putting water supply and wastewater disposal measures into practice against the will of those for whom they are designed. The instruments for achieving this goal include individual discussions, neighborhood meetings, round-table talks with interest groups, coordination with legal decision makers and the like. Despite the popularity of this demand, it is still extensively unclear which form of participation is appropriate, effective and efficient. A considerable need for research still exists here.

There is a particular need for participative procedures for involving affected citizens in the planning of their environment in those cases where the political institutions or private organizations necessary to perform planning functions are absent. In many soci-

BOX D 5.3-4

Experience with discursive procedures in the environmental sector in Germany and abroad

In European and North American states as well as Japan, discursive forms of planning and conflict resolution are gaining in popularity – examples are round-table discussions, citizen involvement, alternative procedures for settling conflicts and the like. These participative procedures should offer a guarantee that the demand for proper and competent political planning corresponds with the wishes and preferences of the affected population for an environment in which life is worth living. These discursive procedures include: consensus conferences, political dialogues, sectoral dialogues, mediation and moderation procedures, planning cells as well as various combinations of these (Weidner, 1996a). All these forms of participation feature direct communication. In the political administration systems governed by Anglo-Saxon law, informal procedures of mediation and planning at local and regional level play a major role. In certain cases, authorities can assign decision-making authority to commissions or mediation rounds in a "pre-empting" process. In Germany the situation is different. Here, mediation and participative procedures cannot formulate legally binding conclusions, but can only draw up recommendations. They represent an instrument of voluntary self-organization and self-coordination at the horizontal level.

In the face of mounting environmental problems and the increasing need for regulation, traditional, hierarchical forms of political management and the associated scope for action and decision-making have reached their limits. This means that the basis has been established for a political approach based on cooperative discourse. Despite many reservations regarding direct forms of citizen involvement, participation procedures are nevertheless seen by public administrators and politicians as desirable and essential, because traditional decision-making strategies for solving environmental problems are considered to be inadequate (Renn et al., 1995). There are a number of reasons why mediation procedures should be necessary (Holzinger, 1994). A fundamental change is taking place in the relationship between central government and its citizens, a transformation that is also reflected in the fact that environmentally relevant decisions made by the responsible political bodies are no longer accepted by those concerned without opposition. An indication of the dilemma of political planning in the environmental sector is that expertise has to be available to assess planning projects, but expertise alone is not enough to obtain a democratically and ethically legitimated solution (Renn and Oppermann, 1995). On the one hand, leaving decision-making to experts contravenes the basic principles on which democratic systems are based. On the other hand, when decisions are left to political whims, objective constraints tend to be ignored, with high consequential costs the result. Mediation and participative procedures are strategies that provide for fairness when adopting resolutions, and at the same time a competent solution to the problem at hand.

Experience in Germany and abroad demonstrates that although alternative conflict resolution procedures, such as environmental mediations, are still in an early stage of development and consolidation in many countries, they still represent promising and successful approaches of direct democratic participation in the political decision-making process in environmental conflicts. Comparative studies by the Berlin Scientific Center for Social Research (WZB) show that particularly in three of eleven industrialized countries – namely the USA, Canada and Japan – mediation procedures play a role in quantitative terms in the settlement of environmental conflicts (Weidner, 1996a). In Europe, mediation procedures are more rare; in Switzerland and Austria, about ten mediation procedures can be identified. They are encountered more frequently in Germany, where they are used in the many areas of environmental policy and related policy fields (transportation planning, urban planning, waste treatment).

The USA is to a certain extent the starting point of mediation procedures in environmental policy. Some were already in place in the early 1970s. By the mid-1980s, Weidner (1996b) had listed over 160 major environmental conflicts in which mediation procedures were applied, though more recent estimates indicate that the growth rate is rising. Various types of mediation in environmental conflicts have been legally institutionalized in many states as well as at federal level. The spectrum of environmental mediation at the various political levels ranges from specific locational decisions, planning of major installations, infrastructural measures, legislative procedures to political declarations of principle. In Canada, which has taken its cue from the USA, a growing trend toward environmental mediation has also been noted. Similar to the USA, legal regulations that give environmental mediation a legal foundation in certain cases have been enacted there at the provincial and federal level. Weidner (1996b), too, comes to the conclusion here that the use of environmental mediation will increase. The transboundary waters regime between the USA and Canada shows that mediation not only leads to peaceful and long-term resolution of conflicts, but is also a key, integral element of international agreements (see Section D 4.1.2.4).

In Japan, mediation procedures also have a tradition that can be attributed to the consensus-oriented political culture. As early as the late 1960s, mediation procedures were institutionalized to settle environmental conflicts. At the municipal level, mediation procedures are very widespread. Weidner (1996b) assesses the different institutionalized procedures of conflict resolution positively, and in the case of larger-scale environmental conflicts they are considered to be extremely effective.

eties, the institutions necessary for maintaining national water management or for establishing and monitoring private-sector systems are either non-existent, or fail to perform their function for a variety of reasons (tribal feuds, shortage of money, corruption, lack of authority, etc.). In this situation, it makes sense to deploy instruments such as round-table talks, collective forums or other forms of cooperative action. The GTZ has already recognized this potential and formulated appropriate procedural instructions (GTZ, 1996). Such consensus-oriented discussion groups are frequently better adapted to the political culture of the respective countries than majority systems or individualized market systems. Such discussion groups often make it easier to overcome the difficult problems faced by societies undergoing transition (e.g. in the former communist countries). These are typical transitional phenomena that tend to be surmounted by more stable systems of order based on institutional rule. However, in countries where the government lacks authority, such sub-political decision-making forums are also meaningful as a long-term solution if they acquire a legitimate function within the existing political framework. In particular, problems relating to water resource allocation and collective wastewater treatment, or to flood management, can be tackled on an equitable and results-oriented basis with the help of such bodies. Although the principle of consensus hinders a quick, and sometimes also an efficient solution, it ensures the feasibility of implementing the measures decided on and strengthens the cohesion of the social system and the peaceful approach to conflict settlement.

Apart from planning tasks, discursive forms of communication can also play a role in conflict resolution. Conflicts frequently arise between different water users. Utilization of water resources by one party imposes limits on the use that the respective other party can make of the same resources. The legal system is often incapable of settling such conflicts. In this situation, mediation can offer a way out. This, of course, presupposes that the number of possible solutions is not equal to zero and that an experienced mediator manages the process. The Federal Government could not only provide mediators, but also make an important contribution to the peaceful solution of local and regional water conflicts by training mediators on site.

In real practice, all these discursive procedures in the field of environmental policy must be structured in such a way that they meet two goals: firstly, they must ensure a proper understanding of the problem and adequate solutions; secondly, they should grant everyone potentially affected by the conflict the same chance to input his or her interests and values into the decision-making process (Renn, 1996b). The procedure must ensure that the necessary expertise is provided, that valid standards and laws are complied with, that social interests and values are incorporated in a fair and representative manner, and that objective, emotional and normative statements can be included (Zillessen, 1993). In the German and Swiss environment sector, for example, a three-stage procedure (Renn et al., 1993) involving a combination of mediation of interest groups, discussions with experts and citizens' participation according to the planning cell model has proven effective in practice. In the first step, the concerns and values of interest groups are translated into criteria with the help of the so-called value-tree method in order to make value preferences visible. The various options of the interest groups resulting from this are evaluated by experts in the second step, thus creating a profile for each option that can then be compared to the others. In the third step, the profiles are evaluated and ranked, and the ranking is then passed on as a recommendation to the political decision-makers. The important feature about multi-stage procedures is that randomly chosen citizens make the decision that is recommended in the end, and not established interest groups (Feindt, 1996).

Whatever form of discourse is considered appropriate, applying discursive procedures of participation to solve water problems and for joint planning and conflict settlement is seen by the Council as a key challenge. In addition to efforts at international level, Germany should make its practical experiences and know-how gained from environmental mediations and the findings obtained from mediation research available for consultation purposes in domestic water conflicts in other countries; this should be done in the form of know-how transfer through personnel support involving environmental mediators, scientists and environmental policy experts.

5.3.3 Recommendations

The target groups for the following recommendations include all social actors that (for different reasons) are interested in establishing a new, sustainable form of water resource management on the part of the general populace, or who are compelled to do so (national actors: federal government, the *Länder*, above all municipalities; enterprises: industry, commerce, craft trades; water providers; environmental and development NGOs and consumer associations; environment consultants and educators; development cooperation organizations).

GENERAL

- *Instrument mix:* Changes in the way that people treat water resources cannot usually be achieved with a single strategy or measure, but require the deployment of various instruments in a mix. Technical, economic, legal and administrative measures must therefore be combined with behavior-oriented interventions and public communication.
- *Consideration of the sociocultural context:* All attempts to change the way water resources are managed occur within a specific environmental and sociocultural context. This context, which manifests itself in various dimensions (stocks of knowledge, collective conceptions and norms, economic, political, legal and administrative structures, religious practices, etc.), must be taken into account as a barrier and as a potential.
- *Initiation of a public water discourse:* There is little reflection on the management of water resources, even in regions where water is scarce. To discuss the issues and problems related to the type, state, origin, importance and management of local water resources, therefore, it is necessary to initiate a public water discourse at all levels of society, independently of any acute water crises.
- *Demand-side management:* A sustainable water policy must also be based on demand. Traditional water policies, based in many places on a supply-side approach, must be supplemented and extended through intensified demand-side management, both in regions with existing and in those with potential water problems.
- *Encouraging the differentiated use of water:* The use of water of varying quality (drinking water, industrial water, rainwater) for different functions must be encouraged. To promote this process, appropriate "image campaigns" should be conducted and at the same time courses of action and incentives for behavioral changes created.
- *Deployment of instruments to change behavioral patterns:* In order to effect a change in the management of water resources in a manner that is ecologically effective, economically efficient and socially compatible, individual and social factors influencing environmentally harmful behavior must be identified and more intensive use made of effective intervention instruments for behavioral change. These have to be based on:
  - the perception and assessment of environmental conditions,
  - knowledge and processes of data processing,
  - cultural values, norms and rules,
  - monetary and symbolic incentives,
  - provision and creation of opportunities (infrastructure),
  - feedback on the impacts of personal behavior and changes in behavior,
  - the perception of the actions of others (reference groups, role models).

All intervention instruments are to be adapted to the specific sociocultural contexts, lifestyles and stakeholder groups.

SPECIFIC

- *Perceptibility:* Water, water problems as well as both positive and negative patterns of water-related behavior must be made more perceptible to the population. This can be effected, for example, by opening up flows and channels of information, by providing households with feedback on water consumption, basing water charges on consumption, etc.
- *Information on perception and assessment patterns:* To make people sufficiently aware of water issues, the respective population should be sensitized to culturally related and social patterns of perception and valuation
- *Suitable presentation of information:* Information on the management of water should be presented in the context of other environmental and social issues, in order to illustrate the interrelations between individual problems and related patterns of

behavior. The "binding" principle of sustainable development, now recognized worldwide, could provide a framework for behavioral recommendations.

- *Communicating causal relations:* Causal connections between personal behavior and its consequences, which often take a long time to emerge, should be made understandable and perceptible. This requires the development of educational and training measures for specific target groups.
- *Feedback on the impact of behavior:* Feedback on the impacts of personal behavior and the consequences of a changed behavior, e.g. with the help of technical information aids such as display boards, can lead to effective saving effects when linked to progressive or linear water rates.
- *Effectiveness of behavioral changes:* A personal contribution to a socially desirable environmental goal is often felt to be minimal or marginal by individual citizens. Therefore, the effectiveness of collective behavior must be made clear, such as in the form of publicly visible display boards on the saving achievements of a community or by publicizing local "Top Tens".
- *Symbolic rewards:* The incentive effect of symbolic rewards (honor, public esteem) both for individuals and groups (e.g. school classes) and for municipalities and regions (e.g. by awarding a "sustainability medal") should be used to a greater extent.
- *Integrating existing knowledge and cultural rules:* In the case of measures aimed at preserving or reestablishing water-protecting behavior, locally available stocks of knowledge and cultural rules (religiously motivated norms, taboos, etc.) must be taken into consideration and actively incorporated, instead of thoughtlessly introducing Western value systems and lifestyles.
- *Establishment of suitable communication opportunities:* Water-related communication activities should be designed in line with the characteristics and interests as well as the sociocultural context of the target groups. Examples of this include the creation of local discussion groups ("water circles"), holding "open days" at supply and wastewater treatment facilities, the introduction of water consulting hours or the inclusion of water issues in cultural and church meetings.
- *Enabling participation:* Effective implementation of even the best programs is impossible if there is resistance among the population. The local population must therefore be involved in the planning and execution of water-related measures, and various forms of participation are to be taken into account and applied according to the situation.
- *Promotion of cooperative action:* In societies in which the requisite conditions for water resource management by the state or for the establishment of private-sector systems are (still) absent, forms of cooperative decision-making and appropriate action patterns should be promoted. They include assistance in putting in place and maintaining water management cooperatives where the prerequisites for that are given or can be established.
- *Negotiation-based conflict solutions:* Communal forums for conflict resolution and mediation procedures should be used for peaceful solution of conflicts between different water users. Mediators and moderators are to be provided and/or trained for this purpose.

5.4 Economic approaches to the sustainable management of water resources

Diversity of uses – Quality and quantity – Drinking- water function – Use in agriculture – Valuation issues – Market valuation – Option, bequest and existence value – Water-related risks – Market and government failures – Allocation issues – Methods of community regulation – Water markets – Enforceable property rights – Institutions for allocation – Coverage of minimum water requirements – Costs for water supply and wastewater disposal – Efficiency aspects – Pollutant loads – Comparison between Germany and the USA

5.4.1 Special characteristics of water

The management of freshwater resources exhibits a number of important features. On the one hand, a differentiated solution to water problems is necessary due to the regional differences in supply, quality and demand. On the other hand, it is possible to elaborate certain principles governing the "sound management of water resources" and which lead to certain typical ways of solving water-related problems.

Water's most important features may be crudely summarized in the following terms:

- the multifunctionality and diverse valuation of water resources,
- the divergent properties of water as an economic commodity,
- the regional nature of most water problems and
- the increasing importance of efficiency in water supply and wastewater disposal systems.

It is essential here to realize the problems that arise when water is analyzed in isolation from the wider context, because the boundaries between water and soils as environmental media (riverbanks or wetlands, for example) are indistinct as far as the natural supply of water is concerned. This often affects the measuring and valuation of water's habitat function.

5.4.1.1
Multifunctionality and the diverse valuation of water resources

Diversity of uses

There is hardly a single resource or category of goods in the world – with the exception of soils perhaps – that is characterized by such diversity of use and valuation as water. Water has natural life-sustaining, habitat and regulatory functions, for example. From an anthropocentric perspective, water is an indispensable food (drinking-water function), serves in the preparation of meals and beverages, is used for cleaning, is a raw material, operates as a productive means in the economic value creation process, performs disposal functions by virtue of its self-purification function, is used as a means of transport and performs important esthetic and religious functions.

From this diversity of uses, we have picked out those that are of great importance for developing countries. Many of these countries

- withdraw water from nature for consumptive purposes and must usually treat it prior to final use (e.g. as drinking water or cleaning water) in more or less costly ways (use of withdrawals with specific costs for extraction, treatment and distribution),
- allow water to enter production processes as a decisive input, whereby agriculture is clearly the predominant sector of application (again, use of withdrawals involving high costs for extraction, treatment and distribution),
- use water in its natural state (rivers, lakes, groundwater) for recreational purposes in the context of long-distance tourism, for transporting goods or for dumping waste (usage), often involving interference with nature in order to enhance benefits (e.g. developing lakes infrastructurally to create leisure areas, or making rivers navigable),
- regard water as something that generates benefits through its very presence or outward appearance (cultural function).

Specific functions may be mutually excluding, while others may complement each other. From the perspective of economics, the inherent scarcity of water resources gives rise to the problem of allocation – how much water should be used when, where and for what purposes? Every country and region must find its own solution to this economic problem, a process that always and unavoidably requires valuations – for example a classification into lower-value and higher-value water uses. Multifunctionality and diversity of valuation are therefore inseparably linked, but at the same time can take on different guises at each level.

Solving the valuation problem

For all the heterogeneity of regional water problems, a number of basic alternatives emerge for resolving valuation issues. The most important of these are the use of markets, asking experts and political valuation. Market-based valuation, the idea of determining value according to individuals' willingness to pay, evidently plays a major and perhaps even the decisive role. In this scheme of things, individuals operate as the key valuation authority, since they are best able to measure the individual benefit of a specific water use. Individuals know their preferences and have appropriate notions concerning the "profitable" use of water in production processes. In the context of markets, individuals are prepared to articulate their conceptions of value as a willingness to pay, i.e. in the form of a demand expressed in monetary terms. This type of valuation is efficient and in no sense hypothetical.

In contrast, any form of valuation made by experts or government authorities runs the risk of being divorced from reality, i.e. of incorrect valuation and subjectivity. Thus, even in economic systems in which decisions on allocation are supposed to be made centrally or collectively, there is a repeated tendency towards the development of spontaneous markets – in the gray economy or as informal markets – in which individual willingness to pay can be expressed. This underlines the fact that valuation by markets is the best means by which the whole range of consumptive and productive uses of water resources can be expressed.

Markets first assign a value to the key drinking-water and sanitary function of water. This is usually reflected in very price-inelastic demand. Other consumptive uses – water for private swimming pools, for play or for watering gardens – are also taken into account, however. This willingness to pay for consumption beyond basic needs is normally characterized by a higher price elasticity of demand. This, too, is an economic variable indicative of the utility character of water.

If the focus is confined to water's drinking-water and sanitation function, which cannot be performed for humans by any other resource, then water is seen to be a resource of paramount importance, without which human existence would be unthinkable. Low water availability is reflected in markets by high water prices, placing those groups in society at a disadvantage or indeed at risk who are least able to meet their basic water needs for survival due to their income and property situation. Poor families living in the illegal squatter settlements in the Pueblos Jovenes of Lima, for example, were paying US$ 3 per cubic meter in the early 1990s for water of dubious quality from mobile tankers, equivalent to about 10% of the average household income of this group (Briscoe, 1996). The poor were therefore doubly hit: by inadequate incomes, on the one hand, while the pressure created by low incomes and high prices to save water gave rise to sanitary problems and subsequent costs to health, on the other.

Separating the valuation issue from the allocation issue

Occasionally, experience of this kind generates criticism of market-based valuation and leads to intervention – in the form of subsidies, for example. Such criticism is unfounded, in the view of the Council, because it confuses valuation and allocation problems. Both problems can be solved by markets, but must be approached from different angles. The valuation problem is well resolved by market mechanisms, especially in connection with the notions that people now entertain in respect of water use. Hence, subsidizing water to ensure adequate allocation is obviously a misguided policy in terms of valuation, one that can actually be counter-productive in solving the problem of water scarcity because it increases the demand for water and exacerbates the problem of allocation. If the problem of allocation is to be surmounted – in order to supply the poor and destitute with an adequate basic supply, for example – then other methods need to be utilized (such as providing "water money" – see Section D 5.4.2.3), not price subsidies.

Valuation of agricultural use

Another special aspect that markets are able to demonstrate concerns the productive uses of water. Water is an important element in the value creation chain and in the development process, something that is usually evident in a willingness to pay for water under market conditions. The quantitative demand for water resulting from this willingness to pay is significant and in many cases amounts to more than two thirds of the national demand for water. From the viewpoint of farmers, the valuation of water is based in normal cases on the production factors in agriculture (see Box D 5.4-1), on technical progress as well as the level and/or structure of consumer demand, and can undergo constant modification as a result of changes in incomes and methods (such as increased efficiency of water use). This assessment is likewise reflected in market prices and thus monetizable and/or measurable with cardinal numbers. As shown by the example of subsidized water supplies to agriculture in many countries, government influence on water prices can affect the level of production in a particular industry (see Section D 3.3) and thus operate as a selective form of structural policy. What is normally involved is an artificial cheapening (subsidization) of water as an input variable – with the result that water is consumed more intensively by those sectors that depend on water. The ulterior motives may relate to the distribution of wealth or to security issues (support for agricultural enterprises, or autonomy with regard to food production). But the same principle applies: market interventions to reduce prices cannot be substantiated by reference to the problems associated with the market-based valuation of water.

The benefit represented by water generally exceeds the value assessed by individuals and expressed as demand, and is therefore greater than the value attached to it in the form of market prices. This is especially so with regard to the religious and cultural functions of water, its user, existence, bequest and option values (see Box D 5.4-2) and hence also its ecological significance (habitat function for ecosystems; solidarity with other creatures).

Valuation of ecological functions

Computing the value of water consumed as drinking water in private households or as a production factor in agriculture is comparatively simple, in that the utility function of the water dominates and the water is thought of in terms of its usefulness for individuals. Determining the value of water's ecological functions, especially its habitat and regulatory functions (see Section D 1.1), is a more difficult undertaking. As has already been emphasized, it is no longer possible to perform a valuation in isolation from other environmental media. On the contrary, there are indistinct transitions between water resources and soils that need to be considered.

Aquatic habitats are essential for a large number of flora and fauna, and conversely for the assimilation capacity of water. Usually, however, the functions performed for the ecological system are uppermost; individual benefits are generally involved in an indirect sense only. The habitat function, for example, has direct consequences for fisheries, which are dependent on sufficient fish stocks. Indirect benefits (see Box D 5.4-2) derive, for example, from the value that an individual assigns to an aquatic landscape without seeing and experiencing that value (the existence value), or the value that results from a potential utilization of genetic resources in the future (the option value). Such assessments of value, which are often based on intrinsic motivation, are rarely manifested as observable demand. To put a figure on the habitat function it is therefore necessary to apply dif-

BOX D 5.4-1

Economic valuation of agricultural water use

Two methods in particular are used for the valuation of water use in the agricultural sector, each of which can be used to determine a demand function for water (Gibbons, 1986).

Economic valuation based on a production function. The relationship between the deployed production factors and the goods produced – in this case agricultural produce – can be described mathematically on the basis of a production function. If other input factors such as fertilizer or working hours are kept constant, it is possible to calculate the marginal product that can be obtained by increasing water input. The data for such a production function can be acquired from experiments in which other inputs remain constant and the change in productive output induced by a variation in the quantity of water is determined.

Economic valuation based on a budgetary analysis. Since it is usually difficult to calculate the physical increase in production, recourse is often made to the budgetary analysis method, which enables these difficulties to be circumvented. The starting point of budgetary analysis is the total monetary yield that can be achieved with a certain kind of produce. The costs for procuring all the inputs, with the exception of water, are deducted from that amount. The remainder is the sum that the farm can spend on water without making a loss. If this sum is divided by the quantity of water used, the result obtained is the maximum average value of the water. This figure can be increased if the farm tries to reach a maximum sales volume for given input prices and capacity limits. With the help of linear programming, the optimum volume of water can be calculated for different water prices. The relevant demand curve can then be derived from these results.

An example is now given for a valuation using a production function. Ayer and Hoyt (quoted in Gibbons, 1986) begin their study by calculating the physical marginal productivity of water input in Arizona cotton farming, and then working out the marginal product value of the water in the case of a 10% reduction in water input. To do this, they took the selling price for cotton in 1975 and the higher price in 1980. The chart in Figure D 5.4-1 shows the results they obtained.

The chart shows two clear features: firstly, the marginal product value of water declines as water input increases due to the decrease in marginal productivity. Secondly, it is evident that the marginal product value of water is greater when the selling price is higher. The conclusion that can be drawn from these results is that government subsidies for farm produce, such as those provided by the Common Agricultural Policy in the EU, influence the economic value of water and therefore distort the allocation of the resource. The exaggerated value of water induces an excessive demand for water in the cultivation of subsidized produce, while incentives to develop water-efficient methods of cultivation are reduced accordingly.

In addition to distortions induced by manipulated sales prices, there are also those caused by water subsidies as such. There are many examples of how governments influence water prices in an attempt to control the level or the structure of production in sectors that are heavily reliant on water inputs (see Section D 3.3). The consequence of this state-induced undervaluation of water re-

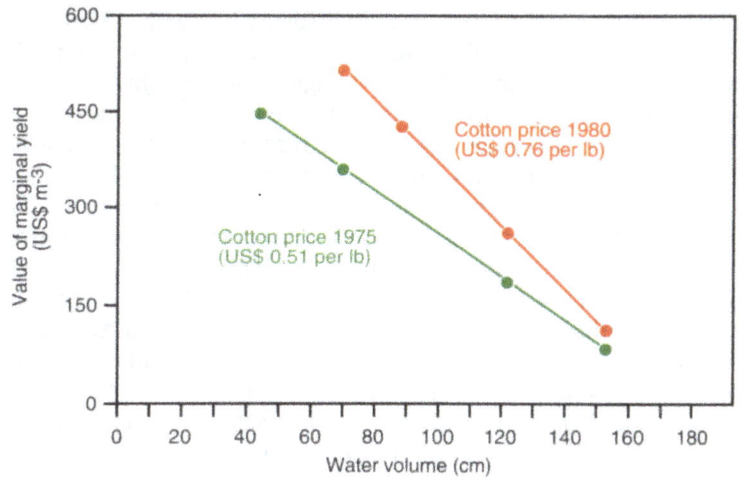

Figure D 5.4-1
Marginal yield value for water in Arizona cotton farming (1975 and 1980). 1 lb = 453.59 g.
Sources: adapted from Ayer and Hoyt, 1981, and from Gibbons, 1986

sources is that water is consumed more heavily than in the case of non-subsidization. This can be summed up by saying that market valuation produces good results as long as there is no distorting of prices for either farm produce or water as a result of market interventions.

In any economic valuation of agricultural water use, one must realize that the latter usually has a negative influence on the drinking-water function (deterioration in the quality of natural water resources) and that intensive irrigation often leads to land consumption due to salinization of soils and to loss of property. These consequential costs would have to be taken into account when valuating agricultural water use.

ferent methods than for determining the value of drinking water or of water inputs as a production factor in agriculture, where reference can be made to market data.

Indirect methods for measuring such values include the travel expenses approach, market simulations and the contingent valuation method (Pommerehne and Romans, 1992; Cansier, 1996 and Ewers and Rennings, 1996 for a discussion of these approaches). Applications of the travel expenses approach include surveying the leisure and recreational benefit of local recreation areas, waterbodies or forests. The basic idea behind this approach is to draw conclusions about an individual's willingness to pay from the actual expense that he or she must incur in order to consume a particular good, an "aquatic landscape", for instance. Another way to determine value is the direct valuation method, according to which individuals are surveyed with regard to their potential willingness to pay – e.g. to preserve an aquatic landscape. This approach can be used in any context, in theory, but it acquires special significance when there is no direct link to market variables. However, applying the approach is bound up with a number of significant problems regarding measurement, information, generation and distribution (Schulz, 1989). Generational problems, in particular, limit the usefulness of this valuation method. The loss of species diversity is an irreversible loss that is unlikely to be reflected to any adequate extent in individuals' willingness to pay. Reverting to expert knowledge is no alternative in this case, because it is generally difficult to assess the economic value that a diversity of genetic resources will have in the future. Too little is known about the potential uses of these resources, or about the future preference for the preservation of biodiversity. In valuing natural resources, it is therefore necessary to examine in each case the extent to which the cost-benefit approach can be applied as the basis for economic valuation. The essentiality of benefits and the likelihood of irreversible loss of a benefit have been specified as filter criteria in this context (Fromm, 1997).

Similar problems are encountered with the regulatory functions of water as with the habitat function. Water is a critical regulator of the Earth's energy and mass balance. Weather is greatly influenced by the evaporation and condensation of water, for example. Functions such as these are not included in economic variables either. The fact that the regulatory function of water has economic significance as well does not become fully apparent until these functions are impaired, or lost entirely. An example here is the clearing of floodplain forests, which adversely affects the regulatory function of water. In the process, the ecological system loses its capacity to absorb discharges of varying size. The value of this regulatory function becomes apparent when floods occur, and can then be measured directly in the form of flood damages. The example highlights the fact that economic valuation of water's natural functions cannot be confined to water alone, but must also take into account the neighboring ecosystems and landscape components of surface waters. A complex analysis of this kind is essential if the role of water in the overall ecological system is to be considered and evaluated to any adequate extent. It is at this juncture that the limitations of the economic valuation of ecological functions become patently obvious. Whatever the reservations, the complementary results obtained with different evaluation methods can provide important data for the sustainable management of freshwater resources.

Economists normally try to measure the value added or the national assets in terms of monetary variables, using systems of national accounts. There are major imperfections in the way that the value of water is determined, as the problems encountered in evaluating the ecological functions of water amply illustrate. It is therefore possible that Systems of National Accounts will underestimate rather than overestimate the importance of water to a society.

The critique of market valuation

Valuation by the market on the basis of individual assessment is occasionally criticized for making water prices too low, because people underestimate the

BOX D 5.4-2

Categories of value not normally manifested as willingness to pay

- The user value represents the benefit that an individual derives from the enjoyment of an aquatic landscape and possibly from its ecological diversity (e.g. by observing it, photographing it, etc.). The attempt is usually made to measure it indirectly using the travel expenses approach.
- The existence value describes the benefit accruing to people who are glad at the very fact that a landscape or living organism exists, without seeing or experiencing it (Krutilla, 1967). This type of value is manifested indirectly, for example in people's financial and personal commitment to nature conservation organizations.
- The bequest value results from the desire to bequeath a part of nature or a particular resource (e.g. a wetland area) to future generations on account of its symbolic or identification value, and to pay for same (e.g. in the form of a voluntary donation to an association for the conservation of a certain lake).
- The option value refers to the potential benefit obtained by preserving some form of future access (a gene pool, or a minimum reserve in the case of fossil groundwater stocks), the price for which is non-utilization in the present.

Source: adapted from Hampicke, 1991

importance of water to a society and are restricted in their horizons by narrow-mindedness and short-sightedness when assessing the value of water (for the debate on this issue, see Gawel, 1996; Endres and Finus, 1996; von Knorring, 1996; Stroebele, 1987). In response to such criticism, it should firstly be pointed out that most political interventions in the valuation process are directed against prices that are seen as too high, not too low. However, the logical conclusion for a critique based on market failures would be to demand higher prices. Leaving aside cases in which there is a lack of water-related rights, one speaks of market failures when insurmountable barriers prevent the development of markets (price-less conditions), when high transaction costs make market pricing more difficult and/or when major external effects lead to low prices (Benzler et al., 1995; Karl, 1997). Of these various types of market failure, externalities are the only one that plays a certain role in the field of water resource management. External effects are mostly a product of government failures (inadequacies in the legal framework) and would have to produce higher market prices to be successfully internalized. The political practice of many countries is to combat precisely these price increases by means of subsidies or price approval regulations. Indeed, in many developing countries the "private costs" are not even recovered in full, let alone the additional social costs. A recent study by the World Bank comes to the conclusion that, on average, the revenues of municipal utilities for water supply in developing countries cover only 35% of the extraction, processing and distribution costs, and that inadequate supplies to the poor are attributable to precisely this lack of cost recovery (Haarmeyer and Mody, 1997). This is quite obviously a case of government rather than market failure in solving the problem of valuation.

The worrying aspect is the fact that individuals – especially in countries that have to fight for their economic survival and/or have a low level of education – have a shorter and narrower planning horizon than experts or government decision makers. Such concerns are often mentioned in connection with the sustainability principle, which requires a long-term focus in planning and decision making, and that thought be given to the many side effects of (problematic) water use. When the myopic or blinkered perspective behind individual decisions is the result of poverty or lack of education and training, the appropriate response would be to combat poverty and disseminate information to individuals. Moreover, there is no assurance that government agencies are able to value resources more comprehensively and with a longer-term perspective than private-sector bodies. The answer to the question of who thinks more on a long-term basis, politicians with elections on their minds or private individuals, varies even in industrialized countries (Gerken and Renner, 1996; Kurz et al., 1996; Rennings et al., 1996; Klemmer et al., 1996). Caution should therefore be exercised when criticizing the decisions taken by individuals. It quickly leads to the legitimation of the state's valuations, which – as the example of many developing countries shows – gives rise to water prices that are counter-productive from the environmental perspective.

Another concern that is voiced in connection with valuation by markets and/or individuals is that the failure of competition could lead to a monopolist control of water resources and therefore to high prices. There are various ways of minimizing this risk, ranging from regulative controls on market transactions to a prohibitive tax on non-usage of ownership rights (Simpson, 1994). In addition, the specific properties of water permit a considerable degree of market transparency that facilitates not only the identification of power being misused, but also interventions against market manipulations and misuse. Another aspect is that the inefficiency of public-sector suppliers normally leads to higher prices than result from the abuse of power by private-sector suppliers. There is nothing to stop the state from remaining the actual owner of natural water stocks, however.

Summarizing what has been said so far, we can see that the multifunctionality of water produces a considerable diversity of valuation that is primarily an expression of different individual preferences and/or diverging assessments regarding the gainful use of water in the production process. Since any decision on allocation necessarily involves valuation of some kind, economic valuations are ultimately unavoidable. In this connection, the valuation of consumptive and productive uses by the members of society is best performed by markets. Market processes represent non-authoritarian and depoliticized methods of valuation which produce monetizable results that can be compared over time and between regions. Prices convey information in concentrated form and act as impersonal signals. Knowledge relating to the different uses of water and the associated benefits for individuals is focused in such prices. They are a collection point not only for past experience, but also for divergent expectations about the future – something that is often overlooked. Efforts should therefore be made to ensure that this method of valuation applies to all forms of regional water use.

5.4.1.2
Divergent properties of water as an economic commodity

Water scarcity necessitates not only a solution to the problem of valuation, but also resolution of the allocation issue. Who may receive how much water, of what quality, when, where and for what purposes? The decision as to how to solve this allocation problem depends on water's characteristics as a commodity. It should be noted here that water can have different commodity properties in the course of its utilization, so the recommended allocation mechanism can vary within the "water cycle" (see Section D 5.4.2.2). The following thoughts are therefore aimed at a better understanding of the various methods for solving the problem of allocation, which are examined in a later section.

The fact that water changes its characteristics as a commodity and that this leads to different conclusions regarding the allocation problem and how to manage it (private- or public-sector organizations, solutions based on user associations or cooperatives, a market or state function) means that the issue at stake involves more than mere definitions of terms. Different conceptual conclusions and political recommendations are inevitably produced.

Water can be an open-access resource (see Box D 5.4-3). This refers to the way that access is regulated. Water is classed as an open-access resource if there is free access for all. Free access generally means the absence of vested titles, or property without restrictions or controls on access. In the course of human development, nearly all natural resources were initially open-access resources. This was not a problem as s long as resources were available in abundance – relative to the number of users and level of withdrawals. Problems did not arise until open access, combined with rivalry over utilization (the unit of resources used by A is no longer available to B) in connection with renewable resources (animals and plants, as well as most water stocks), resulted in excessive use, in the sense of non-compliance with regeneration rates, and hence to declining stocks. The next consequence, especially when users exhibited little willingness or ability to cooperate, was competition over utilization with no incentives to consider the costs of use (e.g. declining regenerative capacity) and needs of sustainability (Fischer, 1997). Under certain conditions, therefore, open access can lead to loss of stock (use rates vs. regeneration rates) or to deteriorating quality, with all the associated risks that are all too visible in the case of water resources.

Similar problems occur when ownership rights exist – mostly with the state being the owner of major water resources – but the owner is unable to control or limit access to the resource. Nationalization of water resources is no guarantee that the common problem of excessive use or declining quality (e.g. adverse impacts on groundwater stocks due to agricultural practices) will actually be solved. The crucial requirement is rather the will and the energy to prohibit open access and to tackle the problem of surveillance. Problems are also generated in the case of transboundary resources when there is little willingness to cooperate internationally and competitive struggles over utilization in which regeneration needs are neglected.

State assignment of limited rights, cooperation, or the granting of private property rights can be ways of solving the problems of open access. However, all of these approaches involve an assignment of property rights. In the first case, the owner is the state, in the second the members of a certain community (com-

BOX D 5.4-3

Important categories of goods

- Open-access resources, as the term suggests, are those that are freely accessible.
- Economic goods are goods that form the subject matter of business negotiations and which generally have a price tag.
- Private goods are goods (individual goods) that are no longer available to others when used by someone with a demand for those goods, so there is rivalry for use.
- Collective property, unlike private property, does not lose its specific benefit for others once it has been used by somebody – "clean air can be breathed by everybody". The exclusiveness principle does not operate here, which is why existing demand is not manifested as an individual willingness to pay. This is because it is economically advantageous for the individual to wait until demand by others induces production of the goods, thus enabling one to participate in the benefit as a matter of course.
- Meritorial goods are private goods that political decision-makers believe are not supplied to an adequate extent.
- Commons property are goods that are characterized – mostly in the local or regional context – by open access and rivalry over use, at least in theory. The latter two aspects do not operate in practice, however, when a user's association is formed or when a collective solution limits access.

mon rights) that exhibits a willingness and capacity for cooperation, while in the latter case, the owners are individuals or legal entities (Fischer, 1997; Ostmann et al., 1997). This turns water into a commodity, an object of valuations, allocation decisions, negotiations and contracts, i.e. an economic commodity. The fact that water is scarce in most regions, indeed extremely scarce in some, and is therefore a factor that can unleash social conflicts, makes it necessary to value water in some way, as argued in the previous section. A "price" (weighting, prioritization) must be attached to water resources in respect of its particular functions. This is an essential characteristic of commodities. This price can develop through the operation of markets, or it can be set by collectives. The issue is also raised here as to who may use water and how the benefits to society, including the non-monetizable types of benefit and the various notions of equity, can be increased through the assignment of rights.

If water is private property, as in many cases (see Section D 5.4.2.2), utilization by one user inevitably excludes others from such use. It is then possible to allocate private or individual rights of disposal or withdrawal, and people can then vie over use. Such rights are exclusive and transferable. Private willingness to pay can then express itself, and a market-based solution involving the highest bidders is possible. Private property therefore enables the exclusion of others, in most cases. If an association of users is able to exclude third parties from use, this set-up will also permit partnerships. To that extent, local or regional cooperation on the basis of common property rights is quite compatible with market-based solutions.

These contrast with so-called collective property. Natural characteristics mean that such property is non-excludable. This means in the case of most surface waters and groundwater resources that when the natural assimilation capacity is exceeded, local deposition of pollutants can cause a loss of quality elsewhere, and others cannot be "excluded" from such impacts. Many natural water resources are connected hydrogeographically, which means that the activities of individuals, once they reach a certain intensity, can result in external impacts spreading over a wider area, affecting animals and plants as well. However, the quality of many natural water resources is an "asset" over which rivalry cannot ensue, similar to the quality of the air we breathe or noise protection, and which can therefore be characterized as collective property or public property. Sewage purification plants, for example, are producers of the public property referred to as "water quality", whereas agriculture is a stakeholder that adversely affects this quality in many cases.

As long as there is no consensus among the collective of users, the result can be false valuation and/or unwise use. In theory, a private willingness to pay may then develop (e.g. for water purification), but due to the lack of excludability this is not usually articulated as market demand, making it more difficult to solve the problem of valuation and allocation of "water property" as a good. Water users are not prepared to disclose their marginal benefit or disadvantage curves. Negotiated solutions in the sense of the Coase theorem are conceivable, but the negotiation costs rise quickly to very high levels as the number of participants increases. Responsibility for solving the valuation and allocation issue can then be transferred to government, or one can look for ways that enable a market solution after all. An example is the introduction of tradeable emission rights in conjunction with a stipulated maximum volume of emissions.

"Common property" resources are a special case. There is a basic rivalry over their use, to which private rights may be allocated. However, in contrast to individually assigned rights of disposal, there is an identifiable "community of interdependent users" here (Fischer, 1997) that regulates the rights (and obligations) of use within the community and which may exclude non-members from using the resources. This internal regulation can ensure efficient and sustainable use, as well act to preserve other functions of water.

Water can take on different forms of property, depending on the context of use and the particular stage in the water cycle. In the case of a waterbody used to preserve a biotope it is a public good, in the case of withdrawals from a free-flowing river an open-access resource, in the case of cult usage it is a meritorial good, while after treatment it can be an economic good on a market for drinking water. Certain water resources can also represent a good that yields a benefit beyond the narrower monetizable value. Such resources can acquire the character of a regional common good or a global common good. The desire to preserve and protect a certain condition usually dominates. The state assumes responsibility for protecting resources in most cases, which in the long run involves the non-utilization of resources. This diversity of uses and the corresponding types of property make it unlikely that only one class of measures, whether based on economic, political-administrative, social or cultural principles, will be sufficient to gain control over all water problems in their manifold utilization contexts. Different allocation mechanisms are available as options, depending on the specific problem and type of property in each case (see Section D 5.4.2.2).

Variety of valuation methods

The greater the scarcity of water resources – in terms of quantity and/or quality – the greater the competition between different uses of water. It has also been shown for market systems that the allocation of scarce resources and hence a separation of these competing uses occurs through the mechanism of price, in which individual willingness to pay is expressed. If water were merely private property, just like bread for example (another important source of nutrition), and not a crucial component of real ecological capital, then one could leave the question of valuation to market forces alone, whose functioning would have to be secured, however. But given the fact that water is also a public good, as seen in its habitat function, there are additional uses at stake that require supplementary or corrective valuations.

These, too, are economic valuations. They are supposed to estimate the value of water in those cases where individual assessments are too costly or where specific components of use have no (market) price. The intention here is to improve the allocation of water, in particular the allocation of this resource to various designated purposes. This usually leads to the demand for comprehensive benefit-cost analyses (including all external effects). In the 1996 amended version of the Safe Drinking Water Act in the USA, for example, more cost-benefit analyses are required than previously, and each year the limit values for five pollutants are to be (re)defined with their assistance (Davies, 1996). If one were to ignore the methodological problems associated with such a benefit-cost analysis, this would have to be endorsed as the most comprehensive form of economic valuation.

Critics of this economic approach demand the protection of human life and ecosystems, detached from human notions of utility and without the assessment of monetized values. As soon as water becomes scarce, however, valuations are made as a matter of necessity, so dispensing with this aspect would be counter-productive from the viewpoint of environmental policymaking. A total renunciation of pricing would imply, for example, that utilization rights be allocated by the state, initially at no cost. This would necessarily stimulate water consumption if there were no price tag on water, or if water prices were too low. The inevitable result would be a substitute valuation in the gray economy. The non-utilization of certain resources that would have to be ordained by the state if it is responsible for valuation would also be valued elsewhere instead. For example, if the state prohibits the use of a watercourse in order to protect a plant or animal species, this means formally spending the amounts of money that would otherwise have been generated by present or future uses (the opportunity costs). As can be seen, it is not possible to dispense with (economic) valuations. However, as will be shown in the following, they are extremely complex in nature because of the diversity of information and valuation aspects that have to be included in the reckoning.

The crucial issue in all valuations concerns who is to evaluate and/or which preferences are regarded as relevant. Are the values to be derived on the basis of individual assessments, or on the assessments made by environmental experts or politicians? In individualistic and democratically organized social systems, individual preferences are an important, if not the most important source of all value assessments. If behavior is rational, individuals will choose the collection of goods that provides the maximum benefit.

This is also the basic philosophy underlying the free-market economy. Benefits cannot normally be measured by direct means, so this is done indirectly instead by measuring the willingness to pay. Market-based valuation, within the limits referred to above, is therefore of enormous significance.

As previously mentioned, this approach based on the free market and individual preferences is the focus of repeated criticism. It is argued, for example, that many forms of environmental degradation, implications of resource use and long-term aspects are not directly perceived by individuals due to a certain narrowness of perspective, bias and short-sightedness, and that these aspects are not taken into account in any adequate way through the disclosure of preferences on markets. Contamination of groundwater cannot be experienced, for example, while rare biotopes may be so unattractive and noxious that people spontaneously reject them (Bonus, 1994). This is compounded by the "human defect" referred to by Platt, namely the tendency to give spontaneous preference to things that are directly accessible and short term (Platt, 1973; Ostmann et al., 1997), as well as by a certain inclination to turn a blind eye to crises (Ostmann et al., 1997). Reference is made again and again to inefficient and counterproductive patterns of individual behavior in a complex motivational context, with the demand that valuation be made by other agencies instead. The argument is sometimes advanced, especially from an expert or ecocentric perspective, that the valuation of natural resources should not be effected via markets, but by select committees that can think long and hard on these issues and base their assessments on preserving the functional capacity of ecological systems.

The assumption made here is that experts possess knowledge about nature's special characteristics. The question therefore emerges as to whether such expert committees are able to arrive at "better" valuations, who selects the experts and who is recognized as an expert in the first place. In view of the methodological problems involved in producing comprehensive benefit-cost analyses, it is legitimate to query the superiority of other valuation methods. Moreover, the criticism expressed about valuation by individuals is often too simplified, because a low willingness to pay for enhanced environmental quality is very often due to the fact that the individuals in question are inadequately informed. The criticism is therefore leveled more at the inadequate level of information among individuals, rather than the economic valuation method as such. Experts have the important task here of informing the public about pollution that cannot be perceived directly and its consequences, especially long-term and irreversible damage. The items of information communicated are frequently contradictory, so it is essential to ensure lively competition in the development and dissemination of hypotheses, i.e. scientific redundancy must be striven for, and the duplication of research is worthwhile.

5.4.1.3 The regional character of most water problems

Most water problems have a regional dimension. This results, on the one hand, from hydrogeographical conditions. Groundwater flow leads, for example, to local pollutant loads (e.g. from agriculture) affecting the quality of groundwater over a much larger area. Land-use changes in the catchment area of a waterworks can affect the quality of the collected water to be treated, and hence also the costs for treatment. Nitrates in particular, as well as some pesticides are deposited in groundwater – especially from intensively cultivated or special crops such as sugar beet, vegetables and vines (SRU, 1985 and 1987). The type and duration of agricultural use affects the regional water balance. Extending the utilization period by multiple cropping reduces runoff and infiltrate by about 10%, while grassland needs 20% more water on average than cropland due to the longer transpiration periods (SRU, 1985). This can lead to lowering of the water table and thus to escalation of water conflicts. Surface waters are also threatened. Nutrient-rich drainage water and eluviated soil increase the high nutrient loads of waterbodies still further. This can lead to excessive weed growth, algal blooms and to the risk of standing waters "dying" of eutrophication, have unfavorable effects on other water functions (recreation and bathing function) and cause water purification costs to rise.

In many large cities, the supply of water is dependent on the development of water reserves considerable distances away. This automatically leads to rising costs for water extraction and water transport, sometimes also for water treatment, and in any case to increased costs for water purification and wastewater disposal, as well as disadvantages for the areas from which the water is drawn. The construction and maintenance of a complex infrastructure for water supply and wastewater disposal are necessary, whereby a substantial proportion of these costs depends on distance. If these costs are not included in the price, there are no checks on agglomeration processes, and the areas needing water supply and wastewater disposal services expand uncontrollably. If costs are dependent on distance, the logical conclusion is to regionalize water supplies or to seek a regional solution to water problems. This leads in addition to regional disparities in valuation, and hence to regional price differentials in the case of market-based valuation and full costing.

In summary, then, water problems should be solved within the regional context. Only on this level is it possible to tackle conflicts over use in an suitable manner and master the problems of freeriding. This complies as well with the classical fiscal principle of equivalence. If an efficient solution is to gain sway, there should be a maximum spatial congruence between the beneficiaries of a (public) goods supply and those paying for the service. The liability principle obtains greater force in the process, and the subsidiarity principle comes to the fore. The latter aspect needs emphasizing in light of the problems that can arise when the state neglects to develop the infrastructures for public supply and disposal, or for local public transport, as exemplified by the rapidly growing Asian economies (see Section D 3.2). Disposal (of sewage, municipal waste, etc.), especially, requires that a specific infrastructure be created and enlarged in a time-consuming and capital-intensive process. If this is left out because of one-sided support for the generation of private capital, bottlenecks can develop that hamper development. This is particularly evident today in the Asian megacities. Regional governments and state should not be allowed to escape from their responsibility to consider these subsequent effects.

France, for example, has drawn the consequences and introduced a regional system of water resource management. The latter is organized into six major hydrogeographical areas, combined with an appropriate form of national policy control (Feder and Le Moigne, 1994). These regions correspond to the country's four large water catchments and the two densely populated and industrialized areas. Each region is managed by a committee (a kind of regional water parliament) and an implementing body, the river basin agency. The water parliament is a forum that brings together as many regional stakeholders in water resources as possible, i.e. a kind of regional assessment panel that ensures maximum participation. However, water supply and wastewater disposal are organized primarily by the private sector. A kind of pollution charge is levied at regional level, while the resultant revenue is also allocated within the region. Above all, water supply and wastewater disposal aspects are considered simultaneously. There are many reasons for transferring this model to developing countries, whereby the system could be started up in a series of "islands" (e.g. in megacities) (see Section D 3.5).

5.4.1.4 Growing importance of water efficiency

In economic terms, regional water problems are becoming more and more expensive to manage. In the early 1990s, it was assumed that investments of at least US$ 600–700 billion would be needed in the coming decade for the developing countries alone (for water supply, sewage systems, etc.). Recent studies conclude that approximately US$ 60 billion a year will be needed over the next ten years purely for drinking water supply. The colossal investments needed in central and eastern Europe were left out of the reckoning here. Thus, solving regional water problems involves the issue of cost to an ever greater degree, whereby it is not just a question of how to finance planned projects but above all the enhancement of efficiency.

It is true to say that major effects are often achieved by small-scale measures – in an attempt to reduce per capita water consumption, Mexico City replaced about 350,000 toilets with smaller and more efficient models, saving enough water to cover the needs of 250,000 inhabitants (Feder and Le Moigne, 1994). In the majority of cases, however, it is necessary to make large-scale investments and to establish and develop a very capital-intensive infrastructure for supply and disposal. Mounting sanitation problems require large investments in sewage systems, for example. The trend to large-scale projects in the growth regions of Asia, Latin America and Africa is based on the idea of integrating the twin services of drinking water supply and sewage disposal. The planning authority no longer invites tenders for "construction of a sewage treatment plant" (for financing reasons alone), but comprehensive "build, operate and transfer" models (Rudolph and Gärtner, 1997).

Yet it is precisely here that efficiency acquires special relevance. In particular, it is evident that many economic problems (inefficiencies, excessive costs) – this applies to Germany as well in part – result from the fragmentation and/or the segmentation of the various administrative tasks in water resource management, and from oversized government administrations. It is increasingly obvious that, in view of the high costs involved, major efforts are needed to improve efficiency and to exploit the potential for cutting costs. The World Bank, which channels a large proportion of its loans to water development projects, is becoming more and more active in this area (World Bank, 1992; Ayub and Kuffner, 1994; Feder and Le Moigne, 1994; Engelman and LeRoy, 1995; Briscoe, 1992).

In view of the investments needed, the search is on for incentive systems that attract more foreign capital and know-how and promise greater efficiency and return on investment, in cases where the lag in development and the poverty of the population exclude any self-financing. It can be assumed on the whole that enough private-sector corporations in the world are in a position to perform these tasks in a profitable manner. French, British and North American companies are ready to take over not just the planning and construction of water supply and wastewater disposal systems, but also the financing and operation (temporarily, at least). What is surprising is the minimal involvement of German companies in this field of activity. This contradicts the high (technical) reputation enjoyed by German water companies and is attributable in large measure to the fact that the German model is not particularly convincing for others (Rudolph and Gärtner, 1997). The latter results to a considerable extent from the inefficient organization of water resource management and from the excessive costs for municipal sewage disposal (Rudolph, 1990; Gellert, 1991). Entrepreneurial failure (unwillingness to get involved in activities abroad) seems to play a role here as well, however. This is regrettable, because a convincing German model could turn into an export hit, not only for environmental but also for economic reasons. For example, there are innovative ideas relating to disposal infrastructures (accessible pipes, for example), that could prove interesting for the rapidly growing megacities of Asia (Stein, 1997). Support could be given to pilot projects in this area.

Various conditions need to be met before private capital will flow into the water resources sector (solution to the financing problem) and lead to greater efficiency. Some of the most important are: permitting full recovery of costs with a reasonable profit margin, the renunciation of price subsidization policies and guarantees of calculable long-term contracts granting certain utilization and/or property rights. As previously mentioned, there are ways to counteract the formation of exploitative monopolies. Competitive tendering is of particular importance in this context.

Various formulae are possible with regard to ownership (of natural water resources, for example), investments, operation, cost accounting and cost recovery, which allow for a variety of water supply systems (Table D 5.4-1). As can be seen from the analysis of models already in operation, natural water resources remain the property of the state in the majority of cases; the state need only ensure that these do not become open-access resources, with all the quality and overexploitation problems that this involves. Property rights pertaining to investments are best left in the private sector, it would seem. Putting operations into the hands of private-sector enterprises generates benefits above all in the form of enhanced efficiency. Transferring the business risk to the private sector creates incentives for efficient management (Haarmeyer and Mody, 1997). Full-service concessions provide the broadest scope for improvements in the business administration and financial areas. In order to reduce the risks for private operators, it is necessary in many cases to arrange credit guarantees from international organizations (the World Bank, for example) or from governments, or other forms of partial risk assumption.

Table D 5.4-1
Distribution of responsibilities in alternative approaches to water supply: evaluation of case studies.
BOO = Build-Own-Operate; BOT = Build-Operate-Transfer.
Source: Haarmeyer and Mody, 1997

	Management contract	Rental agreement	BOO/BOT Concession	Full-Service Supply company	Sale of shares
Rights transferred to					
Ownership:	State	State	State	State	Private sector
Investment:	State	State	Private sector	Private sector	Private sector
Operation:	Private sector	Private sector	Private sector	Private sector	Private sector
Collecting charges:	State/Private sector	Private sector	State	Private sector	Private sector
Recent cases					
	Puerto Rico Mexico Trinidad and Tobago Antalya, Turkey	Guinea Gdansk, Poland North Boehmen, Czech Republic Chihuahua, Mexico	Johor Sydney, Australia Izmir, Turkey Côte d'Ivoire Macao	Buenos Aires Malaysia Liberia Brazil	England and Wales

5.4.2 Solving the allocation problem

5.4.2.1 Basic options

Once the qualitative and quantitative problem of scarcity has been recognized and the valuation issue has been resolved, the uppermost question then concerns who is allowed to use water resources to what extent (for which purposes, when, where and how) and/or who is to assign the utilization rights, i.e. how the problem of allocation is to be managed giving due consideration to equity and sustainability aspects. Economic science and social research have identified various organizational and regulatory systems for solving the allocation problem (see Reuter, 1994 on the social scientific research and Wagner and Lorenz, 1995 on evolutionary economics). The institutions-theoretical approach proceeds on the basis of rational decision-making by individuals acting primarily in their own self-interest, regardless of whether these are consumers, entrepreneurs, politicians or bureaucrats (Brennan and Buchanan, 1985; Williamson, 1985). On the basis of these individual goals, the approach analyzes decision-making and coodination procedures to identify those that appear acceptable to all concerned on account of the benefits they provide (on the criterion of acceptability see, *inter alia*, Buchanan, 1975 and 1987; Knight, 1992). Observing the acceptability for rational individuals maximizing their self-interest provides for greater stability of agreements and security of expectations for the stakeholders involved (see Williamson, 1985; Ostrom, 1986).

Institutions can thus be understood as generally recognized rules. These rules can arise spontaneously, or be laid down by government. Individuals will behave in conformity with rules if they can assume that others will also do so, or if non-compliance with rules leads to sanctions being imposed. In repetitive situations, this leads to mutual expectations about behavior and thus to reduced complexity in decision-making situations. Institutions are therefore able to steer individual behavior in certain directions.

Although there is some controversy regarding the classification of regulatory and organizational systems, the Council assumes the following types of regulatory systems:

1. *Anomie, or the total failure of institutions:* This condition is characterized by the disintegration of all institutions and principles of order, or by the unenforceability of (traditional or state-defined) rules. Rule-less conditions are usually temporary in nature, often dominated by the ruthless enforcement of particular individual interests coupled with disregard for the harm caused to others. In a situation of this kind, where conflicts are fought out without the constraints of rules, the probability that arms will be used to resolve the conflict is very high. Transboundary environmental problems and the management of transboundary resources must clearly be resolved by negotiations and the establishment of institutions, especially when these resources are collective property.
2. *Problem-solving options within institutional frameworks:*
   a) Government solutions: In this regime, the allocation of utilization rights is decided in simple societies by a person recognized as an authority by all (the shaman, the medicine woman or the chieftain, for instance), while in modern societies this decision is usually the responsibility of the state. This may take the form of quotas (a certain amount per user), administrative prices, or restrictions on use (e.g. the establishment of water protection zones). A government solution presupposes that problems concerning information and evaluation have been surmounted, and that there is sufficient enforceability and monitoring authority.
   It is important in this context that the allocation targets relevant for assigning utilization rights must normally be formulated within a decision-making complex in which the self-interest of politicians, as well as the aims and ideas of political competitors, voters, bureaucracies and key interest groups all play a role (see Endres and Finus, 1996). The result can be a markedly short-term orientation. Current thinking is focused on the question of the best way to institutionalize long-term responsibility (Rennings et al., 1996; SRU, 1994; Maier Rigaud, 1994). The idea of setting up Environmental Councils or giving added weight to the subsidiarity principle are controversial. Research is needed here, precisely with respect to water resources.
   b) Market-based solutions under competitive conditions: In this free-market approach, the members of society compete with each other for scarce goods in accordance with certain "rules of the game". The granting of private property rights is constitutive for this solution. Property and utilization rights with respect to water resources provide owners with the incentive to maintain over the long term the potential of the resources to satisfy the needs of water customers and to manage the resource carefully and wisely. Property rights to resources can be sold

and acquired on the market in the same way as land or other goods.

Here, too, it is possible that an inadequate sense of long-term responsibility may develop. A relevant issue in that case is how to instigate an enhanced long-term orientation in individual planning and stimulate intrinsic motivation (especially in terms of allocation aspects) (Frey and Oberholzer-Gee, 1996). There is still a considerable need of research in this area.

c) Community or common property solutions: The lack of national solutions based on privatization for the management of natural resources is awakening interest in community or common property solutions. Such solutions are well able to produce good results under certain circumstances. This, too, is an institutional approach, in that the members of the solidaristic community are governed by commonly defined rules for dealing with the common property (Ostman et al., 1997). The commons dilemma (Hardin, 1968) need not arise, because these are not open-access goods and because others can be excluded from use due to the fact that the resources are the common property of the community. In this regime, collective rights to property or utilization are granted to groups of individuals; the resources are viewed as common pools that may be used jointly by the members of a community according to predefined rules (Ostrom, 1990; Fischer, 1997, Frey and Oberholzer-Gee, 1996; Ostmann et al., 1997). A typical form of such communities are cooperatives, in which business operations are conducted jointly.

Such solidaristic communities are social entities, usually with a limited number of members, clearly defined spatial boundaries, certain joint interests, a minimum degree of interaction among members, common cultural norms in most cases and a separate endogenous system of authority (Fischer, 1997). The economic benefit frequently consists in the fact that the net benefit for a community can be raised by eliminating externalities and adopting a long-range orientation. The latter succeeds above all when the time preference rate can be lowered, because, within the community, having to wait (in the sense of non-utilization) is associated with lower risks. In many countries with "traditional" social structures and less focus on private property, such communities can prove to be interesting solutions to water-related problems. More research needs to be conducted into the basic conditions necessary for such solutions to develop.

It is clear from the above that adhering to a fundamental maxim such as *the more market, the better* or *private property is always preferable to common property* makes neither sense nor promises success. It depends on the circumstances in each case. If private property exists, or privatization could occur without major transaction costs, the obvious principle to apply is to give preference to free-market regimes, just as it is obvious to strive for national solutions in the case of public goods. In all mixed forms, and where the respective regime has little chance of success, an appropriate solution has to be developed from the specific context. A matching solution of this kind can be a mixture of different organizational concepts or a change in the conditions such that a single organizational form is adequate for the desired solution.

In the past, many uses of water were regulated by the state even when water resources were mainly private property. However, it is becoming increasingly clear that the costs associated with this usually inefficient solution are higher than regulation by a system of competition and that the concerns associated with the state regime could have been solved similarly or even better by the market. Today's imbalance between market and state should thus be equalized in most cases by increasing the role of market forces. However, it would be wrong to conclude from this that market-based systems can be applied to all property regimes for water resources.

Nor should one jump to the conclusion, from the obvious failure of community ownership in most highly developed countries, that cooperative solutions or forms of community property are necessarily pointless in other cultures as well. Just as many commons solutions have proved highly stable over the centuries in Europe, Asia and Africa, many present-day cultures exhibit all the prerequisites to suggest that, in the case of goods that are still predominantly open-access resources, common property solutions will be superior to state-regulated and free-market solutions. Ostrom (1992) showed with reference to irrigation systems that self-governing systems based on shared use of a local or regional resource can respond to changes in the environment (such as drought) more flexibly, allocate water more equitably and counteract the abuse and neglect of irrigation systems. The management of common pool resources shows that users are quite prepared under certain conditions to agree on lasting rules for long-term, efficient use of resources. The crucial factor on which the success of collective agreements depends is that all users are prepared to recognize previously established principles governing the allocation and management of water resources, and also to act according to such principles even in the absence of state controls. However, it seems necessary to have a decentralized system for monitoring compliance and sanctioning those who infringe against the rules (Ostrom, 1990).

These conditions are satisfied only in social systems in which a high degree of social or cultural cohesion, consensus-oriented decision making and little mobility prevail. Social systems fitting this description can be found in a number of developing countries. As mobility increases (leading to inflows and outflows), even successful common pool arrangements will tend to break apart, because newcomers are rarely willing to pay the "entry fee", while those moving away demand that part of their deposit be returned. That said, the common pool arrangement can be the preliminary stage for a later solution based on ownership by a private company. A gradual transition seems especially appropriate in societies exhibiting a high level of intrinsic motivation that could be destroyed all too quickly if privatization is too rapid. The effectiveness of common pool solutions is greatest where cultural traditions and the processes of social regulation encourage and reward a sense of responsibility for the common weal. This can occur through cultural obligations and taboos, social networks and external pressure.

5.4.2.2 Water markets as the solution

In view of the problems associated with state management of valuation and allocation, there has been a growing focus in recent debate in the academic literature and among key organizations (the World Bank, for instance) on water markets as a way of solving water problems. The Council shares this view, whereby the issue of how to cater adequately for social concerns, such as securing minimum water requirements for individuals or social groups, is dealt with in Section D 5.4.2.3. Establishing water markets requires a differentiated approach, however, because when allocating enforceable property rights over water, various stakeholders can develop, as well as markets as reference systems (Fig. D 5.4-2), whereby many of these markets can become small-scale monopoly markets, similar to the situation with power supply and waste management. The following comments are therefore intended to clarify the different levels that are affected and where additional activity is needed on the part of the state.

Rough distinctions can be drawn between the following kinds of stakeholder:

- owners of natural water resources,
- water withdrawers,
- water distributors,
- water users,
- wastewater collectors,
- wastewater recyclers.

If the natural water supply in a water region (river basin, lake, aquifer) is no longer sufficient to meet a growing demand for water, then scarcity and its related problems will ensue. In the case of standing waters, every form of withdrawal from natural water stocks is made at the expense of other withdrawals. With running waters, withdrawals in the upper reaches are usually to the detriment of downstream riparians. If no limits are placed on withdrawals, natural water resources will adopt the character of open-access resources in contexts where property rights and cooperative solutions are absent: although individual users are not excluded from the resource, all users are rivals with each other. Such a situation must lead to overexploitation of the resource – due to open access, renunciation of withdrawal by some individuals will produce no pay-off if use is dictated by self-interest, because everyone must assume that quantities not withdrawn will be used by others instead. Running waters can also turn into overexploited open-access resources if the natural supply of water is no longer sufficient to satisfy all users along the watercourse.

The spatial distribution of negative impacts is uneven in such cases, however. Whereas the excessive use of a standing waterbody harms all riparians, it is primarily the lower riparians who are affected by overexploitation of running waters.

In both cases, it is imperative to limit withdrawals in order to protect the waterbodies. This has two implications: firstly, limits must be imposed on the total periodic withdrawals from the water region in accordance with the resource-specific renewal periods. Secondly, it may be necessary to allocate withdrawal rights to particular individuals (riparians of a surface waterbody, or to landowners in respect of groundwater stocks). This makes it necessary in water regions to specify a proprietor of the water resources who defines the maximum permissible withdrawal levels, and to create an organization that has the capacity to enforce limits on withdrawals. The latter is a serious problem in many developing countries, and reaches major proportions when large areas are owned by the state. If one considers the historical land use patterns in water regions, which are often the result of decisions made before water scarcity became a problem, then the quantitative withdrawal rights that need to be introduced in water regions will have to be geographically differentiated (according to the doctrine of prior appropriation – "first in time means first in right"; Hirshleifer et al., 1972).

Granting withdrawal rights to the riparians of surface waters or the owners of land situated over groundwater stocks is the first way to introduce wa-

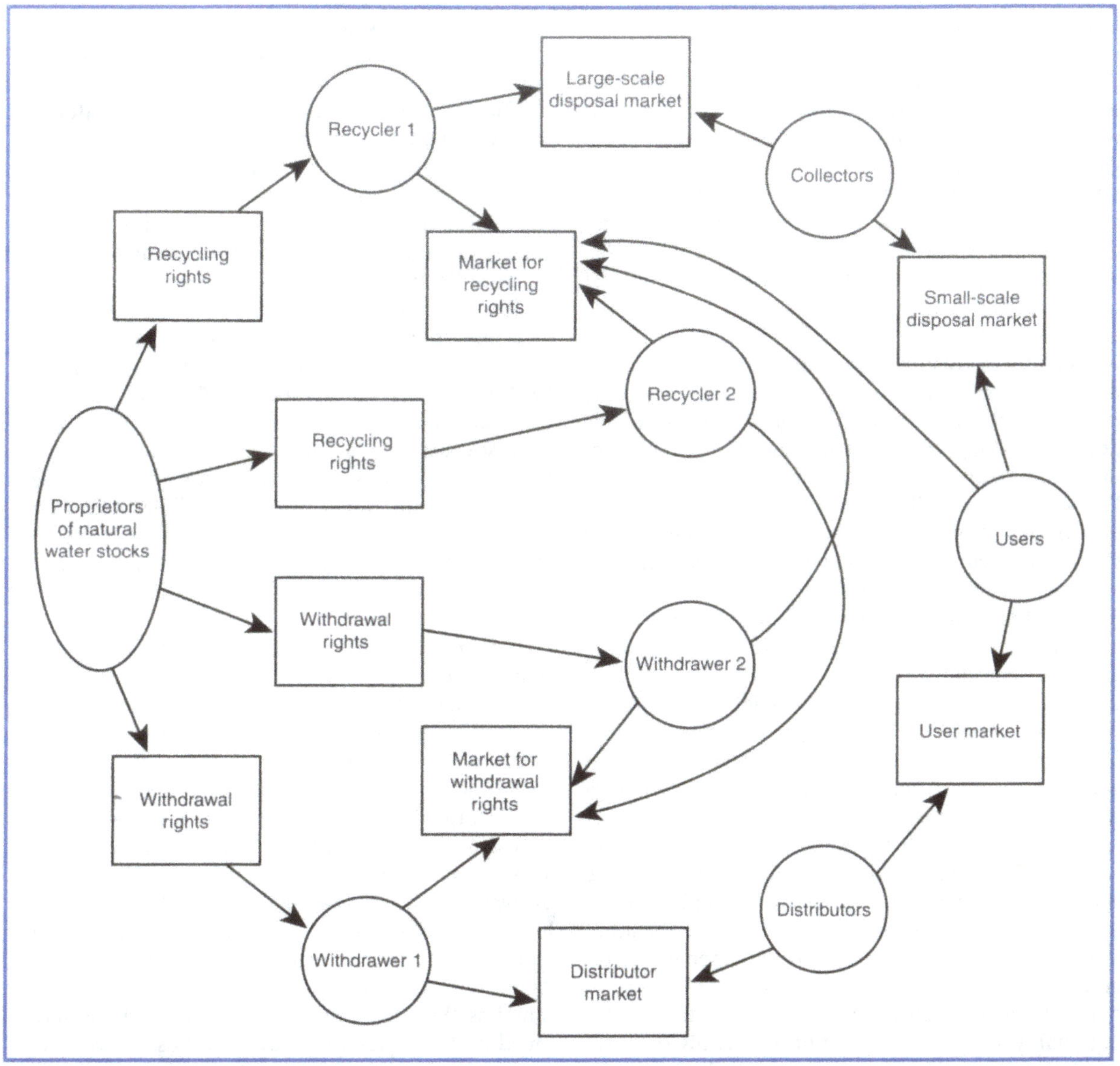

Figure D 5.4-2
Water markets. Explanations in the text.
Source: WBGU

ter markets. These players are thus entitled to trade their water withdrawal rights. Barter deals may be made with the aim of acquiring water withdrawal rights on a long-term basis, or they may serve merely to cope with temporary regional shortages. The price will also incorporate estimates about future trends. This market can be called a market for withdrawal rights. The issue that needs clarifying here is where there may be limits to the market that make it necessary to stipulate when rights may be traded between regions and when not. If unrestricted interregional trade is permitted, a scarcity of water rights may induce regional adjustments on a large scale. Enormous importance is therefore attached to the holders of withdrawal rights, their responsibilities and their goals. In order to prevent stockpiling, it would also be conceivable to revoke water utilization rights if they are not made use of by their owner.

Establishing this system of valuation and allocation requires the granting of property rights. If this is done in accordance with the fiscal equivalence principle, the subsidiarity and liability principles will also operate. To counteract the misuse of power that is liable to occur when such monopolies are created, checks and controls must be included, or additional measures implemented (temporary invitation to tender, price approval systems); water's natural characteristics allow for considerable transparency in this regard.

This market for withdrawal rights is coupled to a second market for distribution rights. The particular way in which the quantities of water to which a region is entitled, on the basis of the original rights and after barter deals, is then distributed within the region to individual users, is an issue that must be decided within the region itself. Withdrawers and distributors need not be identical, however. In contrast to the market for withdrawal rights, the total volume of which must usually be specified for a water region collectively or by government authority, it is neither necessary nor appropriate to decide on the specific uses to which water may flow by means of collective decision-making. Beyond ensuring the minimum individual requirement for water (see Section D 5.4.2.3), the distribution of water as private property once it has been withdrawn from the natural stock can occur through the market and hence be controlled by the willingness to pay. Two markets are relevant here: markets on which the withdrawals for a region are traded in order to distribute them among different sectors (distribution markets), and markets through which the water volume is delivered to users (utilization markets). The willingness to pay on the part of the individual users determines the uses for which the water is supplied. The general tendency will be for willingness to pay to be highest where water has or creates the highest relative values. If individual users are faced with the full costs of using the water, then efficient incentives will be in place to reduce water use. Individual distributors have the additional option of buying water from other water regions. Governments can limit their involvement in particular water regions to checks on whether future regional water needs and the future supply of water are being properly ascertained, and whether steps have been taken to ensure that supply and demand are equalized.

Users return the water they have been supplied back to the water cycle via air (evaporation), the soil (infiltration, runoff) and collection systems. If the quality of the water has not changed to such an extent that subsequent uses are impaired, then scarcity will only arise if – for sanitary reasons, for example – the used water must be collected and disposed of locally using technical facilities (sewage pipes, pumps). Depending on circumstances, two markets may develop here: a market for single disposal facilities (often necessitated by sanitary factors), and a market on which disposal agencies can connect small sub-sections of a region to large sewage systems. It is not necessary to have special recycling rights enabling the discharge of qualitatively unchanged water into natural water resources. In order to establish efficient incentives to save water, the costs for recycling water must be charged to water users on the disposal markets.

The situation is different when the used water is of inferior quality – as is normally the case – and subsequent uses are therefore limited. Scarcities appear here due to water's limited capacity for self-purification. Costs no longer arise purely because scarce resources must be deployed for building and operating water collection systems, but also because contamination of water by one user reduces the options available elsewhere for using the water. Water pollution also results in restrictions on water withdrawals. Withdrawal rights defined purely in quantitative terms can thus become worthless if the quality of water resources is modified so strongly by returned water that the options for using the water – still present in the relevant quantities – are eliminated. If the quality of the water for which withdrawal rights exist is to be safeguarded, there must be a way of influencing the quality of the water that is discharged to natural water resources.

If withdrawal rights are intended to ensure not only the access to certain quantities of water, but also to water of a certain quality, it is necessary to specify the total permitted contamination in recycled water, whereby the parameter can be the water quality at certain withdrawal sites (or utilization sites). As long as this target is not exceeded, then recycling rights assigned to specific players can be exchanged on markets for recycling rights. In a manner similar to withdrawal rights, the granting of recycling rights will have to be based at first on historically acquired utilization rights. In the longer term, trading of recycling rights will lead to a regional utilization structure that matches the qualitative scarcities and the situation faced by the water withdrawal bodies. If scarce and tradeable recycling rights exist, then increasing qualitative demands on the water supply will be reflected in higher prices because recycling rights will have to be purchased elsewhere. This raises the incentives for avoiding or reducing water pollution from the outset. Recycling rights, too, must be subject to rules specifying when such rights may be traded and when not.

The manner in which scarce recycling rights are distributed within a water region is something that markets for large-scale and small-scale disposal services can determine: pollution rights will tend to shift towards those polluters for whom the cost of preventing water pollution are highest. For the water region as a whole, the only important factor is that the pollution caused to the natural water resources does not exceed the legally approved level. It is essential, however, that such a control system induces searches for better solutions or for better institutional arrangements.

Water withdrawers may operate on the market for recycling rights as well, in theory, just as recyclers may become involved in the market for withdrawal rights. Withdrawers can acquire recycling rights, for example. This is of interest to them if it means that the quality of the water they supply can be improved and there is a corresponding willingness in the region to pay for improved quality. This is where free rider problems may develop: every reduction in the discharges to a standing waterbody benefits all withdrawers, while all downstream riparians will benefit from reductions in the volume of wastewater discharged upstream. On the other hand, interested recyclers may acquire withdrawal rights – to the extent that these are linked to "quality" rights – and in this way acquire the license to pollute water more. Problems can arise in this context as well, because increased water pollution can affect all withdrawers from standing waters and all downstream riparians of running waters. Thus, it is necessary as a rule that both withdrawal rights and discharge rights within a region be allocated initially by a single authorized rightholder (usually the state). An exclusive allocation of withdrawal rights that also guarantee a certain level of water quality would probably result – if the number of users in a water region is relatively large – in a system of allocating rights that is not optimal in economic terms, and which cannot be improved by exchange transactions either. Problems occur because an inflow of polluted water into the natural water system is only possible if all affected withdrawers relinquish their claims regarding the quality of their water withdrawals. Conversely, if recycling rights alone are issued, free rider problems will arise among those interested in water of a qualitatively high standard, because every purchase of recycling rights by a withdrawer would benefit other withdrawers.

Finally, there is a special problem that needs addressing. Water pollution as a result of inflows occurs not only through channeled inflow to and channeled disposal from various types of use, but also through the unchanneled discharge from channeled inflows and precipitation. In the latter case, the contamination of water resources is not due to the water being polluted by use, but to the fact that land uses – in agriculture, for example – result in the deposition of substances in soils and that these substances are leached by inflows of water from irrigation schemes or precipitation before leading to water pollution. Depending on the regulations in force, either the land users have to acquire such rights on the market for recycling rights – because, despite not withdrawing and polluting water directly, the contamination of water emanates from their land – or else withdrawers must acquire recycling rights from the land users and take them out of circulation.

5.4.2.3 Securing minimum water requirements

A fundamental objection against water markets as an institutional arrangement for solving water problems is often that water is an essential good for humans, that is, people are not able to survive unless they have access to a basic minimum quantity (of a certain quality). This minimum quantity is needed for drinking, for the preparation of food and for sanitation purposes, and is largely defined by natural determinants. Water is also an essential good for production, because certain vital production processes, food production being the most important, are not possible without an adequate supply of water. The following section examines how securing such requirements can be linked to market-based solutions.

A person's minimum water requirement is therefore shaped by two main factors:

- the essential needs, according to the laws of nature, that must be met in order to maintain the vital functions of the body and
- the minimum amount of water needed to produce essential goods, above all food.

The first type of need is inseparable from the individual person, whereas this is not true for the second. In principle, it is possible for essential goods requiring water as an indispensable production factor to be supplied by third parties and purchased on markets – a system that is also practised in many countries. The individual's minimum water requirement rises if a such an exchange cannot take place, because

- there are no institutional incentives to produce goods under division of labor,
- due to institutional deficits, goods produced under division of labor are not supplied in order to secure individual minimum requirements,
- goods produced under division of labor cannot be demanded by everyone to the extent necessary to cover their minimum needs (mainly because incomes are too low).

Two issues are therefore raised in connection with minimum water supplies: to what extent is it possible and economically advantageous to reduce the direct, minimum water requirement of individuals by ensuring that the production of essential goods requiring water for production is organized in a system of division of labor, and how can the remainder of the minimum water requirement be secured? The acceptance of the division of labor system in production requires that the supply of essential goods to individuals appears sufficiently secure that they are prepared to dispense with producing their own goods. This, in turn, requires institutions that enable and support an economic system based on the division of labor. A supply of the relevant goods must be assured, and there must be guarantees that people demanding these goods will also receive at least a minimum quantity of them. This is true for both essential goods involving water inputs during production, as well as the essential water needs of individuals.

Aside from communities and cooperatives, there are various options in large societies for securing the minimum supply of water for individuals:

- allocation of certain quantities, e.g. in the form of coupons,
- fixing maximum prices,
- subsidizing supply,
- making transfer payments to consumers.

In small societies, it is quite possible that the decision is made on a case-by-case basis according to the severity of need, and that quantities are allocated accordingly. This is because it is relatively easy to acquire knowledge about special circumstances, needs, services etc., and because it easy to monitor the behavior of individuals. In large societies, it is virtually impossible to operate a system enabling the consideration of individual circumstances. Gaps in knowledge and monitoring problems compel the use of other coordination methods for securing minimum requirements that produce satisfactory results in the majority of cases. All such alternatives presuppose a more or less intact and functioning system of public administration, because reallocation should and must be made only for those who are unable with their own resources to secure a minimum supply of water.

Setting maximum prices for water supplies – a method commonly used in developing countries for ensuring supplies to the poor – is bound up, on the one hand, with the problem that either costs are not fully covered on the supply side – meaning that financing supply must be arranged by other means as well – or that the supply is quantitatively and qualitatively inadequate. On the other hand, there is a risk (albeit limited in the case of water supplies) of black markets developing, where water is traded at a higher price, thus making the situation for the poor even worse. Moreover, those profiting from price ceilings are not just people on low incomes, but also the relatively rich members of society, which should not be the intention of allocation policies.

A generalized subsidization of water supplies is problematic for two reasons – no distinction is made with regard either to the purpose for which water is used, or to the people who have a demand for water. The result is that water supplied too cheaply is consumed to an extent far beyond the essential minimum requirement. The development and use of water-saving techniques and organizational forms produces no gains, while switching to less water-intensive applications is prohibitively expensive. Subsidization benefits all those with a demand for water, as long as it is not dependent on individual income. Relatively excessive water withdrawals can then exacerbate scarcities and cause social costs elsewhere (due to the increase in water-related conflicts, for example). If at all, subsidization is appropriate only when it is possible to differentiate between different categories of users and between different types of water use.

In contrast, the target-group approach of paying "water money" to the needy makes it possible, in theory at least, to improve the income situation of the latter to such an extent that they have access to a minimum water supply. Benefit of this kind can be conditional on the particular situation of the recipient in such a way that people earning higher incomes are excluded. If these payments cover more than the absolutely essential minimum requirement, the individuals in question will also have an incentive to save water, because the benefit thus acquired is then available for other goods. All that needs to be financed is access to the minimum water requirement – defined in terms of physical or social needs. Water then remains at the price dictated by scarcity, and significant incentives are generated for reducing water needs by means of measures to reduce waste and reuse water resources. At the same time, the opportunity to operate profitable water supply systems will act as an incentive for foreign investors to intensify their involvement.

5.4.3 Water resource management in Germany and the USA – a comparison

5.4.3.1 Preliminary remarks

One cannot generalize about how states should best cope with water scarcity with respect to water quality and water quantity (Wolff, 1996). While abstract statements can be made about the economic factors that have to be present if scarce resources are to used efficiently, it is not possible to generalize about the institutions (also institutions in the sense of organizations) that are best able to bring about such conditions. Existing institutions may already be in a position to cope with water scarcities, or be suitable – following appropriate adjustments – for performing this task, to such an extent that the main problems to be solved relate to information (e.g. on the presence and the effects of substances, or the connection between water quality and health). In many cases, however, major institutional adjustments will be necessary, sometimes requiring the creation of entirely new institutions. Since these adjustments and provisions involve costs, in addition to the actual operating costs of the institutions themselves, the most economical solutions for managing water scarcities can vary enormously from one state to the next.

The only thing that can be defined in general terms is what has to be available, namely information about water reserves and replenishment rates, about substances and their effects, about sources of loads, about different technical alternatives for water supply and wastewater disposal, about financial and real capital, human resources, etc.

How this knowledge and capital is then deployed for maximum benefit must be ascertained on a mainly country-specific basis, taking into account the legal or traditional institutions that exist, because the use and consideration of existing institutions produce cost benefits – new institutions do not have to be established, and friction between new and old institutions can be minimized.

Two countries are briefly examined in the following: firstly Germany, because the efficiency of water resource management in one's own country is always of interest, and secondly the USA, where experiments are being conducted there with innovative approaches in the form of water markets. This is followed by a look at the legal set-up for water resource management in many countries.

5.4.3.2 Water resource management in Germany

In Germany, the quantity and quality of the water available for and allocated to the various uses are specified by government authorities. This means that the delivery of essential goods (minimum supply of drinking water, protection of important biotopes), which must be managed by public authorities, is largely secured from the outset. Alongside these essential services, however, the state is also responsible for those elements that can be organized through competition. This is most evident in the case of water consumption and water withdrawals.

At first sight, the water supply system in Germany is organized along private-sector lines, i.e. by autonomous companies or joint-stock corporations. In fact, however, governments are able to exert a substantial degree of executive control or at least influence by virtue of their holdings in such private-sector enterprises. The borderline between public-sector and private-sector enterprises is by no means clear-cut. Competition is virtually non-existent due to government constraints. Above all, so-called license agreements were used to establish a set of territorial monopolies in order to comply with minimal technical standards and to prevent the duplication of infrastructural networks.

In Germany, wastewater disposal is likewise the responsibility of the state and, despite the collective property argument having much less weight in this connection, devolves entirely to local government. Bureaucratic administrative structuress akin to public authorities are highly prevalent (complex system of departments, public-service employment law).

This way of organizing the water supply and wastewater disposal systems in Germany is based on the collective property argument and may explain why the technical standards and reliability operational safety are relatively high by comparison and why the German system is greatly acknowledged worldwide. The system creates a poor impression in terms of cost structure, however (Briscoe, 1995; Rudolph and Gärtner, 1997), with the result that the overall system of water resource management is unsuitable as an "export". This is particularly so in the case of potential target countries where the legal and administrative culture are less sophisticated. Although the need for cost-effective solutions is particularly urgent in view of the low level of per capita income, it is questionable whether the German model is a viable proposition given the lack of suitable administrative structures.

5.4.3.3 Water resource management in the USA

Preliminary remarks

A variety of water legislation systems exist alongside each other in the 50 federal states of the USA. The development of competitive regimes is probably favored to a major degree by a specific feature of these systems, namely the fact that water rights in the USA are private property rights that can be sold, while at the same time they represent certain community rights. As individual rights they can be sold. Hence, new users may buy up the additional quantities of water they require, if they can out-bid users with older utilization rights.

Initial experience with water markets

There has been a considerable amount of experimentation with water markets in the western United States. Water transfers are effected either in the form of final transfers of waters from agricultural use to urban use or for the conservation of nature, or in the form of temporary transfers from one type of use to another.

The greatest difficulties in selling water rights are the high transaction costs that often arise when determining the volume of water available for transfer, as well as legal barriers in the form of classical prior appropriation rights (see next section), according to which unused water rights may be cancelled. However, in the western states, transaction costs do not seem to be a major obstacle in practice. A larger study on water transfers in six states comes to the conclusion that – with the exception of "attorney-dominated" Colorado – the current restrictions on transfers are justified by the benefits they produce (McDonnell, 1990). Some states have passed laws that credit "saved water" wholly or partially to the saver. The real barriers are political in nature, not legal. Water districts, which control a large proportion of water resources, are hesitant to approve water transfers because they lose political influence as a result (Thompson, 1993).

Temporary transfers are generally effected by "leasing" or banking water rights. The leasing of water rights increases during dry periods. The protracted drought of the late 1980s and early 1990s stimulated interest in temporary transfers and the creation of water banks. One risk resulting from the American principle of prior appropriation is that water rights may be forfeited under this principle if they are not used. Water banking is an attempt to do away with this "use it or lose it" rule by allowing rightholders to bank water so that their water rights are not cancelled.

Water banking was first introduced for the Snake River in Idaho and taken up by California during the drought in the late 1980s / early 1990s. 1991 was the fifth year of drought in succession for California. Water reserves shrank to 54% below average. In order to make up the difference between supply and demand, the state established the Drought Water Bank. Emergency legislation authorized water suppliers to sign agreements with the bank, and stipulated that further that temporary transfers would have no consequences for the water rights of suppliers. In the last years of the drought, the bank played a major role in reallocating water from agricultural to urban uses. One study concludes that the bank achieved the aim of mobilizing emergency reserves, but that valid procedural rules for the protection of third parties were circumvented in the process (Gray, 1994).

There are numerous schemes aimed at generating greater incentives to induce agricultural users to sell their water rights. Water is power, and many rural areas fear that transfers could transform their land into desert. Although the Congress of the United States has begun to remove some of the barriers to transfers in order to create incentives, it has not gone so far as to establish a deregulated market for state-subsidized water. The Central Valley Project Improvement Act (CVPIA) of 1992 authorizes some of those connected to make any transfers to a Californian Water Association, to state or federal authorities, indigenous peoples or private non-commercial organizations. The transfers are subject to certain conditions being met, including

- review by the Secretary of the Interior,
- the use proposed by the recipient of the transfer must relate to a CVP project or be classified under Californian law as producing benefits for society,
- the transfer may not run contrary to the laws of California,
- the transfer may not restrict the ability of the Secretary to grant allocate water to other CVP beneficiaries.

The most interesting feature is the definition of an upper limit. In order to protect the agricultural areas dependent on the CVP, a maximum 20% of the water that a district receives from the CVP may be transferred without permission of the district and the Secretary of the Interior.

The concept of water markets can also be used also to reduce water withdrawals by agriculture so that the resources are available for ecological functions. A major reallocation experiment is currently taking place at the Truckee and Carson Rivers in eastern Nevada. The Carson River flows into a basin now designated a World Heritage Wildlife Refuge and which was polluted by the runoff from a small irrigation area. In 1990, Congress directed the Secretary of the Interior to conserve 25,000 acres of wetlands in the area, partly by purchasing irrigation rights. A total of 125,000 acre feet of water were required (US Department of the Interior, 1996). An environmental impact assessment in 1996 estimated that an additional 75,000 acre feet would have to be obtained by purchasing existing water rights from farmers in the district. The procedure for such acquisition is very complex, because the authorities must establish whether the purchased water rights are "wet", i.e. that the proprietor acquired the rights under the doctrine of prior appropriation, by channeling the water for a socially useful purpose. Steps are also necessary to ensure that the water genuinely reaches the natural areas to which the water rights are transferred, and that it is no longer used.

The transferability of water marketing

In the following, the pro and cons revealed by practical experience with water markets in the USA are described in order to arrive at tentative conclusions as to whether water markets could be an effective policy instrument in other countries.

The concept of water markets in the USA is based on the assumption that market allocation of water resources results in water flowing into those uses where it has the highest monetary value. Those stakeholders who can achieve the greatest benefit with the water are also prepared to pay the highest price for it. Water markets mean that irrigated agriculture in the western states of the USA must be evaluated according to economic yardsticks, because its purpose is not to provide the population with vital food supplies and because it is a lower-value use in monetary terms.

More generally, the question is raised as to the transferability of water markets to other countries facing water scarcity. The answer is "yes" – because for many countries in arid and semi-arid regions of the planet, the more efficient allocation of developed water resources is an attractive option to developing new resources. Another factor is the caution now exercised, especially by the World Bank on account of the adverse environmental impacts, when granting international assistance for large-scale water development projects (see Section D 3.4). In many developing countries that strive to cover domestic food requirements by promoting irrigated agriculture, state subsidization of water prices to this sector have often led to a relatively careless and wasteful use of water (see Section D 3.3). Water markets could be the answer to such problems.

However, it should also be considered that water markets were established in the USA in response to the dysfunctional effects of prior appropriation, a legal principle with specific historical roots. Prior appropriation may have been a useful principle when the western states of the USA were first being settled, but it leads to problems in the present. The specific legal context implies that caution be exercised regarding the use of transferability of the water markets concept to other countries. Nor should cultural values be ignored. For example, the high value attached to water resources in Islamic countries could generate wise responses to the problem of water scarcity. By attaching a purely monetary value to water resources, water markets could even be detrimental to culturally-specific and time-specific approaches aimed at saving water. Another aspect to consider is that irrigated agriculture in the western United States does not serve to meet essential food requirements, so parallels should not be drawn to countries that operate irrigated agriculture as a vital source of food unless the respective situation is subjected to careful analysis. Nevertheless, as stated above, water markets are possible under such conditions as well.

Water markets demonstrate much greater efficiency than government-controlled systems. Maximum efforts should therefore be made to introduce them in other countries. This applies in particular to countries with severe water scarcity.

5.4.4 Recommendations

A better solution to the valuation problem

Given the diversity of water uses, solving the valuation problem is a matter of particular importance. It has been shown, above all, that many water problems are the result of state-induced undervaluation. Governments have usually pursued specific objectives with such policies, for example to support the agricultural sector, or to provide lower income groups with an inexpensive supply of water. The Council strongly advocates strict separation of the valuation issue from the allocation issue. At the same time, it draws attention to the fact that when governments stipulate maximum utilization rights in natural water resources, lay down quality standards and externalities are largely internalized (e.g. by introducing tougher liability laws), the market-based valuation produces good results that in most cases lead to price rises. The definition of maximum utilization rights can be based on complementary valuation methods that ensure a greater focus on environmental concerns.

Regional responsibility

Many water resources (lakes or aquifers) are connected by surface and underground links to their surrounding environment. In other words: it is possible to demarcate hydrogeographical regions that even behave as self-contained areas in some cases. This has economic implications as well. Something that appears as a problem of externalities at the local level because of functional interdependencies need not be an externality at regional level. At the regional level, the connection between use and pollution is more distinct. It is easier and makes more sense, as far as the use of natural resources is concerned, to grant property rights at the regional level and above all to apply the liability principle there as well. This would also be compliant with the fiscal principle of equivalence (spatial congruence between beneficiaries and those bearing the costs, in order to strengthen the liability principle) that is important for federal reasons (subsidiarity principle). The way in which wastewater is disposed of has a major influence on the quality of regional water resources, so the obvious solution is to combine the responsibilities for water supply and wastewater disposal in one organization. This would enable the negative externalities from wastewater disposal to be internalized much better.

An integrated approach covering sanitary services as well is urgently needed in the densely populated areas of Asia, Africa and South America, above all in the "megacities", where water problems are particularly dramatic. According to a World Bank study, the only way to solve the sanitation problems in Calcutta, for example, is to place water supply, wastewater disposal and sanitation services (emptying pit latrines) in the hands of one organization, and by controlling the liquid, solid and gaseous emissions from industry at regional level (Rudolph, 1990). The latest studies by the Asian Development Bank draw similar conclusions (ADB, 1997).

Cost aspects are another reason for taking a regional approach. The development of infrastructure for water supply and wastewater disposal tends to occur within discrete areas, almost like islands, while expanding water supplies and tapping remote water resources is bound up with disproportionately high costs as a rule. If these costs are not passed on in some adequate form, the result is an effective subsidization of the agglomeration process. Country-specific problems add to the situation in many developing countries. There are limitations on the extent to which the systems for water supply and sewage disposal, many of which were built during the colonial period, can be expanded to wider areas (e.g. sewage collection channels relying on gravity, or branched networks for transporting fresh water). Extending such system involves high threshold costs that sometimes lead separate policies for supply and disposal in the core district and the surrounding areas. This can lead to price and quality differentials that accelerate suburbanization as well as external effects beyond the specific region that can then lead water conflicts (see Section D 4.1). A regional approach is all the more important, while interregional collaboration is essential in the case of watercourses.

The subsidiarity principle imposes an obligation here to take responsibility, accept regional liability and to act accordingly. The newly industrializing countries are a case in point. Their environmental problems, especially those relating to sewage and waste disposal, arise because regional interdependencies are ignored and because timely action is not taken to develop a good infrastructure for supply and above all disposal. This imbalance in the stock of real capital is now avenging itself. In view of their growing prosperity, these countries should first meet their own obligations to solve the problems they have allowed to develop. Assistance from the international community should be provided only in those cases where problem-solving capacities are inadequate.

Efficiency and economy

The water issue has one major implication for environmental policy. In contrast to other environmental media, the important thing is not only to protect resources, but also to achieve certain productive outputs. If a quantitatively and qualitatively adequate water supply is to be provided to cover drinking water needs and other consumptive and productive uses, then it is necessary to develop water resources, to treat the water, distribute it and finally to dispose of the resultant wastewater. This means that a comprehensive range of water-related services must be delivered. This applies to sewage disposal in particular, a field that has been terribly neglected in many developing countries. Polluted water must be collected, treated and returned to the environment in a controlled manner (Karl and Klemmer, 1994). These are value added activities requiring the management of typical entrepreneurial tasks. Technical facilities must be planned, financed and operated. Services of this kind must be adjusted to meet the wishes of final users (production to suit demand) and should be provided at minimum cost – in other words, they have to be efficient. Innovation – the permanent search for new technical and organizational methods – plays an important role in this context. Experience shows that efficiency in the innovation process can be achieved if it takes place decentrally, in the private sector and within a competitive environment. This explains why efforts are increasing worldwide to organize water supply and wastewater disposal more along private-sector lines, with a primary focus on market forces and competition.

The dissemination of certain solutions – "the German model", for instance – is mostly through private-sector enterprises exporting methods and organizational models. This normally occurs through the profitable marketing of entire service packages, or with various forms of participation. Before this can happen, certain minimum sizes have to be present, as well as the commitment that is mostly generated by economic motivation. It therefore comes as no surprise that the topic of *marketing water and wastewater services* is becoming more and more important. It is precisely in this context that deficiencies in the German approach are revealed. Water supply and wastewater disposal are normally separate in Germany; wastewater disposal is organized by government agencies, thus reducing cost-effectiveness and encouraging bureaucratism, and is the responsibility of organizational units that in most cases are far too small. Approximately 80% of all water supplies are furnished by around 5,000 small enterprises and microenterprises that pump groundwater, and while more than three quarters of the 10,000 sewage systems are designed for less than 10,000 inhabitants (Rudolph and Gärtner, 1997). Private-sector involvement is generally only found in the fields of planning (private engineering companies), financing (private banks), building and plant construction (building contractors, plant fitters and electrical engineering companies) as well as service and maintenance (Rudolph, 1987). In contrast to this set-up, about three quarters of the country's water supply and wastewater disposal system have been operated for decades by three major water companies (Compagnie Générale des Eaux, CGE; Lyonnaise des Eaux-Dumez, LED; Societé d'Amenagement Urbain et Rural, SAUR). About twenty years ago, a multitude of different water organizations in Great Britain were combined to form large water authorities, which were privatized and reorganized ten years ago to form ten public limited companies. Powerful private-sector corporations dominate also in the United States, Australia and Japan. All of these enterprises were able to establish the logistical and financial structures necessary to survive on the world market, with the costs of a protracted start-up period, until contractual agreement of a "Build, Operate and Transfer" model. Public-sector enterprises and fragmented utilities as found in Germany have no chance against these major corporations and are therefore globally insignificant. The German water resources and wastewater industry may enjoy a good reputation worldwide for its technical soundness and operational reliability, but it is also considered to be cost-ineffective and too bureaucratic. In some respects one can even speak of entrepreneurial failure.

The Council therefore believes that efforts are needed to turn the German model into an exportable product. In the supply and wastewater disposal field especially, Germany can offer excellent technology and efficient solutions. This applies in particular to the disposal field and recent developments there – such as accessible pipe systems (Stein, 1997). This requires more intensive cooperation (industry as a supplier of technology, the construction industry, service providers and banks) and a more global orientation on the part of German utility companies. The Federal Government could support this global orientation and cooperation in the form of pilot projects, whereby it would suffice in many cases to cover the default risks.

Adequate coverage of minimum needs

It is now obvious that ensuring adequate coverage of the minimum human requirement for water is a matter of great urgency. The failure to achieve this has led in many cases to reservations against market-based methods of valuation and allocation. In societies with small populations it is quite possible to decide on a case-by-case basis, according to the severity of need, and to allocate water supplies accordingly In societies with large populations or with megacities and areas of high population density, such methods are totally inappropriate. The Council advocates payment of "water money" in such cases. Benefit of this kind can be conditional on the particular situation of the recipient in such a way that people earning higher incomes are excluded. In individual cases, consideration could be given to granting an inexpensive basic supply. However, both responses presuppose a functioning system of public administration that is able to solve the problem of monitoring application. A solution to the problem of measuring and invoicing consumption is a prerequisite for granting an inexpensive basic supply.

5.5 Legal principles and instruments pertaining to water resource management

Water management legislation in Germany – Transboundary watercourses – The "equitable utilization" rule – Ban on activities with serious transboundary environmental effect – Convention on the Protection of the Danube – The Mekong Agreement – Water Resources Committee of the International Law Association – UN Convention on the Law of the Non-Navigational Uses of International Watercourses - Integration in the UNCED process – Inclusion of confined groundwaters – Ecosystem approach essential – Global solutions imperative – Information and monitoring – Consultation – Enhancement of dispute settlement mechanisms – Global Plan of Action, World Water Charter or International Framework Convention on Freshwater Resources

5.5.1 Introduction

The institutional structures required for secure and effective allocation of available water resources must first be established by national and international legislation. Productive and consumptive processes must be managed in such a way that the role of water within the economic system is compatible with the hydrological cycle in the ecosphere (Meissner, 1991). The key objective must be the rational use of water resources, especially in water-stressed countries where water shortages are greatly exacerbated by unwise allocation practices. The allocation of water among the direct users takes place within a country, whereas the purpose of international law is to contain the potential conflicts that may arise between neighboring states over the use of transboundary water resources.

In this chapter, the German law on water resource management is presented as one example of a national allocation mechanism and evaluated with respect to its transferability to other countries. Special attention is given to the coverage of basic water needs. The efficient allocation of water resources also requires technical-logistic and planning concepts for a functioning system of water resource management. These are dealt with elsewhere in this Report (Section D 4.5) in connection with the exportability of German technology for water supply and wastewater treatment as a form of technical assistance and environmental protection. The latter chapter also examines the water markets that have recently developed in the USA, a further example of a national allocation mechanism (Section D 5.4.3.3).

The second section of this chapter deals with international freshwater law. The main focus is on recent developments in this area, in particular the adoption of the Convention on the Non-Navigational Uses of International Watercourses by the United Nations. The convention is aimed at demarcating the powers of individual states in respect of transboundary waters. A second section provides an overview of possible mechanisms for settling water-related conflicts.

The final section attempts to identify forms of international cooperation that go further in protecting and conserving water resources. Thoughts in this direction are based on the "Global Consensus" on freshwater resources called for on behalf of the UN organizations by the UN Secretary-General, which should involve better international cooperation over and above what already exists between neighboring states.

5.5.2 Water resource management in Germany

5.5.2.1 Legal regulation of water utilization in Germany

Utilization of water resources is governed in Germany by the Federal Water Act (the Water Management Act – (WHG) as amended in 1996, in association with the water laws at a state level that specify the details for implementing the national framework law.

The foundations for today's system of waterbody management in the Federal Republic of Germany were laid in the 19th century, when the fragmentary ownership of water resources under a multiplicity of private- and public-sector water laws was replaced by common public ownership under the law of public property. The German water laws are thus an example of "the deindividualization of a legal order" (Kloepfer, 1989). Waters are subordinated to public-law regulations on utilization that require all significant forms of water use to be approved. This means that the use of water is not a matter of personal choice, but that the state has authority to manage the country's water resources. The Federal Constitutional Court has confirmed the constitutionality of the Federal Water Act in the so-called Naßauskiesungsbeschluß (BVerfGE 58, 300ff).

Pursuant to Section 1 WHG, surface waters, groundwater and territorial waters are subordinated to the public-law regime for water use. Those forms of utilization requiring approval are listed in Section 3 WHG. All the main categories of water use are included – with the exception of indirect discharges of wastewater, which for historical reasons are covered by local authority bylaws (Brockhoff and Salzwedel, 1978). Some insignificant water uses do not require approval. A special feature of the German system is that there is no legal entitlement to approval (repressive prohibition with conditional grant of permission). The competent authorities are thus granted considerable discretionary powers in approving a specific type of water use, powers that enable them to observe public interests in addition to the interests of those applying for approval.

In the interest of protecting water resources, administrative discretion to grant approval is limited by mandatory grounds for refusal pursuant to Sections 6 and 7 WHG. The legislature has thus reserved the right to make wastewater discharges conditional on certain requirements being met and to prohibit those uses that are likely to be detrimental to general public welfare, in particular those that threaten the public water supply.

The approval procedure distinguishes between permission ("Erlaubnis") and authorization ("Bewilligung"). The difference between the two types of approval consists in the type of legal status that is granted (Breuer, 1997), rather than the specific water resources that may be used and the volume involved. Permission granted under Section 7 WHG is a revocable licence to use a waterbody for a specified purpose in a specified manner and to a specified extent. Authorization, on the other hand, is an irrevocable right, albeit limited in time, to use water resources in a certain way. Authorization thus provides the user with a higher measure of security. Pursuant to Section 8, sub-section 2 WHG, such authorization may only be granted if it would be unreasonable to expect the entrepreneur to carry out his project without an assured legal status, and the use in question serves a specific, regular purpose. An irrevocable grant may not be granted for wastewater discharges.

The administrative authority for water resource management and hence the decision on approval in specific cases is assigned to the individual federal states, the Länder. The water authorities established by laws at federal state level follow the general structure of public administration and are organized into lower, intermediate and higher water authorities. Approvals are generally granted by the lower water authorities, but when the scope of the project exceeds a certain significance the decision rests with the intermediate water authorities. Applications for approval are decided on decentrally.

One could summarize by saying that the allocation of utilization in Germany exhibits a high measure of flexibility. This is achieved in the substantive sense by placing water utilization under public law regulations, and formally by decentralized approval procedures administered by regional and local authorities. However, the drawbacks of such a system are the substantial costs for administration. Moreover, efficient administrative structures are essential if public law regulations on resource utilization are to function properly. This is a critical obstacle when it comes to the transferability of Germany's utilization regulations to other countries. To quote the World Bank: "Where there is a long-established and deeply entrenched tradition of sound governance (as in Botswana, Korea, and Singapore), it is evident that autonomous, accountable public sector agencies can provide efficient and equitable service. For many countries, however, such levels of governance are not attainable in the short run (World Bank, 1992).

Despite the conceptional strengths of public-law regulations on utilization, they are dependent to a substantial degree on the smooth functioning of public administration if they are to prove effective. Where this is not the case, other ways of regulating utilization must be considered.

5.5.2.2 Public supply of drinking water

In Germany, people's basic water needs are secured by public authorities. The central water supply system is traditionally the responsibility of local authorities, in the context of their constitutional guarantee of self-government and in accordance with the federal state's water laws; it is understood as part of the sovereign duty of the state to provide for its citizens. About 98% of the population obtain their drinking water from the public water supply system (BMU, 1996a). The vital coverage of basic requirements is fully safeguarded by the state.

Although local authorities may join forces and form larger utilities to ensure that water supplies are organized in a technically and economically efficient manner, in most cases water supplies are provided by small organizational units: 80% of water supplies are furnished by around 5,000 small and micro-sized enterprises that draw predominantly on groundwater resources.

The specific way in which a local authority organizes water supplies may take a variety of forms: it can set up bodies under its own direction, establish wholly-owned utility companies, or delegate responsibility to private-sector companies. The private-sector arrangement serves the purpose of covering costs through market-oriented management, in that all the costs of supplying drinking water must be covered by the price charged for water. However, responsibility under public law for performing this task remains with the local authority (BMU, 1996a), so responsibility for meeting basic requirements and for actual provision of water supplies may be divided.

The central public water supply covers not only the basic requirements of the population; a fifth of the water provided by utilities is supplied to industrial enterprises (BMU, 1996a). These have either no water supply of their own, or cover part of their demand by resorting to the utilities, for example because they require some of their water to be of the high quality provided by the latter. In addition, private water consumption in Germany is "luxurious" to a considerable extent and therefore exceeds basic requirements by far.

In many countries, meeting basic water needs is considered the responsibility of the state, so water is obtained from a central public supply system. Nationalized water supply and efficient allocation are often incompatible, however. In Caracas and Mexico City, an estimated 30% of all connections are not registered, which makes it impossible for utilities to cover their costs. Unaccounted water consumption amounts in most Latin American cities to approximately 40%. The resultant loss of income has been estimated at US$ 1–1.5 billion for Latin America as a whole. In Western Europe, the number of people employed by water supply companies per 1,000 connections is only two to three, whereas in most Latin American utilities the figure is ten to twenty (World Bank, 1992). Efficient operation in accordance with the cost-coverage principle is imperative, however, if a nationalized water supply system is to provide more than the population's vital basic needs.

5.5.3 International water law

International cooperation regarding the utilization of shared water resources has a longer tradition than other areas of international environmental law. For more than hundred years, states have tried to settle conflicts in this area by concluding bilateral and multilateral agreements. Back in 1978, the FAO had already counted over 2,000 international treaties pertaining to the utilization of international waters (FAO, 1978). Updated figures published in 1993 show that this institutionalization is a continuing proces (FAO, 1993). However, the number of treaties in place is misleading to a certain extent, because in many regions their actual content produces an inadequate level of cooperation. The commitments entered by the parties go little beyond the existing rules of customary international law.

These customary regulations constitute as it were a "a minimum standard" of cooperation, to which neighboring states are bound even in the absence of treaties and other forms of agreement. Such rules develop when states consistently observe certain principles over an extensive period of time, while at the same time assuming that this practice is a legal obligation (opinio iuris) and not a mere "custom". Identification of the norms of customary international law in this area is mainly due to the efforts of international committees. They include the "Salzburg Resolution on the Use of International Non-Maritime Waters" of the Institut de Droit International (Institute de Droit International, 1961), the "Helsinki Rules on the Uses of the Waters of International Rivers" adopted in 1966 by the International Law Association (ILA) and the work of the International Law Commission of the United Nations, which worked from 1974–1994 on "Draft articles on the Law of the Non-Navigational Uses of International Watercourses" (UNGA Official record A/49/10). Pursuant to the UN Charter, the UN International Law Commission shall "encourage the progressive development of international law and its codification". The work of the Commission in the field of water law involved collecting and assessing existing treaties. The idea was that work on codification would be based on actual state practice and therefore meet with greater acceptance on the part of states (McCaffrey, 1996). In addition to the codification of existing rules in international customary law, the International Law Commission also attempted in its draft articles to develop rules going beyond those of customary law.

On December 9, 1994 (Res. 49/52), the UN General Assembly decided to elaborate a Framework Convention on the basis of the draft articles drawn up by the International Law Commission. Negotiations in the Sixth Committee of the General Assembly were completed in April 1997, and the Convention was adopted as Resolution 51/229 by the General Assembly on May 21, 1997. With the planned convention, the "minimum standard" of cooperation among states will rise – in global terms – and in some regions will extend beyond existing commitments. This could prove a major contribution to the protection of transboundary waters and to the prevention of conflicts between riparians over the use of water resources.

In the following section, the relevant rules of customary international law will be outlined and more recent tendencies in international law, in particular "the UN Convention on the Law of the Non-Navigational Uses of International Watercourses", will be examined.

5.5.3.1
Rules of general international law regarding the use of transboundary watercourses

THE PRINCIPLE OF EQUITABLE AND REASONABLE UTILIZATION

The principle of "equitable and reasonable utilization" enjoys almost universal acceptance in customary international law on transboundary surface waters. However, application of the principle to groundwater stocks connected to surface waters is a matter of some controversy, while the principle does not apply at all to confined groundwaters because there is no evidence of commonly accepted practice on the part of states.

Equitable utilization correlates with the reconciliation of the conflicting territorial rights of riparian states: although states have the basic right to use resources on their national territory in whatever way they see fit, they are also under an obligation not to interfere in the way that other states use their resources. Equitable use of international waters means here that riparians strike a balance between these rights and duties. What equitable utilization means in the individual case is therefore dependent on the specific circumstances. The following factors play a role:

- the geographical conditions of the waterbody,
- the "historical" use of the resource in the past,
- the economic and social needs of the riparians,
- the number of people dependent on the waters and
- the costs that would arise if the economic and social needs of the states were covered in a different way.

This list of factors is neither complete nor hierarchically organized. This may seem rather unsatisfactory, in that the scope available to states remains vague, but there is virtually no alternative method of regulating use, because the geographical situation of specific watercourses and the socioeconomic positions of the respective neighbors are all too divergent from one region to the next (Heintschel von Heinegg, 1990).

The rule of equitable utilization of international waters must therefore be specified by contractual provisions for the specific watercourses and the riparians concerned. Customary law stipulates only that parties are required to reach agreement, and specifies a number of relevant factors without weighting them. This does not necessarily imply that customary law performs no function alongside contractual arrangements. In practice, the principle of equitable utilization marks the framework within which new treaties or amendments of existing treaties are negotiated. In diplomatic exchanges during real disputes, states generally base their arguments on the rules of customary law as elaborated by expert bodies.

Utilization must be not only equitable but also "reasonable". It goes without saying that each state has the sovereign right to determine its needs independently of other states (provided international laws are not violated). Other states do not have the right to question this definition and impose their own assessment of needs, even though negotiations on the sharing of water resources implicitly require the mutual evaluation of respective needs. The principle of "reasonable" utilization is supposed to protect individual riparians against exaggerated demands on the part of other states, i.e. that a state need only accept demands raised by another that any state in a comparable situation would be reasonable in making. In practice, the same criteria must be applied as with equitable utilization (Hafner, 1993).

The third aspect to be discussed concerns the "optimal utilization" of a watercourse by states. However, the "Helsinki Rules"of 1966 contain no reference to this principle, which first appears in the draft articles drawn up by the UN International Law Commission (Art. 5 (1)). And yet the basic idea was discussed as far back as the 1972 Stockholm Conference on the Human Environment, where it was adopted as Recommendation 51 of the Plan of Action. It is self-evident that the optimal utilization of the resource will limit environmental damage, to the extent that no more water than is necessary will be degraded. However, the rule of optimal utilization is not yet established on the basis of international treaties and documents in this area (Hafner, 1993).

The ban on activities with significant adverse transboundary environmental effect

The prohibition of "activities with significant adverse transboundary environmental effects" under international law is the central rule governing relations between neighboring states. According to the rule, states must ensure that they cause no "significant harm" to human life, health or to objects used by people in other states.

The prohibition of significant environmental harm is based on an additional environmental concept that protects not only the environmental media as such but also (and primarily, from the traditional perspective) the human activities relating to those media, such as the agricultural and industrial use of water resources (Epiney, 1995). To that extent the use of the word "environmental harm" is misleading. The ban does not protect the ecology of a state, but its sovereign interests. A state has an interest not only in the integrity of its environment but also in the management of its environmental resources. Traditionally, international law has not distinguished whether protected management of resources may itself cause harm to the environment, because this is the immediate responsibility of the state. Reference to "transboundary" pollution does not mean that states have to share a frontier – any state that can be affected by pollution emanating from another is deemed a neighboring state (Epiney, 1995). As far as watercourses are concerned, all riparians are referred to alike.

The harm caused must always be "significant", however. The greatest problems in applying the principle derive from the "significance" criterion – quite apart from proving the causal linkage between activities in one state and harm to another. When disputes erupt, it is not the validity of the principle that is disputed in practice, but the causality, the significance and the attribution of harm (Kunig, 1992). The difficulty consists in defining the indeterminant legal term "significant". Given the necessarily general nature of customary international law, it is not possible to derive exact limit values and fixed criteria for delineation and demarcation from it. Instead, the states concerned must rely on negotiations in order to arrive at a consensual solution and to specify this either informally or in a treaty. The prohibition of significant harm must therefore be understood first and foremost as a requirement to reconcile interests through negotiation (Kunig, 1992). This explains why the Convention on the Law of the Non-Navigational Uses of International Watercourses, which includes the obligation not to cause significant harm, refers primarily to consultations and cooperation in this context, while compensation is dealt with cautiously as a secondary issue.

Procedural obligations

The substantive rules of equitable and reasonable utilization, on the one hand, and the prohibition of significant harm, on the other, amount effectively to an obligation on the part of states to reach agreement. The implication is that this obligation is supplemented by formal obligations to exchange information and consult with each other. Such consultation can only be based on information provided by the respective other party. Without consultation, the basic consensus between states that is required for the two principles above to operate would amount to nothing. The obligation to provide prior and timely notification and relevant information on activities that may have a significant adverse transboundary environmental effect was enshrined in Principle 19 of the Rio Declaration. It can also be found in various international treaties.

In the Convention on Non-Navigational Uses of International Watercourses, the principles of "equitable utilization" and the "obligation not to cause significant harm" are supported by the general obligation to cooperate and to exchange data and information on a regular basis. By lending its support to the Convention, the Federal Government could help establish as rules of international law the obligation to exchange information and provide notification in the case of activities that may have a significant adverse transboundary environmental effect, and the obligation to cooperate for equitable and reasonable utilization.

5.5.3.2
Recent treaties at regional level

As described above, the rules of customary law as they pertain to freshwater resources are usually specified in treaties, the content of which may differ enormously. They may range from simple promises of consultation prior to changes in the pattern of utilization, to the promise not to intervene in the way the other party uses its water resources, to attempts to share water resources and the related benefits, to cooperative management of watercourses, culminating in the highly developed instrument of integrated watershed management (Dellapenna, 1996). Integrated watershed management should not be viewed as a fixed concept, but as the instrument exhibiting the highest level of integration among participant states, based on the current state of scientific understanding. Existing regimes for the protection of water resources must therefore be modified in response to scientific advances and the specific challenges that are posed.

New conventions in Europe

The most recent example is the Convention on Cooperation for the Protection and Sustainable Use of the River Danube, which replaces the Bucharest Declaration of 1985. The Convention, which was signed in 1994, was implemented as national law by the German legislature on June 12, 1996 (Federal Gazette 1996, Part II, p. 875ff.). The Convention is exemplary for the degree of regulation that can currently be provided by an agreement on shared use of a watercourse based on the idea of sustainable management. The geographical scope of the Convention covers the entire hydrological catchment of the Danube and is aimed, in addition, at mitigating pollution of the Black Sea. The aims, which include the prevention of lasting damage to environmental resources and protection of ecosystems, are supported by a set of powerful instruments. The Convention is based on the precautionary and the "polluter pays" principles (Article 2 IV). It is planned to establish an "International Commission" as a permanent joint commission (Article 18, 19). Combating pollution under the terms of the Convention ranges in detail from the differentiation of specific industrial sectors and hazardous substances to the definition of emission limits that are to be developed through joint consultation (Article 7, in association with Annex II). The Parties are supposed to develop common or to harmonize national targets and criteria relating to the quality of water resources (Article 7 IV and Article 9 I). Another aim of the Convention is to produce an inventory of diffuse and point sources. Procedural obligations ensure the close cooperation necessary for carrying out the planned measures. Disputes are to be settled by means of the arbitration procedures specified in Annex V. According to Article 14 of the Convention, Parties must also ensure that the public is provided access to information about the condition and the quality of running waters. The Convention on the Protection of the Danube bears major similarities to the Convention on the Protection and Use of Transboundary Watercourses and International Lakes adopted in Helsinki on March 17, 1992 (the Helsinki Convention; Federal Gazette 1994, Part II: p. 2333ff.). The same is true of the treaties on the Maas and the Schelde agreed in 1994 by France, Belgium and The Netherlands (Teclaff, 1996; ILM, 1995). The Helsinki Convention was drafted by the UN Economic Commission for Europe (ECE) and can be thought of to a certain extent as "a convention for Europe".

The new Mekong Agreement

There are other – non-European – regions where environmental protection has acquired greater significance by virtue of international treaties. On April 5, 1995, the riparians of the lower Mekong – Cambodia, Laos, Thailand and Vietnam – concluded an "Agreement on the Cooperation for the Sustainable Development of the Mekong River Basin" (ILM, 1995), which reorganizes the cooperation that has existed since 1957. The most important body for cooperation is the Joint Commission for the lower Mekong that has been in existence since 1957. The Treaty is basically founded on the notion of equitable and reasonable utilization (Article 5). However, Article 3 assigns a dominant role to protection of the environment, natural resources, aquatic life and the balance of nature in the Mekong as a factor to be considered. According to Article 6, utilization by the states may not lead to the river level falling below the minimum acceptable monthly streamflow, except when there are droughts or floods, and during the flooding season utilization may not lead to average daily rises exceeding normal conditions. All in all, major importance is attached in the Treaty to the goal of protecting the environment. Yet it is questionable whether these lofty aims can be achieved unless the Joint Commission is made a powerful institution. As a rule, states are only obliged to notify each other of water withdrawals. Another major obstacle is the fact that the upper Mekong riparians, China and Myanmar, have refrained from any form of cooperation with the lower riparians except in the technical and scientific research fields. In 1993, China commissioned the first of its planned dams on the Mekong, the Man-Wan dam, without consulting the lower riparians (Chomchai, 1995). The ecological consequences are unforeseeable as yet. Worst of all, the absence of any collaboration on the part of the upper riparians hinders cooperation among the downstream states and reduces its effectiveness.

5.5.3.3 Progress in the work of the International Law Association

The International Law Association (ILA) is the oldest and largest private association of lawyers in the world. According to Article 2 of its statutes, its aim is "the study, eludication, and advancement of international law, public and private, the study of comparative law, and the furthering of international goodwill and understanding". The ILA compiles reports in specialized committees, on the basis of which it may then adopt resolutions. ILA reports focus on issues of customary international law, draft articles for treaties or conventions, the discussion of recent trends in law and jurisdiction, and on the elaboration of rules and principles of international law. The reports of the individual committees and the drafts proposed for adoption exert a major influence on the development of international law (Stoedter, 1995). In the field of water law, for example, the work of the ILA played a crucial role, in the form of the so-called Helsinki Rules, in which equitable and reasonable utilization was recognized as a principle of customary law.

Since 1954, three ILA committees have concerned themselves with the law governing international water resources. The latest to be established was the Water Resources Committee (WRC), set up in 1990 and consisting of 22 members. The WRC has concentrated its attention on four areas:
- estuaries,
- the transfer of water from or into an international drainage basin,
- legal recourse,
- inter-medial pollution.

In addition, the WRC has set up a working party for assessing the work of the International Law Commission in this area. The topic of estuaries has not been addressed by the Committee as yet. Water transfers from or into an international drainage basin was dropped. The working party assigned to this issue came to the conclusion that water transfers out of or into a drainage basin is not subjected to any special rules, but to general principles instead (Bourne, 1996). Efforts have been more fruitful with regard to the issues of inter-medial pollution and legal recourse. The articles drawn up by the WRC on these topics were adopted by the ILA in August 1996 at its Helsinki conference.

The WRC has drafted four articles on the question of inter-medial pollution emanating from water drawn from an international drainage basin. One new approach in the field of environmental law is that pollution of a particular medium, e.g. water, is no longer viewed in isolation, but that the impacts on other environmental media, i.e. soil and air, are also taken into consideration by the legal regulations. The traditional method is to place the individual media under the protection of specialized regimes. The natural linkages between the media are largely ignored in the process. An approach involving the inter-medial analysis of pollution therefore makes a great deal of sense. However, apart from having to create a new regulatory framework, there is also the problem of weighing up the integrative approach against the existing regimes for the respective media. Within the European Union, an important step forwards is the Directive adopted in September 1996 on Integrated Pollution Prevention and Control (the IPPC Directive).

The Council sees a need for research in this area. In debate among international legal experts, considerable attention has been given for some time to the IPPC Directive, and the implementation of the directive that Member States must now ensure will certainly be followed with great interest. However, there is a lack of research with an adequate focus on the international legal context of inter-medial pollution. Interdisciplinary research is therefore essential in this respect.

The ILA also recognizes these research needs. The articles adopted by the EU do not represent a comprehensive body of regulations. The ILA has modified the title to "Supplemental Rules on Pollution".

This is meant to suggest that the regulations in their present form are too underdeveloped to be compared with the work of the ILA as reflected in Articles IX–XI of the 1966 Helsinki Rules, the articles on land-based marine pollution and the Montreal Rules on Water Pollution in an International Drainage Basin (1982). The WRC came to the conclusion that a comprehensive body of regulations is not possible, because there are no prospects of consensus among members on this complex legal terrain, which is still at an early stage of development.

Nor is there a compulsory global forum at which compensation issues concerning activities with adverse transboundary impacts could be negotiated. As a basic principle, only national governments have the power to grant legal protection to resources. For that reason, protecting the environment and resolving international conflicts over natural resources are mainly tackled with preventive approaches. Regulations are agreed upon in order to avoid adverse impacts from arising in the first place. This is the function performed, for example, by treaties on equitable and reasonable utilization, or on the discharging of substances into international watercourses. If damage is caused despite such arrangements, it is essential that the issue of compensation be clarified.

The articles adopted by the ILA in 1996 do not address the question of state responsibility, i.e. compensation for damages between states. They are more concerned with some aspects of the question pertaining to the legal protection options open to a private individual in the event of transboundary damage (Brandt, 1995). This material is traditionally a matter for international private law, which regulates the jurisdiction of courts and the applicability of procedural and material law. The ILA articles go beyond that, however.

According to Article 3, people who suffer or must fear harm as a result of another state using waters from an international drainage basin are entitled to participate in certain procedures in the latter state to the same extent and under the same conditions as an inhabitant of said state. Examples include: participation in environmental impact assessments, examination of the permissibility of harmful utilization, judicial or administrative proceedings, as well as preventive legal recourse and the right to information. Foreign public-law corporations and private non-governmental organizations should also be entitled to the same procedural and participation rights as domestic public-law corporations and private non-governmental organizations when harm is caused or is likely to be caused. Article 3 para. 2, in association with article 1 para. 2 adds that foreign corporations and non-governmental organizations should be entitled to the same extent as their domestic equivalents to institute action in the interest of their members or the general public, i.e. the possibility to seek legal recourse against environmental damage that does not constitute a violation of the association's own rights.

5.5.3.4 UN Convention on the Law of the Non-Navigational Uses of International Watercourses

The Convention is the result of work carried out by the International Law Commission over a period of twenty years. The 1994 draft adopted by the International Law Commission was accepted by the Sixth Committee of the UN General Assembly as a basis for discussion in order to elaborate a draft convention for the General Assembly. This was then adopted by the General Assembly on 21.5.1997 as Resolution 51/229. The UN Convention sets forth a global framework of "minimum standards", within which states should cooperate in future by concluding more specific watercourse agreements at regional level. The following section examines the history of the convention.

Consideration of the UNCED process

The draft articles drawn up by the International Law Commission were the basis on which the convention was elaborated. Much of the working time of the International Law Commission was taken up with the difficult economic and security policy issues involved. On the other hand, recent documents of major importance in international law, such as the results of the Earth Summit in Rio – in particular Chapter 18 of Agenda 21 – could no longer be considered by the International Law Commission to any adequate extent. The proposals submitted by governments in response to the draft articles of the International Law Commission reflected an acceptance that more recent developments and the acknowledged common tasks would have to be integrated into the planned convention (e.g. UN Doc. A/51/275). However, this attitude was taken by a minority of states only. One positive outcome was that the principle of sustainability was included in Article 5 of the convention, which states the central principle of equitable and reasonable utilization. Centering the convention on an ecosystems approach, as demanded by Portugal, for example, did not meet with general approval.

International watercourses

The definition of the subject matter is crucial to successful integrated water resource management. Leading studies agree that the river basin is generally the most appropriate geographical entity for planning the utilization and development of water resources, and the land resources relating to them (e.g. UN DESA, 1958; President's Water Resources Council, 1962; UN DTCD, 1991; Teclaff, 1996).

The concept of "international watercourse" developed by the International Law Commission is based on a similar approach, and was ultimately included in the Convention text. According to Article 2 (a), a watercourse means "a system of surface waters and groundwaters constituting by virtue of their physical relationship a unitary whole and normally flowing into a common terminus". The implication is that scientific knowledge must be integrated more intensively into the convention process. The definition extends the subject matter in that many regional agreements restrict themselves to the management of transboundary rivers and lakes and leave out inflows entirely, which is not the case with the term international watercourse. Article 3 does away with the idea of the convention being a framework law, by stating that the terms of the convention may be applied and adjusted to the characteristics and uses of the respective international watercourse. As can be seen from the commentaries of the International Law Commission, this also applies to the flexible definition of the subject matter of any watercourse agreement to be concluded on the basis of the Convention, be it a river basin, a watercourse, or only a part thereof (ILC Draft Articles 1994). The Convention therefore allows states to define the specific watercourse when they enter into regional agreements.

However, the statements of position made by governments and the history of the draft articles of the International Law Commission are clear evidence of the substantial resistance put up to the term "international watercourse", notwithstanding the flexible arrangements in this regard. This also shows the importance attached to a convention in this field. Recalling the present-day dimensions of water development schemes, like the diversion of entire rivers (the feeders to the Aral Sea and the Israeli "national water carrier" spring to mind), then even broader concepts such as the river basin and the international watercourse appear too small to be planning entities (Teclaff, 1996). Keeping the term "international watercourse" must therefore be welcomed.

Confined groundwaters

The term confined groundwaters refers to those aquifers that do not have any exchange with surface waters. Although it is difficult to apply the principle of equitable and reasonable utilization to groundwaters to any adequate degree, there is an urgent need of legal provisions with respect to these resources, in that a large proportion of potable water is stored as groundwater stocks and because of the growing conflict potential resulting from technological progress and regional population growth. Reference must be made here to the ILA Groundwater Rules, which were adopted in Seoul in 1986, as well as the elaboration of the Bellagio model agreement by an international committee of experts (Hayton and Utton, 1989). For this reason, the International Law Commission has also recommended in a special resolution on confined groundwaters that states apply the principles in its general articles in disputes over confined groundwater resources as well (GAOR, 49th session, supp. 10).

Stephen McCaffrey, former Special Rapporteur of the UN International Law Commission in this area, recently pointed out how unfortunate it was that the growing importance that confined groundwaters will have for states in the future was only addressed in a separate resolution. He referred to this as "a hasty effort tacked onto the draft articles at the conclusion of the Commission's work". He expressed the hope that the Working Group of government representatives convened by the UN General Assembly would correct this omission. Unfortunately, the convention text does little to clarify this issue (McCaffrey, 1996).

Ecosystem analysis

According to Article 20 of the draft produced by the International Law Commission, States shall protect and preserve the ecosystems of international watercourses. According to McCaffrey (1996), "Article 20 fairly cries out for further elaboration". Within the Sixth Committee, Portugal advocated that neighboring ecosystems should be included in the definition of "watercourse" (UN Document A/51/275). This would have been welcome for two reasons: incorporating the ecosystem approach as part of the subject matter would have had a much greater effect than parceling it within a single article. Moreover, this would have been a more appropriate response to the fact that not only the ecosystem of the watercourse itself, but also the neighboring systems exert a major influence on the sustainable management of watercourses. The proposals made by Portugal with respect to an ecosystemic focus were very far-reaching, but would have been desirable if the challenges are to be tackled from a forward-looking perspective. Since the early 1990s there has been a mass of evidence to show that the water development schemes constructed in the past have caused substantial damage to ecosystems (Teclaff, 1996). Semi-natural water resource development is possible in many fields in Germany today, without detriment to others, and indeed is put into practice to an increasing extent (Binder, 1996). One example is the restoration of canalized rivers to the natural state. If new water resources can be developed without repeating the mistakes of the past, this will obviate the need for such rehabilitation measures and avoid the associated costs. Not only in Germany is this the case, but also and precisely in those regions of the world where new water resources are being developed as an essential factor in development.

Relationship between the "equitable and reasonable utilization" principle and "no significant harm"

A central controversial issue in the work of the International Law Commission concerned the relationship between the principle of "equitable and reasonable utilization" (Article 5) and the "obligation not to cause significant harm" (Article 7) and which norm should obtain priority in cases where the two collide. The various statements made within the Sixth Committee on the Commission's draft articles provide no evidence of a consensus on this question.

It is important to realize the wide scope of the significant harm rule, which embraces not only the environment but also all human uses relating to water as an environmental medium, including industrial uses. It makes little sense from the environmental policy perspective to prioritize the prohibition of significant harm against the principle of equitable use. This would effectively protect all existing uses against any new uses having significant adverse effects on new uses (Bourne, 1992; Brandt, 1997). This would make it more difficult to settle present and prevent future water conflicts. The International Law Commission granted priority in its draft articles to the principle of equitable and reasonable utilization. From the viewpoint of environmental policy, it would have been desirable to qualify this priority only in cases where the harm caused has direct impacts on the environment. This could have been achieved by giving precedence to Article 21, which specifically prohibits any pollution of international watercourses that may cause significant harm to other watercourse states or to their environment (Brandt, 1997). Accepting such consequences is hardly conducive to a long-term solution for water-related conflicts – however equitable and reasonable a short-term solution may be – because future utilization options are inevitably curtailed, which contradicts the paradigm of sustainable development and resources management.

Summary

The Convention on the Law of the Non-Navigational Uses of International Watercourses recently adopted by the General Assembly stems primarily from the efforts of the International Law Commission. The Convention falls short of the level of integration that is already achieved by international agreements in some parts of the world. Nor is this the explicit function of the Convention. Its purpose is to establish a global "minimum standard" for riparians of transboundary watercourses that may already be accepted by some states but not by others. Thus, the Convention represents in large measure a codification of the customary law in this area – with some aspects that go further than the status quo. It would have been desirable if recent developments in state practice had played a role in the elaboration of the Convention, i.e. if international law had been developed and refined to a greater extent rather than just being codified (e.g. Brunee and Toope, 1997). For example, Part IV of the Convention (Articles 21–26) pertains to "protection, preservation and management". What is really needed, however, is a greater emphasis on environmental protection and the integration of these issues into central parts of the Convention. In summary, the Convention does not contain any surprise innovations with respect to environmental protection, but represents a well-founded codification of existing customary law in this field that provides a basis for negotiating progressive regional watercourse agreements in the future.

5.5.4 Strengthening international mechanisms for the prevention of conflicts

In addition to long-term and institutionalized structures in international relations, it is imperative to have direct and effective mechanisms for mediating between states in order to prevent potentially violent situations from escalating into military conflict. Mediation is a conflict management process in which the disputing parties or their representatives seek the support or assistance of individuals, groups, states or organizations to change or influence attitudes or behavior without the use of arms or recourse to the law (Bercovitch, 1992; Kriesberg, 1991).

An arbitrator or mediator can be called in as an impartial third party. Such a mediator role can be performed by a representative of a third state or an international organization, whereby the mediator may not be directly involved in the conflict and must be recognized by all parties to the conflict as impartial. In the field of international relations, in which there is neither adherence to generally valid rules nor a central authority with power to regulate the international conduct of states, mediation offers an effective method for peacefully resolving conflicts between states (Bercovitch, 1991).

If mediation is to succeed, the adversaries may not simply pursue their own interests in an egoistic manner as if they were playing a zero-sum game in which the losses or damages incurred by the other side signify one's own advantage; instead, the parties must strive on a voluntary basis to achieve an amicable, "out-of-court" settlement and a peaceful solution to the dispute. Thus, the crucial components of mediation are voluntary participation, consensus, the search for win-win solutions and, if necessary, the acceptance of autonomous decision-making. It is precisely these elements that enable a satisfactory result to be attained for all concerned.

Kriesberg (1991) differentiates between four phases of conflict de-escalation through mediation. In the first phase, the disposition for conflict is reduced. Negotiations are activated in the second phase, in which the parties are brought around the negotiating table. The third phase involves conducting and managing the negotiation process. Finally, joint agreements are concluded and implemented. International mediation is not feasible unless the conflicting parties are willing to seek and accept a consensual solution, because imposing any heavier obligation on the states would be an infringement of the sovereignty principle enshrined in Article 2 paras. 1 and 7 of the UN Charter (Czempiel, 1994). Accordingly, the prospects of third parties being called in to mediate in domestic conflicts are all the smaller.

If transnational or international mediation is to function as a reliable dispute settlement mechanism in the case of international water-related conflicts, and as one that is not just invoked ad hoc in conflict situations and depending on the willingness of third-party states, then what is needed is a permanently available institutional framework for conflict settlement. Such permanent conflict settlement mechanisms can be provided by international regimes, international organizations as well as by international conventions. The two most important procedures for peaceful settlement of disputes are mediation and good services. In the wider United Nations context, mediation is institutionalized as a graduated system for conflicts management (Frei, 1990). Negotiations and enquiries are first in this procedural sequence.

The next step is conciliation, in which a neutral commission submits proposals for an amicable settlement of the conflict. The final step is mediation by neutral representatives of the United Nations. The regional Organization of American States (OAS) has an additional instrument for dispute settlement, besides mediation and good services, in which a "Commission of Investigation and Conciliation" is set up (Czempiel, 1994). Here, representatives of the disputing parties and a neutral chairperson convene and must report within 6 months. A good example in Europe consists of the dispute settlement mechanisms within the Organisation for Security and Cooperation in Europe (OSCE). In addition to the Court of Conciliation and Arbitration in Geneva, the organization has various other mechanisms for handling conflicts using the discursive style of mediation (Schlotter et al., 1994).

Schlotter (1994) lists the arguments for a regional and a universal codification of these mechanisms. The factors favoring a regional foundation for conflict resolution mechanisms is that the immediate neighbors of a conflict zone are best suited for mediation because they are most affected by the destabilizing impacts of conflict. In many cases, direct neighbours must bear the political, social and economic consequences of military confrontation without themselves causing it. Another aspect is that neighbors can understand the dynamics of specific conflicts better than external third parties, and the ability to draw on personal contacts to the parties in conflict may benefit the mediation process (Robert, 1993). However, being an immediate neighbor does not imply from the outset that mediation will succeed. Direct proximity to the crisis zone and implicit involvement mean that neighbors are potentially vulnerable to extortion. Familiarity with the circumstances may generate a bias towards one side or the other, so the impartiality needed for mediation is no longer present.

International mediation has been applied in a wide variety of cases, from conflicts over national sovereignty and colonial wars of independence to conflicts over the utilization of natural resources (Susskind and Babbitt, 1992). According to quantitative surveys carried out by Bercovitch and Regan (1997), 981 conflicts since the Second World War have been the subject of mediation efforts. In 38.5% of cases, mediation succeeded in achieving a ceasefire and/or partial or full conciliation. In almost 20% of cases, the dispute was at least partly settled, while more than 5% of the conflicts were fully resolved by mediation.

In the 1994 nuclear weapons dispute with North Korea, the former US President Carter acted as mediator. The Deputy UN Secretary-General, Cordovez, mediated between the USSR and Afghanistan, resulting in a withdrawal of Soviet troops since 1982. In the Libya-Chad conflict (Amoo and Zartman, 1992), the mediating role was played by the Organisation for African Unity (OAU). In Bosnia-Herzegovina, US President William Clinton conciliated between Croats, Serbs and Bosnians until signing of the peace treaty in Dayton, Ohio.

One of the more remarkable and successful examples of international mediation are Washington's efforts over many years to mediate in the Israeli-Arab conflict. The current peace between Israel on the one side, and Egypt, Jordan and the PLO on the other is largely attributable to the influence of successful US mediation. Kissinger's efforts to mediate between Israel and Egypt in the 1970s (Mandell and Tomlin, 1991) helped Jimmy Carter, the US President, to conclude the Camp David peace treaty. Although the Middle East conflict essentially revolves around security issues and territorial claims, the question of utilizing water resources has always played a major role. In the ongoing peace process between Israel, Jordan and the PLO that was initiated by the USA in the early 1990s, water is one of the key negotiation issues (Bäechler et al., 1996).

Special mention must be made of the mediator role played by Norwegian sociologist Marianne Heiberg, who persuaded Israelis and Palestinians to sit down at the negotiating table and thus instigated the process of dialogue (Czempiel, 1994). In his study on the water conflict in the Jordan basin, Libiszewski (1996) documents the crucial importance of the water issue in the peace treaty between Israel and Jordan.

These examples show that mediation procedures can be conducted successfully by third parties even when the political issues are national security and territoriality. In the classification of conflicts (see Section D 4.1.1), those over security and territoriality are classed as conflicts about values, which are based on dissension over the normal status of an object. Compared to conflicts over objectives and conflicts of interest, conflicts about values are those most difficult to manage. The prospects of successful mediation are greatest in pure conflicts of interest. Given that mediation can succeed even in difficult conflicts about values, the likelihood that mediation in water conflicts will be successful is relatively high, because these are conflicts of interest.

Germany can help prevent the escalation of conflicts over freshwater resources and contribute towards long-term peacekeeping by supporting international, peace-promoting mediation mechanisms. A constant readiness to arbitrate can have direct, peace-promoting effects in conflict situations. The establishment of institutional structures for conflict resolution operates over the long term to secure peace and prevent military conflicts.

- The Council recommends, firstly, that Germany offer its services as a neutral third-party mediator and arbitrator in international conflicts over transboundary waters. Experience and know-how in the field of environmental mediation and the regulations applying to European watercourses can prove valuable in this context.
- Secondly, Germany should strive, within the United Nations system and in line with the amalgamation and strengthening of environment-related institutions as recommended by the Council, for the creation of dispute settlement mechanisms with the functional, structural and personnel capacities to mediate in international water conflicts at the request of parties or as circumstances require.

5.5.5 Intensifying international cooperation for the protection of freshwater resources

5.5.5.1 "Global Consensus" on freshwater resources

The threat to freshwater resources by excessive use and pollution is a worldwide problem that is gaining in severity. In the year 2025, two-thirds of humankind are likely to be affected by water crisis (UN Document E/CN.17/1997/9, Section 2). In addition, regional water crises are an acute or potential cause of international conflicts. Local water scarcity may induce migration from rural to urban areas and even the abandonment of entire settlement areas, thus triggering migration flows (Favela Syndrome). The attempt to confine water crises by means of large-scale projects such as dams or irrigation systems might unleash or intensify international conflicts if the water needs of the riparian states exceed the total amount of water available and the states are thus compelled to share these limited water resources - quite apart from the threats they pose to the environment (Aral Sea Syndrome). The social science literature refers in this connection to threats to the "environmental security" of states and to "water wars". Egypt, Botswana, Bulgaria, Mauritania, Turkmenistan, Hungary and Usbekistan, for example, each obtain more than 90% of their respective surface waters from abroad (see Section D 4.1).

The enormous importance of freshwater for regions affected by water crisis, and the prevalence of such crises indicate the need for a global solution. In addition to the essential regulation of international conflicts over utilization of transboundary waters as discussed in Section D 5.5.3, the Council deems it imperative to bolster international cooperation on a global scale for the protection and improved use of freshwater resources.

This global cooperation should discharge those functions that cannot be covered by the UN Convention on the Law of the Non-Navigational Uses of International Watercourses. They include coordinated interregional cooperation as well as targeted and more effective support for states suffering from a combination of low per-capita income and severe water scarcity. Around 1.2 billion people live in such countries, where water scarcity can become the major limiting factor to socioeconomic development (UN Document E/CN.17/1997/9, 71ff., 150f.). In other countries as well, however, higher priority must be given to the protection of water resources, e.g. against pollution.

In the "Comprehensive Assessment of the Freshwater Resources of the World", published in 1997 and drawn up by the special organizations of the United Nations, the UN Secretary-General came to the conclusion that it would be "illusory to believe that anything short of a global commitment would provide the means to sustainability. Because some of the water crises could be very severe, the whole world has a stake in averting them." The Secretary-General therefore considered it necessary "to reach a Global Consensus over and above what is contained in the existing principles and agreements on freshwater resources of the world" (UN Document E/CN.17/1997/9, 100, 177).

Such a Global Consensus is necessarily confined to a relatively "weak" international provision, because it is not possible to apply globally binding environmental standards, such as those under the ozone regime, to freshwater resources, given the fact that water problems occur in qualitatively and quantitatively different forms from one country to the next. Another aspect to be considered is that a large number of international schemes have been implemented in the field of water policy, and that regional Water Commissions in industrialized countries and in some developing countries have advanced towards the integrated management of watercourses.

In the view of the Council, there are four functions that would have to be performed by a Global Consensus on freshwater resources, supplementing existing cooperation at regional and sectoral level (Section D 5.5.5.2), and three possible institutional forms that such a Global Consensus could have (Section D 5.5.5.3).

5.5.5.2 Functions

A Global Consensus should fulfil four functions:

- Improved information on the status of global water resources, including an assessment of their "criticality" and regular monitoring of water policy at national level (in the case of Germany, this would cause minimal extra costs due to existing programs and capacities),
- Enhanced consultation between the states on different ways to solve acute water crises, including the various technologies for rational water use and water protection, and appropriate policy instruments,
- Intensified and precautionary support, in particular through technology transfers, for states exposed to the threat of water crisis or already affected by acute water crisis, and
- Improved mediation between and advisory services for states that are in conflict or potentially in conflict over the use of transboundary water resources, whereby synergies are to be sought with the Convention on the Law of the Non-Navigational Uses of International Watercourses (see Section D 5.5.3).

Information and monitoring

Chapter 18 of Agenda 21 contains a comprehensive list of non-binding recommendations for the protection and improved use of water resources. These recommendations should be given added strength by means of improved systems for reporting on implementation. Unlike the conventions on climate, ozone, biodiversity and desertification, where the reporting function is performed by the Conferences of the Parties, or the protection of the forests and the seas, where appropriate non-binding mechanisms exist, reporting on the implementation of water policy recommendations remains the responsibility of the UN Commission on Sustainable Development, which is only able to support the protection of freshwater resources as part of its general activities. In addition, there are the obligations of states under regional regimes to notify and consult with other states, duties which serve cooperation between the riparians of an international watercourse and which are now extended by the new Convention on the Law of the Non-Navigational Uses of International Watercourses (see Sections D 4.1 and D 5.5.3). All in all, these requirements fail to match the scale and severity of the global water crisis.

Reporting is inadequate with regard not only to national water policies, but also to the quality and quantity of freshwater resources. In many developing countries the ability to collect data on freshwater resources and assess them accordingly has actually declined in recent years, despite the various Plans of Action (UN-SG 1997a, 6; UN Doc. E/CN.17/1997/9, 99). The social-scientific analysis of the environmental policies to date has shown that the very obligation to report can have beneficial effects that offset the costs incurred. An obligation on the part of states to submit regular reports on their water policies and the state of their water resources should therefore be a key element in a Global Consensus on freshwater that would have to be linked in appropriate ways to existing, sectoral approaches (e.g. research programs such as the Global Network for Isotopes in Precipitation or the World Hydrological Cycle Observing System).

Regular and improved national reports can provide a basis for identifying immanent or acute local or regional water crises; this could enable governments to modify their national policies for water protection and water use and help international organizations to deliver support more efficiently. As a basic, initial approach, the Council has developed a criticality index (see Section D 3.1) that could be used in this context as a tool for identifying local water crises. On the basis of such a criticality index, "states affected by severe water crises" could be defined politically and legally ("most vulnerable nations"/"most affected nations"). The advisory, support and mediating functions of enhanced international cooperation should then be concentrated on these states. The "full incremental costs" of reporting could be borne by the industrialized countries through additional funding, as is provided for by the Montreal Protocol, the Climate Convention and the Biodiversity Convention.

CONSULTATION

Consultation among states regarding appropriate policy instruments for water utilization and protection should be improved further, especially in light of the successes and failures of the activities listed in Chapter 18 of AGENDA 21.

Where transboundary water resources are involved, an important step would be the targeted integration of regional organizations into the global consultation process, by which regional experiences could be passed on and exchanged through interregional consultation. Partnerships could be formed between individual water boards, for example, as a result of which new and improved channels for technology transfer could be created, especially in the North-South context. The Internet could be used here as an inexpensive and rapid means of communication.

Another possibility would be to set up a Clearing House for water-related technologies, e.g. for irrigation, sewage treatment or for preventing harmful emissions. This Clearing House should have research competence in the fields of social science, economics and law, and evaluate the experience gained with various water policy instruments in different national contexts. As is common practice in the ozone regime, for example, the Clearing House could produce detailed lists of organizations supplying the relevant technologies and distribute this information – by inexpensive electronic means via the Internet. Clearing House agencies have already been established in other environmental fields, or are now in the making (the Biodiversity Convention being one example), so there is a wealth of experience that can be tapped.

SUPPORT

Collecting and disseminating information on regional water crises and advising affected states only makes sense if the latter have adequate resources for implementing an efficient and effective water resources policy. This is rarely the case in precisely those developing countries where water crises are a relatively frequent occurrence (UN-SG 1997b, 14f.). Improved cooperation at international level should therefore aim at providing targeted support to those states identified as acutely or potentially affected by water crises, but which are unable to help themselves to the requisite degree. In the context of a Global Consensus on water policy, this could be achieved with a suitable multilateral financing mechanism, supplemented by bilateral programs.

In view of the enormous significance of water crises for human health and the possible consequences (migration pressure and international "water wars", for example), this support should be provided in the form of additional funding for the countries concerned; this seems an attainable goal that accords with the increase in official development assistance recommended by the Council (WBGU, 1994). Chapter 18 of AGENDA 21 details the annual cost requirements for implementing programs for the sustainable use and protection of freshwater resources (Table D 5.5-1). This financial framework, acknowledged in 1992 by virtually all governments of the world as a legally non-binding global target, does not seem to have been implemented in the intervening period either by donor countries or, for lack of own funds, by developing countries.

As well as stepping up efforts to achieve the global target laid down in AGENDA 21, the Council recommends making greater use of innovative financing instruments that achieve major successes with comparatively modest outlays: consideration should be given to supporting potentially profitable projects – e.g. improving irrigation systems in agriculture – with revolving funds at regional and local level that could become self-supporting after "start-up funding" has been provided by the international community. Such innovative financing instruments, currently under intensive discussion within the UN Commission on Sustainable Development, could promote a more efficient and more effective use of available funds.

Another option involves the Global Environment Facility of the World Bank, UNEP and UNDP. However, its area of responsibility has been confined until now to "global environmental problems", among which only "international watercourses" have been numbered so far. Thus, either the sphere of responsibility of the GEF would have to be expanded, or a separate mechanism would have to be created, within the framework of existing institutions, if necessary.

Table D 5.5-1
Estimated average total annual cost (1993–2000) for implementing Chapter 18 of AGENDA 21.
Source: UNCED, 1992

Program area	Annual cost (US$ million)	
	International community "on grant or concessional terms"	Government investments
Integrated Water Resources Development and Management	115	
Water Resources Assessment	145	210
Protection of Water Resources, Water Quality and Aquatic Ecosystems	340	660
Drinking-Water Supply and Sanitation	7,400	12,600
Water and Sustainable Urban Development	4,500	15,500
Water for Sustainable Food Production and Rural Development	4,500	8,700
Impacts of Climate Change on Water Resources	40	60

ENHANCING CONFLICT RESOLUTION MECHANISMS FOR INTERNATIONAL WATER CONFLICTS

The UN Convention on the Law of the Non-Navigational Uses of International Watercourses is an urgently needed codification of international water law that the Council – with the criticisms expressed above – explicitly welcomes (see Section D 5.5.3). This Convention provides no new mechanisms that in individual cases could promote a peaceful resolution on the basis of these principles, however. Moreover, the issue of actual distribution of scarce water resources remains necessarily open in specific cases; states must still agree among themselves as to what constitutes "equitable" utilization, based on the factors specified in the Convention. And yet it is this very utilization of scarce resources that lies at the core of international water conflicts.

The international community can exert its influence here in order to reduce conflicts using various conventional methods of arbitration – above all, non-confrontational mechanisms for resolving conflicts over transboundary water resources could be strengthened, and institutionalized to a certain degree. One option would be to offer good services and mediation or conciliation by neutral states in a regional water conflict, in order to coordinate negotiations and to support the conciliation process in the form of non-binding proposals for compromise. Germany in particular – as a country that is not directly affected by serious international conflicts over water use – could offer itself wherever it can as a neutral mediator between the parties to such regional water conflicts, for example by providing the infrastructure for negotiation on "neutral territory" (good services) or, if so desired by the parties concerned, through mediation and arbitration. This could include the establishment of a specialized organizational structure in Germany, in close cooperation with the United Nations, such as a permanent mediating body for water conflicts to which disputing parties can turn for help (see Section D 5.5.4).

States are also free to call the International Court of Justice. The courts of arbitration already in existence could also be strengthened, or specialized tribunals created to which states could submit their disputes if necessary. Finally, in light of the UN Security Council's acknowledgement in 1992 that "environmental problems" can constitute a potential threat to world peace, it would be conceivable for the latter to bring its influence to bear in certain cases (UN Document A/47/253).

5.5.5.3
Possible institutional arrangements

The Council considers it urgently necessary that further efforts be made in the field of water policy through improved international cooperation – in the sense of the "Global Consensus" on the protection and conservation of freshwater resources proposed by the UN Secretary-General. This cooperation should fulfil all four of the functions referred to above. As far as institutional arrangements are concerned, there are three main options:

"GLOBAL PLAN OF ACTION"

The international community could adopt a Global Plan of Action containing a set of recommended activities in the field of water policy. However, it would be inadequate in the eyes of the Council if this were the only instrument. Sociological research indicates that plans of action have only minimal effects on the actual behavior of states; a case in point is the relative failure of the Mar-del-Plata Action Plan. Action plans have been replaced in several contexts by framework conventions containing at least the obligation to report on the implementation of recommendations and financing mechanisms (see the Desertification Convention, WBGU, 1996), or which even specify substantive environmental standards in the form of protocols (Montreal Protocol, Geneva Convention on Long-Range Transboundary Air Pollution) or which aim at such standards (Biodiversity and Climate Conventions). However, negotiating conventions is usually a very difficult and protracted process, and rarely promises any short- or even medium-term success.

World Water Charter

A further option would be the adoption of a World Water Charter on the principles for sound management of water resources, which would unite political forces and enable the building of institutions, despite being non-binding under international law. A World Water Charter could be agreed upon relatively quickly. Since it would not be a document of international law, it could be signed by parties that do not otherwise qualify as subjects of international law – local authorities, regional parliaments, environment and development organizations, or indeed business enterprises. The Council considers such a World Water Charter to be absolutely imperative.

In Section E 2 of this Report, the Council presents an draft framework for such a Charter, in which a number of proposals regarding institutional structures are also put forward. The substantive content of the World Water Charter – in other words the specific norms that it should contain – could be filled in on the basis of the detailed recommendations made by the Council and similarly included in Section E 2. The World Water Charter could bring together and consolidate the various sectoral activities of the international community in the field of water policy; it could become the central element of a global reporting system on water policy and assist states in implementing Agenda 21 on a non-confrontational basis, whereby recourse could be made to the experience of other nations as communicated in reports.

The World Water Charter must be linked in appropriate ways to existing institutions and action plans, in order to avoid duplication of effort and to exploit synergies. One of the most important of these would be a linkage to the Convention on the Non-Navigational Uses of International Watercourses. The latter is unable to fulfil all the functions of a global water strategy because of its priority focus on the utilization of transboundary water resources, although this must also include local problems such as drinking-water supply and non-transboundary water resources; by promoting regional watercourse commissions, however, the Convention plays an important complementary role that would have to be supported through appropriate integration.

Another linkage should be established to marine environment policy. The "Global Plan of Action on Protection of the Marine Environment from Land-based Activities" agreed in Washington in 1995 is welcomed as a first step towards the Convention for the Protection of the Marine Environment recommended by the Council (see WBGU, 1996). Reducing emissions from land-based discharges, as aimed at by the Washington Plan of Action, is directly linked, via pollutant loads to rivers, with policies for water use and water protection on land. Accordingly, suitable forms of cooperation would have to be sought between institutions in the fields of marine environment policy and the protection of freshwater resources.

Further synergies could develop through cooperation with biodiversity institutions, above all the Ramsar Convention, the World Heritage Convention and the Biodiversity Convention. An appropriate linkage should also be established with the UN Convention to Combat Desertification. Additional connecting lines would have to be drawn to the International Action Programme on Water and Sustainable Agricultural Development (IAP-WASAD) initiated by the FAO in cooperation with other organizations.

International "Framework Convention on Freshwater Resources"

A third option would be the conclusion of a "Framework Convention on the Protection and Conservation of Freshwater Resources", with binding force under international law. The Desertification Convention provides an excellent model: it provides for global cooperation in combating local or regional problems; there is participation on the part of states that are not themselves affected by the specific environmental problem, and the particularly critical situations are concentrated in the developing countries. The Desertification Convention does not place the Parties under a material commitment to comply with certain environmental standards, but acts primarily to institutionalize a reporting system on the progress achieved in national policymaking, combined with appropriate forms of international consultation and support for needy and severely affected developing countries.

Compared to a Plan of Action or a World Water Charter, a Framework Convention on Freshwater Resources would perform the aforementioned functions of a Global Consensus on a higher level, i.e. with

- regular Conferences of the Parties, with binding obligations to report,
- scientific sub-committees for expert consultation,
- a Convention Secretariat and
- an improved financing mechanism.

This structure corresponds to the three "C"s that even weakly endowed environmental conventions can achieve, namely raising concern for the environmental problem among national governments and international organizations, enhancing the contractual environment by improving international institutions so that more detailed measures can be negotiated through the political process, as well as national capacity-building with international institutions that promote the implementation of international recommendations (Haas et al., 1993).

If obligations to exchange information are laid down in conventions with binding force under international law, the effectiveness and (depending upon the specific convention arrangements) the quality of such obligations are enhanced; confining a framework convention to commitments to inform and consult enables states that are particularly affected (those involved in international conflicts over water resources, for example) to join a universal regime, and lowers the costs incurred by the potential parties.

For all the conceivable benefits, however, a binding framework convention for the protection of freshwater resources does not seem to the Council to be a viable proposition at this point in time. A better and faster way to resolve the present freshwater crisis (which shows every sign of worsening) would be a World Water Charter that represents a strong political commitment by the parties signing it, rather than a legal obligation.

5.5.5.4 Summary

The Council wishes to emphasize that if Germany plays a more committed role in the protection of the world's freshwater and makes best use of its technological and financial resources in this domain, it will be acting not only in the interest of the international community and the people in water-stressed regions, but also in its own long-term interest. The syndrome analysis developed by the Council indicates that water supply is a central factor in the network of interrelations and that many other problems can be caused or intensified by a local water crisis (self-reinforcing loops). Local water crisis can lead to desertification and thus intensify the greenhouse effect and the loss of biodiversity. Water crises can force local populations to leave their home areas and thus generate substantial migration pressure within or between states; from a broader perspective, water crises lead to excessive use of international watercourses and hence to an increase in regional conflicts, which can then escalate into "water wars". All this shows that even local water crises can have immense global significance that make water a problem for the entire international community.

To ensure that assistance is given to developing countries facing acute or potential water crises, the Council recommends that Germany give every support to the UN Secretary-General in his efforts to forge a Global Consensus on freshwater resources; as one of the largest industrialized countries, Germany could pressurize for negotiations on a World Water Charter focusing primarily on an improved reporting system on national water policy, intensified consultation at international level and enhanced support to developing countries affected by water crises. Within the framework of such a World Water Charter, Germany could deploy its technological and financial strengths in support of a Clearing House that could even be domiciled in Germany (Bonn). Another possibility is to create partnerships between the European Water Commissions in which Germany is involved and the corresponding bodies in the developing countries.

With regard to the settlement of international water conflicts, Germany should offer itself wherever it can as a neutral mediator between the parties to regional water conflicts, for example through good services or, if so desired by the parties concerned, through mediation and dispute settlement procedures. This could include the establishment of a specialized institution in Germany, in close cooperation with the United Nations, such as a mediating body for water conflicts to which disputing parties can turn for help.

5.6 Instruments

Instrument mix – Designate protected status – Financial contributions of the international community – Guaranteed minimum supply of water for all – Public information – Local capacity-building – Technical measures for monitoring and control – Strengthening of public health service – Technology and know-how transfer – Participation – Compliance with ecological guard rails by means of laws and agreements – Development of free market or cooperative structures – Extension services and training – Orientation of technology and organization to the yardstick of resilience – Improving the legal framework – Mediation

5.6.1 Preservation of valuable biotopes (World Heritage)

Protection by international organizations

Freshwater biotopes of special value for nature, humankind and science that exhibit features of importance beyond the regional or national context should be given a special internationally recognized status as protected areas. Such characteristics include: a high proportion of endemic plant and animal species (that only occur in the habitat concerned), special (unique) natural features and beauty of landscape, special scientific value, outstanding or even existential economic importance for the population of the region or country concerned, and particular threat due to the economic and/or political conditions (instabilities) in the regions concerned. Lake Baikal, for example, meets many of these criteria (see Box D 1.2-1).

Financial contributions by the international community

An appropriate way of protecting such areas is to include them in the UNESCO World Heritage List. This is also a way of accessing and providing international funds. Germany generously supports the UNESCO World Heritage and could encourage the nomination of other waterbodies, although this would have to be effected by the respective governments and promoted through targeted appropriation of funds. Candidates for special protection include Lake Tanganyika (extremely deep tropical graben lake, of great scientific interest and considerable economic importance), the Three Gorges of the Yangtze Kiang (especially beautiful landscape and importance for the Yangtze dolphin), Lake Titicaca (large lake having the highest altitude in the world; substantial economic importance, of great scientific interest), Ochrid lake (endemism because of old geological age, resulting in considerable scientific interest), Lake Chad (biogeographically isolated large lake in an arid region with high endemism, great importance for the local population, severe threat due to very unstable political conditions).

Know-how transfer

Germany possesses substantial scientific experience which could be utilized both for research in the framework of internationally coordinated or bilateral research projects and for designing targeted environmental protection and rehabilitation programs. The following are suitable instruments that would be extremely effective but not costly:

1. Organizing courses in the respective region for local researchers and environmentalists, as well as for optimization of fishery management.
2. Enabling university education in Germany for highly talented students from the countries concerned and support for doctorate work in the respective region using the "sandwich" approach (special training and familiarization with the necessary research methods in Germany; performance of practical work with German financial and equipment support locally; supported evaluation, analysis and publication of the results in Germany).

Both instruments are used to motivate multipliers who take over responsibility in their home countries for scientific research, protection work and measures for the rehabilitation and optimization of economic use (e.g. for the fishery industry), as well as training of further qualified personnel. There is already a well-developed infrastructure in Germany for the execution of such programs in the framework of the GTZ, the DAAD and in exceptional cases the Alexander-von-Humboldt Foundation, and by numerous private and semi-private foundations. In all programs in which highly qualified persons are sent to Germany, steps must be taken to ensure that they do not stay in Germany, otherwise they would no longer be able to perform their role as multipliers.

5.6.2 Water supply and wastewater disposal

Technology transfer

Technology transfer can take place alongside the exchange of scientific and technical information, training, consultation and the assumption of management functions in connection with direct investments and the export of equipment (e.g. wastewater engineering, sewage treatment plants, pipe technology, etc.), combined with support in the form of training and advisory services. Forward-looking strategies for wastewater treatment, for example, must also address existing drainage systems as well as local, social and cultural conditions (agglomeration or rural area). Innovative facilities for biological wastewater treatment can be used economically in many cases in conjunction with relatively large sewage farms. The health hazards posed by drinking water supplies in many developing countries means it is essential to install simple purification systems first of all that can then be upgraded step by step. There are now improved methods for identifying and eliminating leakages without having to replace entire pipe systems. These methods, which tolerate minor leakages, are less expensive than conventional methods involving investments in new piping.

CREATING MARKETS FOR WATER UTILIZATION

Ownership rights relating to water supply and wastewater treatment companies should be strengthened wherever conditions permit. In this way, use of water is automatically regulated as a scarce commodity by the market mechanisms of supply and demand, thus enabling the efficient allocation of water. It is particularly recommended that responsibility for water supply and wastewater disposal be consolidated within a single corporation.

ESTABLISHMENT OF COOPERATIVES IN REGIONS WITH POOR GOVERNMENT AND ANOMIC CONDITIONS

If the prerequisites for private-law regulations are not met, or if private-sector solutions are likely to fail due to other circumstances, then solutions based on cooperatives are suitable options. These require that users form special-purpose alliances and develop common rules governing the use and control of water resources.

MINIMUM SUPPLY OF DRINKING WATER FOR ALL

Everyone should receive a minimum supply of water, regardless of his or her income. The World Water Charter recommended by the Council commits the international community to cover such minimum needs. Implementation of this demand can be carried out on two levels: either the state distributes the respective minimum quantity to each citizen (through certificates of entitlement, for example) or a transfer takes place in the form of "water money (direct subsidies to the poor). The effectiveness and efficiency of these two measures cannot be determined in advance, so their success is dependent on the respective circumstances. In the case of piped water supply systems, suitable tariff schedules (e.g. the first 20–50 liters a month are free or low-priced) could perform the function of a minimum basic supply.

SUPPORTING EDUCATIONAL WORK AND DISCOURSE WITH THE POPULATION

The introduction of new structures in water resource management makes it necessary to educate the public about these structures. In discussions with various groups, needs and wants can be ascertained and integrated as a conditional framework for economic activities into the new structures.

5.6.3 Health

INFORMING AND EDUCATING THE PUBLIC

Most water-based health hazards are due to pathogens; chemical contamination with organic pollutants and heavy metals may pose problems in some places, but it plays only a marginal role with regard to the mortalities caused worldwide by impure water. The epidemic dimensions of morbidity and mortality due to contaminated water necessitate major efforts to ameliorate the situation. First and foremost, this involves educating the public about sanitary and preventive measures they can take. Educating the population about the health-related impacts of their own behavior is not very successful if well-meaning advisers from rich countries try to tell people how they should behave. Instead, it is essential to discuss water resource management in small groups and develop appropriate sanitation strategies for the given situation.

LOCAL CAPACITY-BUILDING

There is a growing awareness that complex tasks related to health care and hygiene require collective efforts that go beyond educating the individual. The aims of the communication process are to overcome a basic apathy and to develop an awareness that "together we are strong". Supplying pure water, preventing unsanitary conditions and modifying self-hazardous behavior are prime areas where local or community structures need to function. Fostering this with appropriate activities could be a key task for development practitioners at the local level. Procedures requiring or enabling participation on the part of individual community members further consolidate a sense of personal responsibility.

TECHNICAL MEASURES FOR MONITORING AND CONTROL

Apart from the usual technical measures for water purification and wastewater disposal, it is important to develop simple technical aids that permit people to distinguish between polluted and pure water, particularly in countries lacking an adequate infrastructure for water and sewage treatment. For example, one could add a harmless coloring agent whenever untreated industrial effluent or sewage residues are discharged, in order to warn people downstream who use the water for drinking, bathing or washing of the dangers and to keep them away from the polluted flow. It also make sense to install easy-to-operate water filters, either in individual households or in smaller communities.

IMPROVING PUBLIC AND PRIVATE HEALTH CARE

Health education, preventive medicine and treatment are the key elements of an effective health service. Efforts must center on mobile health advisors and medically trained staff who can go to the villages and districts where people live to advise and, if necessary, provide medical service. However necessary it may be to build hospitals, this can be no substitute for mobile medical workers. As soon as patients are dismissed from hospital, they revert to the same behavioral patterns that may have put them in hospital in the first place. Preventive measures and health care at local level, in contrast, provide greater chances of acutally influencing people's behavior. Mobile services such as these should first be offered as part of the public health service. As development level and incomes rise, however, the public health services can be gradually supplemented by private service providers, or even replaced entirely if conditions allow.

CETERUM CENSEO

Poor sanitation and health protection are highly dependent on income – more so than any other resource. The richer a person is, the smaller the chance of contracting an infectious disease. This is the case the world over. To that extent, as with many other development problems, improving productivity and ensuring the equitable distribution of earnings within an economy are the most effective means of preventing damage to health due to impure water. However, there are obvious limits to the contribution that the German Federal Government can make towards accomplishing these two goals.

5.6.4 Irrigation and food

TECHNOLOGY AND KNOW-HOW TRANSFER

Many of the irrigation systems in widespread use today are not very efficient and have unwanted side effects. Know-how transfer is necessary for two reasons: first of all, it is necessary to install irrigation systems that match local conditions (risk of salinization, sedimentation, health hazards due to standing water, groundwater pollution, etc.) and help to achieve the objectives pursued with a minimum of secondary environmental impacts. In order to meet these requirements, traditional irrigation systems in particular must also be included in the planning process, since they have often been adapted to the specific conditions at the respective location in the course of cultural evolution. Israel, for example, uses irrigation systems that have enabled sustainable agricultural use in very arid regions for the last 4,000 years. Secondly, when selecting crops or livestock, it is important to select breeds and varieties whose water needs are in reasonable proportion to water availability in the region. Only if both conditions are met can the negative effects of irrigation be avoided.

PARTICIPATION

All forms of technology and know-how transfer are likely to miss their target if they fail to stimulate interest in such innovations among the users of irrigation systems. However, enthusiasm for these solutions cannot be ordained from above, nor are the means of communication adequate for this purpose. Instead, users must be forged into a local network and provided with options for achieving the set objectives. Only when the users themselves are in a position to take an active part in planning and operating such facilities (empowerment) will they maintain and use the facilities in a proper way.

COMPLIANCE WITH ECOLOGICAL GUARD RAILS

Large-scale irrigation systems cause major alterations to the landscape. For this reason, efforts must be made to ensure that no endangered biotopes or species are destroyed as a result of irrigation. Legal regulations are essential for the conservation of biotopes meriting special protection. In the case of particularly valuable biotopes, the international community should provide funding to conserve them (Lake Baikal, for example). International conventions can also play a major role here. In addition, the respective national governments must be admonished to ensure compliance with the key ecological guard rails.

DEVELOPMENT OF FREE MARKET OR COOPERATIVE STRUCTURES

In arid areas, water for irrigation projects must be treated as a scarce resource. At the same time, free access to the sources of water, as is frequently the case, makes it difficult to regulate the water market through the private sector. In most cases, owners of water resources have few options open to them with which they can prevent free riders from tapping their resources. If this problem can be solved by technical means or with controls, then a free market solution is preferable to any government-controlled system of water distribution. This deserves special emphasis, given that states normally supply water "below cost" and thus encourage wastage. However, if free riding cannot be avoided, or in areas where intensive relations already exist between local users, priority should be given to cooperative solutions. It is essential here that all users commit themselves to the jointly negotiated rules and that they develop ways to monitor and control the use of water. Even among peoples with basic communitarian values, intrinsic motivation alone does not provide a sufficient foundation. Cooperative solutions tend to collapse when the population exhibits a high level of mobility. In such cases, a gradual conversion to private-sector forms of organization is needed.

EXTENSION SERVICES AND TRAINING

Apart from the transfer of know-how and technology, users of irrigation systems must be made aware of the possible threats to soil, crops as well as their own personal health. The same aspects must be kept in mind as with all such educational activities. They must match local conditions, be integrated into the sociocultural context and promote motivation on the basis of learning through personal insight.

5.6.5 Disaster prevention and control

BASING TECHNOLOGY AND ORGANIZATION ON THE CRITERION OF RESILIENCE

In the past, many measures for the regulation of river courses and flood protection were carried out with the aim of greatly reducing risk with a minimum of technical expenditure. A blind eye was turned to the fact that such measures continuously increase the overall disaster potential, i.e. the maximum number of people affected by a disaster. A paradigm shift has meanwhile occurred within international debate. Maximization of benefit has been superceded by the principle of resilience. The latter principle states that even under adverse circumstances or in the case of improbable events, the maximum scale of damage must be within limits. This new perspective implies, for example, that few smaller dams are built one behind the other instead of one large dam, that flood basins (floodplains) are created in addition to structural facilities, that error-tolerant installations are preferred, and that decentralized and flexible forms of organization are chosen rather than giant organizations. However, the principle of resilience does not mean that large-scale projects are to be assessed more negatively than many small projects right from the beginning and in all circumstances, but that the failure of systems or the occurrence of unexpected events should limit the maximum extent of the damage to an acceptable magnitude.

TECHNOLOGY AND KNOW-HOW TRANSFER

The water development schemes commonly implemented nowadays are still predominantly based on the old paradigm. In connection with transfers of technology and know-how, greater attention must therefore be given to compliance with resilience criteria.

PARTICIPATION

With flood protection measures, too, it is imperative to involve the affected population in the decision-making process. Only if residents are put in a position to take an active part in the preparation and planning of water development projects (empowerment) will they accept the inconvenience involved in construction and conduct themselves appropriately in the case of an emergency. Water development activities that require resettlement merit special attention. Here, participation is an indispensable condition that must be fulfilled before construction can begin. Not only should all those concerned be able to have a say in the decision; representatives of the affected population should also be able to plan and execute resettlement themselves, with a reasonable budget at their disposal.

COMPLIANCE WITH ECOLOGICAL GUARD RAILS

Large-scale water development systems cause major alterations to the landscape. To this end, special care must be taken to ensure that water protection activities do not destroy endangered biotopes or species. As with irrigation systems, legal or international regulations are imperative in this context.

5.6.6 Resolving conflicts at national and international level

INTERNATIONAL AGREEMENTS AND REGULATIONS

Institutional arrangements – such as global water programs as well as international conventions and international regimes for the protection of freshwater resources (e.g. under the auspices of the United Nations, for example) – prescribe, standardize and regulate the environmental sound and sustainable management of water resources. These arrangements can make a valuable contribution towards preventing conflicts from arising and settling those that do. However, this depends to a critical extent on the specific content of such arrangements. It is hoped that the "Water 21" Plan of Action proposed by the European Union at the CSD-5 and supported by Germany will contain stringent provisions. This Report is seen by the Council as a source of ideas and guidelines in this respect.

International treaties and conventions are another means by which disputes are avoided and conflicts resolved. This function is clearly evident as far as the substantive provisions in an agreement governing use of a waterbody are concerned (see Section D 5.5). By contrast, the function of the arrangements for cooperation that such an agreement may contain is more complex. Commitments to exchange information and coordinate activities may serve to close certain loopholes in the substantive law. This is particularly necessary if the contracting parties are merely prepared to stipulate abstract rules of utilization and bans on pollution. A higher level of cooperation exists if cooperation involves continuous environmental monitoring, implementation of substantive contractual standards as well as formulation of specific bans on pollution. In general, one can say that cooperation is a forerunner of substantive commitments (Hinds, 1997). If one considers that a direct need for cooperation exists exclusively among the riparians of international watercourses, then the following differentiation applies: substantive obligations are specified above all in the regional agreements between the riparian parties of a transboundary waterbody. At the same time, global agreements such as the pending UN Convention on the Law of the Non-Navigational Uses of International Watercourses do not perform the same function, but serve rather to promote regional agreements and specify a "minimum standard". This can also be inferred from the fact that nearly half of the articles in the planned UN Convention deal with states' obligations to cooperate. Even though extremely divergent regional circumstances with regard to water resources stand in the way of uniform standards, it remains to be hoped that generalizable standards and rules of utilization going beyond regional agreements will be established through the increasing contractual regulation of the use of transboundary water resources.

Improving the legal framework for conflict resolution

Enhancing international mediation mechanisms for the prevention of conflicts were dealt with in Section D 5.5.3. Not only should these "softer" forms of international cooperation be strengthened – so, too, should institutionalized dispute settlement procedures. For instance, not all regional agreements for regulating international watercourses provide for institutionalized cooperation or mechanisms for the settlement of disputes. Major differences exist between those agreements that do contain provisions on institutionalized cooperation. River commissions, for example, may be granted competencies of varying scope. Moreover, the members of such commissions should be independent or bound to directives. In some cases, commission resolutions have to be unanimous, while a majority vote is sufficient in others. The same heterogeneous picture arises on closer inspection of the rules for dispute settlement contained in regional contracts. The obligation to settle disputes may amount to nothing more than a stipulation that, in the event of a dispute, the parties shall seek a solution in good faith. However, a non-binding procedure for settlement of disputes can also be prescribed. A binding mechanism, on the other hand, is very rarely stipulated. Article 24, in association with Annex V of the Convention for the Protection of the Danube will be presented here as an example of a differentiated dispute settlement procedure. As a general principle, it is incumbent on the parties to settle any disputes within a reasonable period through negotiation or any other means. This period extends for a maximum of twelve months from notification of the Danube Commission about the dispute by one of the parties. After that, the dispute should be submitted for a binding decision either to the International Court of Justice or to the ad-hoc arbitration panel provided for under Annex V of the Convention. A distinctive feature of the arbitration procedure in Annex V concerns the provision for ending the dispute settlement procedure if one of the parties refuses to cooperate. What is striking here is the graduated and differentiated procedure, characterized in essence by the fact that if non-confrontational means fail to achieve an acceptable result, binding dispute settlement bodies are called in. Improving the legal framework for conflict resolution cannot be achieved with the ideal of a central and binding decision-making body, something that has not (yet) been possible to implement in practice, but instead by reinforcing all forms of dispute settlement while providing for graduations in their application.

Mediation

The Federal Republic of Germany can offer its role as a neutral third party and mediator in conflict-laden situations involving the distribution of water resources in order to help achieve a consensual and peaceful settlement and thus to avoid a violent escalation. A conflict over the use of water resources can emerge between social groups within a country as well as between states that are riparians of international waterbodies. Nonviolent ways of conducting the conflict should be maintained and ensured on a long-term basis through voluntary institutionalization of standards, regulations and guidelines for behavior within an (international) regime applying to the region in question. In addition, new mediative procedures should be developed and promoted in a targeted manner for regional and local conflicts as well.

Recommendations

E

Key recommendations for research on freshwater resources 1

1.1 Sectoral analysis of the system as a whole

INTERNATIONAL RESEARCH COLLABORATION

Precautionary strategies for averting a global water emergency in coming decades must be based on robust predictions of the future supply of water. Such predictions, in turn, can only be developed on the basis of a much-improved understanding of the cycles within the Earth System and the resilience of these cycles to human-induced perturbations (e.g. enhanced CO_2 concentrations in the atmosphere). In essence, the WCRP and IGBP international research programs are an ideal platform for research activities aimed at acquiring a well-founded understanding of the system as a whole, particularly with reference to the hydrological cycle. Unfortunately, however, the declining willingness of national institutions to provide the requisite funds is liable to deprive these programs of their current momentum. The Council has already drawn attention to the importance of these forms of international research collaboration (WBGU, 1997).

And yet freshwater problems are characterized, like no other environmental issue, by more than purely natural-scientific aspects; as explained in the Annual Report, they are determined to a major degree by the "human dimensions" as well. An integrated research approach connecting the WCRP, IGBP and the international social-scientific research program IHDP is the only way to provide valid analyses and well-founded solutions for this specific context. Such an approach can draw, for example, on the structural logic of the syndrome approach advocated by the WBGU (WBGU, 1995 and 1997). Independently of the latter, the Council strongly recommends that the Federal Government adopt and support, as part of the new environment research program, a broadly based and viable approach to the integrated analysis of freshwater problems against the background of global change.

The network research projects on "The Hydrological Cycle and Urban-Ecological Development" and "Ecology of the Elbe Basin" are good examples of interdisciplinary projects addressing the issue of how people manage water resources at all levels.

CLIMATE AND THE HYDROLOGICAL CYCLE

The principal goal is to gain an understanding of climate variability, especially at time scales of up to 100 years, in order to distinguish between anthropogenic and natural factors. What is needed is an improved description of the coupled atmosphere-hydrosphere-cryosphere-biosphere system, including the effects on that system when biological parameters in climate models are changed.

- Some crucial elements in the climate models (evapotranspiration, cloud formation, sea-ice formation) have not yet been described with the precision or depth necessary for simulating the hydrological cycle.
- Potential changes in regional precipitation patterns are of crucial significance in many regions of the world. In order to predict these changes, it is essential to elaborate global and regional climate models with high geographical resolution.
- Knowledge about the extent and renewal rate of groundwater resources must be greatly improved in many regions, with special reference to fossil reserves.

LINKAGES BETWEEN LIMNETIC AND TERRESTRIAL ECOSYSTEMS

Quality assessment of groundwater, running waters, wetlands and lakes makes little sense unless their interlinkages with neighboring ecosystems and the climate are also analyzed. There is a pressing need for in-depth research into the impacts of land-use changes on the interactions between riparian areas and waterbodies. In addition, it is imperative to examine the potential responses of terrestrial and limnetic ecosystems to natural climate variability and to determine the stress limits in respect of climate change, with specific reference to long-term climate variability.

Chemical processes in surface waters are influenced in their particular environments by a diversity of factors that are not sufficiently known. Research issues in this field include:

- Quantifying biogeochemical processes in groundwater that lead to pressures on surface waters (e.g. interaction between dissolved organic carbon and nitrogen load in groundwater; contamination by pesticides).
- Quantifying the influence of neighboring wetlands on water quality (denitrification, precipitation of heavy metals).
- Economic valuation of the ecological services that wetlands, floodplains and riparian vegetation provide with respect to water quality.

Connections between water and soil degradation

Irrigated farming, which in future will play an even greater role in feeding the world population, is particularly exposed to the threat of the most fertile soils being degraded by erosion and salinization. These soil losses are mostly offset by converting natural ecosystems (e.g. by clearing forests to grow crops). There is an enormous need of research in this area, due to the limited amount of land suitable for farming and the conflicting goals of international conventions on protection of the atmosphere and biodiversity:

- Development of appropriate, water-saving utilization strategies that are beneficial to soils and based on the principle of sustainable land use.
- Research into the succession of plants and animals on degraded land.
- Rehabilitation of degraded soils using the natural capacities of the specific location.

Sociocultural frameworks for water resource management

Research on "water resource management" has exhibited a bias towards the natural sciences and technology, and should now be widened in scope to embrace issues addressed by the separate disciplines of the social and behavioral sciences as well as interdisciplinary and transdisciplinary issues. The key research topics will be:

- Systematic exploration of the normative sociocultural foundations for water resource management by means of inter- and transdisciplinary analysis of water culture in the past and present, in all its manifold dimensions (science, technology, trade and industry, law, administration, religion, symbolism, esthetics, etc.).
- Historical and comparative cultural study of traditional knowledge, rules and techniques in the management of water resources, with special reference to sustainability.

1.2 Specification and application of the guiding principle

Ecological guard rails

Determining critical stress limits and the ecological services performed by aquatic systems

Critical stress limits must be defined in relation to the climatic boundary conditions and the properties of the waterbody being analyzed. These limits can serve as the basis for planners' decisions. To define them, it is necessary to investigate the basic characteristics of standing and running waters (their morphology, hydraulic properties, temperature) with regard to physical and chemical loads (irrigation, drinking water, discharged cooling water, shipping; inputs of plant nutrients and pollutants).

- Studies of contaminant accumulation in aquatic habitats through the operation of physico-chemical and biotic processes, and of the decomposition and conversion of pollutants in waterbodies, soils and riparian habitats (especially through microbial processes).
- Research on pollutant groups inadequately studied to date (including chelating agents, volatile organic compounds, hormonal substances, artificial fragrances, persistent organic pollutants) with respect to their formation, turnover and impacts.
- Inventorizing the biodiversity of aquatic habitats at the levels of genetic diversity, species diversity and ecological diversity. The research goals must involve an advance in our understanding of the role played by biodiversity in ecosystem responses to anthropogenic interventions.
- Investigation of the impacts of introduced species (e.g. of fish) and the invasion of non-native species (e.g. via the ballast water of ships) on the structure and function of aquatic ecosystems.

Global criteria for water quality assessments

Standardized indicators must be developed in order to improve existing monitoring systems. The key areas should be:

- Developing assessment criteria (indicators and combined variables) for aquatic habitats that can be applied regardless of edaphic conditions and biogeographical region.
- Defining water quality standards for agricultural, industrial and other uses, taking regional factors

and especially health aspects into consideration.

Weighing up irreconcilable ecological and sociocultural guard rails

The separately determined sociocultural and environmental criteria defining the "guard rails" delimiting the social action space are not always commensurable with each other. For example, compliance with quality standards for freshwater may necessitate local withdrawal rates that result in irreversible damage to key ecosystems. In cases like that, it is essential to achieve a balance between conflicting norms and interests. Pertinent research into the theory and practice of "multiobjective decision-making processes" is becoming increasingly important in the context of global change and therefore needs continuous support.

Monitoring

Efforts to build a global database on freshwater ecosystems are in general need of support. Geographical and ecological parameters should be included, as should adverse anthropogenic impacts. The database should also provide the results of a globally coordinated waterbody monitoring program, support the production of specialized maps and be available via the Internet to a wide range of users. Special research needs exist with respect to:

- Determining the status of aquatic habitats by expanding waterbody monitoring in regions (e.g. Asia, South America and Africa) and categories (e.g. wetlands, groundwater, lakes) on which little data has been collected hitherto, in order to provide a foundation for the global database.
- Collection of reference data from relatively unpolluted waterbodies, and investigation of the natural variability of factors relevant to quality (e.g. lacustrine sediments) in order to assess global and regional changes. Extending monitoring to include parameters that have not been adequately measured worldwide (e.g. metals, pesticides, organic trace substances).

Institutions for ensuring compliance with the social and ecological guard rails

An important component of integrated environmental protection schemes is the inter-medial analysis of pollution with a central focus on the mutual interdependencies between air, soil and water as environmental media. Such approaches are finding their way into national legislation and the laws of the European Union.

- Approaches of this kind should also be studied with reference to international environmental law.
- In the field of environmental law, the concept of sustainability is being examined intensively. The right to development is also gaining recognition. The underlying purpose of the sustainability debate, namely to integrate the twin issues of environmental protection and development, needs to be focused on much more within the field of law, however.
- The level of cooperation in connection with transboundary watercourses is very high on account of the long traditions in this area. Research needs relate to the systemization and characterization of the various mechanisms of collaboration, which would also be useful in fostering consensus in other areas of global and regional cooperation to protect the environment.
- There is also a need for research into possible ways of predicting international conflicts, and into the conditions and prerequisites for resolving them.

Role of non-governmental organizations and the involvement of non-nationals

Through the Rio process, non-governmental organizations have acquired a greatly enhanced role in international negotiations. The question whether they should be subjects of public international law has gained actuality again. The question of their democratic legitimation and the investigation of possible forms and rights of participation, particularly in the field of international environmental protection, are interesting aspects in this regard.

- Research is needed in connection with the issue of which participatory rights should be granted to non-nationals in administrative procedures, and their entitlement to take legal action in the event that they are harmed or must fear harm in their own country by transboundary environmental impacts originating from a different state. This question could generate some interesting work and be of immediate practical benefit, particularly in the context of the harmonization of laws in Europe.

Tradable emission certificates and international funds in the field of environmental law

- There is a need for research concerning the legal assessment of tradable emission certificates, an instrument deployed in the USA, from the viewpoint of Germany and Europe (see WBGU, 1997). Research in the legal field should focus greater attention on the international debate over joint implementation and tradable emission certificates in the context of climate protection.
- Environmental funds in the field of liability law are already a subject of research; preventive funding models, such as the Global Environment Facility, are still relatively unresearched (WBGU, 1997).

1.3 Specific applications of the guiding principle

FOOD

The cutbacks in funding for international agricultural research should be counteracted with greater commitment on the part of Germany.

- In particular, it is recommended that the BMZ provide more funds than in the past for promoting research on the wide variety of aspects related to the use of water for food production in the developing countries. The strengths of international research programs such as IGBP and IHDP should be exploited to a greater extent than hitherto.
- Integrated strategies for water resource management in connection with land-use practices must be developed further.
- Research is also needed into the development of methods for distributing and dosing irrigation water in order to increase the efficiency of water use, including the re-utilization of water resources and new "water harvesting strategies".
- Research into the breeding of salt- and drought-resistant crop varieties should be intensified.
- More research attention should be focused on the potential offered by aquaculture, the societal conditions for its success, and its environmental impacts on surface waters and coastal ecosystems.

LAND-USE OPTIMIZATION: AGROFORESTRY AND MULTIPLE CROPPING

In order to meet their basic requirements, people need not only cereal crops, but also fodder for their livestock, high-protein plants and firewood. In a situation of limited cultivated area and rising costs for irrigated agriculture, modern agricultural methods such as multiple cropping and agroforestry will become immensely important. Enormous research efforts are still needed in this area, particularly with regard to:

- Increasing the acceptance and implementation of multiple cropping systems.
- Breeding and selecting locally appropriate crops for such cultivation systems (avoiding worldwide cultivation of leucaena and eucalyptus).
- Optimizing complex crop rotation techniques.

"INTEGRATED WATERSHED MANAGEMENT"

Effective and economical utilization of the Earth's water reserves is closely linked to land use. It is therefore essential for water resource management to be carried out at the level of natural catchments. The costs for limiting use (e.g. in steeply sloping source areas) and for negative impacts (e.g. eutrophication) in the catchment basin must be ascertained as the basis for such an integrated analysis, and compared to the resultant benefits. There is a considerable need of research in this area (see above):

- Quantification and evaluation of the adverse effects of different land-use forms on groundwater and surface waters.
- Development of strategies for precautionary action and responses against the negative impacts of land use (e.g. by means of afforestation, preventing erosion when using slopes, preservation of wetlands, floodplains, etc. for retaining and delaying runoff).
- Identifying and quantifying the biogeochemical conversion processes that occur in wetlands and in groundwater.
- Economic valuation of these interactions and measures.

HEALTH

One person in two in the developing countries suffers from a water-related disease. This situation necessitates research in the following areas:

- More intensive development of vaccines against waterborne infections.
- Further investigation into the pathogens of waterborne infections with regard to verification procedures, ecological and epidemiological characteristics, the impact of water treatment techniques, disinfection methods and reproductive conditions.
- Research capacities in malaria-afflicted regions must be strengthened, in order to monitor and identify the ecological, social and economic determinants of disease.
- Establishment of European networks for epidemiology similar to the Centers for Disease Control and Prevention (CDC) in the USA, featuring fast, efficient health information systems (using GIS technologies).

WATER TECHNOLOGY

The storage, distribution, utilization and purification of water involves technical measures that must be refined in future in order to provide water of high quality, also in water-poor regions, to reduce consumption, prevent pollution and to recycle polluted water. Priority should be given to research support in the following fields:

- Development and refinement of biological methods, membrane techniques or chemical oxidation methods for treating water.
- Development of durable pipe systems that minimize the risk of secondary colonization by microorganisms.
- Development of water-saving technologies for private households, industry and agriculture.
- Optimization of cost-effective, decentralized

wastewater treatment facilities for rural areas.

- Further development of land-based wastewater treatment and methods for multiple use.
- Research on inexpensive, easily handled technologies for purification and sanitation, and on biological sewage treatment plants.

FLOOD CONTROL

Floods are the third way, after disease transmission and droughts, in which water threatens human life and health. Research is needed on:

- Integrated analysis and modeling of the entire causal chain, from precipitation, formation and concentration of runoff, the course of flood events (also in flooded areas), to damage assessment.
- Earlier and more precise forecasts of precipitation with the help of mathematical models. Greater use of remote sensing techniques to predict flood events, and refinement of techniques for direct conversion of remote sensing data into streamflow figures.
- Derivation of scenarios for extreme weather situations, both on a regional and local scale, on the basis of global warming scenarios, global and regional climate models and analysis of the influence of cyclonal weather patterns on precipitation in Germany.
- Research into the social processes of perception, communication and response in connection with the handling of flood risks as compared to other risks to which individuals and societies are exposed. Investigation of the role of limit values in the acceptance of risks.
- Research to develop simple flood control technologies for decentralized use.

ECONOMICS

- To what extent does market-based valuation impair intrinsic motivation – e.g. for thrifty and/or wise management of water resources? The interplay between intrinsic and extrinsic motivation is an issue of increasing topicality in the field of game theory, and should therefore be subjected to more detailed analysis.
- How can the damages incurred through degradation of soils (salinization) and groundwater be integrated into the valuation of water resources for agricultural purposes? These, after all, are damages to assets that are often taken into account to an inadequate extent in period-based accounting when the assets are natural resources.
- The interactions between water and its surroundings (e.g. riparian zones) mean that the environmental media of soil and water can be separated to only a certain degree as far as their ecological functions are concerned. How can this be taken into account in the ecological valuation of water resources?
- In many cases, water problems cannot be solved within the regional context unless water supply and wastewater disposal are put in the hands of a single organization. What solutions are there to the problem of monopolization that then ensues? What conditions must be met by the state so that it can discharge its responsibility for monitoring and control to a satisfactory degree?
- The reports of the World Bank often refer to the regional models in France as worthy of emulation. A systematic assessment of the strengths and weaknesses of the French approach would make sense in this context. Do the regional water parliaments guarantee a more comprehensive valuation of the various functions that water performs?
- How can the economic weaknesses of the German model be overcome and the model become "exportable"?
- Cooperation models play an important role in coping with many types of water problems. These communities of interest operate as social units, usually with a limited number of members and clearly defined geographical boundaries. In many countries with strong cultural traditions and less focus on private property, such communities can prove to be interesting solutions to water-related problems. More research needs to be conducted into the basic conditions necessary for such solutions to develop.
- A look at the historical components of land use shows that many decisions on water use were taken when water scarcity was still not an issue. What experience was acquired with various options for introducing water property rights when this resource became scarce (e.g. according to the doctrine "first in time means first in right")?
- Payment of "water money" to the needy is supposed to improve their incomes in such a way that they have access to a minimum supply of water. If these payments cover more than the absolutely essential minimum requirement, the individuals in question will also have an incentive to save water, because the benefit thus acquired is then available for other types of goods. What experience has been gained so far with such "water money" schemes? To what extent are tariff models (involving exemption of minimum requirements) a superior alternative?

EDUCATION AND SOCIAL ORGANIZATION

The most important feature of the Desertification Convention is the recognition of participative strategies, the greater role assigned to non-governmental organizations and hence of bottom-up approaches as a precondition for implementing the national action programs. Research needs relate primarily to identifying the conditions for successful participation in different sociopolitical and cultural contexts. Of particular importance are:

- Investigation of the options for enhancing public awareness of water problems as a contribution towards ecologically beneficial, economically efficient and socially compatible water resource management, e.g. by fostering perception of the problems, information campaigns, price policies, feedback, etc.
- Research into and development of culturally appropriate and integrative strategies for sustainable water resource management, addressed to specific stakeholders and situations ("instrument mix" comprising technical, economic, legal and psychosocial measures) and with specific reference to the respective water culture.

1.4 Integrated system analysis

Intensifying research on water-specific criticality indices

Worldwide, there is a basic inadequacy of data on freshwater resources, a situation that has actually deteriorated in some respects. The downward trend is partly attributable to the deterioration of scientific organization in many developing countries.

If the local significance of the freshwater crisis is to be analyzed in greater detail, it is essential to have data of finer geographical resolution and of greater topical range. The following research issues are of particular importance in this context:

- Problem-solving capacity should not be based solely on the economic strength of a country or region (GNP), but should also contain a regional assessment of water-relevant know-how, the efficiency of the relevant institutions as well as the stability of political structures. Determining these basic indicators could be achieved in a wider program for analyzing the local management capacity for responding to environmental crises, e.g. in the framework of global vulnerability studies.
- The integration of water quality aspects is immensely important when calculating the criticality index. Human-induced contamination of natural water resources is a serious problem throughout the world and in some regions imposes severe limits on water availability. Existing databases in this field (e.g. GEMS/Water) must be improved. For example, many statistical surveys for assessing water quality use the term "safe water", yet there are no precise, standard definitions of the term.
- Despite having a standardized system, many national statistics cannot be compared because they are based on different uses within the various sectors (agricultural, industrial, domestic).
- Large countries such as Brazil, China, Canada or the USA are so heterogeneous in their geography and population distribution that different patterns of water use have developed. Using socioeconomic variables for the country as a whole is inadequate as an approach (e.g. equal treatment of Alaska and the east coast of the USA); what is needed, instead, is a certain degree of regional differentiation.
- Harmonizing the data with respect to content, reference frame and geographical variables is more than just a methodological research issue. It is imperative in order to define a standard for surveys and projections that enable studies to be carried out at the global scale.

Widening and deepening the syndrome concept

Breaking down the global freshwater crisis into the various underlying patterns of cause and effect generates a number of additional research challenges:

- The specific relevance for freshwater resources of the syndromes identified by the Council should be analyzed in greater detail in the framework of interdisciplinary research networks.
- Specialized maps should be developed on the basis of syndromic webs, showing the intensities of syndromes at high geographical resolution.
- The degree to which a region is affected by given syndromes should be evaluated, using systemic indicators, with respect to both the causes and the impacts of the freshwater crisis, taking into account the specific relevance for freshwater resources and the regional syndrome intensities.
- In developing instruments (strategies, technologies, incentives, etc.) for integrated curing of the global water crisis, and in assessing its potential repercussions, it is imperative to address the syndrome-specific mechanisms in each case.

Research recommendations regarding the Green Revolution Syndrome

Research concerning the relationship between food production and freshwater scarcity is needed in the following areas:

- Our knowledge about the trajectory of the green revolution in this age of globalization is still fragmentary. The impacts of the international debt crisis and subsequent structural adjustment measures have been inadequately studied with regard to their interactions with the green revolution.

There is a need for research in this area.
- The public participation frequently called for in connection with regional development programs has not been sufficiently specified. The conditions on which successful participation depend need to be studied in detail.
- The baseline irrigation water requirement for securing nutritional subsistence must be determined for each specific country and region.
- Exploring the opportunities and risks of biotechnologies in agriculture deserves further support (biosafety aspects).

RESEARCH RECOMMENDATIONS REGARDING THE ARAL SEA SYNDROME

A substantial and wide-ranging body of knowledge now exists on the impacts of large-scale water development projects. Translating this knowledge to the context of individual case studies is difficult, however. Research is needed in the following areas in particular:
- Improving methods and data for the assessment of environmental costs and their internalization in cost-benefit analyses.
- Renewed evaluation and longer-term monitoring of major water development schemes erected in the past.
- Fundamental research work for solving the technical and financial problems that arise through the decommissioning of dams.
- Research into technical measures aimed at preventing the migration of animals (e.g. fish ladder).

RESEARCH RECOMMENDATIONS REGARDING THE FAVELA SYNDROME

Infrastructural resources are often inadequate or totally absent precisely in those urban regions that are most densely settled. This situation gives rise to specific research needs:
- The interplay of classical push and pull factors as causal forces behind rural-urban migration flows has been extensively studied. However, the growing importance of the informal sector means that customary analyses and explanations will have to be re-examined. There is a need for research regarding the extent to which the informal sector represents a new quality as a pull factor, and the capacity it provides for coping with urban water problems.
- In the field of urban planning and urban development, there are major deficits in basic knowledge about the conditions for successful water management at local and municipal level. There is a need for research into appropriate ways of increasing the competencies of urban dwellers and devolving responsibility to them, and into the suitability of public private partnerships as cooperative alliances.
- In view of the serious risk of illness due to inadequate sanitation in large parts of the megacities, especially in the developing countries, it is essential to implement simple, fast and cost-effective installation schemes. Investigations must be carried out into ways of organizing, coordinating and implementing such schemes in the short-term and with a minimum of problems.

2 Key recommendations for policy action on water resources

2.1 Elements of a global water strategy

In Section D 5.1, the Council presented its general paradigm for the efficient, equitable and sustainable management of freshwater resources, specified in terms of guidelines. Given a basis consensus on these normative principles, policymakers in the fields of environment and development must try to ensure that the paradigm is specified, shaped and operationalized.

Germany can help resolve global water problems primarily by asserting its influence in various fields of international policymaking. These include international development cooperation, foreign trade, the transfer of knowledge and technology, and support of existing and forthcoming international regimes in the environmental and development field. Furthermore, by implementing a national water policy in the sense of the guiding principles outlined by the Council, Germany can strive for an enhanced role as a "model" of sound water resource management for other regions to follow.

At the international level, three different platforms for the organization of specific measures and activities in this connection would be:

- A "global consensus" on the definition and adoption of the paradigm, especially with regard to specifying the sociocultural and environmental guard rails.
- A "World Water Charter" based on the latter, in which the members of the international community commit themselves to respect the principles and conditional framework of "sound management of water resources" as defined by the paradigm.
- A "Global Plan of Action" for detailed specification and implementation of the agreed action to be taken by societies within their permissible boundaries, in accordance with the principle of maximized efficiency.

In specifying the paradigm, it is essential to observe certain operative guidelines deriving from the geophysical properties of water as an environmental medium, from its specific functions as an environmental and cultural asset, or from the current distribution of access and rights of access to water resources. Such principles are discussed and analyzed in Sections D 5.2 to D 5.6.

2.2 Specification of the paradigm

Sound management of water resources requires a definition of the sociocultural and ecological "guard rails". It is crucially important in this context to take an integrated view of environmental and development standards and to elucidate in sufficient depth the repercussions of water-related projects. Specifically, the Council recommends that:

1. minimum standards be defined for the supply of drinking water and water-related sanitation facilities to individuals,
2. the resultant country-specific and culture-specific demand for freshwater be ascertained with respect to quantity and quality, giving special consideration to health aspects,
3. general safety standards be defined with respect to natural water-related disasters,
4. the geographical and sociopolitical pattern of vulnerability and the resultant need for precautionary action be determined in light of (3.) above,
5. internationally recognized equity principles governing access to or control over national and transboundary freshwater resources be agreed.
6. global groundwater reserves in fossil aquifers and the renewal and self-purification rates of recent groundwater reservoirs be determined,
7. the global stock of ecosystems dominated or influenced by freshwater and in need of protection be identified and classified,
8. the respective stress-bearing limits of the seminatural systems identified under (7.) above be determined in respect of water stocks, water quality and water variability, and that
9. the methods for integrated analysis and assess-

ment of water-relevant projects in the private or public sector be developed further.

However, the Council has no mandate to specify in every detail the recommended "guard rails" for the sound management of water resources. This remains a task for practical policymaking and the legitimated legislative organs.

2.3 Compliance with and application of the model

A fundamental consensus between competing users, societal groups or states on the specific guard rail criteria for sound management of freshwater resources does not automatically mean that these limits will be respected. This would require agreement on institutional regulations that can be enhanced by technical, educational and economic programs. Specifically, the Council recommends:

With regard to the further elaboration of international law and international regime-building, to:

- Make water-relevant standards a more integral part of international trade and credit agreements (WTO, World Bank programs, Hermes credit guarantees, etc.).
- Support the negotiation of a World Water Charter and a comprehensive Global Action Program for "Sound Management of Water Resources".
- Continue to strive in this context for an international regime relating to the problem of soil resources (WBGU, 1995), given that soil conservation in particular is urgently needed to protect water resources and to ensure adequate water supplies.
- Ensure that sound management of water resources is taken into consideration more as a cross-cutting task in sectoral regimes for the transition to sustainable development (examples being the Climate Convention, negotiations on the protection of forests, the Biodiversity Convention and the Desertification Convention).
- Step up international cooperation with regard to water-relevant aspects of the International Covenant on Economic, Social and Cultural Rights and the relevant responsibilities of the United Nations High Commissioner for Human Rights.
- Examine the options for improving and strengthening the coordination of various international organizations and programs operating in the field of "sustainable development" by integrating them into a single "Organization for Sustainable Development", whereby UNEP, CSD and UNDP in particular could be integrated, while closer links would have to be forged to the World Bank, the International Monetary Fund and UNCTAD.
- Regarding the amendment of the United Nations Charter, which Germany supports (German membership of the Security Council), the Federal Government should also examine the possibility of including articles on sustainable development in the Charter; in particular, consideration could be given to the inclusion of environmental protection in Article 55b, and the goal of sustainable development in the Preamble and in Article 1 or 2.
- Take action to ensure the broadest possible support for LOCAL AGENDA 21 initiatives worldwide, given that species protection and soil conservation, for example, are problems that must penetrate public awareness at local level and which require local activities most importantly of all.

With regard to foreign trade policy and development cooperation:

- Ensure that multilateral agreements on development cooperation give greater consideration to securing a basic supply of water for nutrition and sanitation purposes, and to environmental aspects, whereby agreement must be reached with the partner countries in question.
- Acknowledge the priority of water recycling over primary withdrawals and that fossil aquifers should only be tapped as a last resort.
- Place a primary focus on local traditions of environmental and waterbody protection, and to take account of their enabling or obstructive impacts.
- Ensure participation on the part of the affected population at the local level, as a means of guaranteeing the social acceptability and effectiveness of development policies, identifying the real needs of the designated beneficiaries and hence to enhance the acceptance of projects.
- Make the transition to "robust" structures for the "sound management of water resources" and to pay special attention to variabilities (droughts, floods, etc.); the type of system that is usually suitable in this context are large modular systems that are able to compensate more easily for local or temporal insufficiencies (e.g. in storage, purification and retention of water) and which therefore exhibit certain "error tolerant" characteristics.
- Support the development of national "master plans" based on regular inventorization of water supplies in terms of quality and quantity and comparison with ongoing assessments of needs.
- Strengthen integrated water resource management, primarily with a combined analysis of quantity and quality, the coupling of water supply and wastewater disposal issues, and the designation of catchments (as opposed to local authority and national territories) as basic planning units.

- Give support to states affected or threatened by water crises, above all with regard to
  1. modernization of existing irrigation systems,
  2. upgrading and expansion of water supply networks,
  3. establishing or improving systems for distributing drinking water, treating and disposing of wastewater, and for recycling water.

  These activities should be carried out within the framework of bilateral development cooperation as well as through close collaboration with international organizations such as the FAO, the WHO, the UNDP or the World Bank.
- Vigorously promote environmental and development projects advancing the cause of peace in areas suffering from water crisis (the Middle East, for example).
- Transfer technologies and expertise for the maintenance of sociocultural and ecological water standards more intensively, or provide financial support for such transfers,
  1. especially in water-scarce regions and to protect the world's natural heritage,
  2. primarily with the help of water-saving and environmentally, culturally and locally compatible methods.
- Take macroeconomic externalities (such as long-term impairment of waterbody quality as a result of industrial activities) into account by appropriate operationalization of the liability principle, whereby the ecological guard rails can be complied with in an effective way by issuing tradable emission certificates, for example.
- Significantly improve the conditional framework for efficient management of scarce freshwater resources, whereby special weight should be attached to:
  1. securing rights of tenure and disposal as far as possible and subjecting the available water resources to economic valuation,
  2. deregulating water use within the permissible action space as far as possible and imposing limits on subsidies that reduce competition,
  3. encouraging regional and international water markets where effective competition and anti-trust laws are in place to prevent monopolies forming and the poor from being disadvantaged,
  4. ensuring that a basic supply of freshwater in water-scarce countries is secured by appropriate forms of direct assistance ("water money" rather than large-scale water resources projects).
- Advance environmental education in a culturally appropriate manner focusing on the specific stakeholders and situations:
  1. make water, water-related problems, water-related patterns of behavior and their impacts easier for people to perceive,
  2. encourage a change in values towards sustainable water resource management,
  3. exploit the incentives produced by material and symbolic rewards, and
  4. create and provide opportunities for sustainable water resource management.

With regard to international research collaboration:

- Intensify the international transfer of know-how concerning physiological, epidemiological and ecological interrelationships of relevance to water resources and about all other aspects of sound water resource management, communicating above all
  1. scientific and technical aspects (in the fields of hydrology, hydraulic engineering, water treatment or sanitation),
  2. well-established rules for the organization of institutions, and
  3. methods for the efficient management of scarce environmental resources.
- Develop integrated and participatory mechanisms for maintaining water-specific standards in private- and public-sector projects (water impact assessments, etc.), and disseminate the relevant information.

With regard to the financing of measures:

- Undertake greater efforts in order to increase Germany's financial contribution in support of countries with insufficient resources, particularly in light of the UN Secretary-General's estimate of US$ 50 billion per year to meet global drinking water needs over the 1990–2000 period.
- Exploit every opportunity for reducing the debt servicing burden on developing countries threatened by water crisis, and examine possible links to water policy programs (debt for water security swaps).
- Explore the possibility of assisting financially overburdened countries from a global Water Fund replenished via robust international financing mechanisms (for example, a "World Water Penny" levied on water consumption).

The draft text in Box E 2-1 is intended by the Council as a basis on which to negotiate a World Water Charter that is non-binding under international law. The text contains general clauses exemplifying the specific arrangements that such a Charter might contain. Placeholders have been inserted for behavioral standards; if the political will exists, the Council's recommendations for action as listed above could be inserted at these positions (see also Section D 5.5.5).

2.4 Selected key recommendations for preventing a worldwide freshwater crisis

1. Management of water resources by private enterprises and stakeholders, e.g. on the basis of temporary licences for catchments or through the creation of water markets, wherever social and geo-environmental relations permit.
2. Launching of a World Water Charter, in which the Parties commit themselves to comply with fundamental social and ecological guard rails ("right of humans and nature to water resources").
3. Safeguarding a basic supply of freshwater to all people within a given country, for example by providing a government "water money" to those living in poverty. Those countries with insufficient financial resources to set up and expand water supply systems, the possibility should be explored of providing assistance for a limited period from a global Water Fund replenished via robust international financing mechanisms (for example, a "World Water Penny" levied on water consumption).
4. Consolidation of the confusing and fragmented plethora of international institutions and programs within the environment/development complex by establishing an "Organization for Sustainable Development" within the United Nations system. A primary task of this organization should be the planning and management of a global Plan of Action for the management of freshwater problems, because it is in this area that a particularly high humanitarian benefit can be achieved for each unit of capital invested.

BOX E 2-1

Global Code of Conduct for Implementing the Right to Water ("World Water Charter")

PREAMBLE

The undersigned States, international organizations and non-governmental organizations,

IN SUPPORT of the human right to water, which forms an integral part of the human right to food as proclaimed in the Universal Declaration of Human Rights of 1948, the International Covenant on Economic, Social and Cultural Rights of 1966 and numerous declarations issued by international diplomatic conferences,

AWARE of the results of the World Water Conferences held in 1992 in Dublin and in 1997 in Marrakech,

CONCERNED that international and national Action Programmes have not been sufficient to provide every person with adequate access to safe water for food and sanitation,

CONCERNED that drastic water scarcity may cause a threat to world peace in the future,

CONSCIOUS that intensified international efforts on the part of states, international organizations and non-governmental organizations are urgently needed in order to provide every person as quickly as possible with adequate access to clean water,

HAVE AGREED on this Global Code of Conduct for implementing the right to water and recognize this as the guiding principle for their respective programs.

ARTICLE 1 - SCOPE

1. This Global Code of Conduct for implementing the right to water (hereinafter: the "World Water Charter") is a declaration of global consensus, non-binding in international law, on principles governing the human right to water, recognized by the undersigned States, international organizations and non-governmental organizations as a standard of conduct.
2. All States, international organizations and non-governmental organizations, including business undertakings, are called on to sign the World Water Charter and to integrate its principles into their programs and plans of action for implementing the human right to water.

ARTICLE 2 - PRINCIPLES

1. The right to water means that every person has physical and financial access at all times and in sufficient amounts to water of adequate quality in order to cover his or her basic needs for food and sanitation.
2. Safeguarding the right to water is fundamentally the responsibility of the state. Every state must use the maximum of resources at its disposal to secure the right to water pursuant to para. 1 for all people under its jurisdiction, without endangering other major human rights. The state may transfer this responsibility to non-governmental enterprises in accordance with national legislation, provided there are guarantees that the right of every person under its jurisdiction to sufficient water pursuant to para. 1 is safeguarded.
3. Support for states that are unable to guarantee a right to water for people under their jurisdiction is the responsibility of the international community, in conformity with Article 55 of the United Nations Charter and Article 11 para. 2 of the International Covenant on Economic, Social and Cultural Rights. All states and international organizations are called upon to provide assistance, according to their individual capabilities, to those countries suffering from water crisis, in order to realize the right to water according to para. 1.
4. Access to water may not be used as a means of coercion in order to achieve political or military goals.
5. The right to water includes the duty of states to protect waterbodies under their jurisdiction so that no harm is caused to people, and that harm to animals and ecosystems is avoided as far as possible.
6. Utilization of a transboundary watercourse by a state must not impair in any way the right of another state to sufficient clean water for food and sanitation, to the extent that similar rights of the utilizing state are not affected.

ARTICLE 3 - IMPLEMENTATION BY THE STATES PARTIES

1. The undersigned States Parties shall not restrict the existing physical and economic access to adequate clean water on the part of people under their jurisdiction.
2. The undersigned States Parties acknowledge that the implementation of the right to water is a national task of great priority and shall endeavor, to the utmost of their capabilities, that all people under their jurisdiction who do not

yet have adequate access to clean water shall obtain such access as soon as possible.

3. The undersigned States Parties shall comply with the resolutions of the World Water Conferences, in particular the 1992 and 1997 Conferences in Dublin and Marrakech respectively, when implementing this World Water Charter as a minimum standard.
4. Those signatory States Parties in which a basic supply of food and water is not secured, accept the target of allocating at least 20% of their national budget to social security programs, whereby special attention shall be given to providing a basic supply of water to people under their jurisdiction.
5. If water of adequate quality is scarce, governments should consider whether, in the context of national cultural traditions, water should not be supplied free of charge, and that the price to consumers be linked to the cost of supply in order to foster the efficient use of water resources. States must ensure by means of economic compensation mechanisms that all people under their jurisdiction have access to water pursuant to Article 2 para. 1.
6. All government measures should be implemented with maximum participation of the people affected and the local and regional authorities, in accordance with national laws.
7. The undersigned States Parties shall contribute towards realization of the right to water by means of the following activities:
   (a) by ratifying the International Covenant on Economic, Social and Cultural Rights if this has not already occurred;
   (b) by taking account of the right to water in their national legislation;
   (c) by introducing appropriate environmental standards in order to ensure a supply of safe drinking water and the protection of ecosystems;
   (d) by means of a national Plan of Action for Water guaranteeing all people physical access to water on the basis of goals clearly defined in terms of quantities, qualities and time scales;
   (e) by introducing "water money" or similar instruments when responsibility for supplying water has been transferred to profit-oriented business undertakings, in order to guarantee economic access to water on the part of the poor;
   (f) by setting up an efficient system for monitoring the right to water that is suitable for identifying vulnerable groups and developing joint solutions through participation of the latter,
   e.g. by
   (i) enabling complaints to be lodged when the right to water is not fulfilled;
   (ii) having a national Ombudsman for water who reports to the national parliament and international organizations, in conjunction with Article 7, on the implementation of the right to water.
   (...) (see Section E 2.3 of this Report concerning additional specific obligations)
8. The undersigned States Parties shall report regularly on the implementation of measures carried out in accordance with Article 7.

Article 4 - Implementation through international cooperation

1. The undersigned States Parties recognize the importance of international cooperation in safeguarding the right to water and food. All states, according to their respective capabilities, shall support other states in protecting the right to water.
2. The undersigned States Parties commit themselves to the principle of equitable and optimal utilization of transboundary watercourses.
3. The undersigned industrialized countries shall provide additional means for safeguarding the right to water in developing countries. They shall allocate 20% of their official development assistance for social programs and assign special priority to providing a basic supply of water and food.
4. The undersigned newly industrializing countries acknowledge that the level of debt servicing in many developing countries prevents larger programs for implementing the right to water from being carried out to the requisite extent. Industrialized nations and developing countries shall intensify their joint efforts to find suitable solutions, including the feasibility of debt for water security swaps.

Article 5 - Implementation by international organizations

1. The international organizations shall include the right to water in their activities, as far as is possible and reasonable, and shall work towards guaranteeing this right to all people.
2. International organizations shall elaborate an

internal Plan of Action and report regularly on its implementation in accordance with Article 7.

3. In particular, international organizations shall adopt measures aimed at

(...) (see the detailed recommendations for action advocated by the Advisory Council in Section E 2.3 of the Annual Report)

ARTICLE 6 - IMPLEMENTATION BY NON-GOVERNMENTAL ORGANIZATIONS

1. Non-governmental organizations set up as nonprofit-making bodies shall assign high priority in their programs to the human right to water.
2. Non-governmental business undertakings shall shape their policies in such a way that the human right to water is not endangered. Non-governmental business undertakings that are particularly active in the field of water resource management shall
   (a) discriminate against no one when distributing water;
   (b) endeavor in particular to promote the access of vulnerable groups to water;
   (c) ensure full participation by the local population in all decisions pertaining to water development infrastructures;
   (...) (see the detailed recommendations for action advocated by the Advisory Council in Section E 2.3 of the Annual Report)
3. Non-governmental organizations, including business undertakings, shall report to the national Ombudsman or a similar body in accordance with Article 3, para. 7 f (iii) and to the international organizations pursuant to Article 7 with regard to the implementation of the right to water in the work of their organization.

ARTICLE 7 - INTERNATIONAL MECHANISMS

1. The highest monitoring body for the World Water Charter is the Annual Conference of the Parties, at which the international organizations and non-governmental organizations shall also be represented in adequate numbers, and with the right to consultation ("World Water Conference").
2. The United Nations High Commissioner for Human Rights shall be invited to coordinate and monitor the implementation of the World Water Charter as its Permanent Secretariat. A separate unit dedicated to water rights should be established within the office of the High Commissioner.
3. The United Nations High Commissioner for Human Rights shall report to the Human Rights Commission of the United Nations on progress achieved in realizing the human right to water. This report shall contain the following details:
   (a) data on the percentage of people in each state without access to clean drinking water;
   (b) data on the percentage of people in each state who face acute water scarcity;
   (c) identification of states facing acute water scarcity;
   (d) identification of states threatened by water scarcity;
   (e) identification of states that are particularly affected or threatened by droughts or floods.
   (...)
4. An Intergovernmental Panel on the Right to Water shall be established to support the World Water Charter; the Panel shall hold regular sessions to discuss the implementation of the World Water Charter, to consider further measures, to advise the High Commissioner with regard to his responsibilities pursuant to para.. and to put forward recommendations to the World Water Conference.
5. This Intergovernmental Panel shall consist of 14 members, seven of whom shall be delegates from industrialized countries and seven from developing countries. The Panel shall so comprised as to ensure appropriate geographical representation.

(...)

References

F

Achtnich, W. (1980): Bewässerungsfeldbau. Stuttgart: Ulmer.

ADB – Asian Development Bank (1997): Annual Report 1996. Manila: ADB.

Agras, W. S., Jacob, R. G. and Lebedeck, M. (1980): The California Drought: A Quasi-experimental Analysis of Social Policy. Journal of Applied Behavior Analysis 13 (4), 561–570.

Aitken, C. K., McMahon, T. A., Wearing, A. J. and Finlayson, B. L. (1994): Residential Water Use: Predicting and Reducing Consumption. Journal of Applied Social Psychology 24 (2), 136–158.

Aksamit, D. (1996): A Journey to the Three Gorges. World Rivers Review 11 (2), 8–9.

Alcamo, J. (1994): IMAGE 2.0 Integrated Modeling of Global Climate Change. Dordrecht: Kluwer.

Alcamo, J., Döll, P., Kaspar, F. and Siebert, S. (1997): An Overview of the Global Fresh Water Situation. Expertise for WBGU. Mimeo.

Alexandratos, N. (ed.) (1995): World Agriculture Towards 2010. A FAO Study. Rome: FAO.

Alho, C. J. R., Lacher jr., T. E. and Goncalves, H. (1988): Environmental Degradation in the Pantanal Ecosystem. BioScience 38 (3), 164–171.

Alster, J. (1996): Water in the Peace Process. Justice (9), 11–16.

Amoo, S. G. and Zartman, I. W. (1992): Mediation by Regional Organizations: The Organization for African Unity (OAU) in Chad. In: Bercovitch, J. and Rubin, J. (eds.): Mediation in International Relations. Multiple Approaches to Conflict Management. New York: St. Martin's Press, 131–148.

Amsel, A. and Lanz, K. (1992): Ein Klo geht um die Welt. Greenpeace-Magazin 3, 9–13.

Andreae, M. O. (1995): Climate Effects of Changing Atmospheric Aerosol Levels. In: Henderson-Sellers, A. (ed.): Future Climate to the World: A Modelling Perspective, World Survey and Climatology. Amsterdam: Elsevier, 341–392.

Appasamy, P. and Lundqvist, J. (1993): Water Supply and Waste Disposal. Strategies for Madras. Ambio 22 (7), 442–448.

Arceveila, D. (1996): Waste Water Treatment Problematique. Joint GAP-WHO Report from an Internet Discussion to the UN Habitat II Conference. Stockholm: Royal Institute of Technology.

Arthur, J. P. (1983): Notes on the Design and Operation of WSP in Warm Climates of Developing Countries. Washington, DC: The World Bank.

Assmann, J. (1996): Das Leichensekret des Osiris: Zur kultischen Bedeutung des Wassers im alten Ägypten. Mimeo.

Aubert, V. (1972): Interessenkonflikt und Wertkonflikt: Zwei Typen des Konflikts und Konfliktlösung. In: Bühl, W. L. (ed.): Konflikt und Konfliktstrategie. Ansätze zu einer soziologischen Konflikttheorie. Munich: Nymphenberger Verlagshandlung, 178–205.

Axelrod, R. (1987): Die Evolution der Kooperation. Munich: Oldenbourg.

Ayub, M. A. and Kuffner, U. (1994): Wasserwirtschaft im Maghreb. Finanzierung & Entwicklung 31 (2), 28–29.

Azzoni, A., Chiesa, S., Frassoni, A. and Govi, M. (1992): The Valpola Landslide. Engineering Geology 33, 59–70.

Bächler, G., Böge, V., Klötzli, S., Libiszewski, S. and Spillmann, K. R. (1996): Kriegsursache Umweltzerstörung: Ökologische Konflikte in der Dritten Welt und Wege ihrer friedlichen Bearbeitung. Volume 1. Chur, Zürich: Rüegger.

Bähr, J. and Mertins, G. (1995): Die lateinamerikanische Groß-Stadt. Verstädterungsprozesse und Stadtstrukturen. Darmstadt: Wissenschaftliche Buchgesellschaft.

Bäthe, J., Herbst, V., Hofmann, G., Matthes, U. and Thiel, R. (1994): Folgen der Reduktion der Salzbelastung in Werra und Weser für das Fließgewässer als Ökosystem. Wasserwirtschaft 10, 528–536.

Bangs, M. J., Purnomo, N., Andersen, E. M. and Anthony, R. L. (1996): Intestinal Parasites of Humans in a Highland Community of Irian Jaya, Indonesia. Annales Tropical Medicine and Parasitology 90 (1), 49–53.

Bangura, Y. (1996): Economic Restructuring, Coping Strategies and Social Change: Implications for Institutional Development in Africa. In: Lundahl, M. and Ndulu, B. J. (eds.): New Directions in Development Economics. Growth, Environmental Concerns and Government in the 1990s. London, New York: Routledge, 352–393.

Bantle, S. (1994): Schattenhandel als sozialpolitischer Kompromiß: die „Lybischen Märkte" in Tunesien. Informelle Kleinimporte, Wirtschaftsliberalisierung und Transformation. Münster: Studien zur Volkswirtschaft des Vorderen Orients.

Barber, M. and Ryder, G. (eds.) (1993): Damming the Three Gorges: What Dam Builders Don't Want You to Know. London, Toronto: Earthscan and Probe International.

Bárdossy, A. and Caspary, H. J. (1990): Detection of Climate Change in Europe by Analyzing European Atmospheric Circulation Patterns from 1881–1989. Theoretical and Applied Climatology 42, 155–167.

Barney, G. O. (1991): Global 2000 – Bericht an den Präsidenten. Technischer Bericht. Frankfurt/M.: Zweitausendeins.

Barrow, C. J. (1994): Land Degradation. Cambridge: Cambridge University Press.

Barrow, C. J. (1995): Developing the Environment – Problems and Management. London: Longman.

Barsch, H. and Bürger, K. (1996): Naturressourcen der Erde und ihre Nutzung. Gotha: Perthes.

Baumann, W., Bayer, H., Greupner, P., Kraft, H., Lauterjung, E., Lauterjung, H., Mollien, H., Wolkewitz, H. and Zeuner, G. (1984): Ökologische Auswirkungen von

Staudammvorhaben – Erkenntnisse und Folgerungen für die entwicklungspolitische Zusammenarbeit. Munich, Cologne: Weltforum.

Baumgartner, A. and Liebscher, H. J. (1990): Lehrbuch der Hydrologie. Allgemeine Hydrologie. Berlin, Stuttgart: Bornträger.

Baumgartner, A. and Reichel, E. (1975): The World Water Balance. Amsterdam: Elsevier.

Benzler, G., Halstrick-Schenk, M., Klemmer, P. and Löbbe, K. (1995): Wettbewerbskonformität von Rücknahmeverpflichtungen im Abfallbereich. Untersuchungen des Rheinisch-Westfälischen Instituts für Wirtschaftsforschung (RWI). Essen: RWI.

Beran, M. (1995): The Role of Water in Global Environmental Change Processes. In: Oliver, H. R. and Oliver, S. A. (eds.): The Role of Water and the Hydrological Cycle in Global Change. Berlin, Heidelberg, New York: Springer, 1–22.

Bercovitch, J. (1991): International Mediation. Journal of Peace Research 28 (1), 3–6.

Bercovitch, J. (1992): The Structure and Diversity of Mediation in International Relations. In: Bercovitch, J. and Rubin, J. (eds.): Mediation in International Relations. Multiple Approaches to Conflict Management. New York: St. Martin's Press, 1–29.

Bercovitch, J. and Regan, P. M. (1997): Managing Risks in International Relations: The Mediation of Enduring Rivalries. In: Schneider, G. and Weitsman, P. A. (eds.): Enforcing Cooperation. Risky States and the Intergovernmental Management of Conflict. London: MacMillan, 185–201.

Berk, R. A., Cooley, T. F., LaCivita, C. J., Parker, S., Sredl, K. and Brewer, M. (1980): Reducing Consumption in Periods of Acute Scarcity: The Case of Water. Social Science Research 9, 99–120.

Bernacsek, G. M. (1984): Dam Design and Operation to Optimize Fish Production in Impounded River Basins. Rome: FAO.

Berner, E. K. and Berner, R. A. (1996): Global Environment. Water, Air and Geochemical Cycles. New York: Prentice Hall.

Bernhardt, H. (1993): Überlegungen zur Entwicklung der Trinkwasseraufbereitungstechnik. gwf Wasser-Abwasser 134 (14), 196–198.

Berrisch, G. (1994): The Danube Dam Dispute Under International Law. Austrian Journal of Public and International Law 46, 231–281.

Biermann, F. (1994): Internationale Meeresumweltpolitik. Auf dem Weg zu einem Umweltregime für die Ozeane? Frankfurt/M.: Lang.

Biermann, F. (1997): Financing Environmental Policies in the South. Experiences from the Ozone Fund. International Environmental Affairs 9 (3), 179–218.

Biermann, F. and Hardtke, M. (1997): Tod im Korallenriff. Die ‚Regenwälder der Meere' drohen zu sterben. Ökozidjournal 13 (1), 2–13.

Billig, A. (1994): Ermittlung des ökologischen Problembewußtseins der Bevölkerung. Berlin: Umweltbundesamt (UBA).

Binder, J. (1996): Vortrag auf dem Internationalen LAWA-Symposium (28.–29.11.1996 in Heidelberg). Tagungsband (in print).

Birnie, P. and Boyle, A. E. (1992): International Law and the Environment. Oxford: Clarendon Press.

Björk, S. (1985): Scandinavian Lake Restoration Activities. In: European Water Control Association (ed.): Lakes Pollution and Recovery: European Water Pollution Control Association International Congress Proceedings-Reprints. London: European Water Control Association, 293–301.

Black, M. (1994): Mega Slums – A Coming Sanitary Crisis. Water Aid Report. Internet location www.oneworld.org/wateraid/reports/slums.html. WWW: OneWorld Online.

Bliefert, C. (1994): Umweltchemie. Weinheim: VCH.

BMU – Bundesministerium für Umwelt, Naturschutz und Reaktorsicherheit (1994): Umweltpolitik – Wasserwirtschaft in Deutschland. Bonn: BMU.

BMU – Bundesministerium für Umwelt, Naturschutz und Reaktorsicherheit (1996a): Umweltpolitik – Wasserwirtschaft in Deutschland. Bonn: BMU.

BMU – Bundesministerium für Umwelt, Naturschutz und Reaktorsicherheit (1996b): Umweltbewußtsein in Deutschland. Ergebnisse einer repräsentativen Bevölkerungsumfrage im Auftrag des Umweltbundesamtes. Bonn: BMU.

BMZ – Bundesministerium für wirtschaftliche Zusammenarbeit und Entwicklung (1995): Überlebensfrage Wasser – eine Ressource wird knapp. Bonn: BMZ.

BMZ – Bundesministerium für wirtschaftliche Zusammenarbeit und Entwicklung (1996): Wasser als knappe lebensnotwendige Ressource (Langfassung der BMZ-Studie „Wasser – eine Ressource wird knapp"). Bonn: BMZ.

Bohle, H.-G. (1981): Die Grüne Revolution in Indien – Sieg im Kampf gegen den Hunger? Paderborn: Schöningh.

Bohle, H.-G. (1989): 20 Jahre Grüne Revolution in Indien. Geographische Rundschau 41 (2), 91–98.

Böhme, H. (1988): Umriß einer Kulturgeschichte des Wassers. Eine Einleitung. In: Böhme, H. (ed.): Kulturgeschichte des Wassers. Frankfurt/M.: Suhrkamp, 7–42.

Bonus, H. (1994): Vergleich von Abgaben und Zertifikaten. In: Mackscheidt, K., Ewringmann, D. and Gawel, E. (eds.): Umweltpolitik mit hoheitlichen Zwangsabgaben. Karl-Heinrich Hansmeyer zur Vollendung seines 65. Lebensjahres. Berlin, 287–300.

Bos, R. (1997): New Priority Areas on the Interface of Water Resources Development and Vector-borne Diseases. Geneva: World Health Organisation (WHO).

Bosshard, P. and Unmüßig, B. (1996): Der Damm zu Babel: das Drei-Schluchten-Projekt am Jangtse. Bonn: Weltwirtschaft, Ökologie und Entwicklung e.V. (WEED).

Bourne, C. (1992): The International Law Commission's Draft Articles on the Law of International Watercourses: Principles and Planned Measures. Colorado Journal of International Enviromental Law and Policy, 3 (1), 65 ff.

Bourne, C. (1996): The ILA's Contribution to International Water Resources Law. Natural Resources Journal 155, 175 ff.

Bradley, R. S., Diaz, H. F., Eischeid, J. K., Jones, P. D., Kelly, P. M. and Goodess, C. M. (1987): Precipitation Fluctuations Over Northern Hemisphere Land Areas Since the Mid-19th Century. Science 237, 171–175.

Brandt, K. (1995): Grenzüberschreitender Nachbarschutz im deutschen Umweltrecht. Deutsches Verwaltungsblatt (110), 779 ff.

Brandt, K. (1997): Reichweite und Schranken territorialer Souveränitätsrechte über die Umwelt. Die Notwendigkeit einer Umweltpflichtigkeit der Souveränität. Ph.D. dissertation. University of Trier.

Brehm, J. M. M. (1982): Fließgewässerkunde. Heidelberg: Quelle & Meyer.

Breitmeier, H. (1996): Wie entstehen globale Umweltregime? Der Konfliktaustrag zum Schutz der Ozonschicht und des globalen Klimas. Opladen: Leske and Budrich.

Brennan, G. and Buchanan, J. M. (1985): The Reason of Rules. Cambridge, New York: Cambridge University Press.

Bretschneider, H., Lecher, K. and Schmidt, M. (eds.) (1993): Taschenbuch der Wasserwirtschaft. Hamburg: Parey.

Breuer, R. (1987): Öffentliches und Privates Wasserrecht. Munich: Beck.

Briscoe, J. (1992): Armut und Wasserversorgung. Der Weg voran. Finanzierung & Entwicklung 29 (4), 16–19.

Briscoe, J. (1995): Der Sektor Wasser und Abwasser in Deutschland – Qualität seiner Arbeit, Bedeutung für Entwicklungsländer. gwf Wasser – Abwasser 136 (8), 422–432.

Brix, H. (1994): Constructed Wetlands for Municipal Wastewater Treatment in Europe. In: Mitsch, W. J. (ed.): Global Wetlands: Old World and New World. Amsterdam: Elsevier, 325–333.

Brockhoff, K. and Salzwedel, J. (1978): Korrekte Maßstabsbildung für Entwässerungsgebühren. Berlin: Schmidt.

Bronger, D. (1996): Megastädte. Geographische Rundschau 48 (2), 74–81.

Brown, L. R., Lenssen, N. and Kane, H. (1995): Vital Signs. The Trends That are Shaping our Future. London: Earthscan.

Brun, W. (1992): Cognitive Components of Risk Perception: Natural Versus Manmade Risks. Journal of Behavioral Decision Making 5, 117–132.

Brunee, J. and Toope, S. (1997): Environmental Security and Freshwater Resources: Ecosystem Regime Building. American Journal of International Law 91 (1), 1ff.

Buchanan, J. M. (1975): The Limits of Liberty. Between Anarchy and Leviathan. Chicago: University of Chicago Press.

Buchanan, J. M. (1987): Zur Verfassung der Wirtschaftspolitik. Zeitschrift für Wirtschaftspolitik 36 (2), 101–112.

Buck, W. and Lee, K. K. (1980): Effektiver Hochwasserschutz: Vorarbeiten und deren Anwendung. Wasser und Boden 32, 59–67.

Buck, W. and Pflügner, W. (1991): Nutzwertanalytische Bewertung auenökologischer Wirkungen – Pilotstudie für eine Hochwasserschutzmaßnahme. Wasserwirtschaft 81, 578–587.

Buhse, G. (1989): Biotope Impairment Caused by Potassium Polluted Water of the Werra and Oberweser Rivers East Germany. Zeitschrift für Wasser- und Abwasser-Forschung 22 (2), 49–56.

Burchard, G. D., Büttner, D. W., Korte, R., Kretschmer, H. and Meier-Brook, C. (1996): Schistosomiasis (Bilharziose). In: Knobloch, J. (ed.): Tropen und Reisemedizin. Stuttgart: Fischer, 238–256.

Burkhardt, P. (1995): Auswirkungen wasserbaulicher Maßnahmen auf das Hochwasserverhalten im mitteldeutschen Raum am Beispiel von Elbe und Bode. Vortrag am DFG-Rundgespräch „Hochwasser in Deutschland unter Aspekten globaler Veränderungen“ am 9. Oktober 1995 in Potsdam. In: Bronstert, A. (ed.) (1996): Hochwasser in Deutschland unter Aspekten globaler Veränderungen. Bericht über das DFG – Rundgespräch am 9. Oktober 1995 in Potsdam. Potsdam: Potsdam Institute for Climate Impact Research (PIK), 37–45.

Burns, D. A., Galloway, N. J. and Hendrey, G. R. (1981): Acidification of Surface Waters in Two Areas of the Eastern United States. Water, Air & Soil Pollution (16), 277–285.

Burton, I., Kates, R. W. and White, G. F. (1978): The Environment as Hazard. New York, Oxford: Oxford University Press.

Butler, D. (1997): Time to Put Malaria Control on the Global Agenda. Nature 386, 535–540.

Calvert, S. and Calvert, P. (1996): Politics and Society in the Third World. An Introduction. London: Prentice Hall, Harvester Wheatsheaf.

Campbell, T. (1989): Urban Development in the Third World: Environmental Dilemmas and the Urban Poor. In: Leonard, H. J. (ed.): Environment and the Poor: Development Strategies for a Common Agenda. New Brunswick, Oxford: Transaction Books, 165–187.

Cansier, D. (1996): Umweltökonomie. Stuttgart: Lucius and Lucius.

Carpenter, S., Frost, T., Persson, L., Power, M. and Soto, D. (1996): Freshwater Ecosystems: Linkages of Complexity and Processes. In: Mooney, H. A., Cushman, J. H., Medina, E., Sala, O. E. and Schulze, E.-D. (eds.): Functional

Role of Biodiversity: A Global Perspective. Chichester, New York: Wiley & Sons, 299–325.

Carruthers, I. and Clark, C. (1981): The Economics of Irrigation. Liverpool: Liverpool University Press.

Carson, R. (1962): Silent Spring. London: Penguin.

Carter, V., Bedinger, M. S., Novitzki, R. P. and Wilen, W. O. (1979): Water Resources and Wetlands. In: Greeson, P. E., Clark, J. R. and Clark, J. E. (eds.): Wetland Functions and Values: The State of Our Understanding. Minneapolis: American Water Resources Association, 344–376.

Cassel-Gintz, M. A., Lüdeke, M. K. B., Petschel-Held, G., Reusswig, F., Plöchl, M., Lammel, G. and Schellnhuber, H.-J. (1997): Fuzzy Logic Based Assessment on Marginality of Agricultural Land Use. Climate Research (forthcoming).

Chambers, R. (1992): Rural Appraisal: Rapid, Relaxed and Participatory. Sussex: Institute of Development Studies.

Chao, B. F. (1995): Anthropogenic Impact on Global Geodynamics Due to Reservoir Water Impoundment. Geophysical Research Letters 22 (24), 3529–3532.

Chapman, D. (1992): Water Quality Assessment. London: Chapman and Hall.

Charlson, R. J., Lovelock, J. E., Andreae, M. O. and Warren, S. G. (1987): Oceanic Phytoplankton, Atmospheric Sulfur, Cloud Albedo and Climate. Nature (326), 655–661.

Chomchai, P. (1995): Management of Transboundary Water Resources: A Case Study of the Mekong. In: Blake, G. H., Hildesley, W., Pratt, M., Ridley, R. and Schofield, C. (eds.): The Peaceful Management of Transboundary Resources. London, Boston: Graham & Trotman, Nijhoff, 245 ff.

Cichorowski, G. (1996): Rationelle Wasserverwendung. Die Maßnahmen und ihre Auswirkungen in Frankfurt am Main. Wasserkultur 7, 40–46.

Clark, D. (1996): Urban World/Global City. London, New York: Routledge.

Clarke, R. (1993): Water: The International Crisis. Cambridge: Cambridge University Press.

Claus, B. and Neumann, P. (1995): Masterplan, Development Objectives and Agenda for the Natural Reconstruction of the Lower Weser and Her Marshes. Archiv für Hydrobiologie Supplementband 101 (3–4), 615–627.

Collier, M. P., Webb, R. H. and Andrews, E. D. (1997): Die experimentelle Überflutung im Grand Canyon. Spektrum der Wissenschaft (3), 76–83.

Cone, J. D. and Hayes, S. C. (1980): Environmental Problems/Behavioral Solutions. Monterey: Brooks/Cole.

Cooper, A. B. (1990): Nitrate Depletion in the Riparian Zone and Stream Channel of a Small Headwater Catchment. Hydrobiologia 202 (1–2), 13–26.

Costanza, R., d'Arge, R., de Groot, R., Farber, S., Grasso, M., Hannon, B., Limburg, K., Naeem, S., O'Neill, R. V., Raruelo, J., Raskin, R. G., Sutton, P. and van den Belt, M. (1997): The Value of the World's Ecosystem Services and Natural Capital. Nature 387, 253–260.

Coy, M. (1991): Sozio-ökonomischer Wandel und Umweltprobleme in der Pantanal-Region Mato Grossos (Brasilien). Geographische Rundschau 43 (3), 174–182.

Cramer, W. and Leemans, R. (1992): The IIASA Database for Mean Monthly Values of Temperature, Precipitation and Cloudiness of a Global Terrestrial Grid. Laxenburg: IIASA.

CYJV – CIPM Yangtse Joint Venture (1989): Three Gorges Water Control Project Feasibility Study. 13 volumes. Montreal: CYJV.

Czempiel, E.-O. (1994): Die Reform der UNO. Möglichkeiten und Mißverständnisse. Munich: Beck.

Dai Qing (1994): „Jangtse! Jangtse!“ London, Toronto: Earthscan and Probe International.

da Silva, C. J. and Silva, J. F. (1995): No ritmo das águas do Pantanal. São Paulo: NUPAUB/USP.

Davies, T. (1996): Two Steps Forward. Center for Risk Management Newsletter (10), 1–2.

de Haan, G., Kuckartz, U. and Rheingans, A. (1996): Lokale Agenda 21: Der Stand der Dinge November 1996. Berlin: Freie Universität. Forschungsgruppe Umweltbildung.

Dech, S. W. and Ressl, R. (1993): Die Verlandung des Aralsees – Eine Bestandsaufnahme durch Satellitenfernerkundung. Geographische Rundschau (45), 345.

Dellapenna, J. (1996): Rivers as Legal Structures: The Example of the Jordan and the Nile. Natural Resources Journal 36 (2), 217ff.

Denny, P. (1994): Biodiversity and Wetlands. Wetlands Ecology and Management 3 (1), 55–61.

Dickerson, C. A., Thibodeau, R., Aronson, E. and Miller, D. (1992): Using Cognitive Dissonance to Encourage Water Conservation. Journal of Applied Social Psychology 22, 841–854.

Dickson, W. (1981): Freshwater Acidification in Scandinavia and Europe. An Overview. Chichester, New York: Wiley & Sons.

Dierberg, F. E. and Kiattisimkul, W. (1996): Issues, Impacts, and Implications of Shrimp Aquaculture in Thailand. Environmental Management 20 (5), 649–666.

Diesfeld, H. J. (1997): Malaria auf dem Vormarsch? Geographische Rundschau 49, 232–239.

Dombrowsky, I. (1995): Wasserprobleme im Jordanbecken. Perspektiven einer gerechten und nachhaltigen Nutzung internationaler Ressourcen. Frankfurt/M.: Lang.

Donald, C. M. and Hamblin, J. (1976): The Biological Yield and Harvest Index of Cereals as Agronomic and Plant Breeding Criteria. Advances in Agronomy 28, 361–405.

Drijver, C. A. und Marchand, M. (1986): Taming the Floods: Environmental Aspects of Floodplain Development in Africa. Natural Resources 22 (4), 13–22.

Dugan, P. (ed.) (1993): Wetlands in Danger. A Mitchell Beazley World Conservation Atlas. London: Reed International.

Dunhoff, J. L. (1992): Reconciling International Trade With Preservation of the Global Commons. Can We Prosper and Protect? Washington and Lee Law Review 49 (4), 1407–1454.

Dunlap, R. E., Gallup jr., G. H. and Gallup, A. M. (1993): Health of the Planet. A George H. Gallup Memorial Survey. Results of a 1992 International Environmental Opinion Survey of Citizens in 24 Nations. Princeton: The George H. Gallup International Institute.

Durka, W. (1994): Isotopenchemie des Nitrat, Nitrataustrag, Wasserchemie und Vegetation von Waldquellen im Fichtelgebirge (NO-Bayern). Bayreuth: Bayreuther Forum Ökologie.

Durth, R. (1996): Zwischenstaatliche Zusammenarbeit an grenzüberschreitenden Flüssen und regionale Integration. Zeitschrift für Umweltpolitik und Umweltrecht 19 (2), 183–208.

Dynesius, M. and Nielsson, C. (1994): Fragmentation and Flow Regulation of River Systems in the Northern Third of the World. Science 266, 759ff.

Eaton, J. W. and Eaton, D. J. (1994): Water Utilization in the Yarmuk-Jordan, 1192–1992. In: Isaac, J. and Shuval, H. (eds.): Water and Peace in the Middle East. Amsterdam: Elsevier, 107ff.

Efinger, M., Rittberger, V. and Zürn, M. (1988): Internationale Regime in den Ost-West-Beziehungen. Ein Beitrag zur Erforschung der friedlichen Behandlung internationaler Konflikte. Frankfurt/M.: Haag and Herchen.

Ehrmann, M. (1997): Die Globale Umweltfazilität (GEF). Zeitschrift für ausländisches öffentliches Recht und Völkerrecht 57 (2–3), 565–614.

Endres, A. and Finus, M. (1996): Umweltpolitische Zielbestimmung im Spannungsfeld gesellschaftlicher Interessengruppen. Ökonomische Theorie und Empirie. In: Siebert, H. (ed.): Elemente einer rationalen Umweltpolitik. Expertisen zur umweltpolitischen Neuorientierung. Tübingen: Mohr, 35–133.

Engel, H. (1995): Die Hochwasser 1994 und 1995 im Rheingebiet im vieljährigen Vergleich. Proceedings of the DGFZ (6), 59–74.

Engel, H. (1997): Understanding Recent Large River Flooding. 1st European Workshop on River-Basin-Modelling and Flood Mitigation (RIBAMOD) in Delft, February 13–14, 1997. Delft: RIBAMOD.

Engelman, R. and LeRoy, P. (1995): Mensch, Wasser! Die Bevölkerungsentwicklung und die Zukunft der erneuerbaren Wasservorräte. Hannover: Balance.

Enquete Commission „Protection of People and Environment“ (1996): Öffentliche Anhörung: Kommunen und nachhaltige Entwicklung. Beiträge zur Umsetzung der Agenda 21. Bonn: Enquete Commission.

Epiney, A. (1995): Das „Verbot der erheblichen grenzüberschreitenden Umweltbeeinträchtigungen“: Relikt oder konkretisierungsfähige Grundnorm. Archiv des Völkerrechts (33), 332 ff.

Erickson, C. L. (1992): Prehistoric Landscape Management in the Andean Highlands. Raised Field Agriculture and its Environmental Impact. Population and Environment 13 (4), 285–300.

Evans, G. and Cohen, S. (1987): Environmental Stress. In: Stokols, D. and Altman, I. (eds.): Handbook of Environmental Psychology. Volume 1. Chichester, New York: Wiley & Sons, 571–610.

Evenari, M., Shanan, L. and Tadmor, N. (1982): The Negev – The Challenge of a Desert. Cambridge: Harvard University Press.

Ewers, H.-J. and Rennings, K. (1996): Quantitative Ansätze einer rationalen und umweltpolitischen Zielbestimmung. In: Siebert, H. (ed.): Elemente einer rationalen Umweltpolitik. Expertisen zur umweltpolitischen Neuorientierung. Tübingen: Mohr, 135–171.

Ex-Im Bank – Export-Import Bank of the United States (1996): Frequently Asked Questions About the Three Gorges Dam Project. Internet location http://www.exim.gov/3gorges.html. Washington, DC: Ex-Im Bank.

Exner, M. and Gornik, V. (1995): Kryptosporidiose. Sonderdruck 1/97. In: Beck, E. G. and Eikmann, T. (eds.): Hygiene in Krankenhaus und Praxis. Landsberg: Ecomed.

Ezcurra, E. and Mazari-Hiriart, M. (1996): Are Mega Cities Viable? A Cautionary Tale from Mexico City. Environment 38 (1), 6–33.

Falkenmark, M. and Lindh, G. (1993): Water and Economic Development. In: Gleick, P. (ed.): Water in Crisis. A Guide to the World's Fresh Water Resources. New York, Oxford: Oxford University Press, 80–91.

FAO – Food and Agriculture Organization of the United Nations (1976): Water Quality for Agriculture. Irrigation and Drainage Paper. Rome: FAO.

FAO – Food and Agriculture Organization of the United Nations (1978): Systematic Index of International Water Resources Treaties, Declarations, Acts and Cases by Basin. Rome: FAO.

FAO – Food and Agriculture Organization of the United Nations (1987): The Fifth World Food Summit. Rome: FAO.

FAO – Food and Agriculture Organization (1993): FAO Production 1992/1993. Rome: FAO.

FAO – Food and Agriculture Organization of the United Nations (1996a): Food Production – The Critical Role of Water. Rome: FAO.

FAO – Food and Agriculture Organization of the United Nations (1996b): Lessons from the Green Revolution. Towards a New Green Revolution. Rome: FAO.

FAO – Food and Agriculture Organization of the United Nations (1996c): World Food Summit. Technical Background Documents. Rome: FAO.

FAO – Food and Agriculture Organization of the United Nations (1996d): World Food Summit. Technical Fact Sheets. Rome: FAO.

FAO – Food and Agriculture Organization of the United Nations (1996e): The State of the World Fisheries and Aquaculture. Rome: FAO.

FAO – Food and Agriculture Organization of the United Nations, World Bank and UNDP – United Nations Development Programme (1995): Water Sector Policy Review and Strategy Formulation. Rome: FAO.

FAOSTAT – Food and Agriculture Organization Statistical Department (1997): FAOSTAT Statistics Database. Internet location http://apps.fao.org/default.asp. Rome: FAO.

Farquhar, G. D., Hubick, K. T., Condon, A. G. and Richards, R. A. (1988): Carbon Isotope Fractionation and Plant Water-use Efficiency. In: Rundel, P. W., Ehleringer, J. R. and Nagy, K. A. (eds.): Stable Isotopes in Ecological Research. Ecological Studies. Berlin, Heidelberg, New York: Springer, 31–40.

Fearnside, P. M. (1988): China's Three Gorges Dam: Fatal Project or Step Towards Modernization? World Development 16 (5), 615–630.

Fearnside, P. M. (1995): Hydroelectric Dams in the Brazilian Amazon as Sources of „Greenhouse Gases". Environmental Conservation 22 (1).

Feder, G. and le Moigne, G. (1994): Umweltverträgliche Wasserwirtschaft. Finanzierung & Entwicklung 31 (2), 24–27.

Feindt, P. H. (1996): Rationalität durch Partizipation? Das Mehrstufige Dialogische Verfahren als Antwort auf gesellschaftliche Differenzierung. In: Feindt, P. H., Gessenharter, W., Birzer, M. and Fröchling, H. (eds.): Konfliktregelung in der offenen Bürgergesellschaft. Dettelbach: Röll, 169–189.

Fischer, L. (1988): Trank Wasser wie das liebe Vieh. Marginalien zur Sozialgeschichte des Umgangs mit Wasser. In: Böhme, H. (ed.): Kulturgeschichte des Wassers. Frankfurt/M.: Suhrkamp, 314–352.

Fischer, L. (1997): Von der Tragedy of the Commons zu deren Benefits. Rahmenbedingungen für die erfolgreiche Organisation von Gemeingütern. Ökologisches Wirtschaften (2), 11–12.

Fischer, R. A. and Turner, N. C. (1978): Plant Production in the Arid and Semiarid Zones. Annual Review of Plant Physiology 29, 277–317.

Flanagan, W. G. (1990): Urban Sociology. Images and Structure. Boston: Allyn & Bacon.

Flohn, H. (1988/89): Ändert sich das Klima? Neue Erkenntnisse und Folgerungen. Mannheimer Forum 88/89, 135–189.

Förster, U. (1990): Umweltschutztechnik. Berlin, Heidelberg, New York: Springer.

Fraser, A. S., Meybeck, M. and Ongley, E. D. (1995): Water Quality of World River Basins. Nairobi: GEMS und UNEP.

Freeberne, M. (1993): The Three Gorges Project and Mass Resettlement. Water Resources Development 9 (3), 337ff.

Frei, D. (1990): Die Organisation der Vereinten Nationen (UNO). Eine Einführung in 15 Vorlesungen. Grüsch: Rüegger.

Frey, B. S. and Oberholzer-Gee, F. (1996): Zum Konflikt zwischen intrinsischer Motivation und umweltpolitischer Instrumentenwahl. In: Siebert, H. (ed.): Elemente einer rationalen Umweltpolitik. Expertisen zur umweltpolitischen Neuorientierung. Tübingen: Mohr, 207–238.

Froese, B. (1997): Umweltchemikalien mit hormoneller Wirkung. GSF – Information Umwelt. Oberschleißheim: GSF.

Fromm, O. (1997): Möglichkeiten und Grenzen einer ökonomischen Bewertung des Ökosystems Boden. Dissertation. Marburg: Universität Marburg.

Frost, T. M., Carpenter, S. R., Ives, A. R. and Kratz, T. K. (1994): Species Compensation and Complementarity in Ecosystem Function. In: Jones, C. and Lawton, J. (eds.): Linking Species and Ecosystems. London: Chapman and Hall, 224–239.

Fuglestvedt, J. S., Jonson, J. E., Wang, W.-C. and Isaksen, I. S. A. (1995): Responses in Tropospheric Chemistry to Changes in UV Fluxes, Temperatures and Water Vapour Densities. In: Wang, W.-C. and Isaksen, I. S. A. (eds.): Atmospheric Ozone as a Climate Gas. Berlin, Heidelberg, New York: Springer, 145–162.

Gaidetzka, P. (1996a): Wasser – Leben für Süd und Nord. In: Bischöfliches Hilfswerk Misereor e.V. (ed.): Wasser: Eine globale Herausforderung. Aachen: Horlemann, 6–10.

Gaidetzka, P. (1996b): Wasser in den Religionen. In: Bischöfliches Hilfswerk Misereor e.V. (ed.): Wasser: Eine globale Herausforderung. Aachen: Horlemann, 138–145.

Gardner, G. T. and Stern, P. C. (1996): Environmental Problems and Human Behavior. Boston: Allyn & Bacon.

GATT – General Agreement on Tariffs and Trade (1991): Dispute Settlement Panel Report on United States Restrictions on Imports of Tuna, übermittelt an die Parteien am 16. August 1991, GATT Doc. DS21/R. ILM 30, 1594.

GATT – General Agreement on Tariffs and Trade (1994): Dispute Settlement Panel Report on United States Restrictions on Imports of Tuna, übermittelt an die Konfliktparteien am 20. Mai 1994. ILM 33, 839.

Gawel, E. (1996): Neoklassische Umweltökonomie in der Krise? Kritik und Gegenkritik. In: Köhn, J. and Welfens, M. J. (eds.): Neue Ansätze in der Umweltökonomie. Ökologie und Wirtschaftsforschung. Marburg: Metropolis, 45–88.

Gaye, M. and Diallo, F. (1994): Case Study: Programme Assainissement de Diokoul et Quartiers Environnants Rufique, Senegal. Mexico: Habitat International Coalition (HIC).

Geller, E. S., Erickson, J. B. and Buttram, B. A. (1983): Attempts to Promote Residential Water Conservation with Educational, Behavioral and Engineering Strategies. Population and Environment 6 (2), 96–112.

Gellert, M. (1991): Kostensenkungspotentiale in der kommunalen Abwasserbeseitigung unter besonderer Berücksichtigung der Organisationsform. Witten: K.-U. Rudolph publisher.

GEMS/Water database (1997a): Global Risk of Surface Water Acidification. Internet location http://cciw.ca/gems/atlas-gwdq/images/figpg25a.gif. Burlington, Ontario: UNEP Collaborating Centre for Freshwater Quality Monitoring and WHO Collaborating Centre for Surface and Groundwater Quality.

GEMS/Water database (1997b): Global Patterns of Sediment Yield, with River Output of Sediment to the Oceans. Internet location http://cciw.ca/gems/atlas-gwdq/images/figpg10.gif. Burlington, Ontario: UNEP Collaborating Centre for Freshwater Quality Monitoring and WHO Collaborating Centre for Surface and Groundwater Quality.

GEMS/Water database (1997c): Range of TDS in Surface Water. Internet location http://cciw.ca/gems/atlas-gwdq/images/figpg15a.gif. Burlington, Ontario: UNEP Collaborating Centre for Freshwater Quality Monitoring and WHO Collaborating Centre for Surface and Groundwater Quality.

GEMS/Water database (1997d): Nitrate Concentrations for Global Major Watersheds. Internet location http://cciw.ca/gems/atlas-gwdq/images/figpg20c.gif. Burlington, Ontario: UNEP Collaborating Centre for Freshwater Quality Monitoring and WHO Collaborating Centre for Surface and Groundwater Quality.

Gerken, L. and Renner, A. (1996): Der Wettbewerb der Ordnungen als Entdeckungsverfahren für eine nachhaltige Entwicklung. In: von Gerken, L. (ed.): Ordnungspolitische Grundfragen einer Politik der Nachhaltigkeit. Baden-Baden: Nomos, 51–102.

Gettkant, A., Simonis, U. E. and Suplie, J. (1997): Biopolitik für die Zukunft. Kooperation oder Konfrontation zwischen Nord und Süd. Bonn: Stiftung Entwicklung und Frieden (SEF).

Ghassemi, F., Jakeman, A. J. and Nix, H. A. (1995): Salinisation of Land and Water Resources Human Causes, Extent, Management & Case Studies. Sydney: CAB International.

Gibbons, D. C. (1986): The Economic Value of Water. Washington, DC: Resources for the Future.

Giddens, A. (1993): Sociology. Cambridge, Oxford: Policy Press.

Giese, E. (1997): Die ökologische Krise der Aralseeregion. Geographische Rundschau 49 (5): 293–299.

Gitlitz, J. (1993): The Relationship Between Primary Aluminium Production and the Damming of the World Rivers. Berkeley: International Rivers Network (IRN).

Glasauer, H. (1996): „Wer küßt schon den Mann, der auf der Toilette Wasser spart?". Individuelle Blockierungen beim nachhaltigen Umgang mit Wasser. Über Bewußtsein und Verhalten, Wollen und Können – Erste Erklärungsansätze. Wasserkultur 7, 48–58.

Glazowsky, N. F. (1995): Aral Sea. Amsterdam: SPB.

Gleick, P. H. (ed.) (1993): Water in Crisis. A Guide to the World's Fresh Water Resources. New York, Oxford: Oxford University Press.

Goldman, C. R., Jassby, A. D. and Hackley, S. H. (1993): Decadal, Interannual, and Seasonal Variability in Enrichment Bioassays at Lake Tahoe. Canadian Journal of Fisheries and Aquatic Sciences 50 (7), 1489–1496.

Goldsmith, E. and Hildyard, N. (1984): The Social and Environmental Effects of Large Dams. San Francisco: Sierra Club Books.

Gray, B. (1994): The Role of Laws and Institutions in California's 1991 Water Bank. In: Carter, H. O., Vaux, H. and Scheuring, A. (eds.): Sharing Scarcity: Gainers and Losers in Water Marketing. Davis: University of California Agricultural Issues Center, 133ff.

Griffiths, R. W. and Schloesser, D. W. (1991): Distribution and Dispersal of the Zebra Mussel (Dreissena polymorpha) in the Great Lake Region. Canadian Journal of Fisheries & Aquatic Sciences 48 (8), 1381–1388.

Grünewald, U. (1996): Abschätzung des Einflusses von Landnutzung und Versiegelung auf den Hochwasserabfluß. Rapporteursbericht zum DFG-Rundgespräch „Hochwasser in Deutschland unter Aspekten globaler Veränderungen" am 9. Oktober 1995 in Potsdam. In: Bronstert, A. (ed.) (1996): Hochwasser in Deutschland unter Aspekten globaler Veränderungen. Bericht über das DFG-Rundgespräch am 9. Oktober 1995 in Potsdam. Potsdam: Potsdam Institute for Climate Impact Research (PIK), 37–45.

GTZ – Deutsche Gesellschaft für Technische Zusammenarbeit (1996): Konfliktmanagement im Umweltbereich. Instrument der Umweltpolitik in Entwicklungsländern. Eschborn: GTZ.

Gubler, D. J. (1996): Arboviruses as Imported Disease Agents: The Need for Increased Awareness. Archive of Virology 11, 21–32.

Gupta, A. C. (1974): Lakes of Sorrow. Journal of the Indian Institute of Engineers 55, 6–11.

Gupta, H. K. (1992): Reservoir-induced Earthquakes. Amsterdam: Elsevier.

Haarmeyer, D. and Mody, A. (1997): Privates Kapital in der Wasserversorgung. Finanzierung & Entwicklung 34 (1), 32–35.

Haas, P. M., Keohane, R. O. and Levy, M. A. (eds.) (1993): Institutions for the Earth. Sources of Effective International Environmental Protection. Cambridge, Ma.: MIT Press.

Habermas, J. (1981): Theorie des kommunikativen Handelns. Frankfurt/M.: Suhrkamp.

Häckel, H. (1990): Meteorologie. Suttgart: Ulmer.

Hafner, G. (1993): The Optimum Utilization Principle. Austrian Journal of Public International Law (45), 113ff.

Hamilton, L. C. (1985): Self-reported and Actual Savings in a Water Conservation Campaign. Environment and Behavior 17 (3), 315–326.

Hampicke, U. (1991): Naturschutz-Ökonomie. Stuttgart: UTB.

Hanssen-Bauer, I. , Førland, E. J., Tveito, O. E. and Nordli, P. Ø. (1997): Estimating Regional Precipitation Trends – Comparison of Two Methods. Nordic Hydrology 28, 21–36.

Hardin, G. (1968): The Tragedy of the Commons. Science 162 (3859), 1243–1248.

Hardoy, J. E. and Satterthwaite, D. (1989): Squatter Citizen: Life in the Urban Third World. London: Earthscan.

Hardoy, J. E., Cairncross, S. and Satterthwaite, D. (1990): The Poor Die Young. Housing and Health in Third World Cities. London: Earthscan.

Hardoy, J. E., Mitlin, D. and Satterthwaite, D. (1995): Environmental Problems in Third World Cities. London: Earthscan.

Harlan, J. R. (1975): Our Vanishing Genetic Resources. Science 188, 618–621.

Hartmann, J. and Quoss, H. (1993): Fecundity of Whitefish Corgonus-Lavaretus During the Eu- and Oligotrophication of Lake Constance. Journal of Fish Biology 43 (1), 81–87.

Hauglustaine, D. A., Granier, C., Brasseur, G. P. and Mégie, G. (1994): The Importance of Atmospheric Chemistry in the Calculation of Radiative Forcing on the Climate System. Journal of Geophysical Research 99, 1173–1186.

Hay, R. K. M. (1995): Harvest Index: A Review of its Use in Plant Breeding and Crop Physiology. Annales of Applied Biology 126 (1), 197–216.

Hayton, R. and Utton, A. (1989): Transboundary Groundwaters: The Bellagio Draft Treaty 29. Natural Resources Journal, 29 (3), 663ff.

Heckman, C. W. (1994): The Seasonal Succession of Biotic Communities in Wetlands of the Tropical Wet-and-dry Climatic Zone I: Physical and Chemical Causes and Biological Effects in the Pantanal of Mato Grosso, Brazil. Internationale Revue der gesamten Hydrobiologie 79 (3), 397–421.

Hecht, U. (1992): Naturnahe Abwasserentsorgung. Politische Ökologie (special issue 5), 67–70.

Hedin, L. O., Granat, L., Likens, G. E., Buishand, T. A., Galloway, J. N., Butler, T. J. and Rodhe, H. (1994): Steep Declines in Atmospheric Base Cations in Regions of Europe and North America. Nature 367, 351–354.

Heintschel von Heinegg, W. (1990): Internationales Öffentliches Umweltrecht. In: Ipsen, K. (ed.): Völkerrecht. Munich: Beck, 805ff.

Helm, C. (1995): Sind Freihandel und Umweltschutz vereinbar? Ökologischer Reformbedarf des GATT/WTO-Regimes. Berlin: Edition Sigma.

Henderson-Sellers, A. and Hansen, A.-M. (1995): Climate Change Atlas - Greenhouse Simulations From the Model Evaluation Consortium for Climate Assessment. Dordrecht: Kluwer.

Henne, G. and Loose, C. (1997): Gutes Geld für grünes Gold. In: Altner, G., Mettler-von-Meibom, B., Simonis, U. E. and von Weizsäcker, E.-U. (eds.): Jahrbuch Ökologie 1998. Munich: Beck.

Heywood, V. H. and Watson, R. T. (eds.) (1995): Global Biodiversity Assessment. Cambridge: Cambridge University Press.

Hinds, C. (1997): Umweltrechtliche Einschränkung der Souveränität: Völkerrechtliche Präventionspflichten zur Verhinderung von Umweltschäden. Frankfurt/M.: Lang.

Hirshleifer, J., de Haven, J. C. and Milliman, J. W. (1972): Water Supply. Economics, Technology, and Policy. Chicago: University of Chicago Press.

Holtz, U. (1997): Welternährung. Spektrum der Wissenschaft (Dossier 2), 16–23.

Holzinger, K. (1994): Politikwissenschaftliche Grundfragen zur Mediation bei Umweltkonflikten. In: Dally, A., Weidner, H. and Fietkau, H.-J. (eds.): Mediation als politischer und sozialer Prozeß. Loccumer Protokolle 73/93. Loccum: Evangelische Akademie, 63–67.

Homagk, P. (1996): Hochwasserwarnsystem am Beispiel Baden-Württemberg. Zeitschrift für Geowissenschaften 14, 539–546.

Howarth, R. W., Billen, G., Swaney, D., Townsend, A., Jaworski, N., Lajtha, K., Downing, J. A., Elmgren, R., Caraco, N., Jordan, T., Berendse, F., Freney, J., Kudeyarov, K., Murdoch, P. and Zhao-Liang Z. (1996): Regional Nitrogen Budgets and Riverine N & P Fluxes for the drainages to the North Atlantic Ocean: Natural and Human Influences. Biogeochemistry 35, 75–139.

HRW – Human Rights Watch Asia (1995): Three Gorges Dam in China: Forced Resettlement, Suppression of Dissent and Labor Rights Concern. New York: Human Rights Watch Asia.

Hutchinson, G. E. (1957): Geography, Physics and Chemistry. In: Hutchinson, G. E. (ed.): A Treatise on Limnology. Volume 1. Chichester, New York: Wiley & Sons, 1–1015.

Hutchinson, G. E. (1961): The Paradox of the Plankton. American Naturalist 95, 137–146.

Hutchinson, G. E. (1967): Introduction to Lake Biology and the Limnoplankton. In: Hutchinson, G. E. (ed.): A Treatise on Limnology. Volume 2. Chichester, New York: Wiley & Sons, 1115ff.

Hutchinson, G. E. (1975): A Treatise on Limnology. Limnological Botany. Volume 3. Chichester, New York: Wiley & Sons.

Hynes, H. B. N. (1970): The Ecology of Running Waters. Liverpool: Liverpool University Press.

ICLEI – The International Council for Local Environmental Initiatives – European Secretariat (1996): Stellungnahme zur öffentlichen Anhörung „Kommunen und nachhaltige Entwicklung – Beiträge zur Umsetzung der Agenda 21“ der Enquete-Kommission „Schutz des Menschen und der Umwelt“ am 18. November 1996 in Bonn. Freiburg: ICLEI European Secretariat.

ICLEI – The International Council for Local Environmental Initiatives (1997): Local Agenda 21 Survey. A Study of Responses by Local Authorities and Their National and International Associations to Agenda 21. Internet location http://www.iclei.org/la21/la21rep.htm. Toronto: ICLEI World Secretariat.

ICOLD – International Commission on Large Dams (1984): World Register of Dams. Full Edition. Paris: ICOLD.

ICOLD – International Commission on Large Dams (1988): World Register of Dams. Updating. Paris: ICOLD.

Ihringer, J. (1996): Hochwasser aus ländlichen und städtischen Gebieten. Zeitschrift für Geowissenschaften 14, 523–530.

Illies, J. (1961): Versuch einer allgemeinen biozönotischen Gliederung der Fließgewässer. Internationale Revue der gesamten Hydrobiologie 46 (2), 205–213.

ILM – American Society of International Law (ed.) (1995): International Legal Materials: Current Documents. Washington, DC: ILM.

ILO – International Labour Organization (ed.) (1993): World Labour Report. Genf: ILO.

Inkeles, A. (1996): Making Men Modern: On the Causes and Consequences of Individual Change in Six Developing Countries. In: Inkeles, A. and Sasaki, M. (eds.): Comparing Nations and Cultures. Readings in a Cross-Disciplinary Perspective. London: Prentice Hall, 571–585.

Institut de Droit International (1961): Annuaire de l'Institut de Droit International. Basel, Munich: Kargo, 381ff.

IPCC – Intergovernmental Panel on Climate Change (1992): Climate Change 1992. The Supplementary Report to the IPCC Scientific Assessment. Cambridge, New York: Cambridge University Press.

IPCC – Intergovernmental Panel on Climate Change (1996a): Climate Change 1995. Impacts, Adaptations and Mitigation of Climate Change: Scientific-Technical Analyses. Cambridge, New York: Cambridge University Press.

IPCC – Intergovernmental Panel on Climate Change (1996b): Climate Change 1995. The Science of Climate Change. Cambridge, New York: Cambridge University Press.

Ipsen, K. (1990): Völkerrechtliche Verantwortlichkeit und Völkerstrafrecht. In: Ipsen, K. (ed.): Völkerrecht. Munich: Beck, 488ff.

Ipsen, D. (1994): Umweltwahrnehmung und Umgang mit Wasser in Agglomerationsräumen. WasserKultur Texte 5. Kassel: Forschungsprojekt Wasserkreislauf und urban-ökologische Entwicklung.

Ipsen, D. (1996): Haben Wassersparkampagnen einen Sinn? Wasserkultur 7, 47.

IRN – International Rivers Network (1996): US Export-Import Bank Say No to Funding Three Gorges. World Rivers Review 11 (2), 1–10.

Jackson, R. B., Canadell, J., Ehleringer, J. R., Mooney, H. A., Sala, O. E. and Schulze, E.-D. (1996): A Global Analysis of Root Distributions for Terrestrial Biomes. Oecologia 108 (3), 389–411.

Jakobeit, C. (1996): Nonstate Actors Leading the Way. Debt-for-Nature-Swaps. In: Keohane, R. O. and Levy, M. A. (eds.): Institutions for Environmental Aid. Pitfalls and Promise. Cambridge, Ma.: MIT Press, 127–166.

Jansson, A. M., Hammer, M., Folke, C. and Costanza, R. (eds.) (1994): Investing in Natural Capital. The Ecological Economics Approach to Sustainability. Washington, DC: Covelo.

Jassby, A. D., Reuter, J. E., Axler, R. P., Goldmann, C. R. and Hackley, S. H. (1994): Atmospheric Deposition of Nitrogen and Phosphorus in the Annual Nutrient Load of Lake Tahoe (California-Nevada). Water Resources Research 30, 2207–2216.

Jassby, A. D., Goldman, C. R. and Reuter, J. E. (1995): Long-term Change in Lake Tahoe California-Nevada, USA and its Relation to Atmospheric Deposition of Algal Nutrients. Archiv für Hydrobiologie 135, 1–21.

Jungermann, H. and Slovic, P. (1993): Die Psychologie der Kognition und Evaluation von Risiko. In: Bechmann, G. (ed.): Risiko und Gesellschaft. Grundlagen und Ergebnisse interdisziplinärer Risikoforschung. Opladen: Westdeutscher Verlag, 167–207.

Jordan, A. (1994): Paying the Incremental Costs of Global Environmental Protection. The Evolving Role of GEF. Environment 36 (6), 12–36.

Junk, W. G. (1993): Wetlands of Tropical South America. In: Wigham, D. F., Dykyjova, D. and Hejn, S. (eds.): Wetlands of the World I. Dordrecht: Kluwer, 679–739.

Junk, W. J. and da Silva, C. J. (1995): Neotropical Floodplains: A Comparison Between the Pantanal of Mato Grosso and the Large Amazonian River Floodplain. In: Tundisi, J. G., Bicudo, C. E. M. and Tundisi T. M. (eds.): Limnology in Brazil. Rio de Janeiro: Brazilian Academy of Sciences und Brazilian Limnological Society, 195–217.

Kadlec, R. H. (1994): Wetlands for Water Polishing: Free Water Surface Wetlands. In: Mitsch, W. J. (ed.): Global Wetlands: Old World and New World. Amsterdam: Elsevier, 335–351.

Kamarck, M. A. (1996): Three Gorges Dam in China. Transcript of a Press Briefing on Board Meeting on May 30, 1996. Internet location http://www.exim.gov:80/t3gorges.html. Washington, DC: Export-Import Bank.

Kantola, S. J., Syme, G. J. and Campbell, N. A. (1982): The Role of Individual Differences and External Variables in a Test of the Sufficiency of Fishbein's Model to Explain Behavioral Intentions to Conserve Water. Journal of Applied Social Psychology 12 (1), 70–83.

Karger, C. R. (1996): Wahrnehmung von Umweltproblemen – am Beispiel von „Wasser". In: Fischer, W., Karger, C. R. and Wendland, F. (eds.): Wasser: Nachhaltige Gewinnung und Verwendung eines lebenswichtigen Rohstoffs. Jülich: Forschungszentrum Jülich (KFA), 185–201.

Karl, H. and Klemmer, P. (1994): Volkswirtschaftliche Effekte privatwirtschaftlich organisierter öffentlicher Investitionen im Bereich der Abwasserentsorgung. Witten: K.-U. Rudolph publisher.

Karl, H. (1997): Umweltökonomik. Universität Dortmund. Mimeo.

Kartsev, A.D. (1995): The Winter Seasonality of Intestinal Infections. Voenna Medicina Zhurnal 80, 44–48.

Kasperson, J. X., Kasperson, R. E. and Turner II, B. L. (eds.) (1995): Regions at Risk. Comparisons of Threatened Environments. Tokyo, New York: United Nations University Press.

Katalyse e.V. – Institut für angewandte Umweltforschung (1993): Das Wasserbuch. Trinkwasser und Gesundheit. Cologne: Kiepenheuer & Witsch.

Keck, O. (1993): Information, Macht und gesellschaftliche Rationalität. Das Dilemma rationalen kommunikativen Handelns, dargestellt am Beispiel eines internationalen Vergleichs der Kernenergiepolitik. Baden-Baden: Nomos.

Keck, O. (1995): Rationales kommunikatives Handeln in den internationalen Beziehungen. Ist eine Verbindung von Rational-Choice-Theorie und Habermas' Theorie des kommunikativen Handelns möglich? Zeitschrift für Internationale Beziehungen 2 (1), 5–48.

Kelliher, F. M., Leuning, R. and Schulze, E.-D. (1993): Evaporation and Canopy Characteristics of Coniferous Forests and Grasslands. Oecologia 95, 153–163.

Kendall, H. W. and Pimentel, D. (1994): Constraints on the Expansion of the Global Food Supply. Ambio 23 (3), 200–212.

Ketterer, W. and Spada, H. (1993): Der Mensch als Betroffener und Verursacher von Naturkatastrophen: Der Beitrag umweltpsychologischer Forschung. In: Plate, E., Clausen, L., de Haar, U., Kleeberg, H.-B., Klein, G., Mattheß, G., Roth, R. and Schmincke, H. U. (eds.): Naturkatastrophen und Katastrophenvorbeugung. Bericht zur IDNDR. Weinheim: VCH, 73–107.

Kidron, M. and Segal, R. (1996): Der Fischer Atlas zur Lage der Welt. Frankfurt/M.: Fischer Taschenbuch.

Kiehl, J. T. and Trenberth, K. E. (1996): Earth's Annual Global Mean Energy Budget. Bulletin of the American Meteorological Society (forthcoming).

Kinder, H. and Hilgemann, W. (1986): dtv-Atlas zur Weltgeschichte. Volume I. Munich: dtv.

Kirschbaum, M., Kirschbaum, U. F. and Fischlin, A. (1996): Climate Change Impacts on Forests. In: Watson, R. T., Zinyowera, M. C. and Moss, R. H. (eds.): Climate Change 1995. Cambridge, New York: Cambridge University Press, 93–129.

Klee, O. (1985): Angewandte Hydrobiologie: Trinkwasser – Abwasser – Gewässerschutz. Stuttgart, New York: Thieme.

Kleidon, A. and Heimann, M. (1996): Interferring Rooting Depth From a Terrestrial Biosphere Model and its Impacts on the Water- and Carbon Cycle. Nature (forthcoming).

Klemmer, P., Hecht, D., Hillebrand, B., Karl, H. and Löbbe, K. (1994): Grundlagen eines mittelfristigen umweltpolitischen Aktionsplans. Untersuchungen des Rheinisch-Westfälischen Instituts für Wirtschaftsforschung (RWI). Essen: RWI.

Klemmer, P., Wink, R., Benzler, G. and Halstrick-Schwenk, M. (1996): Mehr Nachhaltigkeit durch Marktwirtschaft: Ein ordnungspolitischer Ansatz. In: von Gerken, L. (ed.): Ordnungspolitische Grundfragen einer Politik der Nachhaltigkeit. Baden-Baden: Nomos, 289–340.

Klitzsch, E. (1991): Die Grundwassersituation Nordost-Afrikas. Naturwissenschaften 78, 59–63.

Kloepfer, M. (1989): Umweltrecht. Munich: Beck.

Klötzli, S. (1993): Der slowakisch-ungarische Konflikt um das Staustufenprojekt Gabcíkovo. Zürich: Environment and Conflicts Project (ENCOP).

Knight, J. (1992): Institutions and Social Conflict. Cambridge, New York: Cambridge University Press.

Kobus, H. and de Haar, U. (1995): Perspektiven der Wasserforschung. Weinheim: VCH.

Koenigs, T. (1996): Leben wir im Wasserüberfluß? Das großstädtische Trinkwasser kommt aus dem Umland: Probleme und Lösungen. In: Bischöfliches Hilfswerk Misereor e.V. (ed.): Wasser: Eine globale Herausforderung. Aachen: Horlemann, 43–48.

Kohlhepp, G. (ed.) (1995): Mensch-Umwelt-Beziehungen in der Pantanal-Region von Mato Grosso/Brasilien. Beiträge zur angewandten geographischen Umweltforschung. Volume 114. Tübingen: Geographical Institute of the University Tübingen.

Koppes, S. (1990): Delving into Desert Streams. Arizona State University Research 5, 16–19.

Kozhov, M. (1963): Lake Baikal and its Life. Den Haag: Junk.

Krasner, S. D. (1983): Structural Causes and Regime Consequences: Regimes as Intervening Variables. In: Krasner, S. D. (ed.): International Regimes. Ithaca, London: Cornell University Press, 1–21.

Kriesberg, L. (1991): Formal and Quasi-Mediators in International Disputes: An Exploratory Analysis. Journal of Peace Research 28 (1), 19–27

Krutilla, J. V. (1967): Conservation Reconsidered. American Economic Review 57, 777–786.

Kuby, B. (1996): Stand der Umsetzung der Lokalen Agenda 21 in Europa. Erste Auswertung einer Umfrage. In: Rösler, C. (ed.): Lokale Agenda 21. Dokumentation eines Erfahrungsaustauschs beim Deutschen Städtetag am 29. April 1996 in Köln. Berlin: Deutsches Institut für Urbanistik (Difu), 23–29.

Kulessa, M. E. (1995): Umweltpolitik in einer offenen Volkswirtschaft. Zum Spannungsverhältnis von Freihandel und Umweltschutz. Baden-Baden: Nomos.

Kulshreshtha, S. N. (1993): World Water Resources and Regional Vulnerability: Impact of Future Changes. Laxenburg: IIASA.

Kümmerlin, R. (1991): Long-term Development of Phytoplankton in Lake Constance. Verhandlung der Internationalen Vereinigung für Limnologie 24, 826–830.

Kunig, P. (1992): Nachbarrechtliche Staatenverpflichtungen bei Gefährdungen und Schädigungen der Umwelt. Berlin: Deutsche Gesellschaft für Völkerrecht.

Kunreuther, H. (1978): Even Noah Built an Ark. The Wharton Magazine (Summer), 28–35.

Kurz, R., Volkert, J. and Helbig, J. (1996): Nachhaltigkeitspolitik: Ordnungspolitische Konsequenzen und Durchsetzbarkeit. In: von Gerken, L. (ed.): Ordnungspolitische Grundfragen einer Politik der Nachhaltigkeit. Baden-Baden: Nomos, 115–165.

Kwai-cheong, C. (1995): The Three Gorges Project of China: Resettlement Prospects and Problems. Ambio 24 (2), 98–102.

Lampert, W. and Sommer, U. (1993): Limnoökologie. Stuttgart: Thieme.

Lanz, K. and Davis, J. S. (1995): Forschungsbedarf Wasser. Expertise for WBGU. Mimeo.

Lau, K.-M., Kim, J. H. and Sud, Y. (1996): Intercomparison of Hydrologic Processes in AMIP GCMs. Bulletin of the American Meteorological Society 77, 2209–2227.

LAWA – Länderarbeitsgemeinschaft Wasser (1995): Leitlinien für einen zukunftsweisenden Hochwasserschutz. Stuttgart: Umweltministerium Baden-Württemberg.

Lazarus, R. and Folkman, S. (1984): Stress, Appraisal, and Coping. Berlin, Heidelberg, New York: Springer.

Legates, D. R. and Willmott, C. J. (1990): Mean Seasonal and Spatial Variability in Gauge Corrected Global Precipitation. Journal of Climatology 10, 111–127.

Lehn, H., Steiner, M. and Mohr, H. (1996): Wasser – die elementare Ressource. Leitlinien einer nachhaltigen Nutzung. Berlin, Heidelberg, New York: Springer.

Lelek, A. and Koehler, C. (1990): Restauration of Fish Communities of the Rhine River Two Year After a Heavy Pollution Wave. Regulated Rivers-Research & Management 5 (1), 57–66.

Lenhart, B. and Steinberg, C. (1984): Limnochemische und limnobiologische Auswirkungen der Versauerung von kalkarmen Oberflächengewässern. Informations-Berichte Bayerisches Landesamt für Wasserwirtschaft 4, 210.

Létolle, R. and Mainguet, M. (1996): Der Aralsee. Eine ökologische Katastrophe. Berlin, Heidelberg, New York: Springer.

Levy, M. A. (1993): European Acid Rain. The Power of Tote-Board Diplomacy. In: Haas, P. M., Keohane, R. O. and Levy, M. A. (eds.): Institutions for the Earth. Sources of Effective International Environmental Protection. Cambridge, Ma.: MIT Press, 75–132.

Lewis, W. J., Foster, S. S. D. and Drasar, B. S. (1982): The Risk of Groundwater Pollution by On-Site Sanitation in Developing Countries. Duebendorf: IRCWD.

Libiszewski, S. (1995): Das Wasser im Nahostfriedensprozeß. ORIENT 36 (4), 625ff.

Libiszewski, S. (1996): Water Disputes in the Jordan Basin and Their Role in the Resolution of the Arab-Israeli Conflict. In: Bächler, G. and Spillmann, K. R. (eds.): Kriegsursache Umweltzerstörung. Environmental Degradation as a Cause of War. Volume II. Chur, Zürich: Rüegger, 337–460.

Lichtenstein, S., Slovic, P., Fischhoff, B., Layman, M. and Combs, B. (1978): Judged Frequency of Lethal Events. Journal of Experimental Psychology: Human Learning and Memory 4, 551–578.

Lin, B. N. (1994): Some facts and issues about the Three Gorges Project. International Journal of Sedimentation Research 9, 75–84

Lindsay, S. W. and Birley, M. H. (1996): Climate Change and Malaria Transmission. Annales of Tropical Medical Parasitology 90, 573–588.

Lindstrom-Seppa, P. and Oikari, A. (1990): Biotransformation Activities of Feral Fish in Waters Receiving Bleached Pulp Mill Effluents. Environmental Toxicology & Chemistry 9 (11), 1415–1424.

Lipton, M. (1976): Why Poor People Stay Poor – Urban Bias in World Development. London: McMillan.

Lo, F. and Yeung, Y. (1996): Emerging World Cities in Pacific Asia. Tokyo, Paris, New York: United Nations University Press.

Lovett, G. M., Likens, G. E. and Nolan, S. S. (1992): Dry Deposition of Sulfur to the Hubbard Brook Experimental Forest: A Preliminary Comparison of Methods. In: Schwartz, S. E. and Slinn, R. G. W. (eds.): Precipitation Scavenging and Atmosphere-surface Exchange. Volume 3. New York: Hemisphere, 1391–1401.

Lozán, J. L. and Kausch, H. (eds.) (1996): Warnsignale aus Flüssen und Ästuaren. Hamburg: Parey.

Lüdeke, M. K. B., Block, A., Reusswig, F. and Schellnhuber, H.-J. (1997): Weltweite Abschätzung der regionalen Wasserkritikalität. Arbeitspapier. Potsdam: Potsdam Institute for Climate Impact Research (PIK).

Ludwig, W. and Probst, J.-L. (1996): A Global Modelling of the Climatic, Morphological, and Lithological Control of River Sediment Discharges to the Oceans. Wallingford, Oxfordshire: IAHS.

Lugo, A. E., Brown, S. and Brinson, M. M. (1990): Concepts in Wetland Ecology. In: Lugo, A. E., Brinson, M. M. and Brown, S. (eds.): Ecosystems of the World. Volume 15: Forested Wetlands. Amsterdam: Elsevier, 53–85.

Mahmood, K. (1987): Reservoir Sedimentation: Impact, Extent and Mitigation. Washington, DC: The World Bank.

Maier-Rigaud, G. (1994): Umweltpolitik mit Mengen und Märkten. Lizenzen als konstituierendes Element einer ökologischen Marktwirtschaft. Marburg: Metropolis.

Maltby, E. and Procter, M. C. F. (1996): Peatlands: Their Nature and Role in the Biosphere. In: Lappalainen, E. (ed.): Global Peat Resources. Finland: Saarijärvi, 11–19.

Mandell, B. S. and Tomlin, B. W. (1991): Mediation in the Development of Norms to Manage Conflict: Kissinger in the Middle East. Journal of Peace Research 28 (1), 43–55.

Mansell, M.G. (1997): The Effect of Climate Change on Rainfall: Trends and Flooding Risk in the West of Scotland. Nordic Hydrology 28, 37–50.

Manshard, W. (1992): The Cities of Tropical Africa – Cross-Cultural Aspects, Descriptive Models and Recent Developments. In: Ehlers, E. (ed.): Modelling the City. Cross-Cultural Perspectives. Bonn: Colloquium Geographicum, 76–88.

Margat, J. (1990): Les eaux souterraines dans le monde. Orléans: Bureau de Recherches Géologiques et Minières (BRGM).

Marschner, H. (1990): Mineral Nutrition of Higher Plants. London: Academic Press.

Martens, W. J. M. (1995): Modelling the Effect of Global Warming on the Prevalence of Schistosomiasis. Bilthoven: RIVM.

Martens, W. J. M., Rotmans, J. and Niessen, L. W. (1994): Climate Change and Malaria Risk: An Integrated Modelling Approach. Global Dynamics and Sustainable Development Program, Research for Man and the Environment (RIVM). Bilthoven: National Institute of Public Health and Environmental Protection.

Martikainen, P. J. (1996): The Fluxes of Greenhouse Gases CO_2, CH_4 and N_2O in Northern Peatlands. In: Lappalainen, E. (ed.): Global Peat Resources. Finland: Saarijärvi, 29–36.

Matthews, G. V. T. (1993): The Ramsar Convention on Wetlands: Its History and Development. Gland: Ramsar Convention Bureau.

Matthews, E. and Fung, I. (1987): Methane Emission from Natural Wetlands: Global Distribution, Area, and Environmental Characteristics of Sources. Global Biochemical Cycles 1, 61–86.

Maurice, J. (1993): Fever in the Urban Jungle. New Scientist 140, 25–29.

McCaffrey, S. (1984): Yearbook of the International Law Commission 1984. New York: UN.

McCaffrey, S. (1992): A Human Right to Water: Domestic and International Implications. Georgetown International Environmental Law Review 5 (1), 1–24.

McCaffrey, S. (1996): An Assessment of the Work of the International Law Commission. Natural Resources Journal 36 (2), 297 ff.

McCully, P. (1996): Silenced Rivers. The Ecology and Politics of Large Dams. London, New Jersey: Zed Books.

McDonnell, L. (1990): Transferring Water Uses in the West. Oklahoma Law Review 43, 119.

McMichael, A. J., Ando, M., Carcavallo, R., Epstein, P., Haines, A., Jendritzky, G., Kalkstein, L., Odongo, R., Patz, J. and Piver, W. (1996): Human Population Health. In: Watson, R. T., Zinyowera, M. C. and Moss, R. H. (eds.): Climate Change 1995. Impacts, Adaptations and Mitigation of Climate Change: Scientific-Technical Analyses. Cambridge, New York: Cambridge University Press, 561–584.

Meißner, C. (1991): Instrumente des Gewässerschutzrechtes in der DDR. In: Kloepfer, M. (ed.): Instrumente des Umweltrechts in der früheren DDR. Berlin, Heidelberg, New York: Springer, 62–71.

Melillo, J. M., Prentice, I. C., Farquhar, G. D., Schulze, E.-D. and Sala, O. E. (1996): Terrestrial Biotic Responses to Environmental Change and Feedbacks to Climate. In: Houghton, J. T., Meira Filho, L. G., Callander, B. A., Harris, N., Kattenberg, A. and Maskell, K. (eds.): Climate Change 1995. The Science of Climate Change. Cambridge, New York: Cambridge University Press, 444–481.

Mendel, H.-G., Hermann, A. and Fischer, D. (1996): Hochwasser: Gedanken über Ursachen und Vorsorge aus hydrologischer Sicht. Koblenz: Bundesamt für Gewässerschutz (BFG).

Meybeck, M., Chapman, D. and Helmer, R. (1989): Global Freshwater Quality. A First Assessment. Oxford: Blackwell.

Meybeck, M., Friedrich, G., Thomas, R. and Chapman, D. (1992): Rivers. In: Chapman, D. (ed.): Water Quality Assessment. London: Chapman and Hall, 239–316.

Michael, E. and Bundy, D. A. P. (1996): The Global Burden of Lymphatic Filariasis. In: Murray, C. J. L. and Lopez, A. D. (eds.): World Burden of Diseases. Geneva: WHO.

Miller, R., Williams, J. and Williams, J. (1989): Extinctions of North American Fishes During the Past Century. Fisheries 4 (6), 34–36.

Miller, L. H. and Warrell, D. A. (1990): Malaria. In: Warren, K. and Mahmoud, A. A. F. (eds.): Tropical and Geographical Medicine. New York: McGraw-Hill, 245–264.

Milliman, J. D. and Meade, R. H. (1983): World-wide Delivery of River Sediment to Oceans. Journal of Geology 91, 1–21.

Mitchell, J. F. B. (1989): The Greenhouse Effect and Climate Change. Review of Geophysics 27, 115–139.

Mitsch, W. J., Mitsch, R. H. and Turner, R. E. (1994): Wetlands of the Old and New World: Ecology and Management. In: Mitsch, W. J. (ed.): Global Wetlands: Old World and New World. Amsterdam: Elsevier, 3–51.

Mitsch, W. J. (1994): The Nonpoint Source Control Function of Natural and Constructed Riparian Wetlands. In: Mitsch, W. J. (ed.): Global Wetlands: Old World and New World. Amsterdam: Elsevier, 351–368.

Möhle, K.-A. (1983): Wassersparmaßnahmen. Wasserversorgungsbericht des Bundesministers des Innern. Materialien. Berlin: Schmidt.

Mohr, H. and Lehn, H. (1994): Present Views of the Nitrogen Cycle. In: Mohr, H. and Müntz, K. (eds.): The Terrestrial Nitrogen Cycle as Influenced by Man. Halle: Nova Acta Leopoldina, 11–28.

Mooney H. A., Cushman, J. H., Medina, E., Sala, O. E. and Schulze, E.-D. (1996): What We Have Learned About the Ecosystem Functioning of Biodiversity. In: Perrings, C. A., Mäler, K.-G., Folke, C., Holling, C. S. and Jansson, B.-O (eds.): Biodiversity Conservation. Dordrecht: Kluwer, 475–484.

Moore, S., Murphy, M. and Watson, R. (1994): A Longitudinal Study of Domestic Water Conservation Behavior. Population and Environment 16 (2), 175–189.

Müller, G. (1996): Schwermetalle und organische Schadstoffe in den Flußsedimenten. In: Lozán, J. L. and Kausch, H. (eds.): Warnsignale aus Flüssen und Ästuaren. Hamburg: Parey, 113–123.

Müller, H. (1993). Die Chance der Kooperation. Regime in den internationalen Beziehungen. Darmstadt: Wissenschaftliche Buchgesellschaft.

Müller, H. (1994): Internationale Beziehungen als kommunikatives Handeln. Zeitschrift für Internationale Beziehungen 1 (1), 15–44.

Münchner Rückversicherung (1997): Überschwemmung und Versicherung. Munich: Münchner Rück.

Murray, C. J. L. and Lopez, A. D. (eds.) (1996): The Global Burden of Disease. Geneva: World Health Organisation (WHO).

Nash, L. (1993): Water Quality and Health. In: Gleick, P. H. (ed.): Water in Crisis – A Guide to the World's Fresh Water Resources. Oxford, New York: Oxford University Press, 25–36.

Nentwig, W. (1995): Humanökologie. Fakten, Argumente, Ausblicke. Berlin, Heidelberg, New York: Springer.

Neubauer, I. (1995): „Der Fluß ist ein Knecht, um den sich alle streiten" – Was Flüsse und Menschen miteinander verbindet. Wendekreis 6, 4–5.

Nielsen, G., Meyer, C. G., Mantel, C. and Knobloch, J. (1996): Viruskrankheiten. In: Knobloch, J. (ed.): Tropen und Reisemedizin. Stuttgart: Fischer, 272–350.

Niemeyer-Lüllwitz, A. Z. and Zucchi, H. (eds.) (1985): Fließgewässerkunde: Ökologie fließender Gewässer unter besonderer Berücksichtigung wasserbaulicher Eingriffe. Berlin, Frankfurt/M.: Diesterweg, Sauerländer.

Novitzki, R. P. (1979): Hydrological Characteristics of Wisconsin's Wetlands and Their Influence on Floods, Steam Flow and Sediment. In: Greeson, P. E., Clark, J. R. and Clark, J. E. (eds.): Wetland Functions and Values: The State of Our Understanding. Minneapolis: American Water Resources Association, 377–388.

Nriagu, J. O. (1992): Worldwide Contamination of the Atmosphere With Toxic Metals. In: Verry, E. S. and Vermette, S. J. (eds.): The Deposition and Fate of Trace Metals in our Environment. Philadelphia: U.S. Department of Agriculture. Forest Service North Central Forest Experiment Station, 9–22.

Oberai, A. S. (1993): Population Growth, Employment and Poverty in Third World Mega-Cities. Geneva: International Labour Organization (ILO).

Oberhuber, J. M. (1993): Simulation of the Atlantic Circulation With a Coupled Sea Ice-Mixed Layer-Isopycnal General Circulation Model. Part I: Model Description. Journal of Physical Oceanography (22), 808–829.

Odum, E. P. (1971): Fundamentals of Ecology. Philadelphia: Saunders.

OECD – Organisation for Economic Co-operation and Development (1982): Eutrophication of Waters. Monitoring, Assessment and Control. Paris: OECD.

Oerlemans, J. and Bintanja, R. (1995): Snow and Ice Cover and Climate Sensitivity. In: Oliver, H. R. and Oliver, S. A. (eds.): The Role of Water and the Hydrological Cycle in Global Change. Berlin, Heidelberg, New York: Springer, 189–198.

Oldeman L. R., Hakkeling, R. T. A. and Sombroek, W. G. (1990): World Map of the Status of Human-induced Soil Degradation: An Explanatory Note. Wageningen: International Soil Reference and Information Centre (ISRIC).

Oltersdorf, U. and Weingärtner, L. (1996): Handbuch der Welternährung. Die zwei Gesichter der globalen Nahrungssituation. Bonn: Dietz.

Oltersdorf, U. (1992): Hunger und Überfluß. Geographische Rundschau 44 (2), 74–77.

Omar, M. S., Manfouz, A. A. and Abdel Moneim, M. (1995): The Relationship of Water Sources and Other Determinants to Prevalence on Intestinal Protoual Infections in a Rural Community of Saudi Arabia. Journal of Community Health 20, 433–440.

Ostmann, A., Pommerehne, W., Feld, P. and Hart, A. (1997): Umweltgemeingüter? ZWS-Zeitschrift für Wirtschafts- und Sozialwissenschaften 117 (1), 107–144.

Ostrom, E. (1986): An Agenda for the Study of Institutions. Public Choice 48, 3–25.

Ostrom, E. (1990): Governing the Commons. The Evolution of Institutions for Collective Action. Cambridge, New York: Cambridge University Press.

Ostrom, E. (1992): Crafting Institutions for Self-Governing Irrigation Systems. San Francisco: ICS Press.

PAHO – Pan American Health Organisation (1994): Leishmananiasis in the Americas. Epidemiological Bulletin 15, 8–13.

PAI – Population Action International (1993): Sustaining Water – Population and the Future of Renewable Water Supplies. Washington, DC: PAI.

Patz, J. A., Epstein, P. R., Burke, T. A. and Balbus, J. M. (1996): Global Climate Change and Emerging Infectious Diseases. JAMA 275 (3), 217–223.

Pearce, D. W. and Turner, R. K. (1990): Economics of Natural Resources and the Environment. Hampstead: Hemel.

Pearce, F. (1992): The Dammed – Rivers, Dams, and the Coming World Water Crises. London: The Bodley Head.

Peixoto, J. P. (1995): The Role of the Atmosphere in the Water Cycle. In: Oliver, H. R. and Oliver, S. A. (eds.): The Role of Water and the Hydrological Cycle in Global Change. Berlin, Heidelberg, New York: Springer, 199–252.

Pereira, H. C. (1974): Land Use and Water Resources. Cambridge: Cambridge University Press.

Perez, O. M., Morales, W., Paniagua, M. and Strannegard, O. (1996): Prevalence of Antibodies to Hepatitits A, B, C and E Viruses in a Healthy Population in Leon, Nicaragua. American Journal of Tropical Medicine and Hygiene 55, 17–21.

Perrings, C. A., Mäler, K.-G., Folke, C., Holling, C. S. and Jansson, B.-O. (1995): Biodiversity Conservation and Economic Development: The Policy Problem. In: Perrings, C. A., Mäler, K.-G., Folke, C., Holling, C. S. and Jansson, B.-O (eds.): Biodiversity Conservation. Dordrecht: Kluwer, 3–21.

Peterjohn, W. T. and Correll, D. L. (1984): Nutrient Dynamics in an Agricultural Watershed: Observations on the Role of a Riparian Forest. Ecology 65 (5), 1466–1475.

Petschel-Held, G. and Plöchl, M. (1997): A Global Model for Annual River Discharges. Manuscript. Potsdam: Potsdam Institute for Climate Impact Research (PIK).

Petts, G. E. and Amoros, C. (1996): Fluvial Hydrosystems. London: Chapman and Hall.

Pierre, S. (1987): The Green Revolution Re-examined in India. In: Bernhard, G. (1987): The Green Revolution Revisited. Critique and Alternatives. Boston, Sydney, Wellington: Allen & Unwin, 56–75.

Pilardeaux, B. (1995): Innovation und Entwicklung in Nordpakistan. Saarbrücken: Verlag für Entwicklungspolitik.

Pilardeaux, B. (1997): Desertifikationsbekämpfung im Aufwind? 1. Vertragsstaatenkonferenz der UNCCD. Nord-Süd-Aktuell XI (4), 744–749.

Pimentel, D. (1996): Green Revolution Agriculture and Chemical Hazards. The Science of the Total Environment 188 (Supplement 1), 86–98.

Pinay, G., Haycock, N. E., Ruffinoni, C. and Holmes, R. M. (1994): The Role of Denitrification in Nitrogen Removal in Rivers Corridors. In: Mitsch, W. J. (ed.) (1994): Global Wetlands – Old World and New. Amsterdam: Elsevier, 106–116.

Pinay, G., Ruffinoni, C. and Fabre, A. (1995): Nitrogen Cycling in Two Riparian Forest Soils Under Different Geomorphic Conditions. Biogeochemistry 30, 9–29.

Pitt, D. G. (1989): The Attractiveness and Use of Aquatic Environments as Outdoor Recreation Places. In: Altman, I. and Zube, E. H. (eds.): Public Places and Spaces. New York: Plenum Press, 217–254.

Plate, E. J. (1997): Wasser und Katastrophen. Expertise for WBGU. Mimeo.

Platt, G. (1973): Social Traps. American Psychologist 28, 641–651.

Plotnikov, I. S., Aladin, N. V. and Filippov, A. A. (1991): The Past and Present of the Aral Sea Fauna. Zoologicheskii Zhurnal 70 (4), 5–15.

Pommerehne, W. and Römer, A. U. (1992): Ansätze zur Erfassung der Präferenzen für öffentliche Güter. Jahrbuch für Sozialwissenschaft 43 (2), 171–210.

Ponce, V. M. (1995): Hydrogeologic and Environmental Impact of the Parana-Paraguai Waterway on the Pantanal of Mato Grosso, Brazil. San Diego: San Diego State University.

Portes, A. and Schauffler, P. (1993): Competing Perspectives on the Latin American Informal Sector. Population and Development Review (19), 33–60.

Postel, S. (1984): Water: Rethinking Management in an Age of Scarcity. Washington, DC: Worldwatch Institute.

Postel, S. (1989): Water for Agriculture: Facing the Limits. Washington, DC: Worldwatch Institute.

Postel, S. (1993): Die letzte Oase – der Kampf um das Wasser. Frankfurt/M.: Fischer.

Postel, S. (1996): Dividing the Waters. Food Security, Ecosystem Health, and the New Politics of Scarcity. Washington, DC: Worldwatch Institute.

Postel, S., Daily, G. C. and Ehrlich, P. R. (1996): Human Appropriation of Renewable Fresh Water. Science 271, 785–801.

President's Water Resources Council (1962): US Senate Document 97. Washington, DC: President's Water Resources Council.

Prinn, R. G., Weiss, R. F., Miller, B. R., Huang, J., Aleya, F. N., Cunnold, D. M., Fraser, P. J., Hartley, D. E. and Simmonds, P. G. (1995): Atmospheric Trends and Lifetime of

CH_3CCl_3 and Global OH Concentrations. Science (269), 187–192.

Protopopov, N. F. (1995): New Vermitechnology Approach for Sludge Utilisation in Northern and Temperate Climates all Year Round. Second International Conference for Ecological Engineering for Waste Water Treatment. Waedensville, Switzerland.

Putnam, R. D. (1988): Diplomacy and Domestic Politics. The Logic of Two-Level Games. International Organization 42 (3), 427–460.

Radke, V. (1996): Ökonomische Aspekte nachhaltiger Technologien. Zur Bedeutung unterschiedlicher Ausprägungen des technischen Fortschritts für das Konzept des Sustainable Development. Zeitschrift für Umweltpolitik & Umweltrecht 1, 109–128.

Rapp, J. and Schönwiese, C. (1995): Atlas der Niederschlags- und Temperaturtrends in Deutschland 1891–1990: Meteorologie und Geophysik. Volume 5. Frankfurt/M.: Frankfurter Geowissenschaftliche Arbeiten.

Rasmussen, J. (1994): Floodplain Management Into the 21st Century: A Blueprint for Change – Sharing the Challenge. Water International 19, 166–176.

Reeves, R. R. and Leatherwood, S. (1994): Dams and River Dolphins: Can They Co-exist? Ambio 23 (3), 172–175.

Renger, J. (1995): Die Multilateralen Friedensverhandlungen der Arbeitsgruppe „Wasser". Asien/Afrika/Lateinamerika 23, 149–157.

Renjun, L. (1991): New Advances on Population Status and Protective Measures for Lipotes vexillifer and Neophocaena phocaenoides in the Chiangjian River. Aquatic Mammals 17 (3), 181–183.

Renn, O. (1996a): Kooperativer Diskurs. Kommunikation in der Umweltpolitik. In: Selle, K. (ed.): Planung und Kommunikation. Gestaltung von Planungsprozessen in Quartier, Stadt und Landschaft. Grundlagen, Methoden, Praxiserfahrungen. Wiesbaden, Berlin: Bauverlag, 101–112.

Renn, O. (1996b): Möglichkeiten und Grenzen diskursiver Verfahren bei umweltrelevanten Planungen. In: Biesecker, A. and Grenzdörffer, K. (eds.): Kooperation, Netzwerk, Selbstorganisation. Elemente demokratischen Wirtschaftens. Pfaffenweiler: Centaurus, 161–197.

Renn, O. and Oppermann, B. (1995): „Bottom-up" statt „Top-down" – Die Forderung nach Bürgermitwirkung als (altes und neues) Mittel zur Lösung von Konflikten in der räumlichen Planung. Zeitschrift für angewandte Umweltforschung (Special Issue 6), 257–276.

Renn, O., Webler, T. and Wiedemann, P. (1995): A Need for Discourse on Citizen Participation: Objectives and Structure of the Book. In: Renn, O., Webler, T. and Wiedemann, P. (eds.): Fairness and Competence in Citizen Participation. Evaluating Models for Environmental Discourse. Dordrecht: Kluwer, 1–15.

Renn, O., Webler, T., Rakel, H., Dienel, P. and Johnson, B. (1993): Public Participation in Decision Making: A Three-Step Procedure. Policy Science 26, 189–214.

Rennings, K., Brockmann, L., Koschel, H., Bergmann, H. and Kühn, J. (1996): Nachhaltigkeit, Ordnungspolitik und freiwillige Selbstverpflichtung. Umwelt- und Ressourcenökonomie. Heidelberg: Physica.

Rennings, K. and Wiggering, H. (1997): Steps Towards Indicators of Sustainable Development: Linking Economic and Ecological Concepts. Ecological Economics 20 (1), 25–36.

Reuter, N. (1994): Institutionalismus, Neo-Institutionalismus, Neue Institutionelle Ökonomie und andere „Institutionalismen". Zeitschrift für Wirtschafts- und Sozialwissenschaften 114 (1), 5–23.

Rhode, H. (1989): Acidification in a Global Perspective. Ambio 18 (3), 155–159.

Risse-Kappen, T. (1995): Reden ist nicht billig. Zur Debatte um Kommunikation und Rationalität. Zeitschrift für Internationale Beziehungen 2 (1), 171–184.

Rittberger, V. (ed.) (1993): Regime Theory and International Relations. Oxford: Clarendon Press.

Roberts, A. (1993): The United Nations and International Security. Survival 35 (2), 3–30.

Roeckner, E., Oberhuber, J.-M., Bacher, A., Christoph, M. and Kirchner, J. (1996): ENSO Variability and Atmospheric Response in a Global Coupled Atmosphere-Ocean GCM. Climate Dynamics 12, 737–754.

Römbke, J. and Moltman, J. F. (1996): Applied Ecotoxicology. Boca Raton, New York, London: CRC Lewis Publishers.

Rosegrant, M. (1995): Dealing with Water Scarcity in the Next Century. Washington, DC: International Food Policy Institute.

Rösler, C. (1996): Stand der Umsetzung der Lokalen Agenda 21 in deutschen Städten. Erste Ergebnisse der Difu-Umfrage. In: Rösler, C. (ed.): Lokale Agenda 21. Dokumentation eines Erfahrungsaustauschs beim Deutschen Städtetag am 29. April 1996 in Köln. Berlin: Deutsches Institut für Urbanistik (Difu), 47–55.

Rosseland, B. O., Skogheim, O. K. and Sevaldrud, I. H. (1986): Acid Deposition and Effects on Nordic Europe. Damage to Fish Populations in Scandinavia Continue Apace. In: Acidic Precipitation. Proceedings of the International Symposium on Acidic Precipitation. Water, Air and Soil Pollution 30 (1–2), 65–74.

Rott, U. (1997): Wassertechnologien: Grundlagen und Tendenzen. Expertise for WBGU. Mimeo.

Rott, U. and Minke, R. (1995): Verfahren der innerbetrieblichen Behandlung von Abwässern der Textilveredelungsindustrie. Abwassertechnik, Abfalltechnik + Recycling (4), 15–20.

Rozengurt, M. A. (1992): Alterations to Freshwater Flows. In: Stroud, R. H. (ed.): Stemming the Tide of Coastal Fish Habitat Loss. Proceedings of Marine Recreational

Fisheries Symposium. Savannah, GA: National Coalition for Marine Conservation, 73–80.

Rudd, J. W. M., Harris, R., Kelly, C. A. and Hecky, R. E. (1993): Are Hydroelectric Reservoirs Significant Sources of Greenhouse Gas? Ambio 22 (4), 246–248.

Rudolph, K.-U. (1987): Zur Nutzenberechnung der Wasserversorgung. In: Rudolph, K.-U. (ed.): Projektbewertung von Talsperren. Schriftenreihe Wasser und Umwelt. Volume 1. Witten: K.-U. Rudolph publisher, 45–74.

Rudolph, K.-U. (1990): Technische Maßnahmen zur Kostensenkung – Möglichkeiten während der Planungsphase. In: Rudolph, K.-U. (ed.): Kostenprobleme der kommunalen Abwasserbeseitigung. Witten: K.-U. Rudolph publisher, 67–85.

Rudolph, K.-U., Köppke, K.-E. and Korbach, J. (1995): Stand der Abwassertechnik in verschiedenen Branchen. Berlin: Umweltbundesamt (UBA).

Rudolph, K.-U. and Gärtner, T. (1997): Die deutsche Wasserver- und -entsorgung im internationalen Vergleich. Expertise for WBGU. Mimeo.

Ruttner, F. (1962): Grundriß der Limnologie. Berlin: de Gruyter.

Sands, P. (1989): The Environment, Community, and International Law. Harvard International Law Journal 30 (2), 393–420.

Santos, M. (1979): The Shared Space. London, New York: Methuen.

Savenije, H. H. G (1996): The Runoff Coefficient as the Key to Moisture Recycling. Journal of Hydrology (176), 219-225.

Schellnhuber, H.-J., Block, A., Cassel-Gintz, M., Kropp, J., Lammel, G., Lass, W., Loose, C., Lüdeke, M. K. B., Moldenhauer, O., Petschel-Held, G., Plöchl, M. and Reusswig, F. (1997): Syndromes of Global Change. GAIA (forthcoming).

Scherer, P. and Castell-Exner, C. (1996): Forschung und Entwicklung im Wasserfach. gwf Wasser-Abwasser 137 (11), 594–600.

Schindler, D. W. (1990): Experimental Perturbations of Whole Lakes as Tests of Hypotheses Concerning Ecosystem Structure and Function. Oikos 57, 25–41.

Schlotter, P. (1994): Zwischen Universalismus und Regionalismus: Die KSZE im System der Vereinten Nationen. In: Meyer, B. and Moltmann, B. (eds.): Konfliktsteuerung durch Vereinte Nationen und KSZE. Frankfurt/M.: Haag and Herchen, 96–107.

Schlotter, P., Ropers, N. and Meyer, B. (1994): Die neue KSZE. Zukunftsperspektiven einer regionalen Friedensstrategie. Opladen: Leske and Budrich.

Schmid, C. (1993): Der Israel-Palästina-Konflikt und die Bedeutung des Vorderen Orients als sicherheitspolitische Region nach dem Ende des Ost-West-Konflikts. Baden-Baden: Nomos.

Schmidt-Kallert, E. (1989): Staudämme: Symbole für einen Entwicklungsweg. In: Pater, S. and Schmidt-Kallert, E. (eds.): Zum Beispiel Staudämme. Göttingen: Lamuv-Verlag, 7–14.

Schminke, H. K. (1997): Heinzelmännchen im Grundwasser. Biologie in unserer Zeit 3, 182–188.

Schmitz, W. (1961): Fließgewässerforschung – Hydrographie und Botanik. Verhandlungen der Internationalen Vereinigung für Limnologie (14), 541-586.

Schönhuth, M. and Kievelitz, U. (1993): Partizipative Erhebungs- und Planungsmethoden in der Entwicklungszusammenarbeit. Eschborn: GTZ.

Schramm, E. (1995): Wege zu einem gesellschaftlichen Wasserdiskurs. Expertise for WBGU. Mimeo.

Schröer, S. and Staubli, T. (1995): Die Flüsse des Garten Eden. Die Flußsymbolik im alten Orient. Wendekreis 6, 6–9.

Schua, L. and Schua, R. (1981): Wasser. Lebenselement und Umwelt. Freiburg: Alber.

Schug, W., Leon, J. and Gravert, H. (1996): Welternährung. Herausforderung an Pflanzenbau und Tierhaltung. Darmstadt: Wissenschaftliche Buchgesellschaft.

Schulz, W. (1989): Ansätze und Grenzen der Monetarisierung von Umweltschäden. Zeitschrift für Umweltpolitik & Umweltrecht 12 (1), 55–72.

Schulze, E.-D. (1982): Plant Life Forms and Their Carbon, Water and Nutrient Relations. Physiological Plant Ecology II, Encyclopedia of Plant Physiology: New Series 12 B, 615–676.

Schulze, E.-D. (1994): The Regulation of Plant Transpiration: Interactions of Feedforward, Feedback, and Futile Cycle. In: Schulze, E.-D. (ed.): Flux Control in Biological Systems. San Diego: Academic Press, 203–235.

Schulze, E.-D., Kelliher, F. M., Körner, C., Lloyd, J. and Leuning, R. (1994): Relationships Among Maximum, Stomatal Conductance, Ecosystem Surface Conductance. Carbon Assimilation Rate, and Plant Nitrogen Nutrition: A Global Ecology Scaling Exercise. Annual Review of Ecology and Systematics 25, 629–660.

Schulze, E.-D. and Heimann, M. (1997): Carbon and Water Exchange of Terrestrial Ecosystems. In: Galloway, J. N. and Melillo, J. (eds.): Asian Change in the Context of Global Change. Cambridge: Cambridge University Press (in print).

Schwartz, S. E. and Slingo, A. (1995): Enhanced Shortwave Cloud Radiative Forcing Due to Anthropogenic Aerosols. In: Crutzen, P. J. and Ramanathan, V. (eds.): Clouds, Chemistry and Climate. Berlin, Heidelberg, New York: Springer, 191–236.

Schwoerbel, J. (1987): Einführung in die Limnologie. Stuttgart: Fischer.

Seager, J. (1995): Wasserkraft. In: Seager, J., Reed, C. and Stott, P. (eds.): Der Öko-Atlas – Neuausgabe. Bonn: Dietz, 44–46.

SEF – Stiftung Entwicklung und Frieden (ed.) (1993): Globale Trends 93/94. Daten zur Weltentwicklung. Frankfurt/M.: Fischer.

SEI – Stockholm Environment Institute (1996): The Freshwater Resources of the World – A Comprehensive Assessment. Stockholm: SEI.

Selbmann, S. (1995): Mythos Wasser. Symbolik und Kulturgeschichte. Karlsruhe: Badenia.

Seligman, C. and Finegan, J. E. (1990): A Two-factor Model of Energy and Water Conservation. In: Edwards, J., Tindale, R. S., Heath, L. and Posavac, E. J. (eds.): Social Influence Processes and Preventions. Volume 1. Social Psychological Applications to Social Issues. New York: Plenum Press, 279–299.

Shiklomanov, I. A. and Sokolov, A. A. (1985): Methodological Basis of World Water Balance Investigation and Computation. In: IAHS – International Association for Hydological Sciences (ed.): New Approaches in World Water Balance Computations. Hamburg: IAHS, 77–91.

Shiklomanov, I. A. (1993): World Fresh Water Resources. In: Gleick, P. H. (ed.): Water in Crisis. A Guide to the World's Fresh Water Resources. New York, Oxford: Oxford University Press, 13–24.

Shiva, V. (1991): The Violence of the Green Revolution. Penang: Third World Network.

Shukla, J. (1995): On the Initiation and Persistence of the Sahel Drought. In: Martinson, D. G., Bryan, K., Ghil, M., Hall, M. M., Karl, T. R., Sarachik, E. S., Sorooshian, S. and Tallex, L. D. (eds.): Natural Climate Variability on Interannual and Decadal Time Scales. Washington, DC: National Academy Press, 44–48.

Sick, W.-D. (1983): Agrargeographie. Braunschweig: Westermann.

Simmann, H.-Y. (1994): Die Bedeutung von Saprobiensystemen zur Gewässerbeurteilung. Frankfurt/M.: Hessische Landesanstalt für Umwelt.

Simonis, U. E. (ed.) (1996): Weltumweltpolitik. Grundriß und Bausteine eines neuen Politikfeldes. Berlin: Edition Sigma.

Simpson, L. (1994): Wassermärkte. Ein gangbarer Weg? Finanzierung & Entwicklung 31 (2), 30–32.

Slovic, P., Fischhoff, B. and Lichtenstein, S. (1978): Accident Probability and Seat Belt Usage: A Psychological Perspective. Accident Analysis and Prevention 10, 281–285.

Smith, N. (1985): Mensch und Wasser. Geschichte und Technik der Bewässerung und Trinkwasserversorgung vom Altertum bis heute. Wiesbaden: Pfriemer.

Snedacker, S. C. (1984): Mangroves: A Summary of Knowledge With Emphasis on Pakistan. In: Haq, B. U. and Milliman, J. D. (eds.): Marine Geology and Oceanography of Arabian Sea and Coastal Pakistan. New York: Van Nostrand Reinhold, 99.

Snimschikova, L. N. and Akinshina, T. W. (1994): Oligochaete Fauna of Lake Baikal. Hydrobiologia 278 (1–3), 27–34.

Snow, A. A. and Palma, P. M. (1997): Commercialization of Transgenic Plants: Potential Ecological Risks. Bio Science 47 (2), 86–96.

Soffer, A. (1994): The Relevance of the Johnston Plan to the Reality of 1993 and Beyond. In: Isaac, J. and Shuval, H. (eds.): Water and Peace in the Middle East. Amsterdam: Elsevier, 107ff.

Sommer, U. (1985): Comparison between Steady State and Non-Steady-State Competition: Experiments with Natural Phytoplankton. Limnology and Oceanography 30, 335–346.

Spiegel, E. (1970): Slum. In: Akademie für Raumforschung und Landesplanung (ed.): Handwörterbuch der Raumforschung und Raumordnung. Hannover: Jaenecke, 2952–2959.

SRU – Rat von Sachverständigen für Umweltfragen (1985): Umweltprobleme der Landwirtschaft. Sondergutachten. Stuttgart, Mainz: Kohlhammer.

SRU – Rat von Sachverständigen für Umweltfragen (1987): Umweltgutachten 1987. Stuttgart: Metzler-Poeschel.

SRU – Rat von Sachverständigen für Umweltfragen (1994): Umweltgutachten 1994. Für eine dauerhaft-umweltgerechte Entwicklung. Stuttgart: Metzler-Poeschel.

Stanners, D. and Bourdeau, P. (eds.) (1995): Europe's Environment: The Dobris Assessment. Copenhagen: European Environment Agency (EEA).

Stabel, H.-H. (1997): Vergleichende Bewertung der internationalen und nationalen Standards für Nutzwasser. Expertise for WBGU. Mimeo.

Stapelfeldt, G. (1990): Verelendung und Urbanisierung in der Dritten Welt. Der Fall Lima/Peru. Saarbrücken: Breitenbach.

Stein, D. (1997): Moderne Leitungsnetze als Beitrag zur Lösung der Wasserprobleme von Städten. Expertise for WBGU. Mimeo.

Stephens, C. (1994): Environment and Health in Developing Countries: An Analysis of Intra-Urban Differentials Using Existing Data. London: London School of Hygiene & Tropical Medicine.

Stern, P. C. and Oskamp, S. (1987): Managing Scarce Environmental Resources. In: Stokols, D. and Altman, I. (eds.): Handbook of Environmental Psychology. Volume 2. Chichester, New York: Wiley & Sons, 1043–1088.

Stitt, M. (1994): Flux Control at the Level of the Pathway: Studies With Mutants and Transgenic Plants Having a Decreased Activity of Enzyms Involved in Photosynthesis Partitioning. In: Schulze, E.-D. (ed.): Flux Control in Biological Systems. San Diego: Academic Press, 13–36.

Stitt, M. and Schulze, E.-D. (1994): Plant Growth, Storage, and Resource Allocation: From Flux Control in a Metabolic Chain to the Whole-Plant Level. In: Schulze, E.-D. (ed.): Flux Control in Biological Systems. San Diego: Academic Press, 57–118.

Stödter, R. (1995): International Law Association. In: Bernhardt, R. (ed.): Encyclopedia of Public International Law. Amsterdam: Elsevier, 1207.

Ströbele, W. (1987): Rohstofföokonomik. Theorie natürlicher Ressourcen mit Anwendungsbeispielen Öl, Kupfer, Uran und Fischerei. Munich: Vahlen.

Sundrum, R. M. (1990): Income Distribution in Less Developed Countries. London, New York: Routledge.

Susskind, L. and Babbitt, E. (1992): Overcoming the Obstacles to Effective Mediation of International Disputes. In: Bercovitch, J. and Rubin, J. (eds.): Mediation in International Relations. Multiple Approaches to Conflict Management. New York: St. Martin's Press, 30–51.

Tayler, J. R., Cardamone, M. A. and Mitsch, W. J. (1990): Bottomland Hardwood Forests: Their Functions and Values. In: Gosselink, J. G., Lee, L. C. and Muir, T. A. (eds.): Ecological Processes and Cumulative Impacts: Illustrated by Bottomland Hardwood Wetland Ecosystems. Chelsea: Lewis Publishers, 13–86.

Taylor, A. (1987): Looking at Water Through Different Eyes – The Maori Perspective. Soil and Water (Summer), 22–24.

Taylor, M. B., Becker, P. J., van Rensburg, E. J., Harris, B. N., Bailey, I. W. and Grabow, W. O. (1995): A Serosurvey of Water-borne Pathogens Amongst Canoeists in South Africa. Epidemiology and Infection 115, 299–307.

Teclaff, L. (1996): Evolution of the River Basin Concept in National and International Law. Natural Resouces Journal 36 (2), 359 ff.

Thompson Jr., B. H. (1993): Institutional Perspectives on Water Policies and Markets. California Law Review 81, 671.

Thompson, S. C. and Stoutemyer, K. (1991): Water Use as a Commons Dilemma. The Effects of Education that Focuses on Long-term Consequences and Individual Action. Environment and Behavior 23 (3), 314–333.

Tilman, D. and Downing, J. A. (1994): Biodiversity and Stability in Grasslands. Nature (367), 363–365.

Tilzer, M. and Serruya, C. (1990): Large Lakes. Ecological Structure and Function. Berlin, Heidelberg, New York: Springer.

Tilzer, M., Gaedke, U., Schweizer, A., Beese, B. and Wieser, T. (1991): Interannual Variability of Phytoplankton Productivity and Related Parameters in Lake Constance: No Response to Decreased Phosphorus Loading? Journal of Plankton Research 13, 755–777.

Tobler, W., Deichmann, U., Gottsegen, J. and Maloy, K. (1995): The Global Demography Project. Technical Report TR-95-6. Santa Barbara: University of California, National Center for Geographic Information and Analysis.

Tucci, C., Silveira, A., Sanchez, J. and Albuquerque, F. (1995): Flow Regionalization in the Upper Paraguai Basin, Brazil. Hydrological Sciences Journal/Journal des Sciences Hydrologiques 40 (4), 485–497.

UBA – Umweltbundesamt (1991): Ohne Wasser läuft nichts! Rettet unser wichtigstes Lebensmittel. Berlin: UBA.

UBA – Umweltbundesamt (1994): Stoffliche Belastung der Gewässer durch die Landwirtschaft und Maßnahmen zu ihrer Verringerung. Berlin: Schmidt.

UBA – Umweltbundesamt (1995): Umweltdaten Deutschland. Berlin: UBA.

UN –United Nations (1990): Achievements of the International Drinking Water Supply and Sanitation Decade 1981–1990. New York: UN.

UNCED – United Nations Conference on Environment and Development (1992): Agenda 21. Agreements on Environment and Development. Rio de Janeiro: UNCED.

UN DESA – United Nations Department of Economic and Social Affairs (1958): Integrated River Basin Development. UN Doc. E/3066. New York: UN DESA.

UNDP – United Nations Development Programme (1992): Human Development Report 1992. Oxford, New York: Oxford University Press.

UNDP – United Nations Development Programme (1993): Human Development Report 1993. New York, Oxford: Oxford University Press.

UNDP – United Nations Development Programme (1994): Bericht über die menschliche Entwicklung. Bonn: Deutsche Gesellschaft für die Vereinten Nationen (DGVN).

UNDP – United Nations Development Programme (1995): Bericht über die menschliche Entwicklung 1995. Bonn: Deutsche Gesellschaft für die Vereinten Nationen (DGVN).

UNDP – United Nations Development Programme (1996): Bericht über die menschliche Entwicklung. Statistischer Anhang. Bonn: Deutsche Gesellschaft für die Vereinten Nationen (DGVN).

UNDRO – United Nations Disaster Relief Coordinator (1991): Mitigating Natural Disasters: Phenomena, Effects, and Options. New York: United Nations (UN).

UN DTCD – United Nations Department of Technical Cooperation and Development (1991): Integrated Water Resources Planning. New York: UN DTCD.

UNEP – United Nations Environment Programme (1995): Global Biodiversity Assessment. Cambridge: Cambridge University Press.

UNFPA – United Nations Population Fund (1995): Weltbevölkerungsbericht 1995. Welt im Wandel: Bevölkerung, Entwicklung und die Zukunft der Stadt. Bonn: Deutsche Gesellschaft für die Vereinten Nationen (DGVN).

UNICEF – United Nations International Children's Fund (1997): Statistik der Unicef über Wasserversorgung und Abwasserreinigung im Jahre 1996. Korrespondenz Abwasser 44 (4), 596.

UN INSTRAW – United Nations International Research and Training Institute for the Advancement of Women (1991): Women, Water and Sanitation. In: Sontheimer, S. (ed.): Women and the Environment. A Reader. Crisis and Development in the Third World. London: Earthscan, 119–132.

UNPD – United Nations Population Division (1994): World Urbanisation Prospects: The 1994 Revision. New York: UN.

UN-SG – United Nations Secretary-General (1997a): Comprehensive Assessment of the Freshwater Resources of the World. Report of the United Nations Secretary–General to the Fifth Session of the Commission on Sustainable Development (5.–25. April 1997). UN-Doc. E/CN.17/1997/9: New York: UN.

UN-SG – United Nations Secretary-General (1997b): Protection of the Quality and Supply of Freshwater Resources. Application of Integrated Approaches to the Development, Management and Use of Water Resources. New York: UN.

US Department of the Interior (1996): Final Environmental Impact Statement. Water Rights Acquisition for Lahontan Valley Wetlands. Lahontan Valley News 1, 1–5.

Usera, M. A., Echeita, A., Aladuena, A., Alvarez, J., Carreno, C., Orcau, A. and Planas, C. (1995): Investigation of an Outbreak of Water-borne Typhoid Fever in Catalonia in 1994. Enfermedades Infecciosa Microbiologia Clinica 13 (8), 450–454.

Uvin, P. (1993): The State of World Hunger. The Hunger Report. Providence, RI: World Hunger Program.

Vermetten, A. W. M., Hofschreuder, P., Duyzer, J. H., Bosfeld, F. C. and Bouten, W. (1992): Dry Deposition of SO_2 Onto a Stand of Douglas Fir: The Influence of Canopy Wetness. In: Schwartz, S. E. and Slinn, R. G. W. (eds.): Precipitation Scavenging and Atmosphere-surface Exchange. Volume 3. New York: Hemisphere, 1403–1414.

Victor, D. G. (1996): The Early Operation and Effectiveness of the Montreal Protocol's Non-Compliance Procedure. Laxenburg: International Institute for Applied Systems Analysis (IIASA).

Vigarello, G. (1988): Wasser und Seife, Puder und Parfüm. Geschichte der Körperhygiene seit dem Mittelalter. Frankfurt/M., New York: Campus.

Vischer, D. (1996): Hochwassergefahr im Gebirge. Tagungsband des internationalen Symposiums „Klimaänderung und Wasserwirtschaft" am 27. und 28. November 1995 im Europäischen Patentamt in München. Mitteilungen des Instituts für Wasserwesen der Universität der Bundeswehr München (56b), 293–306.

von Knorring, E. (1996): Das Umweltproblem als ökonomisches Problem? Eine nachdenkliche Bestandsaufnahme. List Forum 22, 25–42.

von Thun, L. (1984): Application of Decision Analysis Techniques in Dam Safety Evaluation and Modification. Coimbra, Portugal: International Conference on the Safety of Dams.

von Tümpling, W., Wilken, R. D. and Einax, J. (1995): Mercury Contamination in the Northern Pantanal Region Mato Grosso, Brazil. Journal of Geochemical Exploration 52, 127–134.

Vought, L. B.-M., Dahl, J., Pedersen, C. L. and Lacoursiere, J. O. (1994): Nitrogen Retention in Riparian Ecotones. Ambio 23 (6), 342–348.

Wagner, G. (1976): Simulationsmodelle der Seeneutrophierung, dargestellt am Beispiel des Bodensee-Obersees. Archiv für Hydrobiologie 78, 1–41.

Wagner, A. and Lorenz, H.-W. (1995): Studien zur Evolutorischen Ökonomik III. Berlin: Duncker & Humblot.

Walker, J. T., Mackerness, C. W., Mallon, D., Makin, T., Williets, T. and Keevil, C. W. (1995): Control of Legionella Pneumophila in a Hospital Water System by Chlorine Dioxide. Journal of Industrial Microbiology 15, 384–390.

Wantzen, M. (1997): Siltation Effects on Benthic Communities in First Order Streams in Mato Grosso, Brazil. Verhandlungen Internationale Vereinigung für Limnologie. Mimeo.

WBGU – German Advisory Council on Global Change (1994): World in Transition: Basic Structure of Global Human-Environment Interactions. 1993 Annual Report. Bonn: Economica.

WBGU – German Advisory Council on Global Change (1995): World in Transition: The Threat to Soils. 1994 Annual Report. Bonn: Economica.

WBGU – German Advisory Council on Global Change (1996): World in Transition: Ways Towards Global Environmental Solutions. 1995 Annual Report. Berlin, Heidelberg, New York: Springer.

WBGU – German Advisory Council on Global Change (1997): World in Transition: The Research Challenge. 1996 Annual Report. Berlin, Heidelberg, New York: Springer.

WCMC – World Conservation Monitoring Centre (1992): Global Biodiversity. Status of the Earth's Living Resources. New York, London: Chapman & Hall.

Wehrli, B., Wüest, A., Bührer, H., Gächter, R. and Zobrist, J. (1996): Überdüngung der Schweizer Seen – erfreulicher Trend nach unten. EAWAG News (42D), 12–14.

Weidner, H. (1996a): Umweltkooperation und alternative Konfliktregelungsverfahren in Deutschland. Zur Entstehung eines neuen Politiknetzwerkes. Berlin: WZB.

Weidner, H. (1996b): Umweltmediation: Entwicklungen und Erfahrungen im In- und Ausland. In: Feindt, P. H., Gessenharter, W., Birzer, M. and Fröchling, H. (eds.): Konfliktregelung in der offenen Bürgergesellschaft. Dettelbach: Röll, 137–168.

Weller, D. E., Correll, D. L. and Jordan, T. E. (1994): Denitrification in Riparian Forests Receiving Agricultural Discharges. In: Mitsch, W. J. (ed.): Global Wetlands – Old World and New. Amsterdam: Elsevier, 117–132.

World Bank (1992): World Development Report 1992. Washington, DC: The World Bank.

World Bank (1993): World Development Report 1993. Washington, DC: The World Bank.

World Bank (1995): Learning from Narmada. Washington, DC: The World Bank.

World Bank (1996): Vom Plan zum Markt. Bonn: UNO Verlag.

Wetzel, R. G. (1983): Limnology. Philadelphia: Saunders.

Whitford, P. (1997): The Aral Sea Disaster: Turning the Tide? Environment Matters (Winter/Spring), 20–21.

WHO – World Health Organization (1990): Potential Health Effects of Climate Change: Report of a WHO Task Force. Geneva: WHO.

WHO – World Health Organization (1993): A Global Strategy for Malaria Control. Geneva: WHO.

WHO – World Health Organization (1994): Progress Report 1994: Second Meeting of Interested Parties on the Control of Tropical Diseases, Geneva September 1–4, 1994. Geneva: WHO.

WHO – World Health Organization (1995): The World Health Report 1995: Bridging the Gaps. Geneva: WHO.

WHO – World Health Organization (1996): The World Health Report 1996. Geneva: WHO.

Wichmann, K. (1996): Entsorgung von Wasserwerksrückständen in Deutschland. gwf Wasser-Abwasser 137 (14), 131–136.

Wicke, P. W. (1993): Irrigation in Highlands. Case Study Peru. In: DVWK – Deutscher Verband für Wasserwirtschaft und Kulturbau (ed.): Ecologically Sound Resources Management in Irrigation. Hamburg: Parey, 133–172.

Wilber, D. H., Tighe, R. E. and O'Neil, L. J. (1996): Associations Between Changes in Agriculture and Hydrology in the Cache River Basin, Arkansas, USA. Wetlands 16 (3), 366–378.

Wilke, A. (1995): „In die Ganga tauchen heißt, im Himmel zu baden". Wendekreis 6, 10–12.

Williams, P. B. (1993a): Dam Safety Analysis. In: Barber, M. and Ryder, G. (eds.): Damming the Three Gorges. London, Toronto: Earthscan and Probe International, 126–132.

Williams, P. B. (1993b): Flood Control Analysis. In: Barber, M. and Ryder, G. (eds.): Damming the Three Gorges. London, Toronto: Earthscan and Probe International, 100–117.

Williams, P. B. (1993c): Sedimentation Analysis. In: Barber, M. and Ryder, G. (eds.): Damming the Three Gorges. London, Toronto: Earthscan and Probe International, 133–145.

Williamson, O. E. (1985): The Economic Institutions of Capitalism. New York: The Free Press.

Winkler, R. C. (1982): Water Conservation. In: Geller, E. S., Winett, R. A. and Everett, P. B. (eds.): Preserving the Environment: Strategies for Behavior Change. New York: Pergamon Press, 262–287.

Wissenschaftlicher Beirat beim BMZ (1995): Vernachlässigung der Agrarförderung – Gefahren für die Zukunft. Bonn: Wissenschaftlicher Beirat beim BMZ.

WMO – World Meteorological Organization (1997): Natural Disasters and Human Settlements: A Statistical Survey. Geneva: WMO.

Wöhlcke, L. (1994): Brasilien. Diagnose einer Krise. Munich: Beck.

Wolff, P. (1994): Ist die Vernachlässigung des Bewässerungssektors in der Entwicklungszusammenarbeit verantwortbar? Zeitschrift für Bewässerungslandwirtschaft 29 (1), 115–120.

Wolff, P. (1996): Zur Nachhaltigkeit der Wassernutzung. Der Tropenlandwirt (Supplement 56), 103–126.

World Bank and UNDP – United Nations Development Programme (1990): A Proposal for an Internationally Supported Programme to Enhance Research in Irrigation and Drainage Technology in Developing Countries. Washington, DC: The World Bank.

WRI – World Resources Institute (1990): World Resources 1990–91. Special Focus on Climate Change, Latin America plus Essential Data on 146 Countries. New York, Oxford: Oxford University Press.

WRI – World Resources Institute (1992): World Resources 1992–93. Toward Sustainable Development. New York, Oxford: Oxford University Press.

WRI – World Resources Institute (1994): World Resources 1994–95. A Guide to the Global Environment. New York, Oxford: Oxford University Press.

WRI – World Resources Institute (1996): World Resources 1996–97. The Urban Environment. New York, Oxford: Oxford University Press.

WWI – Worldwatch Institute (1996a): Zur Lage der Welt. Frankfurt/M.: Fischer.

WWI – Worldwatch Institute (1996b): Worldwatch Database Disk. Frankfurt: Umwelt Kommunikation.

Yatsuyanagi, J., Saito, S., Kinouchi, Y., Sato, H., Morita, M. and Hoh, K. (1996): Characteristics of Enterotoxigenic Escherichis Coli and E. Coli Harbouring Enteroaggregative E.coli Heat-stable Enterotoxin-1 (EAST-1) Gene Isolated From a Water-borne Outbreak. Kansenshogaku Zasshi 70 (3), 215–223.

Young, O. R. (1994): International Governance. Protecting the Environment in a Stateless Society. Ithaca, London: Cornell University Press.

Young, R. A. and Haveman, R. H. (1985): Economics of Water Resources: A Survey. In: Kneese, A.V. and Sweeney, J. L. (eds.): Handbook of Natural Resource and Energy Economics. Volume II. Amsterdam, New York, Oxford: North-Holland, 465–529.

Yudelman, M. (1994): Demand and Supply of Foodstuffs up to 2050 With Special Reference to Irrigation. International Irrigation Management Review 8 (1), 4–14.

Zangl, B. and Zürn, M. (1996): Argumentatives Handeln bei internationalen Verhandlungen. Moderate Anmerkungen zur post-realistischen Debatte. Zeitschrift für Internationale Beziehungen 3 (2), 341–366.

Zauke, G. P. and Meurs, H. G. (1996): Kritische Anmerkungen zum Einsatz des Saprobiensystems bei der Gewässerüberwachung. In: Lozán, J. L. and Kausch, H. (eds.): Warnsignale aus Flüssen und Ästuaren. Hamburg: Parey, 329–330.

Zhou, K. (1986): A Project to Translocate the Baiji Lipotesvexillifer From the Main Stream of the Yangtze River China to Tongling Baiji Nature Reserve. Aquatic Mammals 12 (1), 21–24.

Zilleßen, H. (1993): Die Modernisierung der Demokratie im Zeichen der Umweltpolitik. In: Zilleßen, H., Dienel, P. C. and Strubelt, W. (eds.): Die Modernisierung der Demokratie. Opladen: Westdeutscher Verlag, 17–39.

Zürn, M. (1992): Interessen und Institutionen in der internationalen Politik. Grundlegung und Anwendungen des situationsstrukturellen Ansatzes. Opladen: Leske and Budrich.

Glossary

Agenda 21 is the legally non-binding Plan of Action for sustainable development that was adopted in 1992 at the United Nations Conference on Environment and Development (UNCED, the "Earth Summit"). Agenda 21 comprises 40 chapters, each containing recommendations, expressed in politically committing language, on sectoral or cross-sectoral issues (Chapter 18 on protection of freshwater resources, for example, or the sections on "Finances", "Youth" and "Institutions").

Agroforestry is a form of land management involving a combination of agriculture and forestry, in which the cultivation of trees or shrubs is integrated with crop production or animal husbandry. Agroforestry, now being "rediscovered", used to be a widespread practice in the tropics and subtropics.

Alkalization refers to the formation of alkaline soils; these arise primarily in arid and semi-arid climates, for example on clay substrates with poor drainage characteristics due to inadequate irrigation systems.

Aquatic classification system classifies the quality of surface waters; in Germany, classifications are assigned using a system of indicator organisms, oxygen concentration and bacteriological variables. Four classes of water quality are distinguished: no pollution or very slight pollution (oligosaprobic), moderate pollution (beta-mesosaprobic), severe pollution (alpha-mesosaprobic) and extremely severe pollution (polysaprobic).

Aquifers are porous rocks that can feed wells and springs. These rock layers contain →groundwater, which they are able to transmit in quantities significant enough to be economically exploited.

Arid climates are those in which mean annual evaporation exceeds mean annual precipitation.

Aridity index is calculated as the ratio of precipitation to the moisture loss of the soil; hyper-arid, arid, semi-arid, sub-humid and humid regions are distinguished according to the degree of their moisture deficit.

Autotrophic organisms are living organisms that gain their energy from sunlight or chemical reactions; these include all green plants.

Biodiversity refers to species diversity, genetic diversity and ecological diversity, i.e. the diversity of functional groups and the linkages within and between biological communities. Biodiversity thus embraces much more than just the variety of species, which can be calculated as the number of species per unit area.

Biological oxygen demand (BOD_5) is a measure of the amount of degradable organic matter in a waterbody. From the oxygen remaining after five days in an airtight container, the quantity of decomposed organic matter can be deduced.

Chemical oxygen demand (COD) is the measure of the quantity of poorly degradable organic matter in a body of water, for example organochlorine compounds, tensides or naturally occurring substances such as humic matter; it designates the amount of oxygen required to oxidize poorly degradable organic matter by means of a strong oxidant.

Core problems of global change are those global change phenomena of central importance in the →syndrome approach. In the syndrome approach, they appear either as particularly dominant →trends of global change, such as climate change, or consist of several interrelated trends. One such "megatrend" is the core problem of "soil degradation", which comprises several different trends such as erosion, →salinization, contamination, etc.

Criticality index is a composite indicator of the susceptibility of a region or its inhabitants to crises, especially environmental or development crises. In this Annual Report, the criticality index is used for regionally specific assessments of future freshwater crises.

Debt swaps, as the term suggests, refer to the "swapping" of debts (usually those of developing countries) against counter-performance of some kind, for example a certain environment policy (debt for nature swaps) or a certain food security policy (debt for food security swaps). The specific transaction involved depends on the type of debts. In the case of debts to foreign banks, for example, debt for nature swaps provide a means to combat the debt crisis and protect the environment.

Degradation of ecosystems involves changes that lead to the loss or impairment of ecosystem functions.

Desertification is the degradation of land resources in →arid, →semi-arid and dry sub-humid zones.

Disposition refers in →syndrome analysis of the vulnerability of a region to a specific syndrome. The "disposition space", meaning the geographical distribution of the disposition, is determined by natural and anthropospheric conditions that can only change over long time-scales.

Endemic species are organisms which occur within a confined geographical space.

Endocrine system refers to the system of glands in the body that secrete hormones regulating the activity of cells elsewhere in the body.

Eutrophication is the process of nutrient enrichment leading to enhanced primary production, resulting in modifications to natural processes and colonization structures and to increased biological decomposition.

Evaporation is the release of water vapor from waterbody surfaces and soils.

Evapotranspiration refers to the sum total of evaporation resulting from the release of vapor from surface waterbodies and soils (→evaporation) and from plants (→transpiration).

Exposition refers in →syndrome analysis of natural and anthropogenic events and processes, mostly of short duration (e.g. suddenly occurring natural disasters or exchange rate movements), and which can trigger a particular syndrome in susceptible regions (those with a →disposition to the syndrome).

Global network of interrelations refers in syndrome analysis of a qualitative network embracing all →trends of global change identified by the →syndrome concept, as well their respective interactions. The global network of interrelations provides a highly aggregated description of the global change system in terms of its specific phenomena.

Groundwater is water beneath the ground that fills interconnected pores in the upper part of the Earth's crust and whose movement is exclusively or almost exclusively determined by gravitational and frictional forces.

Guide rails demarcate, in →syndrome analysis, the domain of free action for the people-environment system from those domains which represent undesirable or even catastrophic developments and which must be avoided at all costs. Pathways for sustainable development run within the corridor defined by these guard rails. In the framework of this Report, the essential properties of water define the sociocultural and ecological framework and the non-sustainable limits (the "guard rails") within which water must be used efficiently in order to optimize the general welfare of humans everywhere. The Council sees the "guard rail" model as a useful instrument that, by setting clear priorities, enables the dilemma between social, ecological and economic goals to be resolved.

Heterotrophs are living organisms that feed on plants, animals, micro-organisms or organic substances. All animals belong to this category.

Humid climates are those in which the total annual precipitation is greater than the annual potential evaporation. Precipitation that is not evaporated flows off the surface or infiltrates into the →groundwater.

Interception is the proportion of precipitation that is stored in leaves and branches and which eventually evaporates. Interception can amount to 60% or more of precipitation.

LOCAL AGENDA 21 is the process by which cities and local communities consider ways (with citizen involvement) for developing and implementing a local plan of action on →AGENDA 21, as adopted at the Earth Summit in Rio de Janeiro.

Receiving waters are waterbodies that take up and convey water flowing on or under the ground surface.

Resilience refers to the capacity of a system to return to a stable equilibrium or a local equilibrium after a perturbation of some kind (syn. elasticity).

Salinization refers to the accumulation of soluble salts in or on soils or waters. A distinction is made between natural salinization (especially in →arid and →semi-arid climates) and anthropogenic salinization (caused, for example, by artificial irrigation and land-use changes).

Semi-arid is the term used to designate climates in which annual total precipitation is generally less than the total annual evaporation, even though precipitation exceeds evaporation over three to five months of the year.

Structural adjustment is the designation for a political program for stabilizing economies, deregulating markets and liberalizing foreign trade. The aim is to restore or improve international competitiveness and creditworthiness on the basis of a balanced budget (to combat inflation), to raise the internal rate of saving and investment, and to generate a better investment climate for foreign investors. Two institutions – the IMF and the World Bank – are responsible for implementing this program.

Subsistence farming refers to the production of food to satisfy the needs of the farm household, whereby little or no food is supplied to larger markets. It therefore remains outside the monetary economy. Subsistence farming is now considered part of the informal sector, so-called.

Sustainable development is generally understood to be an environment and development policy concept, that was popularized by the 1987 Brundtland Report ("sustainable development is development that meets the needs of the present without compromising the ability of future generations to meet their own needs") and further detailed by the 1992 Earth Summit in Rio de Janeiro. The German Advisory Council on Global Change has created, with its syndrome concept, an approach for operationalizing the concept of sustainable development.

Syndromes of global change are functional patterns of crisis-ridden people-environment relations, characteristic constellations of natural and anthropogenic →trends of global change and their respective interactions. Each syndrome – or "clinical profile", to use a medical analogy – represents an anthropogenic cause-effect complex involving specific environmental stresses, and thus forms a specific pattern of environmental degradation. Syndromes are transsectoral in nature, i.e. they affect several sectors (such as the economy, the biosphere or population) or environmental media (soils, water, air), yet they are always related, directly or indirectly, to natural resources. Syndromes can usually be identified in different forms in many regions of the world, whereby several syndromes may occur simultaneously.

Transpiration refers to the physical and physiological process by which plants lose vapor to the atmosphere, principally through the leaves.

Trends of global change are identified analytically prior to →syndrome analysis, and refer to anthropospheric and natural phenomena that are relevant for and characterize global change. They represent changeable or processional variable factors that can be determined qualitatively. Examples include "population growth", "enhanced greenhouse effect", "growing environmental awareness" or "advances in medicine".

Vapor pressure deficit is the measure of the volume of water that can still be absorbed by air until it reaches the saturation vapor pressure (100% saturation). Also referred to as the air saturation deficit.

Waterbody pollution refers to pollutant, pathogen and thermal loads, causing impairment of ecosystems and of the potential utilization of freshwater resources. The effects of pathogens on ecosystem structure are less significant than the impacts of pollutants. Thermal stress leads, in a manner similar to eutrophication and organic loads, to enhanced turnover rates of substances. For this reason, "thermal pollution" is a fitting term.

Water culture refers to a given society's sociocultural value context, within which people grow up and act with a certain relationship to water. It is generated by a multiplicity of interactions between environmental conditions and sociocultural spheres such as politics, economic activity or religion.

Water supply in a given area is the net total resulting from precipitation, water inflows from upstream and water losses. Exploitation of the water supply by water resource management takes water quality and ecological aspects into consideration.

The German Advisory Council on Global Change

H

THE COUNCIL
Prof. Hans-Joachim Schellnhuber, Potsdam (Chairperson)
Prof. Juliane Kokott (Vice Chairperson)
Prof. Friedrich O. Beese, Göttingen
Prof. Klaus Fraedrich, Hamburg
Prof. Paul Klemmer, Essen
Prof. Lenelis Kruse-Graumann, Hagen
Prof. Ortwin Renn, Stuttgart.
Prof. Ernst-Detlef Schulze, Jena
Prof. Max Tilzer, Konstanz
Prof. Paul Velsinger, Dortmund
Prof. Horst Zimmermann, Marburg

STAFF TO THE COUNCIL MEMBERS
Dr. Arthur Block, Potsdam
Dipl.-Geogr. Gerald Busch, Göttingen
Dipl.-Psych. Gerhard Hartmuth, Hagen
Dr. Dieter Hecht, Bochum
Andreas Klinke, M.A., Stuttgart
Dr. Gerhard Lammel, Hamburg
Referendar-jur. Leo-Felix Lee, Heidelberg
Dipl.-Ing. Roger Lienenkamp, Dortmund
Dr. Heike Mumm, Bremerhaven
Dipl.-Biol. Martina Mund, Bayreuth
Dipl.-Volksw. Thilo Pahl, Marburg
Dipl.-Biol. Helmut Recher, Plön

THE SECRETARIAT OF THE COUNCIL, BREMERHAVEN*
Prof. Dr. Meinhard Schulz-Baldes (Director)
Dr. Carsten Loose (Deputy Director)
Dipl.-Pol. Frank Biermann, LL.M.
Dipl.-Phys. Ursula Fuentes Hutfilter
Vesna Karic-Fazlic
Ursula Liebert
Dr. Benno Pilardeaux
Martina Schneider-Kremer, M.A.

* Secretariat WBGU
c/o Alfred-Wegener-Institute for Polar and Marine Research
PO Box 12 01 61
D-27515 Bremerhaven
Germany

Phone: ++49-471-4831-723
Fax: ++49-471-4831-218
Email: wbgu@awi-bremerhaven.de
Internet: http://www.awi-bremerhaven.de/WBGU/

Joint Decree on the Establishment of the German Advisory Council on Global Change (April 8, 1992)

Article 1

In order to periodically assess global environmental change and its consequences and to help all institutions responsible for environmental policy as well as the public to form an opinion on these issues, an Advisory Council on "Global Environmental Change" reporting to the Federal Government shall be established.

Article 2

(1)
The Council shall submit a report to the Federal Government by the first of June each year, giving an updated description of the state of global environmental change and its consequences, specifying quality, size and range of possible changes and giving an analysis of the latest research findings. In addition, the report should contain indications on how to avoid or correct maldevelopments. The report shall be published by the Council.

(2)
While preparing the reports, the Council shall provide the Federal Government with the opportunity to state its position on central issues.

(3)
The Federal Government may ask the Council to prepare special reports and opinions on specified topics.

Article 3

(1)
The Council shall consist of up to twelve members with special knowledge and experience regarding the tasks assigned to the Council.

(2)
The members of the Council shall be jointly appointed for a period of four years by the two ministries in charge, the Federal Ministry for Research and Technology and the Federal Ministry for the Environment, Nature Conservation and Reactor Safety, in agreement with the departments concerned. Reappointment is possible.

(3)
Members may declare their resignation from the Council in writing at any time.

(4)
If a member resigns before the end of his or her term of office, a new member shall be appointed for the retired member's term of office.

Article 4

(1)
The Council is bound only to the brief defined by this Decree and is otherwise independent to determine its own activities.

(2)
Members of the Council may not be members either of the Government or a legislative body of the Federal Republic or of a Land or of the public service of the Federal Republic, of a Land or of any other juristic person under public law unless he or she is a university professor or a staff member of a scientific institute. Furthermore, they may not be representatives of an economic association or an employer's or employee's organisation, or be permanently attached to these through the performance of services and business acquisition. They must not have held any such position during the year preceding their appointment as member of the Council.

Article 5

(1)
The Council shall elect a Chairperson and a Vice-Chairperson from its midst for a term of four years by secret ballot.

(2)
The Council shall set up its own rules of procedure. These must be approved by the two ministries in charge.

(3)
If there is a differing minority with regard to individual topics of the report then this minority opinion can be expressed in the report.

Article 6

In the execution of its work the Council shall be supported by a Secretariat which shall initially be located at the Alfred Wegener Institut (AWI) in Bremerhaven.

Article 7

Members of the Council as well as the staff of the Secretariat are bound to secrecy with regard to meeting and conference papers considered confidential by the Council. This obligation to secrecy is also valid with regard to information given to the Council and considered confidential.

Article 8

(1)
Members of the Council shall receive all-inclusive compensation as well as reimbursement of their travel expenses. The amount of compensation shall be fixed by the two ministries in charge in agreement with the Federal Ministry of Finance.

(2)
The costs of the Council and its Secretariat shall be shared equally by the two ministries in charge.

Dr. Heinz Riesenhuber
Federal Minister for Research and Technology
Prof. Klaus Töpfer
Federal Minister for Environment, Nature Conservation and Reactor Safety
May 1992

— Appendix to the Council Mandate —

TASKS TO BE PERFORMED BY THE Advisory Council Pursuant to Article 2, para. 1

The tasks of the Council include:

(1)
Summarising and continuous reporting on current and acute problems in the field of global environmental change and its consequences, e.g. with regard to climate change, ozone depletion, tropical forests and fragile terrestrial ecosystems, aquatic ecosystems and the cryosphere, biological diversity and the socioeconomic consequences of global environmental change. Natural and anthropogenic causes (industrialisation, agriculture, overpopulation, urbanisation, etc.) should be considered, and special attention should be given to possible feedback effects (in order to avoid undesired reactions to measures taken).

(2)
Observation and evaluation of national and international research activities in the field of global environmental change (with special reference to monitoring programmes, the use and management of data, etc.).

(3)
Identification of deficiencies in research and coordination.

(4)
Recommendations regarding the avoidance and correction of maldevelopments.

In its reporting the Council should also consider ethical aspects of global environmental change.

Index

A

B

C

Y

Z

GPSR Compliance
The European Union's (EU) General Product Safety Regulation (GPSR) is a set of rules that requires consumer products to be safe and our obligations to ensure this.

If you have any concerns about our products, you can contact us on

ProductSafety@springernature.com

In case Publisher is established outside the EU, the EU authorized representative is:

Springer Nature Customer Service Center GmbH
Europaplatz 3
69115 Heidelberg, Germany

www.ingramcontent.com/pod-product-compliance
Lightning Source LLC
Chambersburg PA
CBHW080535250926
51846CB00016B/1013
9783642641718